Lecture Notes in Computer Science 5932

Commenced Publication in 1973
Founding and Former Series Editors:
Gerhard Goos, Juris Hartmanis, and Jan van Leeuwen

Heung Youl Youm   Moti Yung (Eds.)

Information Security Applications

10th International Workshop, WISA 2009
Busan, Korea, August 25-27, 2009
Revised Selected Papers

 Springer

Volume Editors

Heung Youl Youm
Soonchunhyang University
Department of Information Security Engineering
646 Eupnae-ri, Shinchang-myun
Asan-si, Chungnam-do, 336-745, Korea
E-mail: hyyoum@sch.ac.kr

Moti Yung
Google Inc. and Columbia University
Computer Science Department
Room 464, S.W. Mudd Building
New York, NY 10027, USA
E-mail: moti@cs.columbia.edu

Library of Congress Control Number: 2009941782

CR Subject Classification (1998): E.3, C.2, D.2, D.4.6, K.6.5

LNCS Sublibrary: SL 4 – Security and Cryptology

ISSN   0302-9743

ISBN   978-3-642-10837-2 Springer Berlin Heidelberg New York

springer.com

© Springer-Verlag Berlin Heidelberg 2009

Typesetting: Camera-ready by author, data conversion by Scientific Publishing Services, Chennai, India
Printed on acid-free paper   SPIN: 12811359   06/3180   5 4 3 2 1 0

Preface

The 10th International Workshop on Information Security Applications (WISA 2009) was held in Busan, Korea during August 25–27, 2009. The workshop was hosted by the Korea Institute of Information Security and Cryptology (KIISC), supported by the Electronics and Telecommunications Research Institute (ETRI) and the Korea Internet and Security Agency (KISA), sponsored by the Ministry of Public Administration and Security (MoPAS) and the Korea Communications Commission (KCC), financially sponsored by the ST. Ltd.

The aim of the workshop was to serve as a forum for presenting new research and experimental results in the area of information security applications from academic communities as well as from industry. The workshop program covered a wide range of security aspects, from cryptography to systems and network security and experimental work as well.

It was our great pleasure and honor to serve as the Program Committee Co-chairs of WISA2009. This year, too, the proceedings of the workshop were published in the LNCS series of Springer. The WISA 2009 Program Committee received 79 papers form 16 countries. This year the submissions were exceptionally strong, and the committee accepted 27 papers for the full paper presentation track. All the papers were carefully evaluated through blind peer-review by at least three members of the Program Committee. We would like to say that acceptance is a great achievement since the selection process was highly competitive, and many good papers were not accepted.

In addition to the contributed papers, the workshop had three invited talks. Moti Yung, Hideki Imai and Amardeo Sarma gave us distinguished special talks entitled "The Evolution from Protection of Other to Self-Protection," "Future Direction on Second Round CRYPTREC in Japan" and "Making Identities Work - SWIFT Approaches and Solutions," respectively.

Many people helped and worked hard to make WISA2009 successful. We would like to thank all the people involved in the technical program and in organizing the workshop. We are very grateful to the Program Committee members and external referees for their time and efforts in reviewing the submissions and selecting the accepted papers. We should also express our special thanks to the Organizing Committee members and the General Chair, Kwangjo Kim, for their hard work in managing the workshop.

Finally, on behalf of all those involved in organizing the workshop, we would like to thank the authors of all the submitted papers, for sending and contributing their interesting research results to the workshop, and the invited speakers. Without their submissions and support, WISA2009 could not have been a success.

November 2009

Heung Youl Youm
Moti Yung

Organization

Advisory Committee

Mun Kee Choi	ETRI, Korea
Hideki Imai	Chuo University, Japan
Dae-Ho Kim	ETRI, Korea
Sehun Kim	KAIST, Korea
Hong-sub Lee	Soonchunhyang University, Korea
Pil Joong Lee	POSTECH, Korea
Sang-Jae Moon	Kyungpook National University, Korea
Kil-Hyun Nam	Korea National Defense University, Korea
Bart Preneel	Katholieke Universiteit Leuven, Belgium
Man-Young Rhee	Kyung Hee University, Korea
Min-Sub Rhee	Dankook University, Korea
Joo-Seok Song	Yonsei University, Korea
Dong-Ho Won	Sungkyunkwan University, Korea

General Chair

Kwangjo Kim	KAIST, Korea

Steering Committee

Hyun-Sook Cho	ETRI, Korea
Sang-Choon Kim	Kangwon National University, Korea
Jae-Cheol Ryou	Chungnam National University, Korea
Hyung-Woo Lee	Hanshin University, Korea
Jae-Kwang Lee	Hannam University, Korea
Dong-Il Seo	ETRI, Korea
OkYeon Yi	Kookmin University, Korea
Kyo-Il Chung	ETRI, Korea
Kiwook Sohn	ETRI, Korea

Organizing Committee

Chair	Kyung-Hyune Rhee	Pukyong National University, Korea
Finance	Ji-Young Ahn	KIISC, Korea
	Kyoung Sik Min	KISA, Korea
Publication	Weon Shin	Dongmyung University, Korea

Publicity	Jae-Cheol Ryou	Chungnam National University, Korea
	Jongcheol Moon	The Attached Institute of ETRI, Korea
	MooSeop Kim	ETRI, Korea
Registration	Howon Kim	Pusan National University, Korea
Treasurer	Daehyun Ryu	Hansei University, Korea
Local Arrangements	Sang-Uk Shin	Pukyoung National University, Korea
Webpage Management	Rack-Hyun Kim	Soonchunhyang University, Korea

Program Committee

Co-chairs	Heung Youl Youm	Soonchunhyang University, Korea
	Moti Yung	Columbia University, USA
Members	Frederik Armknecht	Ruhr-University Bochum, Germany
	Joonsang Baek	Inst. for Infocomm Research, Singapore
	Feng Bao	Inst. for Infocomm Research, Singapore
	Rodrigo Roman Castro	University of Malaga, Spain
	Julien Cathalo	UC Louvain, Belgium
	Jianyong CHEN	Shenzhen University, China
	Kefei Chen	Shanghai Jiaotong University, China
	Dooho Choi	ETRI, Korea
	Hyoung-Kee Choi	Sungkyunkwan University, Korea
	Kilsoo Chun	KISA, Korea
	Debbie Cook	Columbia University, USA
	Ed Dawson	Queensland University of Technology, Australia
	JaeCheol Ha	Hoseo University, Korea
	Stefan Katzenbeisser	Technical University Darmstadt, Germany
	Dong Kyue Kim	Hanyang University, Korea
	Howon Kim	Pusan National University, Korea
	Hyung Jong Kim	Seoul Women's University, Korea
	Hyong-Shik Kim	Chungnam National University, Korea
	Seok Woo Kim	Hansei University, Korea
	Young Baek Kim	KISA, Korea
	Brian King	Purdue University of Indianapolis, USA
	Hong Seung Ko	Kyoto College of Graduate Studies for Informatics, Japan
	Chiu Yuen Koo	Google Inc., USA
	Jin Kwak	Soonchunhyang University, Korea
	Deok Gyu Lee	ETRI, Korea
	Dong Hoon Lee	Korea University, Korea
	Pil Joong Lee	POSTECH, Korea
	Chae Hoon Lim	Sejong University, Korea
	Ji-Young Lim	Korean Bible University, Korea

Table of Contents

Multimedia Security

Device Security

HW Implementation Security

Applied Cryptography

Side Channel Attacks

Cryptograptanalysis

Anonymity/Authentication/Access Control

Network Security

Protecting IPTV Service Network against Malicious Rendezvous Point

Hyeokchan Kwon, Yong-Hyuk Moon, Jaehoon Nah, and Dongil Seo

Electronics and Telecommunications Research Institute(ETRI), Korea
{hckwon,yhmoon,jhnah,bluesea}@etri.re.kr

Abstract. In this paper, we present security mechanism to protect IPTV service network from malicious Rendezvous Point. The IPTV service network considered in this paper is overlay network that is constructed in application layer. The overlay-based IPTV service network has several advantages such as cost-effectiveness, dynamicity and scalability. However, there are several security threats against overlay network such as malicious rendezvous point attack, routing interference attack, DoS(Denial of Service) attack and so on. In this paper we analyze the security threats of overlay-based IPTV service network, and we present the brief security guidelines against it. And we present detailed security mechanisms to protect IPTV service network from malicious Rendezvous Point. For this, we design the security mechanism to guarantee trust of rendezvous point and distribute security keys such as self-generated public key of each node and group key of rendezvous point safely manner. This approach is very simple, lightweight and implementation friendly.

Keywords: IPTV, Security, Overlay network, Malicious rendezvous point.

1 Introduction

Traditional IPTV mainly serviced on the special purpose premium network. Therefore, in case of open IPTV environment, the IPTV video is delivered on public network, in this case network load and management cost is a big issue. So, recently there is the attempt to use overlay-based IPTV service to reduce network load and management cost.

Figure 1 shows the example of traditional media delivery network – AOL webcast Live 8 concert[1]. In figure 1, CDN(Content Delivery Network) consisting 1500 distribution server is used to deliver the 300kbps media streaming data to 175,000 users. In this case, total 50Gbps bandwidth is needed for provide media streaming service. Figure 2 shows the example of overlay-based media delivery service. In this case the only 300Kbps bandwidth is needed at the server side, and the streaming media is delivered by p2p-based overlay network. The overlay-based IPTV service network has several advantages such as cost-effectiveness, dynamicity and scalability.

But there exists several security threats against overlay-based IPTV network such as malicious rendezvous point attack, routing interference attack, DoS attack and so

H.Y. Youm and M. Yung (Eds.): WISA 2009, LNCS 5932, pp. 1–9, 2009.

on. Among them malicious rendezvous point attack is especially serious. Generally rendezvous point has a very important role such as group management, root of multi-cast routing and so on.

In this paper, we present security mechanism to protect IPTV service network from malicious Rendezvous point. The IPTV service network considered in this paper is overlay based network that is constructed in application layer. The overlay based network has several security vulnerabilities. In this paper we analyze the security vulnerabilities of overlay-based IPTV service network. And we present the brief security solutions against it. And we present detailed security mechanisms to protect IPTV service network from malicious Rendezvous Point.

We consider the target IPTV service network as pastry-based overlay network. Pastry[2], one of the overlay networks, is very suitable for constructing overlay-based IPTV service network because it has scalable and self-organizing properties. [3] proposed the pastry based overlay multicast mechanism. It is very efficient solution, but it doesn't have sufficient security function to protect multicast network from various security attacks.

In this paper, in order to protect IPTV service network against malicious rendezvous point, we design the security mechanism to guarantee trust of rendezvous point and distribute security key such as self-generated public key of each node and group key of rendezvous point in a safe and efficient manner. The proposed approach is very simple, lightweight and implementation friendly.

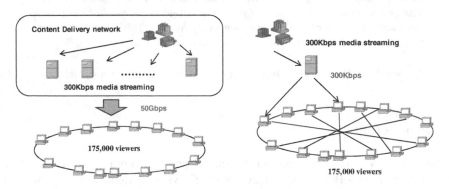

Fig. 1. Server-based media streaming   **Fig. 2.** Overlay-based media streaming

The contents organized as follows. Section 2 presents overview of overlay-based IPTV service network. And in section 3, we present the mechanism of protecting IPTV service network against malicious rendezvous point. Finally conclusion is given in section 4.

2   Overview of Overlay-Based IPTV Service Network

In this section, we provide overlay-based IPTV service network. The main issue of overlay-based IPTV service is a how to make and manage dynamic overlay network. To do this, the virtual overlay tree is used. Generally overlay tree members are constructed by the peers that view the same IPTV channel. Recently structured P2P overlay is used to make overlay tree.

Structured P2P overlay networks like CAN(Content Addressable Network), Chord, Pastry and Tapestry - those are DHT(Distributed Hash Table) based overlay network - provide a self-organizing infrastructure, so they can establish and maintain the overlay topology by themselves. The structured overlay network guarantee location of content if it exists, within a bounded number of hops; for example, Chord guarantees O(logN) messages per lookup in case that no malicious peers[4]. The target overlay considered in this paper is pasty-based overlay network.

2.1 Pastry-Based IPTV Service Network

In this section, we analyze pastry-based IPTV service network that is target network in this paper. Pastry[2] is a self-organizing network and it is one of the DHT(Distributed Hash Table)-based overlay networks. The node ID of pastry is generated by secure hash of the node's public key or IP address. Pastry reliably routes the message to the pastry node with the node ID that is numerically closest to the key. Pastry uses the prefix routing. In the prefix routing, the node forward the messages to a node whose node ID shares longer prefix with the given key comparing the present node, if no such node, forward it to a numerically closer node.

The node IDs and keys are thought as a sequence of digits with base 2^b. The tables required in each Pastry node have only $(2^b - 1) * \lceil \log_{2^b} N \rceil + l$ entries. L is a number of entries in Leaf set. Leaf set contains information of several neighbor nodes that is physically adjacent to it. A node's routing table organized into $\lceil \log_{2^b} N \rceil$ rows with base 2^b -1entries each. The 2^b -1 entry in row n of the routing table each refer to a node whose nodeId matches the present node's nodeId in the first n digits. Figure 3 shows the pastry routing table examples. For example node 02133 whose routing table is shown in figure 3 routes message 02202 to node 02230. Figure 4 shows this routing example.

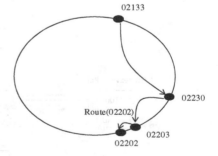

Routing Table of Node **02133**			
0	-1-1212	-2-2301	-3-1203
1-0111	1-1133	2	1-3022
1	02-110	02-230	02-323
021-00	021-10	021-22	3
0213-0	0213-1	0213-2	3

Fig. 3. Pastry routing table example   **Fig. 4.** Pastry routing example

SCRIBE overlay multicast network is build on top of pastry. SCRIBE[3] is a large-scale, decentralized application level multicast infrastructure built upon pastry[2], a scalable, self-organizing peer-to-peer object location and routing substrate overlayed on the Internet. There exist 4 types of API that is used by SCRIBE[3].

(1) create(credentials, groupId) (2) join(credentials, groupId, messageHandler) (3) leave(credentials, groupID) (4) multicast(credentials, groupId, message)

Each group of SCRIBE has a unique groupId. The group id is generated by hash of group's name and creator's name. The figure 5 shows the process of group creation and Join.

In figure 5, node 1001 send the group creation message to node 1100(1100 is the groupID), and message is routed to node 1100. In figure 5, the group creation message is routed through the node 1101. This routing path is decided by prefix-based pastry routing table for each node. Each node in this route can be a member of this multicast tree automatically. These nodes store the information of parent peer and child peer during the deliver group creation message. In this case the node 1100 is the rendezvous point of the group. Rendezvous point is a root of multicast tree. In figure 5, node 0111 join request to group 1100. This join message is routed through the node 1101 by pastry routing. So, node 0111 is a child node of 1101, and node 1101 is parent node of 0111.

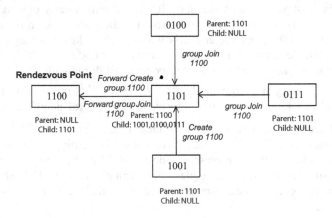

Fig. 5. The process of create & join the overlay group

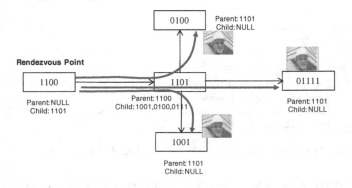

Fig. 6. The process of IPTV video delivery

Figure 6 shows the multicast routing process. The media data is multicasted to the group members through the multicast tree. In case of overlay-based IPTV service, the group (ID: 1100) can be a channel and the members of the group can be a viewer of that channel.

3 Protecting IPTV Service Network

3.1 Analysis of Security Threats of Overlay-Based IPTV Service Network and Brief Solution

In this subsection, we analyze security threats of pastry-based IPTV service network - SCRIBE. In this architecture, the peers in which there is no access right of media but in the pastry transmission path can be a member of a multicast tree. So it is needed to protect multicast media data from these unauthorized members. Moreover, because the arbitrarily selected peer can be a rendezvous point, the security threat by malicious rendezvous point is a big problem. And in the SCRIBE model, it is not possible to verify the multicasting routing is done successfully or not. So, it is possible that some malicious nodes forward the multicast message to incorrect node or drop it. The table 1 shows the vulnerabilities of overlay-based IPTV service network and brief solutions against them.

Table 1. Security threats & brief solutions

Security Threats	Brief Solution guideline
Attacks from malicious rendezvous point: If malicious peer plays the role of Rendezvous Point, a media delivery service disturbed and it is possible to deliver the media data to a non-member node by malicious rendezvous point, and overlay tree itself could not be trust. And it is possible that a set of malicious nodes can make malicious rendezvous point. And they forward the join message from a new node to a malicious rendezvous point	Introduce the new authentication mechanism to authenticate rendezvous point. Another solution is that the system selects the nodes that can be a rendezvous point in advance and send rendezvous point certificate to them. In this case the rendezvous point selection mechanism is needed.
Routing interference attack : It is possible that an malicious group member intentionally forward multicast data to an incorrect node or drop it	Introduce the functions to monitor delivery status of multicast data to rendezvous point, and introduce incentive mechanism.
Information leakage: Some members of the multicast tree have no right for seeing multicast data. These members only perform pastry routing. These member could sniff, store, re-distributes the multicast data illegally.	Introduce the group key management functions to rendezvous point. The rendezvous point issues the group certificate for group member, and delivers it to the group member directly not using pastry routing. And then the rendezvous point encrypts the media data by using group key.
DoS attack: A set of malicious node forwards the multicast data to a specific target node cooperatively.	Add the functions to monitor delivery status of multicast data to rendezvous point, and introduce incentive mechanism

3.2 Security Mechanisms against Malicious Rendezvous Point

In this paper, we present security mechanism to protect IPTV service network from malicious Rendezvous point. For this mechanism, we introduce the rendezvous point candidate and authentication server. The system component is shown in figure 7.

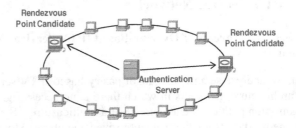

Fig. 7. Rendezvous point candidate & Authentication Server

We assume that auth_server knows the node id and public key of rendezvous point candidate in advance. rendezvous point candidate is pre-selected by the system and the rendezvous point is selected to numerically closest to groupID from rendezvous point candidates. In this application, we assume that each pastry node knows the public key of auth_server. In our approach, each node self-generates public/private key pairs. Auth_server is not a certificate authority server in public key infrastructure, it only manages rendezvous point candidates and provides the information of rendezvous point candidate.

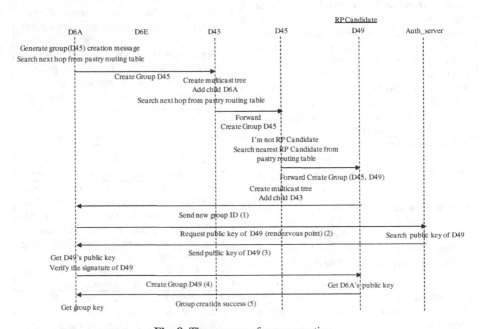

Fig. 8. The process of group creation

The notations used in this paper are as follows.
- R : Rendezvous Point
- AC : Auth_server
- U_k : Public key of peer k
- R_k : Private key of peer k
- G_g : Group key of group g
- $E_K(m)$: Encryption function. Encrypt the message m by encrypt key k
- $D_K(c)$: Decryption function. Decrypt the cipher text c by decrypt key k
- $S_K(m)$: Signature function. Signing to message m by key k

The group creation process is shown in figure 8 and group join process is shown in figure 9. The equations which are used in figure 8 and 9 are shown in table 2 and 3 respectively.

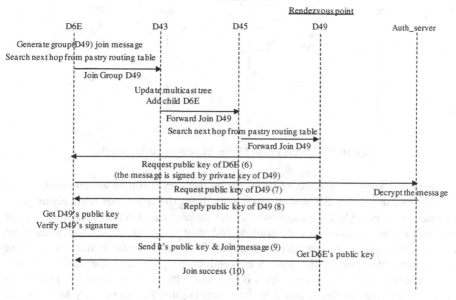

Fig. 9. The process of group join

Table 2. Equations in figure 8

Equation #	Equation
(1)	$\{new\,GroupID\,(D\,49)\} \mid S_{R_{D49}}(m)$
(2)	$E_{U_{AC}}("Request\ public\ key\ of\ D49")$
(3)	$\{U_{D49}\} \mid S_{R_{AC}}(m)$
(4)	$E_{U_{D49}}(ID_{D6A} \mid U_{D6A} \mid "request\ group\ creation\ D49")$
(5)	$E_{U_{D6A}}("group\ creation\ success" \mid G_{D49})$

Table 3. Equations in figure 9

Equation #	Equation
(6)	$\{"request\ public\ key\ of\ D6E"\} \mid S_{R_{D49}}\ (m)$
(7)	$E_{U_{AC}}\ ("Request\ public\ key\ of\ D49")$
(8)	$\{U_{D49}\} \mid S_{R_{AC}}\ (m)$
(9)	$E_{U_{D49}}\ (ID\_D6E \mid U_{D6E} \mid "join\ group(D49)")$
(10)	$E_{U_{D6E}}\ ("group\ join\ success" \mid G_{D49})$

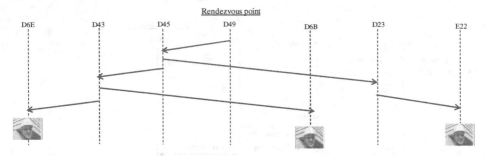

Fig. 10. The process of IPTV media delivery using overlay

Figure 10 shows the process of IPTV media delivery using proposed overlay. This mechanism can protect IPTV service network from the malicious rendezvous point, because it can provide the mechanism to the authenticated rendezvous point candidate, and in this mechanism only pre-selected trust node can be a rendezvous point. And this mechanism can protect information leakage, because it introduces group management functions to the authorized rendezvous point and the multicast video data is encrypted by group key which is generated by rendezvous point. During the process of creation (figure8) and join the group (figure 9), the security keys such as self-generated public key of each peer and group key of rendezvous point are distributed in a safe manner without any heavy security infrastructure.

4   Conclusions

In this paper, we present security mechanism to protect IPTV service network from malicious rendezvous point. The IPTV service network considered in this paper is overlay based network that is constructed in application layer. The overlay network has advantages on dynamicity, scalability, cost-effective but, it has several security vulnerabilities. In this paper we analyze the security threats against overlay-based IPTV service network. And we provide the brief security guidelines against it.

And we present detailed security mechanisms to protect IPTV service network from malicious Rendezvous Point. In order to protect IPTV service network against malicious rendezvous point, we design the security mechanism to guarantee trust of rendezvous point and distribute security key such as self-generated public key of each node and group key of rendezvous point in a safe and efficient manner. The proposed mechanism is very simple, lightweight and implementation friendly. It doesn't need any other heavy security infrastructure such as public key infrastructure and so on.

Currently, we plan to analysis the proposed mechanism as security aspect and design more detailed architecture and operations.

Acknowledgement

This work was supported by the IT R&D program of MKE/KCC/KEIT [2008-S-006-01, Development of Open-IPTV (IPTV2.0) Technologies for Wired and Wireless Networks].

References

1. Girod, B.: Transport of Real-Time Traffic over The Internet, Presentation slides (2005)
2. Rowstron, A., Druschel, P.: Pastry: Scalable, decentralized object location, and routing for large-scale peer-to-peer systems. In: Guerraoui, R. (ed.) Middleware 2001. LNCS, vol. 2218, p. 329. Springer, Heidelberg (2001)
3. Castro, M., Druschel, P., Kermarrec, A., Rowstron, A.: SCRIBE: A large-scale and decentralized application-level multicast infrastructure. IEEE Journal on Selected Areas in Communications 20(8) (October 2002)
4. Stoica, I., Morris, R., Karger, D., Kaashoek, M.F., Balakrishnan, H.: Chord: A Scalable Peer-to-peer Lookup Service for Internet Applications. In: SIGCOMM 2001, San Diego, California, USA (2001)
5. Dhungel, P., Hei, X., Ross, K.W., Saxena, N.: The Pollution Attack in P2P Live Video Streaming Measurement Results and Defenses. In: Proc. of the workshop on peer-to-peer streaming and IPTV, pp. 323–328 (2007)
6. Androutsellis-Theotokis, S., Spinellis, D.: A survey of peer-to-peer content distribution technologies. ACM Computing Surveys 36(4), 335–371 (2004)
7. Lian, S., Sun, J., Liu, G.: Efficient Video Encryption Scheme based on Advanced Video Coding. Multimedia Tools and Applications 38(1) (May 2008)
8. Sethom, K., Masmoudi, K., Afifi, H.: A Secure P2P Architecture for Location Management. In: Proceedings of the 6th international conference on Mobile data management, Ayia Napa, Cyprus, May 09 - 13 (2005)
9. Xiaoyun, L., Tiejun, H., Longshe, H., Luntian, M.: A DRM Architecture for Manageable P2P Based IPTV System. In: ICME 2007 (2007)

Design and Implementation of SIP-aware Security Management System

KyoungHee Ko[1], Hwan-Kuk Kim[1], JeongWook Kim[1], Chang-Yong Lee[1],
Soo-Gil Cha[2], and Hyun Cheol Jeong[1]

[1] Korea Information & Security Agency, Seoul, South Korea
{khko,rinyfeel,kjw,chylee,hcjung}@kisa.or.kr
[2] IGLOO Security, Seoul, South Korea
chascal8@igloosec.com

Abstract. SIP is a signaling protocol used for establishing sessions in multimedia services such as VoIP, instant messaging, and video conferencing. As SIP-aware security devices are emerging for protecting SIP-based services, it is needed for security management system to manage heterogeneous SIP-aware devices. In this paper, SIP-aware security management system is proposed. Design considerations and overall design of the system is described. And implementation and performance test is presented.

Keywords: Session Initiation Protocol, VoIP Security, IP Telephony Security, Security Management System, Security Event and Information Management, Enterprise Security Management, Security Event Correlation.

1 Introduction

SIP (Session Initiation Protocol) is an application-layer control protocol that can establish, modify, and terminate multimedia session [1]. SIP-based services are IP multimedia services such as VoIP (Voice over Internet Protocol), presence, instant messaging, video conferencing, and unified communication.

As SIP-based services are becoming popular, threats and countermeasures in SIP-based services are being studied. As a result, three groups are providing countermeasures for protecting SIP-based services. First, existing security devices such as firewall, intrusion detection system, and intrusion prevention system are extended to detect SIP-based attacks. Second, SIP network devices such as SIP proxy server, session border controller, and IP-PBX provide access control mechanisms. Third, dedicated SIP-aware firewalls or vulnerability scanners are emerged.

As more SIP-aware security devices are deployed in organization's network, it is needed for security management system to manage these devices. Security management system collects information from each SIP-aware device. And then the system analyzes collected information. Also the system controls heterogeneous devices in a uniform way.

In this paper, SIP-aware security management system is proposed. In order to collect security information and event from heterogeneous devices, message format and

H.Y. Youm and M. Yung (Eds.): WISA 2009, LNCS 5932, pp. 10–19, 2009.
© Springer-Verlag Berlin Heidelberg 2009

message exchanging protocol are defined based on existing standards. In order to analyze collected information, rule-based security event correlation is used. The proposed system provides event processing performance for middle and large size organizations

This paper is structured as follows: in section 2, we overview related works briefly. In section 3, we describe design considerations and design for SIP-aware security management system. In section 4, we present the results of implementation and performance test. In section 5, this paper ends with conclusions and future works.

2 Related Works

Related works of SIP-aware security management system are divided into two groups.

The first group is classified as commercial products for managing VoIP devices.

NetIQ added VoIP Security Solution to NetIQ AppManager[2]. AppManager for VoIP manages VoIP solutions, analyzes call usage patterns, and provides performance monitoring and reporting.

Q1 Labs added VoIP module to QRadar products. QRadar combines network behavior analysis and security event correlation to monitor across the network protocol, application and security services layer of a VoIP network [3].

Acme Packet Net-Net Element Management System manages Acme VoIP devices [4]. It provides configuration, performance management and security management for Acme VoIP devices.

Until now, these commercial products support limited kinds of VoIP devices. This is because there is no standard protocol between management system and security devices. And these products are more focused on managing and monitoring VoIP devices than analyzing security information and event.

The second group is research projects. [5] proposed holistic approach for VoIP intrusion detection and prevention system. Open source and platform independent software, SEC (Security Event Correlation) was adopted in [5]. SEC is a lightweight event correlator that can serve different applications ranging from log file and system monitoring to fraud detection, network management and intrusion detection. SEC is written in Perl and no performance result was reported in large size organizations.

3 Design of SIP-aware Security Management System

In this section, design considerations and overall design of the system is described. Detailed design for SIP-aware security management system was presented in [6].

3.1 Design Considerations for SIP-aware Security Management System

SIP-aware Security Management System (SSMS) needs to collect SIP specific security information and event from SIP-aware devices. SIP specific information and event includes as follows:

- Packet payload inspection at application layer: SIP-aware devices capture and inspect packets at application layer. SIP header and body are parsed according to protocol specifications. If RTP (Real-time Transport Protocol) [7] is used for transferring media data, RTP header is parsed according to the protocol specification.

- Cross protocol detection: signaling and media channel is separated in SIP-based services. SIP-aware security devices can detect cross protocol
- Dynamic port filtering: SIP-aware devices can filter ports for media channel which are determined dynamically during session establishment.
- Quality of Service (QoS) metrics such as delay, jitter, and packet loss rate: because SIP-based services are multimedia services, it is important to assure QoS. Many SIP-aware devices measure QoS metrics for traffic control.

Because SSMS collects the above information from heterogeneous SIP-aware devices, it is needed to define message format and message exchanging protocol. The message format and exchanging protocol are implemented as a library, so that any SIP-aware device can communicate more easily with SSMS.

SSMS needs to analyze collected information, although each SIP-aware security device can detect attacks. Because SSMS analyze information collectively based on individual detection result, SSMS can improve accuracy and confidence of detection.

When SSMS is analyzing collected information, SSMS needs to know characteristics of attack in SIP-based services. There are rule-based, codebook, artificial neural network and causality graph approach to represent attack scenario [8]. SSMS uses rule-based approach because it is easy to obtain scenario from experts' knowledge and shows feasibility in many real world systems.

SSMS can be solely used for managing voice network or IMS (IP Multimedia Subsystem) [9] network. But SSMS is designed to be modular and open, so that SSMS can be also used as an add-on module to existing security management system.

SSMS needs to assure performance, because security management system is generally used in middle and large size organizations. Performance metric is defined as the number of events processed per second in SSMS.

3.2 Structure of SIP-aware Security Management System

Fig. 1 shows the system structure for SSMS. SSMS is composed of agent, manager, and console. SSMS agent is installed on the same machine with SIP-aware devices. SSMS agent can also be installed on the different machine, because SSMS agent can communicate with SIP-aware devices using TCP connection. In the latter case, SSMS agent's ability to control SIP-aware devices will be limited.

3.3 Design of SSMS Agent

SSMS agent is responsible for collecting security events from SIP-aware devices. SSMS agent is composed of Client-Side SSMS Interface library, Server-Side SSMS Interface Library, Normalization, Aggregation and Transceiver modules.

Message format for SSMS is defined based on as IETF IDMEF [10]. In order to include SIP specific information, messages for SSMS are defined as follows:

- Alert message: Alert message is defined based on Alert and Service class in IETF IDMEF. These two classes are updated to include SIP specific information such as SIP method, SIP from-URI, SIP to-URI, RTP media port, RTP media delay, RTP media jitter, RTP media packet loss rate, SIP attack class, response action, and so on.

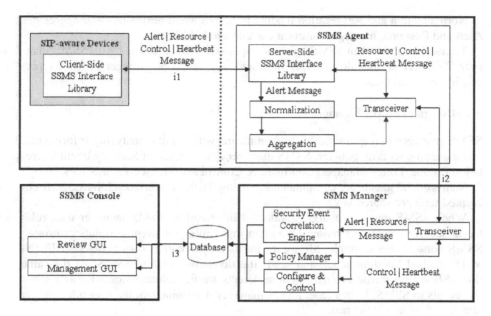

Fig. 1. Structure of SSMS. SSMS is composed of SSMS agent, SSMS manager, and SSMS console.

- Resource message: Resource message is newly defined to include information such as CPU utilization, memory utilization, disk utilization, network interface card, SIP traffic statistics, SIP session statistics, and so on
- Control message: Control message is newly defined to include information such as commands for starting agent, stopping agent, enforcing security policy, returning the results of enforcing security policy, and so on
- Heartbeat message: Heartbeat message is used to indicate current status of devices. Heartbeats are intended to be sent in a regular period. SSMS can use Resource messages as health indicator, because Resource messages are designed to be sent to SSMS in a fixed period.

Message exchanging protocol is defined based on the above message format and [11]. SSMS message exchanging protocol has APIs as follows:

- ksaStartServer() : start server for connecting with agents
- ksaStartClient() : start client for connecting with servers. If disconnected, automatically reconnected.
- ksaCreateEvent() : create event data structure to prepare for exchange events
- ksaWriteEvent() : write data into event data structure
- ksaSendEvent() : send event to server or client
- ksaSendPolicy() : send security policy event to agent
- ksaSendResult() : send the result of enforcing security policy to server

These APIs are implemented as Client-Side and Server-Side SSMS interface library for interface i1 as shown Fig.1

Normalization and Aggregation module in SSMS agent normalizes and aggregates Alert and Resource messages for event correlation.

Transceiver module in SSMS agent is responsible for connecting with SSMS manager. Transceiver module encrypts data and authenticates manager based on pre-configured IP address.

3.4 Design of SSMS Manager

SSMS manager is responsible for communicate with agents, analyzing information, and managing security policies. SSMS manager is composed of Security Event Correlation Engine, Policy Manager, Configure & Control and Transceiver modules.

Manager and agents are communicating using TCP. In interface i2, message is encrypted with pre-shared key.

When SSMS manager analyzes collected information, SSMS manager uses rule-based security event correlation. Users can input rules for event correlation through SSMS console. For example, SIP-aware intrusion prevention system detects SIP INVITE method flooding attack and sends the detection event to SSMS. At the same time, SIP-aware traffic anomaly system detects traffic volume anomalies and sends the events to SSMS. In this case, SSMS manager determine that the network is under attack with more confidence.

Policy Manager and Configure & Control translate security policies and commands generated from GUI into Control Message for agent.

3.5 Design of SSMS Console

In SSMS Console, there are graphical user interfaces (GUIs) for viewing collected information such as system resource, security events, and the result of security event correlation. Also there are GUIs for managing event correlation rules and security policies.

SSMS console sends and receives information to/form database. Database schema and connectivity information for interface i3 in Fig.1 is open. So any GUI programs which are compliant with interface i3 can utilize SSMS manager.

4 Implementation of SIP-aware Security Management System

4.1 Test Environment for SSMS

As shown in Fig.2, test environment for SSMS is composed of four subnets.

The first subnet represents the domain of SIP service providers. On this subnet, there are SIP-aware intrusion prevention system, SIP-aware traffic anomaly detection system, SIP proxy server, session border controller, and SSMS.

The second subnet represents the domain of attackers. On this subnet, there are SIP attack tools and SIP malformed message generators.

The third subnet represents the domain of victims. In the victim domain, there are VoIP hard phones and soft phones.

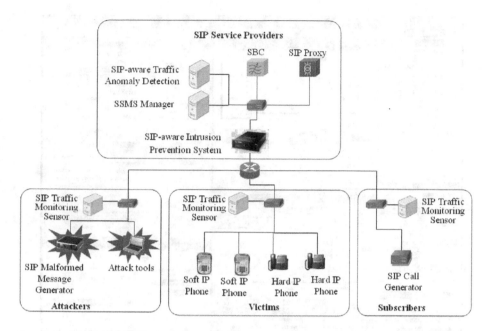

Fig. 2. Test Environment for SSMS. In this test bed, there are four subnets for service providers, attackers, victims, and legitimate subscribers.

The fourth subnet represents the domain of legitimate subscribers. To simulate legitimate calls among users, a SIP call generator is used. The SIP call generator generates 2,000 concurrent calls per second.

4.2 Viewing GUIs of SSMS

Fig. 3 shows main view of SSMS. Users can monitor all the devices that SSMS manages through this view. In the upper left pane, users can select host to watch and monitor. In the lower left pane, users can configure refresh rate for each window, and search information by choosing date, severity, and so on.

Fig. 4 shows the result of security event correlation. Fig.4 shows "SIP scan" alarm were fired. In the lower part of the window, evidences for firing "SIP scan" alarm are displayed. In this case, within predefined period (1 minute), SSMS manager receives three events from a host whose IP address is 10.3.10.100. One SIP_CALL_EVT shows that there are more SIP 5xx responses than predefined threshold count. Two SECURITY_EVTs show that the host detects SIP scanning attack. Generally, attackers use random SIP to-URIs for SIP scanning attack. In test environment described in section 4.1, when SIP proxy server receives messages with unregistered to-URIs, it responds with 503 response message. Therefore, increasing of SIP 5xx response messages can be supporting evidence to detect SIP scanning attack.

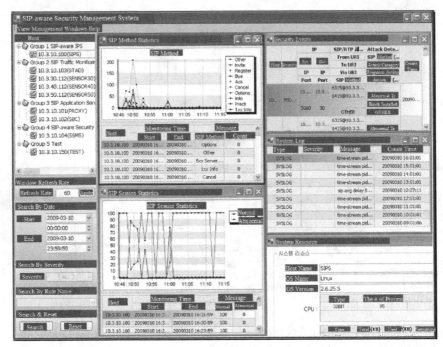

Fig. 3. Main View of SSMS. Users can monitor SIP method statistics, SIP session statistics, security events, system log, system resources, and so on. Because SSMS console was developed in Korean, titles and captions for windows from Fig.3 to Fig.7 were replaced by those of English.

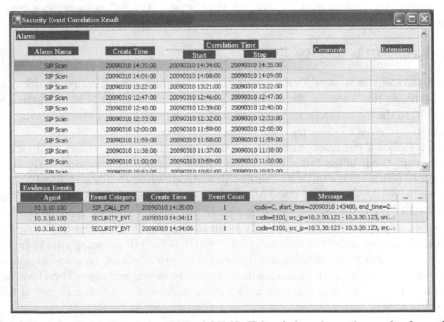

Fig. 4. Security Event Correlation GUI of SSMS. This window shows the result of security event correlations.

4.3 Management GUIs of SSMS

Fig. 5 shows management window for security event correlation. Users can create, retrieve, modify, and delete rules using this window. Fig.5 shows that users can make alarms by using SIP method, SIP From-URI, SIP to-URI, SIP Via-URI as well as attack category, attack severity.

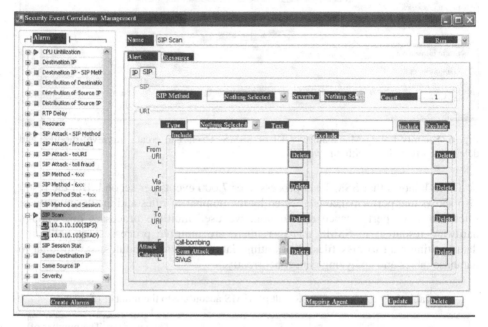

Fig. 5. Security Event Correlation Management GUI of SSMS. Users can create, retrieve, modify, and delete rules with this window.

As shown in Fig. 6, users can edit security policies for SIP-aware devices. User command to SIPS (SIP-aware intrusion prevention system) that SIPS block packets based on SIP from-URIs. In particular, any packets from attacker@10.3.10.123 must be blocked. If user click "Yes" button, SSMS manager sends Control messages to SIP-aware devices.

4.4 Performance Test Result

For middle and large size organization, average 5,000 events can be processed per second in commercial products in Korea.

For testing performance, we increased the number of security event correlation rules from 0 to 20. And we made rules at various level of complexity. Simple rule has a form like "for any Alert messages, if severity is high and the number of same Alert messages within predefined period is over five, then alarm". Complex rule has a form like "If any 10 Alert messages have same source IP and source port, but different destination IPs, then fire alarm".

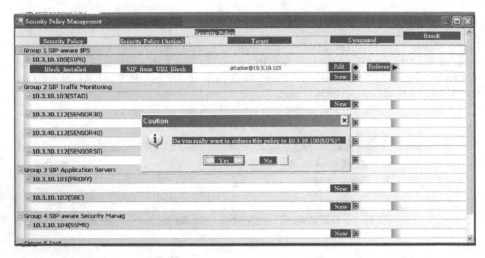

Fig. 6. Security Policy Management GUI of SSMS. Users can create, retrieve, modify, and delete security policies with this window.

Table 1 shows that SSMS can process over 7,000 events per second. But the result also shows that performance is affected by the number of rules and the complexity of rules. To solve performance degradation, we used multiple processes and enough hardware resource. In SSMS manager, there are multiple processes for collecting data, writing data to disk files, distributing data to internal modules, saving data to database tables and analyzing data concurrently.

Table 1. Performance Test Result of SSMS according to the number of rules

No.	The number of rules	The number of test events generated	Elapsed time to generate test events	Elapsed time to save events into database	The number of events saved in database	The number of events processed per second
1	0	500,000	57 sec	50 sec	500,000	10,000/ sec
2	5	500,000	57 sec	49 sec	500,000	10,204/ sec
3	10	500,000	57 sec	66 sec	500,000	7,575/ sec
4	20	500,000	61 sec	70 sec	500,000	7,142/ sec

5   Conclusions and Future Work

In this paper, security management system which can collect and analyze SIP specific information was proposed. We described design considerations and overall design of the system. And then we showed the result of implementation with captured images of our system. We also showed the result of performance test according to the number of rules. As a result, SSMS could collect security events from heterogeneous SIP-aware devices by using standardized message format and message exchanging protocol. Message exchanging APIs were implemented as a Linux library. With this library, it is expected that we can collect information from other devices with fewer efforts.

Until now SSMS uses rule-based event correlations with static threshold count. Rule-based event correlations with adaptive threshold without affecting performance will be studied further.

Acknowledgement

This work was supported by the IT R&D program of MKE/IITA. [2008-S-028-02, The Development of SIP-Aware Intrusion Prevention Technique for protecting SIP base application Services].

References

1. IETF RFC 3261, SIP: Session Initiation Protocol,
   http://www.ietf.org/rfc/rfc3261.txt
2. http://www.netiq.com/
3. http://www.q1labs.com/
4. http://www.acmepacket.com
5. Nassar, M., et al.: Holistic VoIP Intrusion Detection and Prevention System. In: IPT-COMM 2007 (2007)
6. Ko, K., et al.: The Design of SIP-based Enterprise Security Management System. In: Conference on Information Security and Cryptology – Winter, Written in Korean (2008)
7. IETF RFC, RTP: A Transport Protocol for Real-Time Applications (1889),
   http://www.ietf.org/rfc/rfc1889.txt
8. Jiang, G., et al.: Temporal and Spatial Distributed Event Correlation for Network Security. In: American Control Conference (2004)
9. http://www.3gpp.org
10. IETF RFC 4765, The Intrusion Detection Message Exchange Format (IDMEF),
    http://www.ietf.org/rfc/rfc4765.txt
11. Internet Security Technical Forum, Telecommunications Technology Association, Written in Korean (2006)

Application Management Framework in User Centric Smart Card Ownership Model

Raja Naeem Akram, Konstantinos Markantonakis, and Keith Mayes

Information Security Group Smart card Centre, Royal Holloway,
University of London Egham, Surrey, United Kingdom
{R.N.Akram,K.Markantonakis,Keith.Mayes}@rhul.ac.uk

Abstract. The predominant smart card ownership model is the issuer centric, and it has played a vital role in the proliferation of the technology. However, recent developments of multi-application smart card technology lead to new potential ownership models. One of the possible models is the User Centric Smart Card Ownership Model. In this model, the ownership is with smart card users. To support user's ownership, we require a framework that can assist cardholders to manage applications on their smart cards. In this paper, we present such a framework for managing application securely on a smart card.

1   Introduction

Historically, the smart card ownership resides with organizations (card issuers) that provide smart card based services. Smart cards issued by the card issuer will have pre-installed applications, and they cannot customise to suit customer's requirements. This ownership model lacks flexibility, ubiquity and is inconvenient to cardholders.

In last two decades, the smart card technology evolved to support multiple applications. The adoption of multi-application smart cards was hindered by card issuers concerns over the ownership of the card and customer relationship along with branding issues. A possible solution to these issues is to delegate the ownership to users. This proposal is referred to as the User Centric Smart Card Ownership Model (UCOM), which is based on providing the complete control over the choice of applications on a smart card, securely and efficiently, to its cardholder. To do so, cardholders would require a secure and practical mechanism to perform application management tasks efficiently. In this paper, we discuss the need for the new ownership model and describe how it is different from the existing models. The main focus of the paper is the procedures and functions performed by a smart card and a service provider to install or delete an application in the UCOM.

In section two, a short description of the UCOM is provided along with the motivation for the new ownership model. Section three describes the architecture of the Application Management Framework (AMF) that supports the application installation and deletion process on the UCOM-based smart cards. The

H.Y. Youm and M. Yung (Eds.): WISA 2009, LNCS 5932, pp. 20–35, 2009.

application management processes (e.g. install, delete, etc) are described in section four. Section five provides an analysis of the proposed framework. Section six briefly looks on future research directions and finally, section seven draws the conclusion.

2   User Centric Smart Card Ownership Model

In the following sections we provide the motivation behind the User Centric Smart Card Ownership Model (UCOM) proposal along with its architectural overview.

2.1   Motivation

The multi-application smart card technology, except for the initial popularity it never took off. However, recent developments mainly driven by the technologies like Near Field Communication (NFC) [1] and Secure Element (SE) [2] in mobile phones have revived again the concept of having multi-applications on a smart card (chip).

The NFC enables a contactless data exchange between a chip (i.e. smart card) and the terminal. It is also extended to include the mobile phones that enable them to emulate the contactless smart cards. As a result, the existing infrastructure deployed in the different industries (i.e. banking, transport, access control) to support contactless smart card can be utilised. There are many organisations around the world that are currently engaged in the field trials [3, 4, 5], and they are fostering new business models to actively manage the multi-applications through mobile phones.

To support the initiative, there are many different proposals to manage the SE in the NFC based mobile phones. One proposal is to keep the traditional ownership model so that the card issuer (i.e. Telecom) will have the ownership. This model has traditional issues related to the ownership of smart cards and customer relationship. Another model is to delegate the control to a third party that does not use the SE to provide any services to end users. Such a model is referred to as the "Trusted Service Manager" (TSM) based model [6]. In this model, the trust relationships with Telco operators and other service providers are maintained by the TSMs. Eventually it enables the SE to host multiple applications from different companies). Each company only has to establish an individual trust relationship with a TSM.

However, the UCOM goes further by giving choice of applications on a card to its user. The card assures a Service Provider (SP) of its underlying security state and if satisfied the SP's is satisfied; it can lease its application(s). The difference between the TSM and UCOM is that TSM still requires trust relationship between service providers and a TSM that may involve business and financial agreements. This may discourage small businesses (e.g. public library, health centre, leisure club). In the UCOM the small companies only require to develop their applications, and they can be installed onto their customer's SE in a cost effective way.

The multi-application smart cards platforms (i.e. Java [7], Multos [8]) support the installation of applications remotely (after issuance of the card). The standardisation efforts to manage the application remotely like the GlobalPlatform [9] have been effective in the Issuer Centric Smart Card Ownership Model (ICOM). In the ICOM, the control of the card is with a single organisation and they manage the relationship with other organisations that may wish to share the smart card. In these situations, there is always an entity (i.e. card issuer) that has a pre-issuance secure binding with the smart card. The security measures are implemented by card issuers and they provide the security assurance. The pre-issuance secure binding and control of security measures implemented on smart cards provides a secure and reliable model. This notion is based on the presumption that the ICOM is a closed environment and applications are rarely installed and deleted from a card.

The ICOM based frameworks including the GlobalPlatform are proposed with the assumption that the ownership will be either with a card issuer or a third party. This assumption is not necessarily constructive when dealing with the user's ownership of the smart card. The ownership gives the provision to install and delete any application that also brings new security and privacy issues that are not present in the ICOM. The presented proposal is designed with a basic principle that the underlying platform is open, dynamic and in the control of its user that may act as adversary.

2.2 Overview of the User Centric Smart Card Ownership Model

The User Centric Smart Card Ownership Model (UCOM) focuses on the delegation of the ownership (control) to its users. The term "Ownership" in the UCOM does not imply that users own the application(s) installed onto their smart card(s). It only means the freedom of choice to install or delete any application(s). The ownership of applications will always remain with their corresponding SP. The SPs will only lease their applications, after specific security, privacy and operational requirements are satisfied by the UCOM-based smart card. The provision to install or delete an application cannot be performed without the prior authorisation of the relevant SP.

The UCOM-based smart cards should support the ownership of the cardholder and provide adequate functionality for the application management tasks. In addition, it should provide security assurances to SPs who lease their applications. As a crucial design requirement a UCOM-based smart card should be an impartial, secure and robust platform. The impartiality in the UCOM refers to providing assurance that the card does not favour any application or particular set of applications. The following figure illustrates the architectural overview of the UCOM.

In the UCOM, a cardholder acquires a smart card from UCSC supplier. A smart card that supports UCOM is referred to as User Centric Smart Card (UCSC) and a UCSC supplier can be a smart card manufacturer, an SP or a third party vendor. After acquiring the UCSC, the cardholder can request an SP to lease its application(s). The SP will decide the lease based on its Application

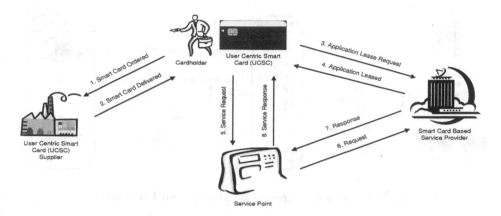

Fig. 1. Illustration of the User Centric Smart Card Ownership Model (UCOM)

Lease Policy (ALP). If the requesting UCSC meets the ALP, the application is leased, otherwise the request is denied. In addition to requesting the lease of an application, the cardholder could also request the removal.

An Application Lease Policy (ALP) defines the minimum requirement of an SP that an UCSC has to satisfy. The APL not only governs the lease of the application(s), but also the terms of the lease. The terms of the lease stipulate the minimum security, privacy and operational requirements of an application while it is installed onto an UCSC. The UCSC will provide adequate measures to enable an application to verify the execution environment before executing. Furthermore, the lease of an application can be temporary (time/execution constraint) as defined by the ALP. The UCSC or the application will initiate the deletion command once it reaches the expiry. After application(s) is leased, the cardholder can request the SP's associated services that are entitled to the cardholder (application) via a service point. A service point is a point of service device (i.e. ATM, Access Controllers) where a user presents his/her smart card to utilise certain services. The basic function of a service point is to connect an application to the relevant SP, so the application can authenticate itself before the user is being facilitated by the service point in accessing the SP's services.

SPs will make their application for installation ubiquitously accessible to their customers by offering them through a web server, referred as an Application Management Server (AMS). In addition to the AMS, SPs also have an Application Services Authentication Server (ASAS). These two servers are essential to support the UCOM from SPs perspective. SPs will provide their customers with the AMS credentials (i.e. AMS web address) and user's credentials (i.e. Account ID, login/password) that they can use to access and authenticate to the AMS.

An AMS typically deals with the application management processes (i.e. installation, deletion). The application management processes also include enforcing the ALP, ensuring that the application is transmitted and installed securely

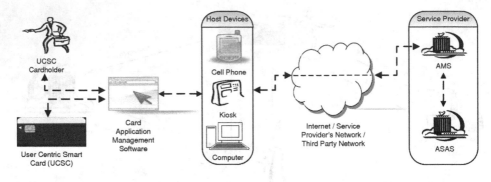

Fig. 2. Illustration of Application Management Framework

.onto a smart card, and managing a user's profile. The user's profile keeps record of the registered smart cards and card(s) that hold the active lease. Depending upon an SP's ALP, a user can have the application installed onto multiple cards; therefore, the AMS keeps track of all cards of a particular user that hold/held the lease.

3   Application Management Framework

An overview of the framework is provided in section 3.1. Section 3.2 explains the basic UCSC configuration required to support the framework. The establishment of a secure connection between an UCSC and an SP's AMS is described in section 3.3.

3.1   Application Management Framework Overview

The UCOM Application Management framework (AMF) that stipulates the mechanism of application installation and deletion is illustrated in Figure 2. To initiate the AMF processes a cardholder presents his/her UCSC to a host device. A host device (i.e. cell phones, kiosks, and computers) acts as the connection bridge between the smart card (i.e. UCSC) and the AMS. The cardholder will provide his/her account credentials for an AMS to the Card Application Management Software (CAMS). The basic functionality of the CAMS is to provide an interface (between a cardholder, UCSC and AMS) and protocol conversion (if required). The protocol conversion addresses any incompatibilities between a smart card and an AMS supported protocols. For example, a smart card may not support TCP/IP protocol so protocol conversion will provide the TCP/IP support. After a cardholder's authentication with an AMS, a secure channel is established (section 3.3) between the UCSC and AMS. The next phase involves the initiation of the required tasks (i.e. installation, deletion, etc) that are discussed in section 4.

3.2   Basic User Centric Smart Card Configuration

The basic design principle of the UCOM is to be independent of underlying Smart Card Operating System (SCOS) [10] or platform. However, for practical and security reasons we have to define the minimum requirements for different components of the UCOM. The minimum requirement that a UCSC should satisfy for the AMF is to have an SSL/TLS public key pair and public key certificate [11]. A UCSC will have a SSL/TLS public key pair and certificate, irrespectively of the underlying protocol (i.e. TCP/IP [12] and SSL/TLS [11]) handling. If an UCSC supports a web server [16] along with the TCP/IP and SSL/TLS protocols, the secure communication channel would be established entirely by the UCSC, otherwise the CAMS should provide the protocol conversion functionality. In any situation, all cryptographic functions are only handled by the UCSC.

The AMF uses both symmetric and asymmetric cryptography [13] to provide security and privacy services. The cryptographic keys used beside the SSL/TLS keys are generated by the AMSs and smart cards. These keys are lease specific and when the lease expires or the cardholder requests the deletion of the application, all cryptographic keys associated with the application will also be deleted. The UCSC supports the domain mechanism for post-installation application lifecycle management as in the GlobalPlatform (GP) [9]. The subtle difference between the GP and UCOM domain mechanism is the non-availability of Issuer's Domain. In addition, no entity (i.e. card manufacturer, SP and cardholder) has ownership of the security domain of the UCSC. The reason for not giving the control of the security domain is to avoid the possibility of indirect control of the UCSC and also to ensure SPs that there will not be any over-riding privileges for an entity.

The UCSCs will have adequate mechanisms to ensure SPs that they satisfy their ALP. One of the integral parts of the APL requirement verification is the validation of the security of an UCSC. The existing security validation is based on initiating the security evaluation of the smart card according to the Common Criteria (CC) [14]. At the end of the Common Criteria Evaluation, the smart card is given the Common Criteria Security Evaluation Assurance Level (EAL). The EAL determines how thoroughly the evaluation is performed and the security of the underlying hardware and software. In the ICOM environment, the EAL is not present on the smart card and it is a certificate that is mostly kept off-card by the organisation (card issuers). However, in the UCOM the Common Criteria Security EAL can play an important role to certify the level of security assurance that an UCSC provides. This can be done by the on-card Common Criteria Security Evaluation Certificate (CC-Certificate). The certificate is cryptographically protected in order to provide the EAL of the platform. As the certificate can only provide assurance of the state of the security at the time of evaluation and the card manufacture may opt to deploy weaker security. To avoid this, the certificate also contains an image (created by cryptographic hash function [13]) of the underlying hardware and software. The smart card itself or an SP's application can request the self-test of the card to gain the assurance of the security. The self-test basically generates the image of the underlying software (Smart Card Operating System) and optional hardware configurations. This image is then verified with the image associated with CC-Certificate. If both match, it can be safe to assume that the platform is at similar state as it was when the CC evaluation was carried out. To perform the image measurement and then comparison, the possible solution can be a Trusted Platform Module (TPM) [15] for the smart card. The scope of exact solution to provide the assurance for the security of an UCSC and how different components will interact with each other is beyond the scope of this paper.

3.3   Secure Channel Establishment between an UCSC and an AMS

A cardholder will initiate the connection with an AMS through the CAMS interface. The user provides the AMS details (e.g. web address) to the CAMS. that initiates a connection. The AMS establishes a secure connection (i.e. SSL/TLS [12]) with the requesting CAMS. After establishing the connection, the AMS requests the user's credentials. The user provides his/her credentials (i.e. account ID, login/password) through the CAMS interface. The details and type of the credentials are on the SP's sole discretion. The AMS verifies the credentials, if it is successful, it will allow the access to its services, and otherwise the connection is terminated. After the authentication the two-way SSL/TLS [12] session is established between the smart card and the AMS. There are well tested and secure protocols already in the public domain; therefore, this paper does not focus on designing a new protocol. However, the secure channel protocol established between the smart card and the AMS should be based on Public Key cryptosystem (e.g. SSL/TLS). After a secure channel protocol between a smart card and an AMS is established then cardholders can request application installation or deletion, which will be discussed in the next section.

4   Application Management Processes

In this section the crucial process of installation of an application is described in section 4.1.

4.1   Installation Process

In this section, the processes that support the secure transmission and installation of an application are discussed. In the ICOM based environment, there are many secure and robust application delivery mechanisms, most notably by the GlobalPlatform.

Most of these mechanisms rely on the assumption that the smart card is in a closed environment and under the total control of the card issuer. The card issuer has a secure binding with their smart cards before they are issued to their customers. Therefore, there is an implicit trust on the smart card in the ICOM, and most of the protocols are based on it. However in the UCOM, there is no implicit trust on the smart card. Therefore, the installation process has not only to take this into account but also that the smart card can be under the control of a malicious user, or it may not be a real smart card (card emulator running on a personal computer). The installation process discussed in this section builds the additional checks around the existing application installation protocols (without preferring anyone) that can provide the assurance of secure and reliable application installation.

The installation request will initiate the process of acquiring an application from an AMS and install it on a UCSC. The entire process can be divided into six sub processes listed below.

1. Requirement Verification
2. Domain Creation
3. Downloading
4. Application Verification by card
5. Localisation (Installation)
6. Personalisation
7. Application Registration by AMS

Each of these sub-processes is explained in sections 4.1.1 to 4.1.7. The application deletion process is similar but the steps will be performed in reverse order.

4.1.1 Requirement Verification

Before the lease of an application, the AMS will verify the compliance of an UCSC with its APL. This verification is illustrated by the flowchart shown in figure 3.

A UCSC creates the Application Request message that contains the UCSC details. The details include the CC-Certificate, UCSC manufacturer certificate, details of the SCOS/runtime environment (i.e. Java Card [7], Multos [8], etc), supported cryptographic algorithms, and communication interfaces (e.g. T1, T2 or CL [10], web [16]]). The manufacturer certificate validates the cryptographic public keys pair and hardware tag. The hardware tag is unique sequence that identifies the UCSC. The length of the tag and how it is generated is on the sole discretion of the UCSC manufacturer. Requirements on the hardware tag by the UCOM are that it does not violate the privacy of the cardholder and actively verifies that the AMS is communicating with the real card (not an emulator).

The AMS will verify whether the requesting UCSC satisfies the ALP. If so, it continues, otherwise process terminates. To verify the CC-Certificate and UCSC manufacturer certificate, the AMS can communicate with either the entity that has issued these certificates, or a third party that plays the role of intermediary between the AMS and the UCSC manufacturers. To validate that the AMS is communicating with a real smart card (not an emulator), the AMS can request the UCSC manufacturer to verify the claim of their card. The details of these processes are beyond the scope of the paper.

After validation of the APL, the AMS generates application requirement details. This contains the application space and on-card security policy requirements. Application space requirement stipulates the memory required for the

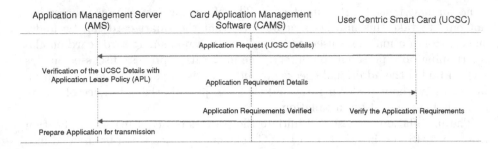

Fig. 3. Illustration of Requirement Verification Phase

application and the on-card security policy requirement includes the required firewall and application access configuration. The firewall configuration defines the mechanism through which an on-card application(s) can access (share) the requested application. The application access configuration details the mechanism through which an off-card application communicates with the application. In addition, the AMS specifies the generation requirement for the domain management key, application download key and algorithm used to encrypt the application for transmission.

The AMS and UCSC can negotiate the application communication protocol. The AMS can decide whether to use any of the UCSC implemented (open) protocols. Once the trust relationship is established and domain keys are generated, any protocols (including the GlobalPlatform) can be used to download the application on to the smart card. The AMS can also opt for their proprietary protocol to download the application. This can be achieved by first using the UCSC supported protocols to download a proprietary (small) application referred to as Application Download Manager (ADM), to a least privilege domain; the ADM will manage the download of the request application. The lease privilege domain is controlled by the UCSC, and it is a temporary domain. Applications installed in this domain are not allowed to communicate with any other applications on the smart card, which means they are in an isolated domain. The security measures will ensure that the ADM will abide by the policy of the UCSC. Before the ADM starts the execution, it will be subjected to security tests on the card to achieve the assurance that the download manager is secure and reliable to execute. During the download process, if the ADM performs any unauthorised action, the UCSC will terminate its execution and deleted it. After the application is successfully downloaded, the ADM will be deleted.

The smart card will examine the application requirement sent by the AMS. If it meets these requirements, it will send an acknowledgement to AMS and proceed to the domain creation process; otherwise, it will terminate the process.

4.1.2 Domain Creation
After the AMS and UCSC have verified each other's requirements, the next phase is to create a domain (SP's Domain), involving the following steps:

1. Allocate memory space for the SP's Domain in the EEPROM [10] (Electrically Erasable Programmable Read Only Memory).
2. After the allocation of memory space, a domain manager is installed in the allocated memory. A domain manager maintains the security aspects of the domain. Its functions are similar to a security domain in the GlobalPlatform [9]. An SP will have a view of their domain as a complete smart card.
3. After a domain is created, the Domain Delegation keys are generated. These keys can be generated in one of the following ways:
   - Either an UCSC or an AMS will generate the key and exchange it.
   - Alternatively, an UCSC and AMS can mutually generate them.

   Which of the above methods is going to be used is negotiated in the requirement verification process (i.e. section 4.1.1). Any requirements regarding the generation of the keys is solely based on SP's discretion and UCSC will follow these guidelines.

4.1.3   Downloading

After generation and mutual authentication of the Domain Delegation keys, the AMS and UCSC will start the application downloading process, shown in figure 4. An AMS will prepare the application(s) for transmitting it to the requesting UCSC. The application consists of multiple modules with varying security and operational requirements. The grouping of the application into modules of varying security and operational requirements is referred to as Application Level Modularity (ALM). Each application (i.e. banking, telecom, transport) can be divided into small modules with vary security requirement. Each application have some operational and security program code and data. These modules simply represent these logical divisions but on the line of sensitivity to the service provider. The exact framework and implementation guidelines of ALM are beyond the scope of this paper. However, for the application download process, each of the modules (i.e. group, level) of an application is encrypted with different key, and these keys are only revealed to the UCSC in incremental fashion after it satisfied the module's security and operational requirements.

The AMS will digitally sign the application with the corresponding SP's signature key, then encrypt with the transmission key. The transmission key is generated during the step three of section 4.1.2. After this the application it is transmitted to the UCSC.

The Application Download Handler (ADH) module in the UCSC handles the incoming packets. The ADH supports different application download protocols (implemented by card manufacturer). The AMS either selects one of the supported protocols or opts for its own protocol. If the SP's opt for its own protocol then the ADM (section 4.1.1) handle the application download process. The function of the ADH is to efficiently download the application in a secure and reliable fashion. The received packages of the application are not installed, because they first require the application signature validation and decryption. Therefore, downloaded applications are stored in a temporary space (in either EEPROM or RAM [10]).

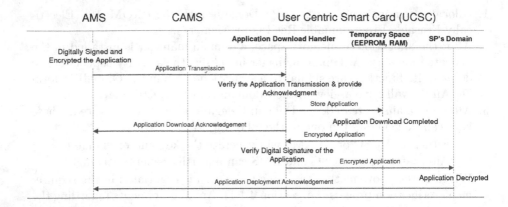

Fig. 4. Downloading an Application on an UCSC

After the download is completed, the digital signature of the encrypted application is verified. After verification of the digital signature the application is transferred to the SP's Domain and the application is decrypted there. A decrypted application is not a fully installed application. It is the equivalent of copying an application in a memory location. The decrypted application cannot be executed unless it satisfies the application verification test, discussed in the next section.

4.1.4   Application Verification by Smart Card

After an application is downloaded into the SP's domain, the next step is to verify whether the application complies with operational and security policy of the UCSC or not. A UCSC's operational and security policy defines the sanctioned operations, privileges and runtime environment restrictions on the SP's domain. To verify whether an application code conforms to specification and standards (i.e. Java Card [7], Multos [8], etc), a byte code verification is performed [17].

The byte code verification will take place on the smart card for security reasons. Performing byte code verification on the CAMS will be much faster, because in most cases it would be hosted on computationally faster machines. However, this violates the security requirement of the SP, because for a CAMS to perform byte code verification, the decrypted application would have to be transferred out of the UCSC.

The scope of this paper is not to define a byte code verifier; however, there are several well defined on-card byte code verification proposals [18, 19, 20].

4.1.5   Localisation

The application is allowed to execute on the UCSC only after it is properly verified by the UCSC. On its first execution, the application registers its security policy details with the card's security services (i.e. firewall, access manager, SP's domain manager, cryptographic services etc.). Furthermore, it may require access to specific logical or physical (i.e. contact, contactless or web server) channels. The application will register with the communication handler (service that handles communications in and out of an UCSC) and the UCSC's application manager that allows application to be selected by an off-card entity's. Once the registration is complete, is the application is considered installed and could be accessed by an off-card entity.

4.1.6   Personalisation

After localisation is completed, the SP's application will initiate the personalisation process. The personalisation data (i.e. user's specific data) is downloaded with the application; however, it is separately encrypted. The process is as listed below.

1. An SP's application creates a message that contains on-card test and localisation process response. In addition, it generates a message for application personalisation request.
2. The AMS verifies the on-card test and localisation message. If verified it will generate the message containing the cryptographic key and digital signature on the encrypted personalisation data and it sent to the smart card. On failure the AMS will terminate the process and the smart card will delete the application.
3. The UCSC decrypts the personalisation data and verifies the digital signature. If the verification fails then the UCSC request the download again. However, if the signature verification fails after multiple tries (depending upon the UCSC's policy) then the process is terminated and the application is deleted.
4. An acknowledgement message is generated to verify to the AMS that the application is personalised successfully.
5. The AMS verify the acknowledgment message and initiate the next phase (i.e. application registration).

4.1.7   Application Registration by an AMS

The final stage of an application installation on an UCSC is the application registration by the AMS. In this stage, the AMS will register the UCSC as authorised card to an Application Lease Database (ALD) hosted on the Application Services Access Server (ASAS). After the completion of this process, the UCSC will be ready to access the SP's services.

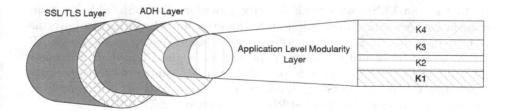

Fig. 5. Secure Communication Channel between an AMS and a Smart Card

In the UCOM, applications can be installed on one or more cards. Actually, the SP will decide whether they will allow the user to keep their application on multiple cards or not (i.e. Application Lease Policy). For certain applications, being on multiple cards would not be an issue like banking application (as it requires PIN to use the card. Therefore, if a person provides the correct PIN and it posses the appropriate application in his UCSC that means the owner was present at the point of transaction).

5   Critique of the Framework

In this section, we will critically analyse the Application Management Framework in terms of its feasibility, practically and overall security perspective.

5.1   Security Analysis of the Framework

The most crucial and sensitive operation in the Application Management Framework is the application installation process. In this process, the application is transmitted over an insecure network from an AMS to a User Centric Smart Card (UCSC). The figure 6 illustrates the security envelopes on the application in transit over an insecure network.

The top envelope is provided by the two-way SSL/TLS session established directly between an UCSC and an AMS. During the second phase (i.e. Domain Creation) of the installation process, the Domain Delegation keys are generated. Along with these keys the Application Installation key (i.e. transmission keys) are also generated that are used by the Application Download Handler (ADH) to securely download the application from its respective SP. The Application Installation keys are only used once, at the time of installation and then they are securely discarded. During the deletion process, an SP only needs to use their Domain Delegation keys to instruct the deletion command to their application. The final layer of protection in the application installation process is provided by Application Level Modularity (ALM). In ALM each group of modules with the same security association and requirements would be encrypted separately with different keys, and each of the keys are only provided to an UCSC after it satisfies the associated requirements for the module level. Therefore, an application has at least three security layers for the secure communication on the insecure network during the installation process, as shown in the figure 6. An obvious attack can be to reset an UCSC with weak security provisions (i.e. SSL/TLS key pair, domain key generation mechanism etc). However, an SP will always have the right to deny any UCSC that cannot satisfy its requirements. Therefore, the SP has to be sure of security measures implemented by the UCSC before it leases its application. This assurance is provided by the Common Criteria Security Evaluation Certificate, UCSC manufacturer certificate and self testing or state assurance mechanism (i.e. Trusted Platform Module in SE).

In addition, attacks like fault attacks or Side-channel attacks [21] on a smart card can be mounted against an UCSC. As a protection measure against the fault attacks during the installation/deletion process (either to corrupt the UCSC or the application to retrieve sensitive data), the UCSC will be in defensive mode. The defensive mode enables an UCSC to save the secure, operational and reliable state of the UCSC before proceeding with the installation process. In addition, the UCSC intercepts each instruction and determines whether the execution of the instruction will violate the safe state of the UCSC or not. If it is safe to execute, it will allow the instruction to be carried out; otherwise, it will terminate the process. If anything goes wrong, the UCSC aborts the installation process and reset to the safe state (saved before the process initiated. The safe state of the smart card represents the state of the operational, and security modules (i.e. Firewall, Access Controller, Communication Channels, Execution Environment) of the UCSC that are responsible for the smart card platform reliability and security. The defence against the side channel attacks is mostly implemented on the hardware layer. Therefore, Common Criteria Security Evaluation Certificate will test the security mechanisms that provide protection against the Side-Channel attacks. If a service provider does not accept the certificate, the process will be terminated. The SPs are in total control of their applications and they have the sole discretion whether to lease or deny the lease request, this assures the service providers that their application will only be installed on an UCSC that meets their ALP, after their authorization.

5.2   Operational Critique of Framework

The User Centric Smart Card Ownership Model (UCOM) emphasise on delegating the ownership of the smart card to its user. Therefore, the framework that supports the UCOM proposal has to be user friendly and less complex. The application installation, especially on the smart card, is technically challenging. Therefore, the proposed Application Management Framework (AMF) performs majority of the processes without the cardholder's interaction. It will be less prone to errors if the user interaction during the application management processes is limited.

As a result of the recent technological developments in mobile handsets (i.e. Near Field Communication), there is a renewed interest by large scale horizontal industries (i.e. banks, transport and telecom operators) in the multi-application smart card initiative. The UCOM capitalises on the development and gives the opportunity to small organisation to develop applications that can be deployed on their user's UCOM supported NFC enabled mobile phones. The framework proposed in the paper does not rely heavily on telecom operators. The basic requirement is to establish an internet connection with the service provider's Application Management Server (AMS). This can be achieved through connecting to the internet by wireless internet, Bluetooth or cable connection (by connecting the mobile phone with a personal computer). The overall framework would not require any change to cope with different intermediary networks (protocols) that establish a connection with the service provider's AMS. In certain situations (i.e. small organisations, home environment), there is no need to set up AMS and connect the UCOM-based smart card through internet. In such cases the user can actually install the application, by directly connecting with the AMS without an internet. The only change to the framework is on the Card Application Management Software (CAMS) that has to connect to a local computer through any of the supported bearers (i.e. Bluetooth, wireless, USB). This makes the UCOM capable of being deployed in local environments that are difficult to achieve in the SE management framework that relies on Trusted Service Manager (TSM). Possible applications in the local environments can be access controller (i.e. doors, car, computers), home appliance management application (controls the intelligent kitchen, home appliances), utility meters payment/management, local library, school/college and local grocery stores application etc.

6   Future Research Directions

This paper should not be considered as one that has solved all of the issues relating to the UCOM or the AMF. The issues that are still not resolved are listed in this section.

- Common Criteria Security Evaluation Certificate: The Common Criteria do not stipulate a security certificate on the smart card itself. The certificate discussed in the paper provides an assurance of security from the neutral security evaluators. SPs can rely on their evaluations to verify the security claim of the UCSC. Further research is required in the security evaluation certificate mechanism and how they can be integrated with other components (i.e. TPM) on an UCSC to provide security assurance to SPs.
- Firewall: Firewall designs of modern smart cards are based on the presumption that the smart card is under complete control of the card issuer. Due to this assumption, the firewall is designed from the point of view of what can be shared. Therefore, traditional firewall mechanism in the smart cards may not be adequate for the UCOM.

7   Conclusion

In this paper, a framework is presented for application management in the User Centric Smart Card Ownership Model (UCOM). The operations preformed by the smart card and Application Managment Server (AMS) are provided and such operations are possible with the present state of the technology as most of these measures are already implemented to support other mechanisms. In addition this paper also provides the associated issues that are required to be resolved to fully explore the user's ownership model. Further work will be conducted in order to attempt to answer the pointes mentioned in the future research directions.

References

1. Near Field Communication and the NFC Forum: The Keys to Truly Interoperable Communications, White Paper (November 2006)
2. The GlobalPlatform Proposition for NFC Mobile: Secure Element Management and Messaging, GlobalPlatform, White Paper (April 2009)
3. Near Field Communications (NFC). Simplifying and Expanding. Contactless Commerce, Connectivity, and Content, ABI Research, Oyster Bay, NY (2006)
4. Pay-Buy-Mobile: Business Opportunity Analysis, GSM Association, White Paper 1.0 (November 2007)
5. Co-Branded Multi-Application Contactless Cards for Transit and Financial Payment, Smart Card Alliance, NJ, USA, White Paper (March 2008)
6. Mobile NFC Services, GSM Association, White Paper Version 1.0 (2007)
7. Java Card Platform Specification; Application Programming Interface, Runtime Environment Specification, Virtual Machine Specification, Sun Microsystems Inc. Version 2.2.2 (March 2006)
8. Multos, http://www.multos.com/
9. GlobalPlatform: GlobalPlatform Card Specification, Version 2.2, GlobalPlatform (March 2006)
10. Rankl, W., Effing, W.: Smart Card Handbook, New York, NY, USA. John Wiley & Sons, Inc., Chichester (2003)
11. Dierks, T., Rescorla, E. (eds.): The Transport Layer Security (TLS) Protocol Version 1.1. RFC 4346 (2006), http://tools.ietf.org/html/rfc4346
12. Comer, D.E.: Internetworking with TCP/IP Vol.1: Principles, Protocols, and Architecture, 4th edn. Prentice Hall, Englewood Cliffs (2000)
13. Menezes, A.J., van Oorschot, P.C., Vanstone, S.A.: Handbook of Applied Cryptography. CRC, Boca Raton (1996)
14. Common Criteria for Information Technology Security Evaluation, Part 1: Introduction and general model, Part 2: Security functional requirements, Part 3: Security assurance requirements, Common Criteria. Version 3.1 (August. 2006)
15. Trusted Module Specification 1.2: Part 1- Design Principles, Part 2- Structures of the TPM, Part 3- Commands, Trusted Computing Group, Rev. 103 (July 2007)
16. Smartcard-Web-Server, Smartcard Web Server Enabler Architecture, Smartcard Web Server Requirements, Open Mobile Alliance (OMA) (2008)
17. Casset, L., Burdy, L., Requet, A.: Formal Development of an Embedded Verifier for Java Card Byte Code. In: DSN 2002: Proceedings of the 2002 International Conference on Dependable Systems and Networks, pp. 51–58. IEEE Computer Society, Washington (2002)
18. Barthe, G., Dufay, G., Jakubiec, L., Melo de Sousa, S.: A Formal Correspondence between Offensive and Defensive JavaCard Virtual Machines. In: Cortesi, A. (ed.) VMCAI 2002. LNCS, vol. 2294, pp. 32–45. Springer, Heidelberg (2002)
19. Deville, D., Grimaud, G.: Building an "impossible" verifier on a java card. In: WIESS 2002: Proceedings of the 2nd conference on Industrial Experiences with Systems Software, p. 2. USENIX Association, Berkeley (2002)
20. Leroy, X.: Bytecode verification on Java smart cards. Softw. Pract. Exper. 32(4), 319–340 (2002)
21. Kocher, P.C., Jaffe, J., Jun, B.: Differential Power Analysis. In: Wiener, M. (ed.) CRYPTO 1999. LNCS, vol. 1666, pp. 388–397. Springer, Heidelberg (1999)

When Compromised Readers Meet RFID

Gildas Avoine, Cédric Lauradoux, and Tania Martin

Université catholique de Louvain
Information Security Group
B-1348 Louvain-La-Neuve, Belgium
{gildas.avoine,tania.martin,cedric.lauradoux}@uclouvain.be

Abstract. RFID-based access control solutions for mobile environments, e.g. ticketing systems for sport events, commonly rely on readers that are not continuously connected to the back-end system. The readers must so be able to perform their tasks even in offline mode, what commonly requires the management by the readers of sensitive data.

We stress in this paper the problem of compromised readers and its impact in practice. We provide a thorough review of the existing authentication protocols faced to this constraint, and extend our analysis with the privacy property. We show that none of the reviewed protocols fits the required properties in case of compromised readers. We then design a sporadically-online solution that meets our expectations in terms of both security and privacy.

Keywords: RFID, Security, Privacy, Authentication, Compromised Readers.

1 Introduction

Radio Frequency Identification (RFID) is getting more and more popular in access control, especially in mobile environments, such as sport events and public transportations. In such applications, the typical framework consists in customers who each holds an RFID ticket, that is a microcircuit with an antenna, called *tag*; some agents who carry an RFID *reader* to control the tags; and a centralized back-end system that manages data about the tickets and customers.

The readers in this context are mobile embedded devices that have an intermittent access to the back-end. For example, the ticket validator of a flying agent in the site of a sport event is connected only when the agent is back to the headquarter; or, the ticket validator in a bus has access to the back-end system only when the vehicle is parked in its lot, usually during the night. Consequently, readers must be able to authenticate offline the customers.

The tickets in this context are reasonably-costed RFID passive tags. They commonly offer some reasonable computation capabilities that allow them to use symmetric key cryptography.

The common RFID-friendly authentication protocols are based on either the ISO/IEC 9798 standard or some dedicated protocols [4,18,22]. They all consider an adversary model where the communication channel between the tag and the

H.Y. Youm and M. Yung (Eds.): WISA 2009, LNCS 5932, pp. 36–50, 2009.

reader is basically not secure, and the tag can be tampered with. The security policies are thus designed following the assumption that the readers and any other infrastructure are secured. This is not enough to enforce strong security. Indeed, readers carry some sensitive information, and the security of the whole system is threatened if an adversary can compromise some of them. For instance, a PDA reader used to check RFID tickets at Beijing Olympic Games [13] could have been stolen. Hence, it is critical for an access control system to be able to restore its integrity upon detection of such an event, without renewing all the delivered tickets.

In those applications, authentication protocols are used to prevent unauthorized entries. But they also require to enforce strong privacy policies to protect the customers. The challenge of designing authentication protocols consists in providing both security and privacy. Today, there is no authentication protocol providing both properties with compromised readers. The goal of this paper is to provide a practical and deployable solution that fits our expectations.

The contributions of the paper are as follows. We raise the problem of compromised readers, and we analyze the security of the existing protocols in this new scenario. We focus on candidates whose security without compromised readers is already well-established. All of them can be considered as a specific instance of the well-known ISO/IEC 9798 standard, from the basic challenge/response to more advanced protocols such as GPS, WIPR and TanSL. We show that none of these protocols preserves all the security properties under the assumption of a compromised reader, and especially tag privacy. We design a solution based on a symmetric-key challenge/response protocol. The secret key shared between a tag and a reader is computed on-the-fly by the tag at each authentication. This computation depends on an attack counter, the identity of the reader, and a long-term key shared by the tag and the back-end system. Our authentication protocol comes along with an update protocol to renew the tag's authentication key through the attack counter when a compromised reader is detected.

The paper is organized as follows: in Section 2, the different assumptions used to analyze the security of RFID protocols are detailed. Section 3 describes the authentication protocols defined by ISO/IEC 9798 and other protocols dedicated to RFID. The security analysis of these protocols is done in Section 4. Section 5 addresses the privacy issue, presents our protocol, and provides its security analysis. We conclude the paper in Section 6.

2   Security Analysis in RFID

In this section, we discuss the architecture, the threat models chosen to analyze RFID authentication protocols, and the different classes of attacks.

2.1   Architecture

We consider an RFID system composed of a trusted back-end, a set of readers, and m tags. The readers can be sealed for their own protection, and are sporadically connected to the back-end through a secure channel. We also assume the use of low-cost tags: they have a low gate complexity ($\approx$ 4000 GE) and tamper-resistance is limited. The latter assumption implies that a unique secret per tag is needed to prevent a large-scale attack where an adversary recovers the secret of all the tags by breaking only one of them.

2.2 Threat Models

We examine two threat models in which an adversary can have different power levels. They mainly differ on the assumption made on the vulnerability of the readers. These two scenarios are described as follows.

Scenario 1. The RFID system does not have any compromised reader. This threat model is the most widespread in the RFID body of literature. The adversary can be either passive or active. She can eavesdrop, delete, swap, or alter the messages sent between the readers and the tags. She can inject her own messages and analyze the tag or reader behavior (e.g. timing attack).

Scenario 2. An adversary can compromise some readers. She is as strong as in *Scenario 1*, but she can additionally obtain all the data stored in these readers which makes her very powerful. We assume that the back-end is able to detect this event: a physical alteration of the reader can be seen (its seal broken or it did not connect with the back-end for a while). This scenario is particularly realistic in applications where the adversary can get access to the readers.

2.3 Attacks

An adversary can use different strategies to undermine an RFID system. The classes of attacks go from tag impersonation to tag privacy. The aim of a well-suitable RFID authentication protocol is to thwart these attacks, especially considering our two threat models.

Tag Impersonation. In this case, an adversary would like to impersonate a legitimate tag T to fool a legitimate reader. The latter should be convinced to interact with an authorized tag. Various methods exist to carry out this attack: a tag can be cloned or an adversary can replay messages previously sent by T. The adversary's chances of success mainly depends on how much information she has.

Denial of Service. In RFID systems, a denial of service (DoS) is when an adversary wants to make the tag unusable by any means. Here, we only consider DoS attacks related to the authentication protocol. The adversary can only exploit the protocol to mount a DoS.

Tag Privacy vs. Reader Complexity. The tag privacy refers to the protection of the owner personal data. This security issue is based on two fundamental properties: the information leakage and the malicious traceability of a tag [3]. The information leakage concerns data exchanged during an RFID communication, which can be inherent to the environment or to the tag particularly. For instance, when Alice validates her transit pass, this latter can send in the clear the type of subscription: one-year or 10-travel ticket. The adversary learns some information about Alice. The malicious traceability arises when an adversary can correlate messages from a given tag over different protocol executions. In mass transportation, if Alice's pass always sends a unique identifier when she checks in, the adversary is able to recognize Alice anywhere. The problem of compromised readers is very critical for tag's privacy: an adversary may be able to track all the tags from the data stolen in a reader.

An efficient protocol preserving the tag privacy should have a low computation complexity for the reader, with respect to the number of tags. Unfortunately, there exist a duality between tag privacy and reader complexity. As explained in [2], there exists in our framework no symmetric-key based protocol that ensures privacy with a reader complexity better than $O(m)$, where m is the number of tags in the system.

3   Available Authentication Protocols

In this section, the content of the ISO/IEC 9798 standard from part 2 to 5 is described. It covers the well-known authentication protocols based on the classical cryptographic primitives. Moreover, three RFID-dedicated protocols are analyzed: GPS, WIPR and TanSL. GPS and WIPR are respectively based on zero-knowledge authentication and on public-key encryption. It has been shown that these two protocols can meet the hardware constraint of RFID. GPS is also mentioned in ISO/IEC 9798. Finally, we study the TanSL protocol because of its interesting secret recomputation mechanism.

3.1   ISO/IEC 9798

ISO/IEC 9798 [12] is currently the international standard for entity authentication and it is widely used in RFID. Parts 2 to 4 of the standard provides four authentication protocols that are respectively based on symmetric encryption algorithms, digital signature, and cryptographic hash functions. Each part describes three mechanisms for achieving authentication: unilateral authentication with timestamps, unilateral authentication with random numbers, and mutual authentication with random numbers. The unilateral authentication with random numbers, designated as Mechanism 2 in the standard, is discussed here. Basically, the reader sends to the tag a nonce n_R called the challenge and the tag responses to this challenge. To do so, it can use:

- a symmetric encryption function (ISO/IEC 9798-2),
- a signature scheme (ISO/IEC 9798-3), or
- a cryptographic hash function (ISO/IEC 9798-4).

The protocol described in Fig. 1 corresponds to a challenge/response based on symmetric-key (SK) encryption. To simplify the notations, the figures and protocols represent the readers data for one tag T. Id_T, s, n_R and n_T denote the

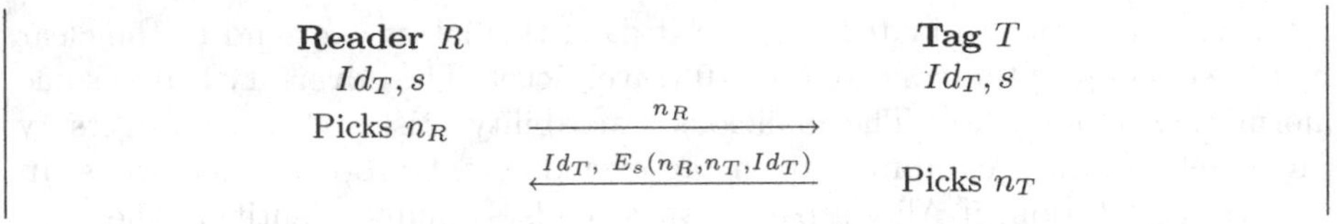

Reader R | | Tag T
Id_T, s | | Id_T, s
Picks n_R | $\xrightarrow{\quad n_R \quad}$ |
| $\xleftarrow{Id_T,\ E_s(n_R, n_T, Id_T)}$ |
| | Picks n_T

Fig. 1. A SK-based challenge/response protocol

identifier of the tag, the symmetric key shared by the tag and the reader, the nonce picked by the reader, and finally the nonce picked by the tag. The function E represents encryption algorithms. Part 5 of the standard provides several mechanisms that are respectively based on integer factorization, discrete logarithms with respect to prime or composite numbers, and asymmetric encryption. Such mechanisms can be zero-knowledge (ZK) proofs, such as Fiat-Shamir [9] protocol. In the next section, GPS, WIPR and TanSL protocols are reminded.

3.2   Protocols Dedicated to RFID

GPS Protocol. The GPS authentication protocol [4] is an interactive zero-knowledge authentication protocol initially proposed by Girault, Poupard, and Stern. It provides provable security based on the composite discrete logarithm problem. It also combines short transmissions and minimal on-line computation, using precomputed "coupons". This protocol has been selected in the NESSIE portfolio [16] and it is mentioned in the ISO/IEC 9798-5 Clause 8 as a reference. Throughout the paper, we will refer GPS as this variant "with coupons".

The parameters used in this protocol are the following:

- S, B, A are public integers, where $|S| \approx 180$, $|B| = 32$ and $|A| = |S|+|B|+80$,
- $n = p \times q$ is a public composite modulus, where p and q are secret primes, $|n| = 1024$, $|p| = |q| = 512$,
- g is an element of $\mathbb{Z}_n^*$,
- $\Phi = (B - 1) \times (S - 1)$,
- $s \in [0, S[$ and $I = g^{-s} \bmod n$,
- a coupon i is a couple $(r_i, x_i = g^{r_i} \bmod n)$, where $r_i \in [0, A[$ is a random number.

At the beginning, the tag T has a unique identifier Id_T, a unique pair of keys (s is the private one and I is the public one) and a set of *coupons* computed by a higher trusted entity (the back-end). Every reader knows the tag's identifier and public key. GPS works as follows:

(1) The tag T chooses a coupon (r_i, x_i), and sends Id_T and x_i to the reader R.
(2) The reader answers a challenge n_R randomly chosen in the interval $[0, B[$.
(3) The tag computes $y = r_i + n_R \times s$, and sends y to the reader.
(4) The reader checks if:
  - $g^y \times I^{n_R} \bmod n = x_i$
  - $y \in [0, A + \Phi[$

WIPR Protocol. This is a variant of the well-known Rabin cryptosystem [20]. WIPR was proposed by Oren and Feldhofer in [18], and improved by the same authors in [19]. It is based on early works of Shamir [21] and Naccache [15]. Recently, Wu and Stinson [23] also presented a version of WIPR with a proven security. We describe here the version of WIPR presented at RFIDSec 2008.

During the system set up, a large number $n = p \times q$ is chosen, $|n| = 1024$ bits, p and q are two prime numbers. n is public while p and q are kept secret by the reader. α and β are two security parameters, such that $\alpha = 128$ and $\beta = 80$. Each tag T has a unique secret identifier Id_T only known by the readers of the system. WIPR is a challenge/response protocol that works as follows:

(1) The reader R sends a challenge n_R to the tag T, where $|n_R| = \alpha$.
(2) The tag picks two random numbers $n_{T,1}$ and $n_{T,2}$, where $|n_{T,1}| = |n| - \alpha - |Id_T|$ and $|n_{T,2}| = |n| + \beta$.
   Then it generates a plaintext $P = BYTE\_MIX(n_R||n_{T,1}||Id_T)$ where $BYTE\_MIX()$ is a classic byte-interleaving operation. Finally, T sends to R the encryption $A = P^2 + n_{T,2} \times n$.
(3) The reader decrypts it with its private key (p, q). Like in Rabin cryptosystem, there are 4 plaintext candidates. Then R checks if one of these contains its challenge n_R. If so, then R also recovers Id_T.

TanSL Protocol. TanSL protocol denotes the first protocol presented in [22] by Tan, Sheng, and Li at PerCom 2007. It is a challenge/response protocol based on a single hash operation. The secret shared between the reader and the tag is computed by the tag at each authentication.

At the initialization of the system, every tag T has a unique identifier Id_T and a secret t_T. Every reader R has an identifier Id_R and a list L containing all the tags' identifiers and a hash value of their secret concatenated with the reader's identifier: for every tag T, $L = [Id_T : h(Id_R||t_T)]$, where h is a cryptographic hash function. TanSL works as follows:

(1) The reader R sends a request to the tag T.
(2) The tag T answers a random number n_T.
(3) The reader sends its identifier Id_R and a random number n_R.
(4) The tag computes a hash $H = h(h(Id_R||t_T)||n_R||n_T)$, where $\ell := |H|$ (e.g. 160 bits for SHA-1).
   Hb and He represent the b first bits and the $\ell - b$ last bits of H, respectively. That is $H = Hb||He$, $|Hb| = b$ and $|He| = \ell - b$.
   Then T sends Hb and a question $ques_R = (ques_R^1, ques_R^2, \ldots, ques_R^k)$, which represents k randomly chosen bit positions from He (notice that $k \leq \frac{\ell - b}{2}$).
(5) For every entry T in L, the reader computes $H' = h(h(Id_R||t_T)||n_R||n_T)$ with $H' = Hb'||He'$, and checks if Hb matches with Hb':
   – If so and $k \leq \frac{\ell - b}{2}$, it sends to T the answer ans_R to the question $ques_R$. ans_R represents the actual bits in positions $ques_R^1, \ldots, ques_R^k$ of He'.
   – Else it sends $ans_R = rand$ where $rand$ is a k-bit random number.
   In turn, it sends $ques_T = (ques_T^1, ques_T^2, \ldots, ques_T^k)$, built like $ques_R$.

(6) The tag T checks if ans_R is correct:
  − If so and $\{\forall x, \forall y : ques_R^x \neq ques_T^y\}$, it answers ans_T to $ques_T$.
  − Else it sends $ans_T = rand$.
(7) The reader R verifies the answer ans_T.

4 Security Analysis

We now study the security of all the previous protocols in the different scenarios defined in Section 2.

4.1 Tag Impersonation

As authentication protocols are designed by nature to be secure in the context of *Scenario 1*, we only focus in Section 4.1 on *Scenario 2*.

For SK-based challenge/response protocols, once the adversary compromised a reader, she knows all the secrets stored by the reader. She is so able to impersonate any tag.

For signature schemes and zero-knowledge protocols (including GPS), the private key used to answer to the challenges is only known by the tag. Thus even if the adversary compromises a reader, she does not know the tags' private keys. She cannot impersonate them.

Regarding WIPR, an adversary who compromised a reader R knows its public and private keys (n and (p, q)) and the tags' identifiers. The result is that she will be able to impersonate any tag to every reader.

For TanSL, an adversary can obtain from a compromised reader R its identifier Id_R and the list L containing all $(Id_T : h(Id_R\|t_T))$, for every tag T. The adversary will not be able to impersonate a tag T in front of any other non-compromised reader R'. Indeed, she does not know the tag's secret t_T, thus she is not able to compute the symmetric key $h(Id_{R'}\|t_T)$ shared between R' and T.

4.2 Denial of Service

The problem of denial of service remains the same for either *Scenario 1* or *Scenario 2*. When an authentication protocol does not modify the content of the tags, no DoS attack is possible in both scenarios. Therefore, all the protocols presented in this paper, except GPS, are resistant to such an attack.

As GPS uses coupons, a DoS attack is feasible. Actually, a tag can perform a limited number of authentications, i.e., one authentication consumes one coupon. The number of coupons available is bounded by the tag memory. As there is no reader authentication, an adversary can ask many authentications to a tag T in a very short time. She can exhaust all the tag's coupons almost instantaneously without T's agreement. T will no longer be able to successfully perform the protocol. If GPS is used "without coupons", there is no DoS attack. However, it increases the number of computations for the tag. This version of GPS has not been considered in [4] for lightweight applications such as RFID.

4.3 Comparison

In practical terms, a first comparison is necessary to find out which protocols can be implemented in wired logic for low-cost RFID passive tags.

In Table 1, we compare the hardware requirements of typical authentication protocols based on challenge/response and of specific protocols (like WIPR and GPS). For typical authentication protocols based on symmetric, asymmetric encryptions and hash functions, the results represent the most costly part of the implementation. As it is quoted, symmetric encryption has good results: AES is efficient in such tags, so is PRESENT. For asymmetric encryption, NTRUEncrypt has been considered as a great candidate in [1], however the parameters of the system achieving good security are not yet known [10]. Also it is commonly admitted that an RSA encryption core cannot fit in low-cost passive tags. The same observation can be made for classical elliptic curves cryptosystems [11]. Actually, the major problem of cryptosystem based on integer factorization or discrete logarithm problems, e.g. signature schemes or zero-knowledge protocols (included GPS "without coupons"), is that their implementation is too costly to fit in less than about 4000 GE. An exception is GPS: it is suitable for low-cost tags, since it requires only 1642 GE [14]. However, the coupons limit the number of authentications. WIPR is a great candidate for asymmetric encryption, as its chip area is reasonable.

In the case of ticketing applications, we consider that the AES encryption time is the reference. WIPR is not fast enough in comparison to the AES ($\times 66$ slower), whereas GPS is faster than the AES ($\times 2.5$ faster). TanSL and SK-based

Table 1. Comparison of different cryptosystem implementations

Type	Algorithm		Frequency [kHz]	Chip Area [GE]	Clock Cycle
Symmetric	AES-128	[8]	100	3400	1032
	PRESENT-80	[5]	100	1570	32
Asymmetric	WIPR	[19]	100	4682	66048
	ECC-163	[11]	100	14976	296000
	NTRUEncrypt	[1]	500	3000	28390
Hash	SHA-1	[17]	100	5527	344
	SHA-256	[6]	100	10868	1128
ZK	GPS	[14]	100	1642	401

Table 2. Comparison of the presented protocols

Scenario	SK Ch./Re. 1	2	Sign. - ZK 1	2	GPS 1	2	WIPR 1	2	TanSL 1	2
Implementation	+		-		+		+		+	
Efficiency	+		-		+		-		+	
No Tag Impersonation	+	-	+	+	+	+	+	-	+	+
No Denial of Service	+	+	+	+	-	-	+	+	+	+

challenge/response are the only protocols relying on a common primitive (SHA-1, AES, PRESENT-80) that achieves a reasonable speed.

We also need a protocol which achieves all the security properties in addition to lightweight implementation and efficiency. For *Scenario 1*, TanSL and SK-based challenge/response provide all the security features. GPS is vulnerable to DoS attacks in any of the scenarios. However, SK-based challenge/response protocols do not prevent from tag impersonation in *Scenario 2*. TanSL and GPS are the most attractive solutions in our context. Table 2 is an overview of the properties studied in this section for the protocols presented till to now. According to the context, the "+" notation denotes the protocol resistance to the attack or its suitability for a specific property.

5   The Problem of Privacy with Compromised Readers

We focus in this section on the problem of tag privacy when a system has compromised readers. We first show that currently none of the remaining candidates (TanSL and GPS) ensures privacy in this context. Then, we propose a new authentication protocol with a key update to preserve tag privacy in *Scenario 2*. The security recovery is done using an update function when the legitimate readers get connected to the back-end after the detection of a compromised reader.

5.1   Privacy Analysis of the Candidates

Scenario 1. The TanSL protocol provides tag privacy because the tag never sends to the reader its identifier in the clear and, more generally, an adversary cannot distinguish the tag's response from a random value. However, the reader does not know which tag it is communicating with and must so carry out an exhaustive search through the list L. The reader complexity is so $O(m)$ where m is the number of tags in the system. The duality privacy vs. reader complexity allows nevertheless to reduce to $O(1)$ the complexity if the privacy is abandoned.

GPS does not provide by design tag privacy because the tag public key I is required by the reader in order to complete the authentication. Since this key I is known by everyone, an adversary is free to perform herself an authentication on a tag. One may think that the duality tag privacy vs. reader complexity can also be applied here in order to get a protocol ensuring privacy at the cost of $O(m)$ computations. This is actually more tricky in this case. In fact, even if I is kept secret by the system and Id_T is not sent in the clear, Bringer, Chabanne, and Icart explain in [7] that an adversary is able to recognize a tag from another by eavesdropping two communications. Indeed, by observing two GPS executions, an adversary cannot recover I but she can determine whether or not the same I has been used in the two executions. Thus, classic GPS does not provide tag privacy, whether the tag public key (and identifier) is disclosed or not.

It is important to notice that the authors of GPS [4] proposed an improvement: the coupons can be pairs of $(r_i, x_i = h(g^{r_i} \mod n))$, where h is a cryptographic hash function; in that case, the reader has to verify if $h(g^y \times I^{n_R} \mod n) = x_i$.

Bringer *et al.*'s attack does not apply to this modification, initially used to reduce the size of the coupons. Thus, the reader complexity is $O(m)$.

We propose below a way to reduce the above-mentioned complexity: the tag can send a random session identifier Id_T^i used once and only known by the reader and the tag. Therefore the tag cannot be recognized by an adversary during the authentication, but the reader who knows in advance the session identifiers can directly identify the public key I to use from Id_T^i. The reader complexity is so $O(1)$. The session identifiers can be integrated into the coupons, such that each coupon is a triple $(Id_T^i, r_i, x_i = h(g^{r_i} \bmod n))$. We will call this protocol the *modified GPS* (mGPS).

Scenario 2. In all the previous protocols, there is no reader's authentication: consequently the tag cannot distinguish a legitimate reader from a compromised reader R. If the system detects such an attack, there is no mechanism to take this into account better than changing physically all the tags and readers.

For TanSL, the adversary knows R's identifier Id_R and the list L that contains all the information required to allow R to communicate with the tag T. The adversary can track every tag, TanSL does not supply tag privacy anymore.

We also notice that, if tag privacy is not provided in *Scenario 1*, it can neither be achieved in *Scenario 2*. GPS has an unchanged security profile. Concerning mGPS, the adversary knows all the hidden tags' public keys I. Thus, she is able to identify which tag is involved in every communication. mGPS no longer provides tag privacy. None of these protocols provides tag privacy in *Scenario 2*.

5.2   Solving the Privacy Issue

In a system with mobile readers, it is more attractive for an adversary to compromise a reader than a tag. We propose a new challenge/response authentication protocol to handle this threat. It is based on a symmetric encryption algorithm to achieve reasonable chip area and efficiency. Our protocol has two steps. First, the secret key shared by the tag and the reader is computed on-the-fly by the tag, as in TanSL. This mechanism is enhanced with an attack counter to allow an update of the system. Second, the tag sends a classical answer in a challenge/response protocol using the previous key. We describe and analyze our protocol: it supplies tag privacy in the context of compromised readers.

Initialization. When the system is set up, each tag T is assigned with the following values:

- a unique identifier Id_T,
- a long-term key K_T,
- three counters c_B, c_R and c_T, initially synchronized and all equal to zero.

And during this set up, each reader R is assigned with the following values:

- a unique identifier Id_R,
- for every tag T, its identifier and an encryption of its secret: $Id_T, k_{TR} = E_{K_T}(Id_R, c_R)$.

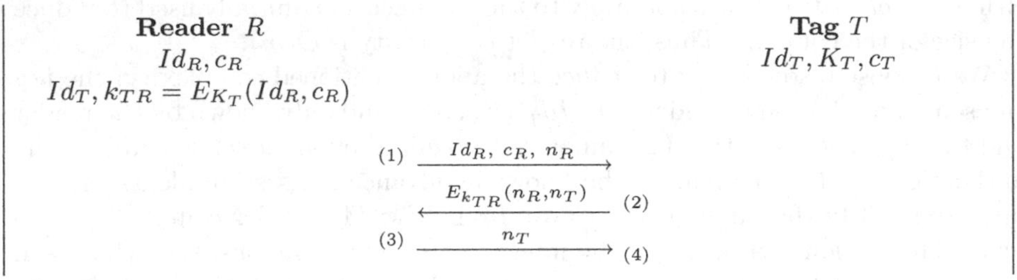

Reader R $\qquad\qquad\qquad\qquad\qquad\qquad\qquad$ **Tag T**
$\qquad Id_R, c_R$ $\qquad\qquad\qquad\qquad\qquad\qquad\qquad\qquad$ Id_T, K_T, c_T
$Id_T, k_{TR} = E_{K_T}(Id_R, c_R)$

$$(1) \quad \xrightarrow{\quad Id_R,\ c_R,\ n_R \quad}$$

$$\xleftarrow{\quad E_{k_{TR}}(n_R, n_T) \quad} \ (2)$$

$$(3) \quad \xrightarrow{\qquad n_T \qquad} \ (4)$$

Fig. 2. Authentication protocol

B stores Id_R, Id_T, K_T and c_B.
R stores Id_R, c_R, Id_T and $k_{TR} = E_{K_T}(Id_R, c_R)$.
T stores Id_T, K_T, c_T.

Authentication. The authentication protocol consists of four steps (see Fig. 2):

(1) The reader sends its identifier Id_R, the counter c_R and a nonce n_R.
(2) The tag checks the value c_R it receives:
  - If $c_R \geq c_T$, it computes the key $k_{TR} = E_{K_T}(Id_R, c_R)$. Then, it picks a nonce n_T and answers the encryption $E_{k_{TR}}(n_R, n_T)$ to the reader.
  - If $c_R < c_T$, the protocol aborts.
(3) The reader decrypts the received message with the symmetric key k_{TR}, and verifies the value n_R. Then, it sends to the tag the recovered value n_T.
(4) T checks the validity of n_T: if so and $c_R > c_T$, it updates c_T to c_R ($c_T \leftarrow c_R$).

Key Update. The update protocol is carried out when a compromised reader is detected (see Fig. 3). We consider that all the readers are synchronized at the same time:

(1) The back-end increments c_B and associates this new value to c_{up}.
  For every reader R and every tag T, it generates a new key
  $k_{TR_{up}} = E_{K_T}(Id_R, c_{up})$.
  Finally, for every tag T, it sends Id_T, c_{up}, and $k_{TR_{up}}$ to every reader R.
(2) Each reader R updates its array as follows:
  - $c_R \leftarrow c_{up}$.
  - $k_{TR} \leftarrow k_{TR_{up}}$, for every tag T.

Our protocol is based on a single cryptographic operation (symmetric encryption) and it consists in two computations for the tag. The first computation is done to generate the secret key k_{TR} used between the tag T and the reader R. Usually in a classical challenge/response protocol, each tag is assigned with a fixed key, which is used to communicate with every reader. In our protocol, the tag computes on-the-fly the key to interact with the reader R. This key k_{TR} is unique for T and R, since it is the encryption of the reader's identity Id_R and the counter c_R with the tag long-term key K_T (shared by the back-end and the tag T). The counter c_R represents the number of updates achieved by the system. At the beginning of the authentication, the reader sends this counter

Back-End B Reader R
Id_R, Id_T, K_T, c_B Id_R, c_R
                                                $Id_T, k_{TR} = E_{K_T}(Id_R, c_R)$

$$(1) \xrightarrow{\quad Id_T,\ c_{up},\ k_{TR_{up}} \quad} (2)$$

Fig. 3. Key update protocol

alongside with its identity Id_R and a nonce n_R to indicate the current state to the tag. The second computation corresponds to the tag's answer in the challenge/response protocol. The nonces n_R and n_T are encrypted with the key k_{TR}. When the reader decrypts successfully this latter message, it replies n_T as an acknowledgment.

We now analyze the security of our protocol against the different classes of attacks. Our protocol introduces a key update for the reader and tag update. The tag update corresponds to the c_T update. These events have an impact on *Scenario 2*. Let consider $\mathcal{E}$ a set of m tags, where at least one tag has already been updated. The set $\mathcal{S} \subset \mathcal{E}$ contains all the non updated tags. We assume that the tag $T \in \mathcal{S}$ recovers its security after it has been updated. We define $\mathcal{P}$ the period spent between the detection of the attack and T's update.

Tag Impersonation. Our protocol inherits from the positive security properties of the classical challenge/response protocol in *Scenario 1*, that is it prevents from tag impersonation. Indeed, the only difference is that the long-term key stored by the tag in a classical challenge/response protocol is replaced in our solution by a key computed on-the-fly by the tag.

The additional values c_R and n_T exchanged during the authentication protocol do not reveal any key material neither in *Scenario 1*, nor in *Scenario 2*: n_T is a random number, and c_R is the system state known by anybody.

For *Scenario 2*, an adversary cannot impersonate a tag in front of all the non compromised readers, since the keys are computed like for TanSL.

Denial of Service. The only value modified into the tag T is the counter c_T in *Scenario 2*. An adversary can try to desynchronize the tag by modifying c_T, impersonating a legitimate reader with a fake update. She cannot forge a fake key $k'_{TR} = E_{K_T}(Id_R, c'_R)$ corresponding to a fake counter $c'_R > c_T$, because she does not know K_T. Therefore, the tag will not accept to update c_T to a fake $c'_R > c_T$, since the adversary will not be able to answer correctly at step (3) of the authentication protocol (see Fig. 2). Our solution is so resistant to DoS.

Tag Privacy vs. Reader Complexity. Our protocol provides tag privacy in *Scenario 1*: no tag's identifier sent in the clear and answers are randomized. But the reader has no clue on the key it should use to check the tags' responses: the reader complexity is so $O(m)$. An improvement for the reader complexity

Table 3. Comparison of our protocol

	GPS		mGPS		TanSL		Our protocol	
Scenario	1	2	1	2	1	2	1	2
No Tag Impersonation	+	+	+	+	+	+	+	+
No Denial of Service	-	-	-	-	+	+	+	+
Tag Privacy	-	-	+	-	+	-	+	+
vs.								
Reader Complexity	$O(1)$	$O(1)$	$O(1)$	$O(1)$	$O(m)$	$O(m)$	$O(m)$	$O(m)$

consists in using session identifiers as in our modification of GPS (mGPS). In this solution, the tag will store a given number of identifiers known by the reader and used once per authentication. The reader complexity is $O(1)$. However, there is only a limited number of authentications: a DoS attack is possible.

In *Scenario 2*, the adversary knows T's key $k_{TR} = E_{K_T}(Id_R, c_R)$ during $\mathcal{P}$. Thus, she can track T. After $\mathcal{P}$, the adversary does not know the new key $k_{TR_{up}} = E_{K_T}(Id_R, c_{up})$ used for the further T's authentications. Thus, she will not be able to decrypt correctly the T's answers and to track T anymore. Such an incorrect decryption by the adversary represents an information leakage: she knows when T has been updated. But this cannot be avoided, since the adversary knows that T will be updated at one time or another. This is the only leakage. However, the adversary is no more able to distinguish any $T \in \mathcal{E}\backslash\mathcal{S}$, since $|\mathcal{E}\backslash\mathcal{S}| > 1$ after the period $\mathcal{P}$. Therefore, the tag privacy is restored. The reader complexity is still $O(m)$.

It should be noticed that the number of messages exchanged in our protocol is variable. Indeed, a legitimate tag answers or not to a reader depending on the attack counter c_R sent in the clear. However, a legitimate tag always answers to an updated legitimate reader. A variation is possible only between a legitimate tag and a rogue reader. If the tag answers $(c_R \geq c_T)$, the rogue reader can use the secrets stolen from the compromised reader to try to decrypt the tag's answer. The adversary knows if the tag has been updated or not. If the tag does not answer $(c_R < c_T)$, the rogue reader learns that the tag has been updated. Thus, we see that any answer of the tag can be exploited to know if the tag has been updated or not. This is the situation described in the previous paragraph.

Table 3 provides a comparison between GPS, GPS with our improvement (mGPS), TanSL, and our protocol.

6   Conclusion

A new issue in RFID systems is presented: the threat of compromised readers. Such an attack is very likely to occur in access control solutions for mobile environments, e.g. ticketing systems for sport events. We present a state of the art of several existing protocols. Our security and practical analysis of these available authentication protocols showed their weaknesses in this context.

We proposed a solution based on a symmetric-key challenge/response authentication protocol with key update. It can face the problem of compromised readers while preventing from tag impersonation and DoS. Our solution provides tag privacy w/o compromised readers. We used symmetric encryption to achieve low chip area and efficiency, such that it is suitable for reasonably-costed tags.

We have made the hypothesis that the period $\mathcal{P}'$ during which the system has updated a single tag T cannot be exploited by an adversary to track T. During $\mathcal{P}'$, the T's answers are the only ones which cannot be decrypted correctly by the adversary. In applications such as ticketing, it is very uncommon to have a single updated user during a long period: our hypothesis is fair in this case. If this hypothesis cannot be verified, new solutions are required to preserve privacy.

References

1. Atici, A.C., Batina, L., Fan, J., Verbauwhede, I., Yalcin, S.B.O.: Low-cost Implementations of NTRU for Pervasive Security. In: International Conference on Application-Specific Systems, Architectures and Processors – ASAP 2008, Leuven, Belgium, pp. 79–84 (July 2008)
2. Avoine, G., Dysli, E., Oechslin, P.: Reducing Time Complexity in RFID Systems. In: Preneel, B., Tavares, S. (eds.) SAC 2005. LNCS, vol. 3897, pp. 291–306. Springer, Heidelberg (2006)
3. Avoine, G., Oechslin, P.: RFID Traceability: A Multilayer Problem. In: Patrick, A.S., Yung, M. (eds.) FC 2005. LNCS, vol. 3570, pp. 125–140. Springer, Heidelberg (2005)
4. Baudron, O., Boudot, F., Bourel, P., Bresson, E., Corbel, J., Frisch, L., Gilbert, H., Girault, M., Goubin, L., Misarsky, J.-F., Nguyen, P., Patarin, J., Pointcheval, D., Poupard, G., Stern, J., Traoré, J.: GPS - An Asymmetric Identification Scheme for on the Fly Authentication of Low Cost Smart Cards. In: A proposal to NESSIE (2001)
5. Bogdanov, A., Knudsen, L.R., Leander, G., Paar, C., Poschmann, A., Robshaw, M.J., Seurin, Y., Vikkelsoe, C.: PRESENT: An Ultra-Lightweight Block Cipher. In: Paillier, P., Verbauwhede, I. (eds.) CHES 2007. LNCS, vol. 4727, pp. 450–466. Springer, Heidelberg (2007)
6. Bogdanov, A., Leander, G., Paar, C., Poschmann, A., Robshaw, M.J., Seurin, Y.: Hash Functions and RFID Tags: Mind The Gap. In: Oswald, E., Rohatgi, P. (eds.) CHES 2008. LNCS, vol. 5154, pp. 283–299. Springer, Heidelberg (2008)
7. Bringer, J., Chabanne, H., Icart, T.: Efficient Zero-Knowledge Identification Schemes which respect Privacy. In: ACM Symposium on Information, Computer and Communication Security – ASIACCS 2009, Sydney, Australia, March 2009, pp. 195–205. ACM Press, New York (2009)
8. Feldhofer, M., Dominikus, S., Wolkerstorfer, J.: Strong Authentication for RFID Systems using the AES Algorithm. In: Joye, M., Quisquater, J.-J. (eds.) CHES 2004. LNCS, vol. 3156, pp. 357–370. Springer, Heidelberg (2004)
9. Fiat, A., Shamir, A.: How To Prove Yourself: Practical Solutions to Identification and Signature Problems. In: Odlyzko, A.M. (ed.) CRYPTO 1986. LNCS, vol. 263, pp. 186–194. Springer, Heidelberg (1987)
10. Gama, N., Nguyen, P.Q.: New Chosen-Ciphertext Attacks on NTRU. In: Okamoto, T., Wang, X. (eds.) PKC 2007. LNCS, vol. 4450, pp. 89–106. Springer, Heidelberg (2007)

11. Hein, D., Wolkerstorfer, J., Felber, N.: ECC is Ready for RFID – A Proof in Silicon. In: Conference on RFID Security, Budapest, Hungary (July 2008)
12. International Organization for Standardization. ISO/IEC 9798 – Information technology – Security techniques – Entity authentication (1997 – 2008)
13. Mathas, C.: Altera CPLDs go to the Beijing Olympics (2008), http://www.eetimes.com/showArticle.jhtml?articleID=208800197
14. McLoone, M., Robshaw, M.J.: Public Key Cryptography and RFID Tags. In: Abe, M. (ed.) CT-RSA 2007. LNCS, vol. 4377, pp. 372–384. Springer, Heidelberg (2006)
15. Naccache, D.: Method, Sender Apparatus and Receiver Apparatus for Modulo Operation. European patent application no. 91402958.2 (1992)
16. NESSIE consortium. Portfolio of recommended cryptographic primitives. Technical report (2003)
17. O'Neill, M., (McLoone).: Low-Cost SHA-1 Hash Function Architecture for RFID Tags. In: Conference on RFID Security, Budapest, Hungary (July 2008)
18. Oren, Y., Feldhofer, M.: WIPR - a Public Key Implementation on Two Grains of Sand. In: Conference on RFID Security, Budapest, Hungary (July 2008)
19. Oren, Y., Feldhofer, M.: A Low-Resource Public-Key Identification Scheme for RFID Tags and Sensor Nodes. In: Proceedings of the second ACM Conference on Wireless Network Security – WiSec 2009, Zurich, Switzerland. ACM Press, New York (2009)
20. Rabin, M.O.: Digitalized Signatures and Public-Key Functions as Intractable as Factorization. Technical report, Massachusetts Institute of Technology, Cambridge, Massachusetts, USA (1979)
21. Shamir, A.: Memory Efficient Variant of Public-key Schemes for Smart Card Applications. In: De Santis, A. (ed.) EUROCRYPT 1994. LNCS, vol. 950, pp. 445–449. Springer, Heidelberg (1995)
22. Tan, C.C., Sheng, B., Li, Q.: Serverless Search and Authentication Protocols for RFID. In: International Conference on Pervasive Computing and Communications – PerCom 2007, New York, USA. IEEE Computer Society Press, Los Alamitos (2007)
23. Wu, J., Stinson, D.: How to Improve Security and Reduce Hardware Demands of the WIPR RFID Protocol. In: IEEE International Conference on RFID – RFID 2009, Orlando, Florida, USA (April 2009)

Coding Schemes for Arithmetic and Logic Operations - How Robust Are They?

Marcel Medwed and Jörn-Marc Schmidt

Graz University of Technology
Institute for Applied Information Processing and Communications
Inffeldgasse 16a, A–8010 Graz, Austria
{Marcel.Medwed,Joern-Marc.Schmidt}@iaik.tugraz.at

Abstract. In the past many coding schemes have been proposed to render arithmetic and logic units fault tolerant. However, most schemes are suited for safety rather than for security applications, i.e. they were not designed to protect against malicious fault injections. Even articles considering an adversary as the source of faults restrict the error-detection discussion to partial fault models.

In this article, we investigate the possibilities of an adversary to inject an undetected fault in different coding schemes. In contrast to other works, we analyze the interaction of erroneous operands and operations. Such an analysis yields quite different results than traditional evaluations. These new results show that each of the schemes has serious weaknesses and neither of them can guarantee a universal protection. Thus, a hybrid approach is favorable to counteract fault attacks.

Keywords: Coding schemes, Error-masking, Fault Attacks.

1 Introduction

In a world that relies on mobile systems, it is vital that security-related devices, like credit cards, protect the data stored on them. These devices must fulfill their security requirements, even if an adversary can access them physically. In 1997, Biham et al. presented attacks that make use of faulty computations [1], so-called *fault attacks*. In order to counteract such attacks, faults have to be detected by the device and no erroneous output must leave the device.

Including methods that ensure fault detection, called *fault countermeasures*, into smart cards is common practice by now. An accurate estimation of the security level provided by fault countermeasures is very important, for example, if a chip should be evaluated against common criteria. On the one hand, a smart-card company wants to ensure a very high probability for the chip passing the test, which includes powerful attacks with sophisticated equipment. On the other hand, the costs per device should be as low as possible, which requires that no inefficient countermeasure is implemented.

Furthermore, since a company does not know the later applications of their product, a method that is independent from the code that runs on the device

H.Y. Youm and M. Yung (Eds.): WISA 2009, LNCS 5932, pp. 51–65, 2009.
© Springer-Verlag Berlin Heidelberg 2009

and which is, in addition, transparent to the user is favorable. Hence, a counter-measure should secure arbitrary operations as well as arbitrary data processed in the device.

Deploying error detection codes, an error detection rate can be stated for a device. This error detection rate is often assumed to be the same as for plain data transmission via a noisy channel. We show that especially for efficient codes, this assumption does not hold. This is because also the performed operations have to be considered in order to allow correct statements about the error coverage of the device.

In this paper, we evaluate different coding schemes according to their fault resistance and their suitability for fault detection mechanisms in cryptographic devices. For each code, we look at the robustness for every supported operation. Our approach allows a fair comparison and yields new results about the robustness of the coding schemes. It turns out that breaking the distance-two barrier is impossible with a single coding scheme unless a redundancy of 100% is exceeded.

Note that the discussed schemes only secure the data path of a device—an adversary that manipulates the program flow remains undetected. However, various methods to ensure the correct execution of the program have been proposed, e.g. by introducing program code signatures [2].

Another problem using codes is the required checking procedure before data is stored or leaves the device. The check may be vulnerable to adversaries that manage to inject a fault in it, making the device output erroneous results [3]. A possible solution is a multi-stage check [4].

The remainder of this paper is organized as follows: Section 2 investigates the strengths and weaknesses of various coding schemes. In Section 3, those schemes are compared in different attack scenarios and also hybrid solutions are taken into account. Finally, we draw conclusion and discuss open problems in Section 4.

2   Coding Schemes

In this section, we investigate several codings schemes. For an arbitrary code, with r bits of redundancy, only 2^{-r} of the codewords are valid and hence the chances of an adversary who injects random faults are also limited by $1 - 2^{-r}$. On the other end of possible faults, there are precise single-bit manipulations. However, also the detection rate for these faults is fixed. This is because coding schemes usually require an adversary to manipulate at least two bits. Thus, those two extreme fault types are not of interest for our investigations. Instead, the main intension of this article is to examine the weaknesses of coding schemes when arithmetic and logic operations are performed on their codewords. One such weakness is error-masking, that is, if an operation renders a detectable error at the input undetectable at the output.

We first look at straight-forward techniques to introduce redundancy, namely time- and space-redundant techniques. Next, we investigate coding schemes which can be used to perform arithmetic and logic operations. Finally, we ex-amine arithmetic codes.

2.1   Time Redundancy

Time redundancy is a well known and straightforward strategy to deal with faults in a system. It is based on executing some algorithm $\mathcal{A}$ several times and checking whether every execution of $\mathcal{A}$ yields the same result. It requires no or little additional hardware, but the performance drops by the number of executions of $\mathcal{A}$.

A major drawback of the method is that it relies on a single piece of hardware. Therefore, a permanent malfunction in the hardware causes the same error on every execution and hence the error cannot be detected in general. An adversary can enforce such a behavior by means of destructive faults. Another way to circumvent the protection mechanism is to introduce the same fault on every execution. Since the timing parameter of an attack is commonly well controllable, the success of such an approach is likely.

In some cases though it is possible to improve the approach. If the algorithm is invertible one can check if $\mathcal{A} \circ \mathcal{A}^{-1}$ yields the original input. For ciphers like AES, this has been proposed in [5]. In the article the authors apply this principle to every round of the block cipher. Also for signature generation algorithms the method was proposed [6]. The approach is especially advantageous in the case of RSA signatures, because usually the computation of $\mathcal{A}^{-1}$ is much faster than the one of $\mathcal{A}$.

Another improvement which might be possible is to alter the algorithm to $\mathcal{A}'$. This new algorithm is identical to $\mathcal{A}$ in the way that their input-output behavior is indistinguishable. However, the performed instructions and the bit-order of the intermediate operands vary. Therefore, the effect of a fault also varies with a certain probability. However, altering an algorithm in such a way is not always possible (or only to a very restricted extend).

From above it becomes clear that time-redundant countermeasures are simple, generic and mostly hardware independent. However, their design space and hence their reliability depend on the algorithm. To improve this, one possibility is to introduce the redundancy into the space or area parameter rather than into the time parameter.

2.2   Space Redundancy

In contrast to time-redundant systems, space-redundant systems use duplicated hardware. Hence algorithm $\mathcal{A}$ is executed several times in parallel. As a result, the approach does not affect the performance, but the hardware costs increase.

The security against destructive faults increases compared to time redundancy. However, inducing the same fault twice in parallel is sufficient to corrupt the system. In general, inducing two faults in parallel is more expensive than inducing two faults one after the other, but the increase of costs is only linear.

Possible improvements are similar to those for time-redundant systems. The functionality implemented by the original system can be implemented in a different way for the redundant part. This can be done statically or dynamically. A static modification manifests itself in the layout for instance. An example for a dynamic modification is bus scrambling. Bus scrambling causes that the adversary cannot predict the position of single bits anymore. However, if the bits in a circuit depend on each other (e.g. an integer-arithmetic unit or a multiplier) scrambling might become rather expensive in terms of hardware.

An advantage of space-redundant systems is that the countermeasure is transparent to the programmer. Every algorithm that can be executed by the original hardware can then be automatically executed as $\mathcal{A}$ and $\mathcal{A}'$ on the space-redundant hardware.

In general, both approaches, time and space redundancy come with an overhead of at least hundred percent, either in terms of performance or in terms of hardware. At the same time, both approaches are susceptible to an adversary who induces multiple faults. In the most straight-forward case, these are two faults which manipulate a single bit. To reduce the overhead and to improve the error detection, coding theoretic approaches can be used.

2.3   Linear Codes

One such coding theoretic approach is to use linear codes to protect a system. Linear codes are known for decades now [7] and hence are well studied. Also the use of linear codes to design fault tolerant systems has already been described in the 80's [8]. However, no previous work investigated the error-masking behavior of arithmetic or logic circuits that deploy linear codes. Also the faults, assumed in previous articles, are not suitable for fault-attack scenarios. Whereas in [8,9] only single-bit faults were assumed (due to radiation in space for instance), much more complex faults have to be assumed if adversary is present. Hence, it is interesting to investigate the behavior of linear codes, if the faults are induced by an adversary. In this section, we revise linear codes and block codes in general. Afterwards, we investigate the error masking probabilities of linear codes when logic, arithmetic or shift operations are applied.

A binary block code maps a datawords of k bits to codewords of n bits with $k < n$. The portion k is called the dimension of the code and $r = n - k$ is the number of redundant bits added by the code. If the code forms a k-dimensional sub vector-space in $\mathbb{F}_2^n$ then the code is called a *linear code over* $GF(2)$. A property of such codes is that the sum of codewords always results in a codeword itself. A common convention to describe a linear code is to write $[n, k]$ *code* or $[n, k, d]$ *code*, where d denotes the distance of the code. The distance is the minimum pairwise Hamming distance of all codewords. Therefore, the distance of a code indicates how many bits have to be changed at minimum in order to transform one valid codeword into another valid codeword. It also states the robustness of a code, since all errors with a Hamming weight smaller d are detected with certainty. For instance, [48,32] linear codes are known up to a distance of 6. The space-redundant approach from Section 2.2 can be seen as a so-called repetition code. In fact, if the hardware is duplicated, it represents a $[2k, k, 2]$ linear code over $GF(2)$. From $d = 2$ it can be seen that two bit-manipulations can already corrupt such a system.

In this section, we only look at systematic linear codes over GF(2). That is, the data is embedded in the codeword and the parity part can be separated from the data part. For such a systematic code[1], every bit of the parity part is the sum of some bits of the data part. The simplest systematic, linear code can be constructed by appending the sum of all data bits as parity. For decoding every bit of the syndrome is calculated as the sum of some bits of the codeword (including the parity bits). An all-zero syndrome indicates, that the codeword is valid. In the above example, the syndrome is the sum of all bits of the codeword.

From linearity it follows that an erroneous codeword $\tilde{A} = A + e_A$ is only valid if the error e_A itself is a codeword. However, this observation does not consider any arithmetic or logic operations. In the following we look at the scenario when one or two operands, involved in an operation, are erroneous. For the implementation of the operations we use the same equations as in [8] and extend them by equations for shift operations.

Logic Operations. To investigate logic operations on linear codes, it suffices to look at the exclusive-or and at the logic-and operation. All other boolean functions can be composed of them. XOR and AND implement the exclusive-or and the logic-and function for codewords. Thus, they operate on codewords and also yield a valid codeword if the input was valid. The bit-wise exclusive-or, that is, the addition in GF(2) is denoted by the $+$ operator. The $\cdot$ operator on the other hand denotes the multiplication in GF(2) or the bit-wise logic-and operation.

The linearity property of the codes states that the sum of two codewords again results in a valid codeword. Since, as stated before, the sum in $GF(2)$ is nothing else than the XOR operation, linear codes over $GF(2)$ are closed under XOR by definition. Therefore, the result of $\tilde{A}$ XOR $B = A + e_A + B$ is only valid if e_A is valid. Also the result of $\tilde{A}$ XOR $\tilde{B} = A + e_A + B + e_B$ is only valid if the sum $e_A + e_B$ is valid. Thus, the XOR operation preserves the distance of the code independent of how many operands are erroneous.

Next, we look at the AND operation. For this, we have to redefine the operation for linear codes, since a bit-wise application does not yield a valid codeword. For that, we introduce the following notation: The k data bits of the dataword a are $a_1, \cdots, a_k$. The r parity bits of the encoded a are $p(a) = p(a)_1, \cdots, p(a)_r$. Enc($x$) encodes x and returns X. $P(x)$ encodes x and returns only $p(x)$. The concatenation operator is denoted by $|$ and Enc(a) = $a \mid p(a)$. Finally, 1 denotes the all-one vector and 0 the all-zero vector respectively. We can now define the AND operation for codewords as

$$C = a \cdot b \mid$$
$$p(a) + p(b) + P(1 + (1 + a) \cdot (1 + b)). \tag{1}$$

As an example we look at the simple [3,2] code, where the parity bit is defined as the sum of the two data bits. Hence, the codewords are

---

[1] Note that every linear code can be transformed into a systematic code.

$$
\begin{aligned}
A &= a_1, a_2 & | \; a_1 + a_2 \\
B &= b_1, b_2 & | \; b_1 + b_2 \\
(A \text{ AND } B) &= a_1 \cdot b_1, a_2 \cdot b_2 & | \; a_1 \cdot b_1 + a_2 \cdot b_2.
\end{aligned}
$$

According to (1), the parity results in

$$
\begin{aligned}
& p(a) + p(b) + P(1 + (1 + a) \cdot (1 + b)) \\
={} & a_1 + a_2 + b_1 + b_2 + \\
& P(a_1 \cdot b_1 + a_1 + b_1, a_2 \cdot b_2 + a_2 + b_2) \\
={} & a_1 \cdot b_1 + a_2 \cdot b_2,
\end{aligned}
$$

which follows exactly the rule of the code. To investigate the effect of one erroneous operand, we add an error $e_a | e_{P(a)}$ to the codeword A. Hence, the erroneous codeword results in:

$$
\tilde{A} = a + e_a \; | \; p(a) + e_{P(a)}.
$$

If such an erroneous codeword is involved in an AND operation, the data part of the result is

$$
a \cdot b + e_a \cdot b.
$$

The parity evaluates to

$$
\begin{aligned}
& p(a) + e_{P(a)} + p(b) + P(a \cdot b + a + b + e_a + e_a \cdot b) \\
={} & e_{P(a)} + p(a \cdot b) + p(e_a) + p(e_a \cdot b).
\end{aligned}
$$

Now we put together the last two results and eliminate all valid codewords. That is, all terms in the data part, where the according parity occurs on the right side of the bar. This leaves us with

$$
\begin{aligned}
& 0 \; | \; e_{P(a)} + p(e_a) \\
={} & e_a \; | \; e_{P(a)}.
\end{aligned}
$$

The last transformation is valid since $p(e_a)$ originated from e_a. It follows that the result is only a correct codeword, if the induced error was already a codeword. Hence, the AND operation is error-masking free and preserves the code's distance if one operand is erroneous.

To investigate the case of two erroneous operands we introduce a second error term $e_b | e_{P(b)}$. Thus, the data part results in

$$
a \cdot b + e_a \cdot b + e_b \cdot a + e_a \cdot e_b.
$$

The parity evaluates to

$$
\begin{aligned}
& p(a) + e_{P(a)} + p(b) + e_{P(b)} \\
& + P(a \cdot b + a + b + e_a + e_b + e_a \cdot b + e_b \cdot a + e_a \cdot e_b) \\
={} & e_{P(a)} + e_{P(b)} + p(a \cdot b) \\
& + p(e_a) + p(e_b) + p(e_a \cdot b) + p(e_b \cdot a) + p(e_a \cdot e_b).
\end{aligned}
$$

Eliminating the valid codewords yields

$$
\begin{aligned}
0 \quad & | \; e_{P(a)} + p(e_a) + e_{P(b)} + p(e_b) \\
= e_a + e_b \; & | \; e_{P(a)} + e_{P(b)}.
\end{aligned}
$$

This is in fact an interesting result since the error terms sum up although the operation is the AND operation and not the XOR operation. As a consequence it is possible to set for instance $e_{P(a)} = e_{P(b)} = 0$ and $e_a = e_b = 1$. In this case the error terms will cancel out and produce a valid codeword. However, the result itself changes by $e_a \cdot b + e_b \cdot a + e_a \cdot e_b$ which renders the result incorrect.

It turns out that logic operations preserve the distance of linear codes if only one operand is erroneous. If both operands are erroneous on the other hand, modifying two bits already corrupts the system.

Arithmetic Operations. As a representative for the arithmetic operations, we take the integer addition. The parity for an integer sum s can be calculated as $p(s) = p(a) + p(b) + P(c)$, where c represents the carry vector. The carry vector consists of the carry-in bit, followed by all carry bits produced internally by the adder. The carry-out bit is not part of c. The main observation, the formula is based on, is that $\text{ADD}(a, b) = a + b + c$. That is, once the carry vector is available, the bits become independent.

Since the carry vector depends on a, an error e_a also causes an additional error e_c in the carry vector. However, e_c also adds directly to the sum s. Hence, we get

$$
\begin{aligned}
a + e_a + b + c + e_c \; & | \; p(a) + p(b) + \\
& \quad\; P(c + e_c) + e_{P(a)} \\
= a + e_a + b + c \quad & | \; p(a) + p(b) + p(c) + e_{P(a)} \\
= e_a \quad & | \; e_{P(a)}.
\end{aligned}
$$

For only one erroneous codeword the operation is distance preserving. However, for two erroneous operands similar problems as for the AND operation evolve. Let $e_{P(a)} = e_{P(b)} = 0$ and $e_a = e_b = 1$. As a result the error cancels out for the check equation. At the same time the error manifests itself in the carry and hence renders the result incorrect. Thus, also for an addition, manipulating two bits suffices to corrupt the system.

Another problem of the integer addition for linear codes is the parity calculation for the carry vector. If a fault affects the carry generation, even a single fault suffices to produce an incorrect but valid result.

Shift Operations. The last family of functions we look at contains typical unary and linear functions, like shifts and rotates. The idea here is to calculate the parity for the sum of the operand and the result. The linear, unary function is denoted by $u(\cdot)$.

$$
\begin{aligned}
u(a + e_a) \quad & | \; p(a) + P(u(a + e_a) + a + e_a) \\
& \quad\; + e_{P(a)} \\
= u(a) + u(e_a) \; & | \; p(a) + p(u(a)) + p(u(e_a)) \\
& \quad\; + p(a) + p(e_a) + e_{P(a)} \\
= 0 \quad & | \; p(e_a) + e_{P(a)} \\
= e_a \quad & | \; e_P
\end{aligned}
$$

For one erroneous operand, the operation is distance preserving and since the operation is unary, this is the only possible case. Note that these considerations only hold for single-position shift and rotate operations where no second operand is involved.

The analysis above shows that a system using linear codes is robust against errors of small multiplicity if only one operand is affected. However, as soon as both operands are erroneous, the minimum distance drops to two.

2.4   Berger Codes

Berger codes present another coding theoretic approach to protect arithmetic and logic operations. Initially, they were introduced by J.M. Berger in 1961 [10]. In 1989, Lo and Rao showed how to implement an ALU which is protected by Berger codes [11]. The equations in their article are similar to those for linear codes and rely on the same principles.

The check symbol for Berger codes is the number of zero-bits in the dataword. As an example, we look at an eight bit word holding the value 14. The number of zeros is 5. Hence, the Berger-encoded word results in $(00001110, 101)$.

What is special about Berger codes is that they have a minimum asymmetric distance of one. The asymmetric distance between two codewords is the minimum of the number of bits which change from zero to one and the number of bits which change from one to zero. For instance, the integers in binary representation have a minimum distance of one, but a minimum asymmetric distance of zero. For Berger codes, this means that only setting or only resetting bits cannot produce a valid codeword. Hence, Berger codes detect all unidirectional errors.

However, the minimum distance of Berger codes is always two. This is because flipping a zero to a one and a one to a zero at the same time always produces a valid codeword. As a result a 2-bit error stays undetected with a probability of 0.5. The redundancy added by Berger codes is limited by $\lfloor log_2(k) \rfloor + 1$ bits.

Berger codes are non-linear codes. As a consequence, the error masking analysis cannot be carried out as straight-forward as for linear codes. This is because if an error cannot be detected does not only depend on the error, but also on the data itself. Also since the distance of Berger codes is only two (which is the minimum for any channel code) a statement about the error masking via the distance does not work. Hence, we simulated the error injection into one and into two operands. The result however was not surprising: An error which is valid for one operand, can be split across two operands (depending on the data). Therefore, it is always possible to find two bits which can be manipulated in order to cause an incorrect but valid result. Furthermore, since the check equations for integer arithmetic also depend on the carry vector, Berger codes have the same problem as linear codes.

Other variants of Berger codes, which have been proposed, are reduced Berger codes [12] and Dong's code [13]. However, the above considerations also hold for these variants.

2.5 Arithmetic Codes

Berger codes as well as linear codes show deficiencies when it comes to arithmetic operations. Therefore, we next investigate arithmetic codes as they are designed towards those operations. In this section we revise some basics of arithmetic codes and discuss the advantages and disadvantages of AN codes and residue codes. Additionally, we discuss the construction of multi-residue codes with a certain distance. Throughout the section, the integer addition is denoted by $A + B$ and the integer multiplication is either denoted by $A * B$ or AB.

Arithmetic Distance. Analogously to the Hamming distance for linear codes over GF(2), the arithmetic distance can be defined for linear codes under integer addition. The arithmetic weight is the Hamming weight of the minimum weight representation $\sum \pm 2^i$ of a binary integer $\sum 2^i$. Such a minimum weight representation is well defined for every integer [14]. The arithmetic distance between two integers is the arithmetic weight of the arithmetic difference. Furthermore, the minimum distance of an arithmetic code equals the weight of the minimum weight codeword of the code.

AN Codes. An AN code is defined by an integer A and a maximum dataword N_0. The codewords are the product of the datawords $n \leq N_0$ times A. A codeword c is error-free if A divides c. Therefore, an error stays undetected if it is a multiple of A and if it can be represented as a sum of powers of ± 2 which is smaller than some $A \cdot N_0$. The minimum number of terms in the sum, needed to inject an undetected error is also the minimum distance of the code.

For a minimum distance of three, there exist ways to determine N_0 for a given A. A minimum distance of three implies that single errors can be corrected and double errors can be detected. Correcting single errors also demands a distinct and non-zero syndrome for every single bit error e smaller $A \cdot N_0$. This on the other hand is given, if for instance 2 is a generator of $GF(A)$ and the order of $GF(A)$ is $\geq \lceil \log_2(AN_0) \rceil$. However, for larger distances and high code rates k/n, the parameters can only be determined by exhaustive search. That is, checking if every codeword $\leq AN_0$ has an arithmetic weight of at least d. In [15], Mandelbaum presented AN codes with a given minimum distance. However, the redundancy is with $r = 2^k - k$ too large for the protection of a processing unit.

Since arithmetic codes have the property that $AN_1 + AN_2 = A(N_1 + N_2)$ with $0 \leq N_1, N_2 \leq N_0$, they support integer addition and furthermore do not mask errors under addition. This is because, for one erroneous operand $AN_1 + e_1$, e_1 must be of the form Ae_1'. For two erroneous operands $(AN_1 + e_1) + (AN_2 + e_2)$, the sum $e_1 + e_2$ must be either of the form $A(e_1' + e_2')$ or $A((e_1 + e_2)/A)$. That is, either each error term is divisible by A or the sum of the error terms is divisible by A. Hence, arithmetic codes preserve their distance even with two erroneous operands.

Disadvantages of AN codes are that they are not systematic and that they only support integer addition. For multiplication the result becomes incorrect since $AN_1 * AN_2 = A^2 N_1 N_2$. However, this problem can be solved by using idempotent AN codes.

Idempotent AN Codes. Idempotent AN codes have been introduced by Proudler in [16]. Gaubatz et al. were the first to consider them in an adversary scenario [17]. Again, A and N_0 are chosen appropriately to achieve a certain arithmetic distance. Addition and multiplication take place in the ring $\mathbb{Z}_{AN_0}$. The difference to AN codes lies in the encoding of datawords. Instead of generating the code with A, an idempotent AN code is generated by an idempotent element $I \equiv I^2 \mod AN_0$. Such an idempotent element exists if A and N_0 are co-prime and can be constructed with $I = \text{CRT}(1 \mod N_0, 0 \mod A)$, where CRT indicates the application of the Chinese remainder theorem.

The masking probability for idempotent AN codes under addition is the same as for AN codes under addition. That is, an error stays undetected if $e_1 \equiv e_2$ (mod A). However, for multiplication, detecting only one erroneous operand is impossible if the code is used like described above. This is because

$$IN_1 * (IN_2 + e_2) =$$
$$\text{CRT}(N_1 \mod N_0, 0 \mod A) * \text{CRT}(N_2 + e_2' \mod N_0, e_2'' \mod A) =$$
$$\text{CRT}(N_1 N_2 e_2' \mod N_0, 0 \mod A).$$

Therefore, the multiplication has to be slightly modified. For instance, it is possible to add $\text{CRT}(0 \mod N_0, 1 \mod A)$ to the operands before multiplication and subtract the same value afterwards. If this approach is pursued, a single erroneous operand stays undetected if $e_1 \equiv 0 \mod A$. Two erroneous operands stay undetected if

$$e_1 \equiv -e_2/(1 + e_2) \mod A. \tag{2}$$

For addition the arithmetic distance of idempotent AN codes stays the same as for standard AN codes. Also for multiplication with one erroneous operand, the code is distance preserving. However, multiplication is not distance preserving if two erroneous operands are involved. To show this, we look at the code with $A = 89$ and $N_0 = 22$. The arithmetic distance of this code is 4. Hence, it is enough to show that it is possible to induce two error terms by manipulating only 3 bits and that further the result after the multiplication is valid. For this, we set $e_2 = 4$ in (2) and get $e_1 = 17$. Since the Hamming weight of 4 is 1 and the Hamming weight of 17 is 2, only 3 bits have to be manipulated.

Nevertheless, for the above code it is not possible to induce an undetected error by only manipulating two bits. The robustness of an idempotent AN code has to be investigated for every case separately.

Idempotent AN codes are suitable for arithmetic operations, but not for logic operations. If also such operations are desired and at the same time a high robustness for arithmetic operations is demanded, the only solution is to use several codes. However, trans-coding a dataword from one code to another is easier if the codes are systematic. Furthermore, comparing two operands needs decoding for non-systematic codes. Therefore, systematic, arithmetic codes with similar properties as idempotent AN codes are desired.

Residue Codes. Residue codes are systematic arithmetic codes. They are composed of a data part N and a parity part P where $P = N$ mod A. For residue codes, A is called the check basis. An advantage of residue codes is that addition as well as multiplication can be carried out without any modifications since

$$N_1 + N_2 \mod A = P_1 + P_2 \mod A,$$
$$N_1 * N_2 \mod A = P_1 * P_2 \mod A.$$

Another advantage is that the complexity of a multiplication is potentially smaller since the operands themselves are smaller. Errors which affect the data part only, need to have the same weight as for AN codes with the same A in order to stay undetected. However, it is always possible to induce an error of the form $(N = 1, P = 1)$, hence the minimum distance is two, independent of A. This also holds for addition. When it comes to multiplication the case is more complex. This is because it also depends on the encoded data if an error stays undetected or not under multiplication. To investigate this behavior we look at the parities of two operands under multiplication.

$$(P_1 + e_1)(P_2 + e_2) = P_1 P_2 + P_1 e_2 + P_2 e_1 + e_1 e_2$$

First, we assume $e_1 \neq 0, e_2 = 0$. In this case the error stays undetected if either $e_1 \equiv 0 \mod A$ or $P_2 \equiv 0 \mod A$ are satisfied. In the latter case, it is impossible to detect any error. If both operands are erroneous then $P_1 \equiv -e_1(P_2 + e_2)/e_2$ mod A must hold. The investigation of that case is as complex as for the idempotent AN codes. However, if an error stays undetected for idempotent AN codes, then it also stays undetected for residue codes (assuming specific values for P_1, P_2). But in general, for inducing a specific and at the same time undetected error into a residue encoded operand, more knowledge about the encoded data is required. Thus, the only low weight errors that can always be injected are low-weight codewords themselves.

Residue codes are attractive because they are systematic. Unfortunately, their minimum distance of two is low. Hence, it would be interesting to have a scheme with similar properties but a larger distance. Suitable candidates for such a coding scheme are multi-residue codes.

Multi-Residue Codes. In [18], Rao presents a bi-residue code capable of correcting single errors. In [19], Rao and Garcia derive connections between AN codes with composite A and multi-residue codes which use the factors of A as their check bases. The authors investigate codes with a distance of 3, that is single-error correcting codes, but do not look at higher distances. Thus, we give requirements for multi-residue codes with larger distance.

A multi-residue code consists of a data part N and multiple check bases $P^1, \cdots, P^l$. The behavior for errors which affect only one operand is the same as for standard residue codes with the exception that larger distances are possible due to the multiple residues. In other words, a trivial error of the form $(N = 1, P^1 = 1, \cdots, P^l = 1)$ has to have at least a weight of $l+1$. However, the number check bases alone is not sufficient to state a certain distance. In the following we state all necessary conditions for a minimum distance d if all datawords are $\leq N_{max}$:

1. If $A = \prod_{i=1}^{l} a_i$ has l factors then l must be at least $d - 1$.
2. Let M be the product of all l' distinct factors of A. Then M can be split in two factors M_1 and M_2. For every possible M_1 with k factors check if an AN code with $A = M_1$ and $N_0 = \lfloor N_{max}/M_1 \rfloor$ has at least a minimum arithmetic distance of $d - (l' - k)$.

Searching for possible As following these conditions allows to guarantee a minimum distance for the multi-residue code. As for AN codes with a high distance, also for multi-residue codes with a distance larger 3, it is necessary to determine A by exhaustive search. However, for multi-residue codes a suitable A for a given d and N_{max} can be found faster than for AN codes.

If an error is induced in both operands before a multiplication, it is data dependent whether the result is valid or not. However, the more residues the error affects the unlikelier an undetected error becomes. On the other hand, an error which affects only one residue, needs to be of weight $\leq d - 1$ according to the conditions above.

Multi-residue codes show the least complexity of all investigated arithmetic codes and they only need a little more redundancy compared to AN codes. This is because the entropy of a t-bit residue is less than t and for a large number of check bases this adds up.

3   Comparison

The analysis of Section 2 is summarized in Table 1. It shows that every coding scheme has its weaknesses or disadvantages. Thus, the optimal choice heavily depends on the assumed adversary. Time- and space-redundant systems provide reasonable security against an adversary with little control on the induced fault. However, the overhead is large and more advanced adversaries can exploit the weaknesses of the two approaches and succeed by manipulating only two bits.

In order to reduce the overhead, coding theoretic approaches have to be pursued. The most general coding schemes, meaning that they support most operations, are Berger codes and linear codes. The disadvantage of Berger codes is their small distance of two. Linear codes provide a higher distance, but show weaknesses when it comes to logic-and and similar operations. Both schemes show problems for arithmetic operations. Additionally, linear codes do not support multiplication.

Arithmetic codes on the other hand can provide a high distance for addition and multiplication (if supported), but do not support logic operations. The major problem of AN codes is that they either support multiplication or allow the comparison of two operands. This is because idempotent AN codes are not systematic. Residue codes solve this problem, but have a low distance. Therefore, multi-residue codes seem to be a good choice for arithmetic operation. However,

Table 1. Properties and weaknesses of the analyzed coding schemes. The supported operations are **L**ogic, **A**ddition, **S**hift/Rotate, **M**ultiplication and **C**omparison. $f()$ describes some bound on the distance, e.g. the Hamming bound for linear codes.

Code	Overhead	Distance	Supp. Op.
Time red.	1	2	L,A,M,S,C
Susceptible to single destructive errors.			
Induction of two equivalent faults stays undetected.			
Space red.	1	2	L,A,M,S,C
Induction of two equivalent faults in parallel stays undetected.			
Linear codes	$\frac{r}{k}$	$f(k,r)$	L,A,S,C
Logic-and is susceptible to two bit-manipulations.			
Manipulation of carry generation stays undetected.			
Berger codes	$\frac{\lfloor \log_2(k) \rfloor + 1}{k}$	2	L,A,M,S,C
Manipulation of carry generation stays undetected.			
AN codes	$\frac{\lceil \log_2(A) \rceil}{\lceil \log_2(N_0) \rceil}$	$f(A, N_0)$	A,C
Idem. AN codes	$\frac{\lceil \log_2(A) \rceil}{\lceil \log_2(N_0) \rceil}$	$f(A, N_0)$	A,M
Multiplication is susceptible to errors with a weight $\leq d$.			
Residue codes	$\frac{\lceil \log_2(A) \rceil}{\lceil \log_2(N_{max}) \rceil}$	2	A,M;C
Some data values inhibit error detection for multiplication.			
Multi-residue codes	$\frac{\sum \lceil \log_2(a_i) \rceil}{\lceil \log_2(N_{max}) \rceil}$	$f(A, N_{max})$	A,M,C
Some data values inhibit error detection for multiplication.			

they have data dependency problems. If an adversary can choose one operand and can attack the other one, he can always succeed in injecting an undetected error. If this presents a problem, heavily depends on the application.

3.1 Hybrid Solutions

Since no code can provide universal protection, depending on the present adversary, a hybrid solution might be desirable. For such a hybrid solution, the various used codes must be easy to trans-code. Furthermore, the trans-coding must not present a weakness itself, for instance due to decoding. If all codes are systematic, the following approach can be pursued:

$$p_2(a) = (p_1(a) + P_2(a)) + P_1(a).$$

Here, the parity of a codeword a under the second code $p_2(a)$ is computed in two steps. First, the parity under the new code $P_2(a)$ is calculated and exclusive-ored to the old parity $p_1(a)$. In the second step, the old parity is calculated and removed from the codeword's parity. Since the codewords stay encoded during the whole process, an injected error must be a valid codeword in order to stay undetected.

Using such a trans-coding technique, one possible hybrid solution would be the combination of linear codes with multi-residue codes. However, even such an approach is not able to preserve a high distance for all operations in the presence of an adversary who is able to manipulate two bits. This is because there is no efficient way to protect logic-and-like operations. Thus, such operations would need to be detected by a compiler which then adds time-redundancy.

4   Conclusion and Open Problems

In this paper, we studied various coding techniques that can be used to design a fault-tolerant environment. In contrast to previous works, we did not restrict the adversary to a specific fault model but rather discussed all possible weaknesses of the coding schemes. It turned out that most schemes possess weaknesses, which can be exploited with only two bit-manipulations. Furthermore, no efficient coding scheme preserved its distance throughout all supported operations, that is, none of them was more robust than straight-forward duplication.

Additionally, we presented parity formulas for linear, unary operations on linear codes and gave conditions for high-distance multi-residue codes.

Comparing the various coding schemes, it turned out that a universally protecting scheme must pursue a hybrid approach. However, not even such a hybrid approach can protect all operations sufficiently. Encoding logic-and-like operations in a robust and efficient way is an open problem and will be subject to future research.

References

1. Boneh, D., DeMillo, R.A., Lipton, R.J.: On the Importance of Checking Cryptographic Protocols for Faults (Extended Abstract). In: Fumy, W. (ed.) EUROCRYPT 1997. LNCS, vol. 1233, pp. 37–51. Springer, Heidelberg (1997)
2. Oh, N., Shirvani, P.P., McCluskey, E.J.: Control-flow checking by software signatures. IEEE Transactions on Reliability 51, 111–122 (2002)
3. Kim, C.H., Quisquater, J.J.: Fault Attacks for CRT Based RSA: New Attacks, New Results, and New Countermeasures. In: Sauveron, D., Markantonakis, K., Bilas, A., Quisquater, J.-J. (eds.) WISTP 2007. LNCS, vol. 4462, pp. 215–228. Springer, Heidelberg (2007)
4. Dottax, E., Giraud, C., Rivain, M., Sierra, Y.: On Second-Order Fault Analysis Resistance for CRT-RSA Implementations. Cryptology ePrint Archive, Report 2009/024 (2009) The final version of this paper will be published in the proceedings of WISTP (2009)
5. Karri, R., Wu, K., Mishra, P., Kim, Y.: Concurrent Error Detection of Fault-Based Side-Channel Cryptanalysis of 128-Bit Symmetric Block Ciphers. In: Proceedings of the 38th Design Automation Conference, DAC 2001, Las Vegas, NV, USA, pp. 579–585. ACM, New York (2001)
6. Lenstra, A.K.: Memo on RSA Signature Generation in the Presence of Faults (1996), http://cm.bell-labs.com/who/akl/
7. Hamming, R.W.: Error Detecting and Error Correcting Codes. Bell System Technical Journal 29, 147–160 (1950)
8. Elliott, I., Sayers, I.: Implementation of 32-bit RISC processor incorporating hardware concurrent error detection and correction. In: Computers and Digital Techniques, IEE Proceedings E., vol. 137, pp. 88–102 (1990)

9. Nicolaidis, M.: Carry checking/parity prediction adders and ALUs. IEEE Transactions on Very Large Scale Integration (VLSI) Systems 11, 121–128 (2003)
10. Berger, J.M.: A Note on Error Detection Codes for Asymmetric Channels. Information and Control 4, 68–73 (1961)
11. Lo, J.C., Thanawastien, S., Rao, T.R.N.: Concurrent error detection in arithmetic and logical operationsusing Berger codes. In: Proceedings of 9th Symposium on Computer Arithmetic (1989)
12. Kim, J., Rao, T., Feng, G., Lo, J.C.: The efficient design of a strongly fault-secure ALU using a reduced Berger code for WSI processor arrays. In: Proceedings of Fifth Annual IEEE International Conference on Wafer Scale Integration, pp. 163–172 (1993)
13. Russell, G., Maamar, A.: Check bit prediction scheme using Dong's code for concurrent error detection in VLSI processors. In: IEE Proceedings of Computers and Digital Techniques, vol. 147, pp. 467–471 (2000)
14. Massey, J.L.: Survey of residue coding for arithmetic errors. ICC Bulletin 3, 195–209 (1964)
15. Mandelbaum, D.: Arithmetic codes with large distance. IEEE Transactions on Information Theory 13, 237–242 (1967)
16. Proudler, I.K.: Idempotent AN codes. In: IEE Colloquium on Signal Processing Applications of Finite Field Mathematics, London, UK, pp. 8/1–8/5. IEEE, Los Alamitos (1989)
17. Gaubatz, G., Sunar, B.: Robust Finite Field Arithmetic for Fault-Tolerant Public-Key Cryptography. In: Breveglieri, L., Koren, I., Naccache, D., Seifert, J.-P. (eds.) FDTC 2006. LNCS, vol. 4236, pp. 196–210. Springer, Heidelberg (2006)
18. Rao, T.: Biresidue Error-Correcting Codes for Computer Arithmetic. IEEE Transactions on Computers C-19, 398–402 (1970)
19. Rao, T., Garcia, O.: Cyclic and multiresidue codes for arithmetic operations. IEEE Transactions on Information Theory 17, 85–91 (1971)

Mechanism behind Information Leakage in Electromagnetic Analysis of Cryptographic Modules

Takeshi Sugawara[1], Yu-ichi Hayashi[1], Naofumi Homma[1],
Takaaki Mizuki[1], Takafumi Aoki[1], Hideaki Sone[1], and Akashi Satoh[2]

[1] Tohoku University
[2] National Institute of Advanced Industrial Science and Technology
{sugawara,homma}@aoki.ecei.tohoku.ac.jp,
yu-ichi@m.tains.tohoku.ac.jp,
tm-paper@rd.isc.tohoku.ac.jp, aoki@ecei.tohoku.ac.jp,
sone@isc.tohoku.ac.jp, akashi.satoh@aist.go.jp

Abstract. This paper presents radiation mechanism behind Electromagnetic Analysis (EMA) from remote locations. It has been widely known that electromagnetic radiation from a cryptographic chip could be exploited to conduct side-channel attacks, yet the mechanism behind the radiation has not been intensively studied. In this paper, the mechanism is explained from the view point of Electromagnetic Compatibility (EMC): electric fluctuation released from a cryptographic chip can conduct to peripheral circuits based on ground bounce, resulting in radiation. We demonstrate the consequence of the mechanism through experiments. For this purpose, Simple Electromagnetic Analysis (SEMA) and Differential Electromagnetic Analysis (DEMA) are conducted on FPGA implementations of RSA and AES, respectively. In the experiments, radiation from power and communication cables attached to the FPGA platform is measured. The result indicates, the information leakage can extend beyond security boundaries through such cables, even if the module implements countermeasures against invasive attacks to deny access at its boundary. We conclude that the proposed mechanism can be used to predict circuit components that cause information leakage. We also discuss advanced attacks and noise suppression technologies as countermeasures.

1 Introduction

There have been many articles published on side-channel attacks [1, 2] and countermeasures against them, mainly focusing on algorithms and implementation (both hardware and software) levels. Consequently, many experimental results have been reported.

In academic context, experiments have been conducted in a laboratory environment where an attacker can directly access the cryptographic modules to measure it. However, such ideal measurements are not always available in practice.

For example, it is common for cryptographic modules to feature countermeasures against invasive attacks such as electromagnetic shielding and tamper-sensing mesh. Such countermeasures can prevent the attacker from intruding to their security

H.Y. Youm and M. Yung (Eds.): WISA 2009, LNCS 5932, pp. 66–78, 2009.

boundaries [3]. At the same time, side-channel attacks also become difficult as such countermeasures prevent attackers from measuring side-channel information. However, the cryptographic modules are still vulnerable to the attacks if other side-channels (e.g. electromagnetic radiation) are available and exploited by attackers.

As stated above, the quality of measurement is vital to the feasibility of the side-channel attacks. Therefore, *the availability of measurements* should be considered to achieve practical security evaluations. For this purpose, it is essential to investigate available side-channels and physical mechanism behind them.

Conventional measurement methods are revisited from the viewpoint of the *availability of measurements*. In the original paper on power analysis [1], a small resistor is inserted between a power pin of a cryptographic chip and a Printed Circuit Board (PCB) to measure the power consumption. Then, Electromagnetic Analysis (EMA), which measures electromagnetic fields near the cryptographic device, is proposed [4, 5]. Many studies have followed these two methods. However, in these methods, an attacker requires close access to the device while conducting the measurements. Therefore, the attacks can be easily prevented by applying countermeasures against invasive attacks to deny close access to the module.

On the other hand, there are reports on successful EMA experiments based on measurements from remote locations [6, 7]. In these cases, side-channel attacks are feasible without close access to the module. These are good examples where the feasibility of the attack is determined by the *availability of measurements*. However, the mechanism behind such radiation was not discussed in these works [6, 7]. As a result, it has been difficult to predict how much and where the radiation occurs.

The problem can be reduced to the following question: how will the electric fluctuation released from an LSI (Large-Scale Integrated circuit) propagate and radiate? This problem has been discussed in the field of Electromagnetic Compatibility (EMC) as unintentional radiation from a digital circuit. From this perspective, the electric fluctuation that can be used to retrieve side-channel information is regarded as noise responsible for interference. Therefore, it is possible to discuss the problem by using a model studied in the field of EMC.

In this paper, the mechanism behind EMA from remote locations is presented. The mechanism behind propagation and radiation is explained by ground bounce, which is studied extensively in the EMC field. The mechanism indicates that electric fluctuations are conducted to peripheral circuits through the ground plane, causing radiation. We demonstrate the consequence of the proposed mechanism by a series of experiments measuring (i) a power cable (ii) a communication cable, and (iii) the surrounding space around the power cable. For the experiments, Simple Electromagnetic Analysis (SEMA) and Differential Electromagnetic Analysis (DEMA) are applied to FPGA implementations of RSA and AES, respectively. We also discuss advanced attacks and noise suppression techniques as countermeasures from the viewpoint of EMC.

2 Leakage Mechanism Based on Ground Bounce

In this section, conventional measurement techniques are described, followed by the mechanism behind the propagation and radiation explained by ground bounce. As conventional techniques, (i) measurements using an inserted resistor and (ii) measurements of electromagnetic fields are described.

2.1 Conventional Measurement Methods

2.1.1 Measurements of Voltage Drop Using an Inserted Resistor

Fig. 1(a) shows an overview of the power measurement method using a resistor [1]. In this measurement, the transient current I released from the power pins of the LSI is assumed to contain information leakage. The mechanism behind the leakage is discussed in [2, 8] based on the switching behavior of CMOS gates.

The current I must be transformed into voltage units as general instruments, e.g., a digital oscilloscope, only accept voltage signals. For this purpose, a small resistor is inserted in series between the pin and the PCB. The voltage across the resistor R is $V = RI$ according to the Ohm's law. Then the attacker can measure the voltage V which is proportional to the current I.

Many studies use the above method as it is simple and easy to reproduce. When we consider the *availability of measurement*, however, the method involves manipulation (i.e., insertion of a resistor) of the PCB. Consequently, it requires close access to the PCB.

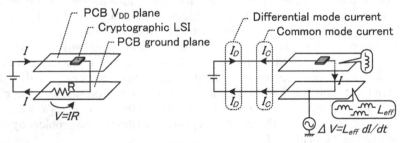

(a) Conventional measurement method   (b) Proposed leak mechanism

Fig. 1. Conventional and proposed leak model

2.1.2 Measurements of Electromagnetic Fields

Side-channel attack based on electromagnetic measurements is referred to as EMA and is proposed in [4, 5]. This method is also widely used. Measurements are conducted by placing a magnetic or electric field probe very close to the chip. Below, we focus on the method that uses magnetic field measurements.

A magnetic probe transforms the magnetic field (or magnetic flux) into a voltage output based on the Electromagnetic Induction. Here, the output voltage V is

$$V = -N \frac{d\Phi}{dt},$$

where Φ represents the magnetic flux within the closed loop comprising the probe; N is the number of the loops. Since $\Phi \propto I$, it follows that $V \propto dI/dt$. A probe that directly outputs V is referred to as a magnetic field probe, whereas one that outputs the value proportional to I, as a result of loop integration, is referred to as a current probe. Although the output of the magnetic field probe is proportional to dI/dt not to I, a number of experiments have shown that the attack using the magnetic probe is feasible and efficient [9].

This method also requires close access to the PCB since the electromagnetic near field decreases in amplitude in proportion to $1/r^3$, where r is the distance between the target and the probe. Therefore, signals measured from a distance suffer from the effects of external noise, resulting in a low S/N ratio. In addition, the measurements become even more difficult when the target module uses electromagnetic shielding as a countermeasure.

2.2 Leak Mechanism Based on Ground Bounce

The proposed mechanism based on ground bounce is described in this section.

In classical circuit theory, the level of the ground plane is assumed to be constantly zero. However, in reality, the ground level can change. Such transient voltage fluctuation in the ground plane is referred to as ground bounce.

Transient current released from a digital circuit is a known cause of ground bounce. Here, the released transient current I is transformed into a voltage fluctuation ΔV through inductance. Such inductance is distributed over the PCB, since conductors with finite length (e.g., pins and lead lines) have parasitic inductance as shown in the Fig. 1(b). When a transient current I is fed into such inductance, an electromotive force occurs [10, 11] due to electromagnetic induction, which results in a voltage fluctuation ΔV in the ground plane. The fluctuation is expressed as the following equation [10]:

$$\Delta V = L_{eff} \cdot M \cdot \frac{dI}{dt},$$

where L_{eff} is the effective parasitic inductance, M is the number of simultaneous switching outputs, and dI/dt is the rate of change of the current. The amount of L_{eff}, which is abstractly illustrated in Fig. 1(b), depends not only on the self-inductance within the cryptographic chip but also on the mutual inductance between the chip and the PCB.

The voltage fluctuation caused by ground bounce can be modeled as an alternate voltage source as shown in Fig. 1(b). The model suggests that the voltage fluctuation can propagate to peripheral circuits through a common ground. Consequently, peripheral circuits, such as attached cables, are driven as antennas, resulting in unintentional radiation from them [12, 13].

Radiation based on ground bounce is even more important since it generates common-mode current [14]. If there is only differential-mode current as shown in Fig. 1(a), then the resulting radiation is limited as the electromagnetic fields radiated from the current pair (current with forward and reverse directions) cancel each other out since they are equal in amplitude and inverse in direction. On the other hand, when common-mode current is considered, the electromagnetic fields from the current pair are not cancelled out since their directions coincide. As a consequence, the common-mode current can cause strong radiation even if it is weak in amplitude.

When we consider a voltage fluctuation ΔV caused by a transient current I released from a cryptographic LSI, ΔV contains information leakage due to $\Delta V \propto dI/dt$.

As a result, EMA is possible by measuring the radiation driven by ground bounce. The mechanism also suggests that peripheral circuits interconnected to a cryptographic module can be an antenna responsible for information leakage.

3 Experiment

In this section, we demonstrate SEMA and DEMA against FPGA implementations of RSA and AES by measuring the radiation of a target module from remote locations.

3.1 Experimental Condition

Fig. 2 and Fig. 3 show a block diagram and overview of the measurement setup, respectively. The measurement system consists of the Side-channel Attack Standard Evaluation Board (SASEBO) [15], a digital oscilloscope, and a PC. The SASEBO is equipped with two Field Programmable Gate Arrays (FPGAs), namely *FPGA1* and *FPGA2*. These two FPGAs work by using power supplied from on-board regulators and clock signal provided by an external function generator. Experiments are conducted by implementing cryptographic cores (circuits) on FPGA1.

Four types of measurements are conducted: (i) the voltage drop across a resistor, the current flowing in (ii) the attached power cable and (iii) the communication cable (RS232C cable), and (iv) the magnetic field around the power cable. Hereafter, these are referred to as (i) *Resistor*, (ii) *Power cable*, (iii) *RS232C cable*, and (iv) *Antenna*. It is important to emphasize that the measuring points (ii) and (iii) are not directly connected to FPGA1; there are circuit components (e.g., voltage regulators and an RS232C level converter) in between. In addition, the locations of the measurements are (ii) 50 cm, (iii) 40 cm, and (iv) 50 cm away from the board.

Measurement instruments are summarized in Table 1. The details of the measurement methods are as follows. In measurement (i), the voltage drop across a 1-Ω resistor inserted between the ground pin of FPGA1 and the ground plane of the PCB is measured using a differential voltage probe. Note that the 1-Ω resistor is short-circuited during the measurements (ii)–(iv) by using jumper pins.

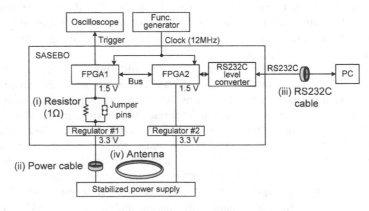

Fig. 2. Block diagram of experimental setup

(i) Resistor   (ii) Power cable   (iii) RS232C cable   (iv) Antenna

Fig. 3. Overview of experimental setup

In the measurements (ii) and (iii), the current flowing in cables is measured by clamping the cables using the probe as shown in Figs. 3(ii) and 3(iii). Then the voltage proportional to the flowing current is measured as described in Section 2.1.2. In measurement (ii), both lines (for V_{DD}: 3.3 V and GND: 0 V) of a twisted pair cable are clamped, while the whole body of RS232C cable is clamped for measurement (iii). Since both of the current pairs are clamped, radiation due to differential-mode current is cancelled out within the probe, and thus the contribution only by the common-mode element is measured.

Finally, in measurement (iv), the magnetic field around the power cable is measured using an antenna. We used an off-the-shelf indoor loop antenna, which is used for amateur radio and priced around \$300. The antenna is used by tuning it to maximize the measured amplitude in the range of 3–40 MHz. The measurement is conducted by placing the antenna over the power cable at a height of about 20 cm.

In each measurement, the frequency bands of measured signals are limited up to 25 MHz by using a low-pass filter equipped with an oscilloscope. This is because the measured raw traces were highly contaminated by high-frequency noise that interfered with the measurements. In addition, a trigger signal generated by the FPGA1 is used in order to align the measured traces in time. As a probe for the trigger signal has a physical contact to the board, the measurements are not exactly remote. However, the setup is enough to examine information leakages from the measuring points (i)-(iv). In practical scenario, an attacker would have difficulty in taking a trigger without

Table 1. Specific measurement instruments

	Probe
Oscilloscope	Agilent MSO6104A (with 25MHz low-pass filter)
Voltage probe for (i)	Agilent 1130A differential (voltage) probe with SMA probe head
Current probe for (ii) and (iii)	Fischer F-2000 current probe [16] (10MHz - 3GHz)
Antenna for (iv)	AOR LA380 Wideband Active Loop Antenna [17] (10kHz - 500MHz)
Pre amplifier for (ii)-(iv)	MITEQ AM-1594-9907 (+51dB, 300kHz-3.0GHz)

contacts to the target, yet it is still possible. One practical way is to observe communication cables and then obtain a trigger from a specific binary sequence on them. In addition, the attacker can consult signal processing techniques to achieve precise waveform alignment [18].

3.2 Simple Electromagnetic Analysis of an RSA Implementation

A 1,024-bit RSA circuit using a left-to-right binary method is implemented on FPGA1. Modular multiplication and squaring are performed by high-radix Montgomery multiplication algorithm using a 32-bit multiplier [19, 20]. In this implementation, one Montgomery multiplication requires 4,386 cycles, and the total number of cycles for the modular exponentiation (1,024-bit RSA operation) is approximately 7 million cycles. The parameters, i.e., the key and the plaintext are embedded into the FPGA1 in order to allow FPGA1 to operate as a standalone module. The goal of the attack is to distinguish between multiplication and squaring [1] in the measured trace. Since we are interested in the difference between the measurements (i)–(iv), a chosen input message of 2^{-1024} [20] is used to enhance the difference between the multiplication and squaring.

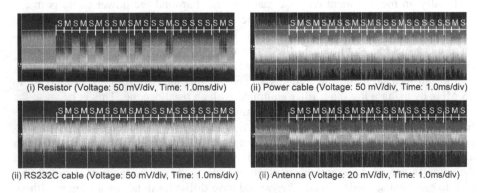

(i) Resistor (Voltage: 50 mV/div, Time: 1.0ms/div)   (ii) Power cable (Voltage: 50 mV/div, Time: 1.0ms/div)

(ii) RS232C cable (Voltage: 50 mV/div, Time: 1.0ms/div)   (ii) Antenna (Voltage: 20 mV/div, Time: 1.0ms/div)

Fig. 4. Result of SEMA on RSA

Fig. 4 shows the results of the SEMA. The traces are measured at a sampling frequency of 400 MSa/s. Each of the traces is aligned in time, in which the modular exponentiation starts at around 1.5 ms. Sequences of symbols 'S' and 'M' shown in the figure represent the corresponding operation (squaring and multiplication).

The result of measurement (i) shows large difference between multiplication and squaring with multiplication having lower spikes compared to squaring. Although the differences are smaller than (i), they are still visible in results (ii)–(iv). The results indicate that it is possible to reveal a secret key by using any of the measurements (i)–(iv).

3.3 Differential Power Analysis of an AES Implementation

In the experiments, an AES circuit, retrieved from the reference[19], supporting a 128-bit key is implemented on FPGA1. The circuit uses a loop architecture, where one round operation is performed every clock cycle. As a result, one encryption takes 10 clock cycles for round operations and an additional clock cycle for data I/O. FPGA2 is configured as the control and communication circuits, and plaintexts are fed from the PC into FPGA1 via FPGA2. During the encryption, the corresponding traces are captured from the four measurement points at a sampling frequency of 500 MSa/s. The measurements are repeated for 30,000 different plaintexts, and the corresponding 30,000 traces are stored for each measuring points. Examples of the measured traces are shown in Fig. 5, where the encryption process starts at around 400 ns and finishes after 11 clock cycles or 916 ns (=11×1/12 MHz).

Correlation Power Analysis (CPA) [21] is applied to these traces. The 128-bit (16-byte) register containing intermediate data is chosen as target. The power estimates are calculated by counting changed bits of the target register in the final round of AES encryption. Here, Hamming distance model [21] is used as power model. Key candidates are searched by byte. In other words, total of 16 power estimates are made correspond to 16 bytes of the round key. In the attack phase, the linearity is evaluated by using Pearson's correlation coefficient.

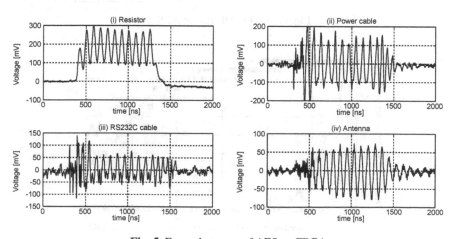

Fig. 5. Example traces of AES on FPGA

The results are shown as error rates in Fig. 6 and Fig. 7. Fig. 6 shows Measurement to disclosure (MTD) graph of each result. On the other hand, Fig. 7 shows error rates. The vertical axis represents the number of incorrectly predicted round-key bytes, and the horizontal axis shows the number of traces. Since the length of the secret key is 16 bytes (=128 bits), the value ranges between 0 and 16, where 0 indicates the successful extraction of the whole key (i.e., completion of the attack).

In the analysis with the measurement (i), the key prediction is regarded as successful when the correlation with the correct key is the highest among 2^8 candidates. On the other hand, in (ii)–(iv), the difference between maximum and minimum

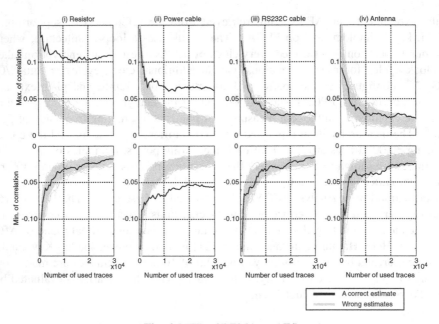

Fig. 6. MTD of DEMA on AES

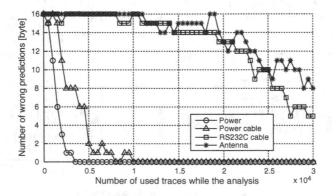

Fig. 7. Results of DEMA on AES

correlations is used instead of maximum. since the results of (ii)–(iv) show correlation in both the positive and negative directions as shown in Fig. 6.

As shown in Fig. 7, the result for (i) goes to zero fastest, using 3,000 traces to extract the whole key. In addition, the attack using the power cable (ii) also shows fast extraction. On the other hand, the attacks using the RS232C cable and the antenna require much larger number of traces. However, the error rate is descending gradually as the number of used traces increases. Therefore, we can say that all the remote attacks can successfully reveal the secret keys.

3.4 Discussion

As shown above, the remote SEMA and DEMA worked well, but the traces from (ii)–(iv) contained smaller amount of information leakage or larger noise in comparison to that from (i). Various disturbing factors between the FPGA and the measuring points are responsible for the results. Such factors include the filtering effect of parasitic circuit, the compensation effect of regulators, and external noise, and so on.

For example, there is an on-board regulator between the chip's power supply pins and the power cable. Since regulators are designed to stabilize their output voltage, they feature buffering and feedback control in order to suppress voltage fluctuation. The results of (ii) indicate that the voltage fluctuation containing information leakage was able to overcome the effects of the regulator.

Measurement (iii) is also affected by such disturbing factors. In the experimental setup, the RS232C cable is connected to FPGA1 via an RS232 level converter and FPGA2. First, the RS232 level converter has the same effect as the voltage regulator. In addition, FPGA1, FPGA2, and the RS232C level converter have their own separated ground plane and they are connected via noise filters (inductors). This feature is used in SASEBO, in order to allow an experimenter to measure the contribution from FPGA1 while avoiding contributions from other elements. The noise filters act as low-pass filters for current through them. However, the success of measurement (iii) indicates that the voltage fluctuation from FPGA1 can propagate to the other part of the board even if such high-frequency current is filtered out.

From these results, we can conclude that the propagation of the voltage fluctuation is rather robust and difficult to suppress.

4 Advanced Attacks and Countermeasures

In this section, advanced attacks as well as countermeasures based on noise suppression are discussed based on the source-path-antenna model [14] from the EMC field.

The source-path-antenna model describes unintentional radiation using three abstract factors. The source is a differential-mode element released from a chip. Then, the differential-mode element is transformed into a common-mode element in the path. Finally, the common-mode element drives an antenna, resulting in radiation. The strength of the resulting radiation is the product of the contributions of these three components. In our experiments, the source is transient current released from FPGA1, the path is ground bounce, and the antenna is the attached cables.

4.1 Advanced Attacks

As the model suggests, an attacker can improve the gain of radiation by enhancing any of the sources, path, and antenna elements. If we assume an attacker who has difficulty gaining close access to a target cryptographic module, it would also be difficult to modify source and path elements. On the other hand, it would be relatively easy to access and manipulate antenna elements, such as attached cables.

A simple enhancing technique of the antenna element is to wind a cable into a loop. Since a cable with a loop structure composes a loop antenna, the magnetic field around the cable is boosted. Improved gain of radiation enables an attack from a more distant position.

A second technique is to tune the antenna element to the frequency band containing information leakage. When the antenna element is tuned (intentionally or unintentionally) to such a frequency band, the radiation becomes more efficient. Consequently, the attacker can conduct attacks from a more remote location [6, 7]. Note here that the resonant frequency of the power cable (75 cm) used for the experiments is 400 MHz for a single wavelength, and the frequency band is not used for the experiments as it is filtered.

4.2 Countermeasures

It is possible to counteract EMA by suppressing each of the source, path, and antenna elements. This implies that existing noise reduction techniques for EMC can be used.

One of the popular noise reduction techniques is the use of bypass capacitors [6]. Voltage fluctuation can be suppressed by placing capacitors close to the pins of the cryptographic chip. This method is regarded as equivalent to noise reduction in the source element. Note here that the board we used for the experiments (SASEBO) does not have bypass capacitors.

A second example is the use of an attached ferrite core on the cables. This method is also widely used to suppress radiation from cables and regarded as equivalent to noise reduction in the antenna element.

It is important to emphasize that noise reduction techniques should be applied carefully. First, noise propagation is robust as we observed in the experiments in Section 3. Second, noise reduction should be applied within the security boundary so that attackers won't remove it. Finally, frequency band containing information leakage should be investigated because a noise reduction usually suppresses a specific frequency band selectively.

5 Conclusion

Leakage mechanism behind Electromagnetic Analysis (EMA) from remote locations is presented. The mechanism based on ground bounce, studied in the EMC field, is described. Consequence of the proposed mechanism is demonstrated through various experiments. In the experiments, measurements are conducted from (i) an attached power cable, (ii) an attached communication cable, and (iii) free space around the power cable. By using the measured traces, SEMA and DEMA attacks on FPGA implementations of RSA and AES successfully revealed secret keys. The results indicate that side-channel information can extend beyond security boundary through unprotected peripheral circuits (e.g., attached cables) even if the module was protected against invasion within the boundary. We conclude that the proposed mechanism can be used to predict circuit components that can cause information leakage. Finally, we discussed advanced attacks as well as noise suppression techniques countermeasures against side-channel attacks.

In this paper, we studied side-channel attacks from the viewpoint of EMC, focusing on the propagation of information leakage. In addition, the malfunction (fault) mechanism due to noise interference is also an important research topic in the field of EMC. We plan to apply these research results to fault injection attacks in order to develop new aspects of physical security for cryptographic modules.

References

1. Kocher, P., Jaffe, J., Jun, B.: Differential Power Analysis. In: Wiener, M. (ed.) CRYPTO 1999. LNCS, vol. 1666, pp. 388–397. Springer, Heidelberg (1999)
2. Mangard, S., Oswald, E., Popp, T.: Power Analysis Attacks: Revealing the Secrets of Smart Cards. Springer, Heidelberg (2007)
3. Anderson, R., Bond, M., Clulow, J., Skorobogatov, S.: Cryptographic Processors-A Survey. Proceedings of the IEEE 94(2), 357–369 (2006)
4. Quisquater, J.-J., Samyde, D.: ElectroMagnetic Analysis (EMA): Measures and Countermeasures for Smart Cards. In: Proceedings of E-smart (September 2001)
5. Grandolfi, K., Mourtel, C., Olivier, F.: Electromagnetic Analysis: Concrete Results. In: Koç, Ç.K., Naccache, D., Paar, C. (eds.) CHES 2001. LNCS, vol. 2162, pp. 251–261. Springer, Heidelberg (2001)
6. Agrawal, D., Archambeault, B., Rao, J.R., Rohatgi, P.: The EM Side-Channel(s). In: Kaliski Jr., B.S., Koç, Ç.K., Paar, C. (eds.) CHES 2002. LNCS, vol. 2523, pp. 29–45. Springer, Heidelberg (2003)
7. Kim, C., Schlaffer, M., Moon, S.: Differential Side Channel Analysis Attacks on FPGA Implementations of ARIA. ETRI Journal. 30(2), 315–325 (2008)
8. Suzuki, D., Saeki, M., Ichikawa, T.: DPA Leakage Models for CMOS Logic Circuits. In: Rao, J.R., Sunar, B. (eds.) CHES 2005. LNCS, vol. 3659, pp. 366–382. Springer, Heidelberg (2005)
9. Peeters, E., Standaert, F.-X., Quisquater, J.-J.: Power and Electromagnetic Analysis: Improved Model, Consequences and Comparisons. Integration, the VLSI Journal 40(1), 52–60 (2007)
10. Sudo, T., Sasaki, H., Masuda, N., Drewniak, J.: Electromagnetic Interference (EMI) of System-on-Package (SOP). IEEE Transactions on Advanced Packaging 27(2), 304–314 (2004)
11. Yang, Y., Brews, J.R.: Design Trade-offs for the Last Stage of an Unregulated, Long-channel CMOS Off-chip Driver with Simultaneous Switching Noise and Switching Time Considerations, Components, Packaging, and Manufacturing Technology, Part B. IEEE Transactions on Advanced Packaging 19(3), 481–486 (1996)
12. Drewniak, J., Sha, F., Van Doren, T., Hubing, T., Shaw, J.: Diagnosing and Modeling Common-mode Radiation from Printed Circuit Boards with Attached Cables. In: Proceedings of the 1995 IEEE International Symposium on Electromagnetic Compatibility, pp. 465–470 (1995)

13. Hockanson, D., Drewniak, J., Hubing, T., Van Doren, T., Sha, F., Wilhelm, M.: Investigation of Fundamental EMI Source Mechanisms Driving Common-mode Radiation from Printed Circuit Boards with Attached Cables. IEEE Trans. Electromagn. Compat. 38(4), 557–566 (1996)

14. Paul, C.R.: Introduction to Electromagnetic Compatibility (Wiley Series in Microwave and Optical Engineering). Wiley Interscience, Hoboken (2006)

15. Side-channel Attack Standard Evaluation Board (SASEBO),
    http://www.rcis.aist.go.jp/special/SASEBO/

16. AOR, LA380 Wideband Active Loop Antenna,
    http://www.aorusa.com/la380.html

17. Fischer Custom Communications, Inc., F-2000 Current Probe,
    http://www.fischercc.com/Quadrary-Pages/Current-Probes/F-2000.htm

18. Homma, N., Nagashima, S., Imai, Y., Aoki, T., Satoh, A.: High-resolution Side-Channel Attack Using Phase-Based Waveform Matching. In: Goubin, L., Matsui, M. (eds.) CHES 2006. LNCS, vol. 4249, pp. 187–200. Springer, Heidelberg (2006)

19. Cryptographic Hardware Project, Computer Structures Laboratory, Graduate School of Information Sciences, Tohoku University,
    http://www.aoki.ecei.tohoku.ac.jp/crypto/

20. Miyamoto, A., Homma, N., Aoki, T., Satoh, A.: Chosen-Message SPA Attacks Against FPGA-Based RSA Hardware Implementation. In: Proceedings of the International Conference on Field Programmable Logic and Applications (FPL 2008), pp.35–40 (September 2008)

21. Brier, E., Clavier, C., Olivier, F.: Correlation Power Analysis with a Leakage Model. In: Joye, M., Quisquater, J.-J. (eds.) CHES 2004. LNCS, vol. 3156, pp. 16–29. Springer, Heidelberg (2004)

EM Side-Channel Attacks on Commercial Contactless Smartcards Using Low-Cost Equipment

Timo Kasper, David Oswald, and Christof Paar

Horst Görtz Institute for IT Security
Ruhr University Bochum, Germany
timo.kasper@rub.de, david.oswald@rub.de, cpaar@crypto.rub.de

Abstract. We introduce low-cost hardware for performing non-invasive side-channel attacks on Radio Frequency Identification Devices (RFID) and develop techniques for facilitating a correlation power analysis (CPA) in the presence of the field of an RFID reader. We practically verify the effectiveness of the developed methods by analysing the security of commercial contactless smartcards employing strong cryptography, pinpointing weaknesses in the protocol and revealing a vulnerability towards side-channel attacks. Employing the developed hardware, we present the first successful key-recovery attack on commercially available contactless smartcards based on the Data Encryption Standard (DES) or Triple-DES (3DES) cipher that are widely used for security-sensitive applications, e.g., payment purposes.

1 Introduction

In the past few years, RFID technologies rapidly evolved and are nowadays on the way to become omnipresent. Along with this trend grows the necessity for secure communication and authentification. RFID-based applications such as electronic passport, payment systems, car immobilisers or access control systems require strong cryptographic algorithms and protocols, as privacy and authenticity of the transmitted data are crucial for the system as a whole. Since severe weaknesses have been discovered in the "first generation" of RFIDs that rely on proprietary ciphers [25,8,7,10], such as Mifare Classic contactless smartcards [21] or KEELOQ RFID transponders [20], future systems will tend to employ stronger cryptographic primitives. This trend can already be observed, as several products exist that provide a (3)DES encryption.

The aim of this paper is to practically evaluate the security of these believed (and advertised) to be highly secure contactless smartcard solutions. Since encryption is performed using well-known and carefully reviewed algorithms, cryptanalytical attacks on the algorithmic level are very unlikely to be found. Thus, we aim at performing a *Side-Channel Analysis* which exploits the physical characteristics of the actual hard- or software implementation of the cipher.

H.Y. Youm and M. Yung (Eds.): WISA 2009, LNCS 5932, pp. 79–93, 2009.

1.1 RFID and Contactless Smartcards

The huge variety of applications for RFID implies that products come in a lot of distinct flavors, differing amongst others in the operating frequency, the maximum achievable range for a query, and their computational power [9].

Passive RFIDs draw all energy required for their operation from the field of a reader and are hence severely limited with respect to their maximum power consumption, i.e., the amount of switching transistors during their operation, which has a direct impact on their cryptographic capabilities. For highly demanding applications, the ISO/IEC 14443 standard for *contactless smartcards* [13,14] has proven to be suitable. A strong electromagnetic field combined with a specified reading distance of only approx. 10 cm provides - contrary to most other RFID schemes - a sufficient amount of energy even for public key cryptography, as realised in the electronic passport [1]. In the standard, a contactless smartcard is also referred to as *Proximity Integrated Circuit Card* (PICC), while the reader is called *Proximity Coupling Device* (PCD). The PCD generates an electromagnetic field with a carrier frequency of 13.56 MHz, that supplies the PICC with energy and at the same time serves as a medium for the wireless communication. All communication is initiated by the PCD, while the PICC answers by load-modulating the field of the PCD [13].

Challenge-Response Authentication Protocol. According to its data sheet, the analysed contactless smartcard uses a challenge-response authentication protocol which relies on a symmetric block cipher, involving a 112 bit key k_C that is shared between PCD and PICC. For the cipher, a 3DES using the two 56 bit halves of $k_C = k_1 \| k_2$ in EDE mode according to [2] is implemented. After a successful authentication, the subsequent communication is encrypted with a session key. We implemented the whole protocol, but however, focus on the step relevant for our analyses as depicted in Fig. 1, where $3\mathrm{DES}_{k_C}(\cdot) = \mathrm{DES}_{k_1}\left(\mathrm{DES}_{k_2}^{-1}\left(\mathrm{DES}_{k_1}(\cdot)\right)\right)$ denotes a 3DES encryption involving the key $k_C = k_1 \| k_2$. The values B_1 and B_2 have a length of 64 bit and are encrypted by the PICC during the mutual authentication. B_2 originates from a random number previously generated by the PICC and is always encrypted by the PICC in order to check the authenticity of the PCD[1]. B_1, a random value chosen by the PCD that serves for authenticating the PICC to the PCD, is mentioned here for completeness only and is not required in the context of our analyses.

1.2 Related Work

Oren and Shamir [22] presented a successful side-channel attack against so-called Class 1 EPC tags operating in the UHF frequency range which can be disabled remotely by sending a secret "kill password". Small fluctuations in the reader

---

[1] The protocol will abort after the encryption of B_2, in case its verification is not successful.

PCD   PICC

$$\text{Choose } B_1, B_2 \xrightarrow{\quad B_1, B_2 \quad} 3\text{DES}_{k_C}(B_2)$$

Fig. 1. Exerpt of the authentication protocol relevant for an attack

field during the communication with the tag allow to predict the password bits. However, the very limited type of RFID tag does not offer any cryptography.

At CHES 2007, Hutter et al. [12] performed an EM attack on their own AES implementations on a standard 8-Bit microcontroller and an AES co-processor in an RFID-like setting, i.e., the self-made devices are powered passively and brought into the field of a reader. On their prototype devices the antenna and analogue frontend are separated from the digital circuitry, while on a real RFID tag, these components are intrinsically tied together. An artificially generated trigger signal before the attacked S-Box operation ensures perfect time alignment. Moreover, the clock signal for the digital circuitry is generated independently from the field of the reader using an external oscillator, hence the carrier is uncorrelated with the power consumption of the AES and can be easily removed.

In contrast, we now face the real-world situation, i.e., have no knowledge on the internal implementation details of the unmodified contactless smartcard to be attacked, cannot rely on artificial help like precise triggering for alignment, and analyse a black box with all RFID and cryptographic circuitry closely packed on one silicon die. In the following, after a brief introduction to power analysis in the context of RFID in Sect. 2.1, we will describe all relevant steps to analyse an unknown RFID device in practise, starting from our special low-cost measurement setup in Sect. 2.3 and including the extensive profiling that is required to gain insight into the operation of the smartcard in Sect. 3, before the results of the actual side-channel attack are presented in Sect. 3.2.

2   Power Analysis of RFIDs

Differential Power Analyis (DPA) was originally proposed in [17] and has become one of the most powerful techniques to recover secret information from even small fluctuations in the power leakage of the physical implementation of a cryptographic algorithm. In this paper, we address the popular *Correlation Power Analysis* (CPA), as introduced in [4].

2.1   Traditional vs. RFID Measurement Setup

For a typical power analysis attack [8] the side-channel leakage in terms of the electrical current consumption of the device, while executing a cryptographic operation, is measured via a resistor inserted into the ground path of the target IC. Since the targeted RFID smartcard circuitry including the anntenna is embedded in a plastic case, lacking any electrical contacts, it is difficult to perform a direct on-chip measurement of the power consumption. Invasive attacks, i.e., dissolving the chip from its plastic package and separating it from the antenna, were not successful [6], maybe due to the strong carrier of the reader that is required for the operation. Anyway, even a successful invasive attack is costly and can be easily detected, hence a non-invasive approach becomes very attractive in the context of RFIDs.

Non-Invasive Analysis with DEMA. A possible source of side-channel leakage that can be exploited in a non-invasive attack scenario is the information gathered from fluctuations of the EM field emanated by a device whilst performing a cryptographic operation. The corresponding side-channel information for this so-called *Differential Electro-Magnetic Analysis* (DEMA) [3] is acquired by means of near-field probes that are positioned close to the chip, and typically require no physical contact to the device, i.e., leave no traces. The analogue signal, i.e., the EM leakage in case of a DEMA, is digitised and recorded as a discrete and quantised timeseries called a *trace*. In practice, several traces for varying input data are collected. In the following, let t_l be the lth trace of one attack attempt, where $0 \leq l < L$, with L denoting the number of traces. Likewise, x_l denotes the associated input challenge for the lth measurement. For simplicity, we consider that all traces have the same length N.

2.2 Correlation DPA

For the actual attack, each *key candidate* K_s, $0 \leq s < S$, where the number of candidates S should be small[2], is input to a *prediction function* $d(K_s, x_l)$, establishing a link between given input data x_l and the expected current consumption for each key candidate K_s. Often, d predicts the power consumption of the output of an S-Box after the key addition, modelled either based on the Hamming weight, i.e., the number of ones in a data word, or based on the Hamming distance, i.e., the amount of toggling bits in a data word.

A CPA essentially relies on calculating the *Normalised Correlation Coefficient* between the predicted and recorded values for one point in time n and a fixed key K_s:

$$\Delta(K_s, n) = \frac{\sum_{l=0}^{L-1} \left(t_l(n) - m_{t(n)}\right)\left(d(K_s, x_l) - m_{d(K_s)}\right)}{\sqrt{\sigma^2_{t(n)}\sigma^2_{d(K_s)}}}$$

with $m_{t(n)}$, $m_{d(K_s)}$ denoting the means of the samples, and $\sigma^2_{t(n)}$, $\sigma^2_{d(K_s)}$ the sample variances of the respective timeseries. Plotting Δ for all n yields a curve indicating the correlation over time that features significant peaks, if K_s is the correct key guess, and has a random distribution otherwise. Thus, by iterating over all K_s and analysing the resulting $\Delta(K_s, 0)\ldots\Delta(K_s, N-1)$, the cryptographic secret can be revealed, given that enough traces have been acquired and that there exists a link between the side-channel leakage and the processed data input.

_____

[2] This is always the case when attacking single S-Boxes with few in- and outputs.

Efficiently Implementing a CPA. Straightforward implementations of a CPA read all L traces, each with a length of N samples, into memory before calculating the correlation coefficient $\Delta(K_s, n)$ (see Sect. 2.2).

This may become problematic for long traces and/or a large amount of measurements, e.g., $L = 10\,\mathrm{k}$ traces with $N = 350\,\mathrm{k}$ data points (stored as 4 byte single precision values) consume $\approx 13\,\mathrm{GByte}$ of memory. Therefore, a *recursive* computation of $\Delta(K_s, n)$ becomes attractive. Instead of first reading and then processing all data, existing values of the correlation coefficient can be updated with every new trace. This approach makes use of an algorithm given in [16], originally proposed by Welford. The update equations are

$$m_{i+1} = m_i + \frac{t_{i+1} - m_i}{i+1}, \quad M2_{i+1} = M2_i + (t_{i+1} - m_i)(t_{i+1} - m_{i+1}).$$

where the initial values are $m_0 = 0$, $M2_0 = 0$, t_i denotes the data points, m_i is the mean and $\sigma_i^2 = \frac{M2_i}{i-1}$ the variance after i samples. Applying this idea for computing the correlation coefficient of a key candidate, it suffices to keep track of N trace means $m_{t(n)}$ and $M2_{t(n)}$. Analogously, $m_{d(K_s)}$ and $M2_{d(K_s)}$ are updated, however, these are independent of n and thus need to be stored only once.

Besides, for evaluating Eq. 2.2, $c(K_s, n) = \frac{\sum_{l=0}^{L-1} t_l(n)d(K_s, x_l)}{L-1}$ is stored for N points in time and updated[3] according to

$$c_{i+1} = c_i + \frac{t_{i+1} \cdot d(x_{i+1}) - c_i}{i}$$

with initial values $c_0 = t_0 \cdot d(x_0)$, $c_1 = t_0 \cdot d(x_0) + t_1 \cdot$. $\Delta(K_s, n)$ after L traces is

$$\Delta(K_s, n) = \frac{(L-1) \cdot c_L(K_s, n) - L \cdot m_{t(n)} \cdot m_{d(K_s)}}{\sqrt{M2_{t(n)}M2_{d(K_s)}}}$$

The application of the recursive approach requires the storage of $\mathcal{O}(N)$ values for each key candidate K_s. In contrast, the traditional two-pass method (read all, then process) needs $\mathcal{O}(L \cdot N)$ memory. Thus, for large L, the memory footprint of the above described computations remains *constant*, while a straightforward algorithm becomes infeasible.

Modelling the Power Consumption of RFID Devices. For a simple model of the frequencies where we would expect the EM leakage to occur, consider a band-limited power consumption $p(t)$ that directly affects the amplitude of the $\omega_0 = 2\pi \cdot 13.56\,\mathrm{MHz}$ carrier, i.e., the amplitude of the field will be slightly smaller in an instant when the chip requires more energy than in an instant when no energy is consumed. This results in possibly detectable frequency components

---

[3] Note that n and K_s have been omitted for readability.

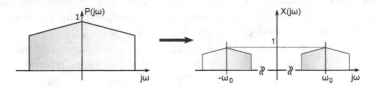

Fig. 2. Frequency spectrum of the carrier signal ω_0 and the assumed information leakage for remote power analysis

in the side bands of the carrier, as depicted in Fig. 2. Equation 1 describes this model more precisely, where $\circ\!\!-\!\!\bullet$ denotes the Fourier transform[4].

$$p\left(t\right)\cos\left(\omega_0 t\right) \circ\!\!-\!\!\bullet X\left(j\omega\right) = \frac{1}{2}\left(P\left(j\omega - j\omega_0\right) + P\left(j\omega + j\omega_0\right)\right) \tag{1}$$

We refer to this approach as *Remote Power Analysis*, as the fluctuations in the power consumption of the device are modulated onto the strong carrier signal of the PCD and may thus be visible even in the far-field[5].

2.3   Measurement Setup

The core of our proposed DEMA measurement equipment for RFIDs, illustrated in Fig. 3, is a standard PC that controls an oscilloscope and a self-built, freely programmable reader for contactless smartcards. These components, a specially developed circuit for analogue preprocessing of the signal and the utilised near-field EM probes are covered in this section.

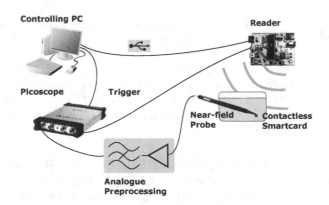

Fig. 3. Measurement setup

---

[4] The Fourier transform is commonly used to transform signals from the time domain into the frequency domain.

[5] For a frequency of 13.56 MHz the far-field begins at approx. 22 m [15].

RFID Reader. The RFID-interface is a custom embedded system both capable of acting as a reader and a transponder [15], whereas in the context of DEMA only the reader functionality is used. The device is controlled by a freely programmable Atmel ATMega32 microcontroller and provides an ISO 14443-compliant analogue front-end at a cost of less than 40 €. Contrary to commercial RFID readers, our self-built device allows for sending chosen challenges during the authentication.

Scope. The *Picoscope 5204* is a dual-channel storage USB-oscilloscope [23], featuring a maximum sample-rate of 1 GHz, an 8 bit analogue-to-digital converter (ADC), a huge 128 MSamples waveform memory and an external trigger input. These conditions are extremely good for side-channel analysis[6], alone the minimum input range of ± 100 mV might pose a problem in the context of DEMA attacks, where small voltage changes need to be detected with a high accuracy.

Probes. For measurements of the EM-field emanated by the contactless smartcard, a *RF-U 5-2* probe [18] is suitable, because it captures the near H-field that is proportional to the flow of the electric current in the horizontal plane. Note that, if no commercial EM probes are at hand, a self-wound coil can be a suitable replacement [5]. The small signal amplitudes (max. 10 mV) delivered by the probe are preamplified with the *PA-303* amplifier [18] by 30 dB over a wide frequency range of 3 GHz.

Analogue Signal Processing. Although to our knowledge there exist no reliable estimations about the exact amplitude of the EM emanations caused by digital circuitry — especially when attacking an unknown implementation — the unintented emanations of the chip are clearly orders of magnitude smaller than the strong field generated by the reader to ensure the energy supply of a PICC. The quantisation error induced by the ADC of the oscilloscope constitutes a minimum boundary for the achievable *Signal-to-Noise Ratio* (SNR), depending on the number of bits used for digitising an analogue value. Following [11], each bit improves the SNR by about 6 dB. Thus, for the best SNR the full input scale should be utilised for the signal of interest, implying that a maximum suppression of the carrier frequency and a subsequent amplification of the small side-channel information must already take place in the analogue domain, before the digitising step.

For minimising the disturbing influence of the carrier frequency on the measurements, we have built and tested several types of active and passive analogue filters. We here present our most straightforward and most unexpensive idea which in fact turned out to be the most effective approach in order to bypass the influence of the field of the reader. A part of the analogue front-end of the reader is a crystal-oscillator generating an almost pure sine wave with a frequency of 13.56 MHz that serves as the source for the field transmitted to the contactless

---

[6] In fact, for a typical side-channel attack such a large memory will never be fully used.

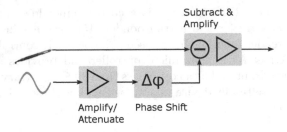

Fig. 4. Block diagram for removing the unwanted carrier frequency of the reader

smartcard. The straightforward principle introduced in the following is to tap
the oscillator of the reader and subtract its signal from the output of the EM
probe. The sine signal has a constant amplitude and a constant shift in time,
compared to the field acquired with the EM probes. Hence, as shown in Fig. 4,
the developed analogue circuitry is capable of delaying and scaling the sine wave
of the crystal, in order to match its amplitude and phase to that of the EM measurements, before substracting the pure sine from the EM measurements. This
approach, based on low-cost circuits employing operational amplifiers, allows to
suppress the unwanted signal component while keeping all possibly interesting
variations. The analoque preprocessing unit can also be used for other types of
RFIDs, such as 125 kHz transponders in car immobilisers.

3   A Real-World EM Attack on Contactless Smartcards

By performing a full authentication and reproducing the responses[7] of the cryptographically enabled contactless smartcard under attack on the PC, we verify
that a standard (3)DES [2] is used for the encryption of the challenge according
to Fig. 1. We further observe that the card unconditionally encrypts any value
B_2 (cf. Sect. 1.1 sent to it, hence we can freely choose the plaintext. For the CPA
described in the following, we will send random, uniformly distributed plaintexts
for B_2 and attack the first DES round.

3.1   Trace Preprocessing

The raw traces recorded between the last bit of the command sent by the reader
and the first bit of the answer of the card do not expose any distinctive pattern,
hence, digital preprocessing is applied in order to identify interesting patterns
useful for a precise alignment of the traces. On the basis of the RFID power
model introduced in Sect. 2.2, we assume that the power consumption of the
smartcard modulates the amplitude of the carrier wave at frequencies much
lower than the 13.56 MHz carrier frequency, which is justified by a preliminary

---

[7] Note that in this context the secret key of the implementation can be changed by
us and is hence known.

spectral analysis and the well-known fact that the on-chip components (such as capacitances, resistors, inductances) typically imply a strong low-pass filter characteristic.

Digital Amplitude Demodulation. In order obtain the relevant side-channel information, we record raw (undemodulated) traces and perform the demodulation digitally, using a straightforward incoherent demodulation approach (Fig. 5, following [26]). The raw trace is first rectified, then low-passed filtered using a *Finite Impulse Response* (FIR) filter. An additional high-pass *Infinite Impulse Response* (IIR) filter removes the constant amplitude offset resulting from the demodulation principle and low-frequency noise. Good values for the filter cutoff frequencies $f_{lowpass}$ and $f_{highpass}$ were determined experimentally and are given in Sect. 3.2.

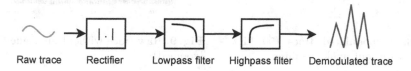

Raw trace   Rectifier   Lowpass filter   Highpass filter   Demodulated trace

Fig. 5. Digital amplitude demodulator

Fig. 7 displays a demodulated trace ($f_{lowpass} = 2\,\text{MHz}$, $f_{highpass} = 50\,\text{kHz}$) in which distinct patterns are visible, especially two shapes at 240000 ns and 340000 ns preceded and followed by a number of equally spaced peaks. For comparision, Fig. 6 shows a zoomed part of the same trace without demodulation. Fig. 8 and Fig. 9 originate from a trace recorded without the analogue prefilter described in Sect. 2.3 and demonstrate that our filter circuit effectively increases the amplitude of the signal of interest and reduces the noise level of the demodulated signal.

Trace Alignment For precise alignment during the digital processing, we select a short reference pattern in a demodulated *reference trace*. This pattern is then located in all subsequent traces by finding the shift that minimises the squared difference between the reference and the trace to align, i.e., we apply a least-squares approach.

For devices with a synchronous clock, the alignment with respect to one distinct pattern is usually sufficient to align the whole trace. However, in our measurements we found that the analysed smartcard performs the operations in an asynchronous manner, i.e., the alignment may be wrong in portions not belonging to the reference pattern. The alignment has thus to be performed with respect to the part of the trace we aim to examine by means of CPA.

3.2 Results of DEMA

The process to perform a DEMA of the 3DES implementation can be split up into the following steps, of which we will detail the latter two in this section:

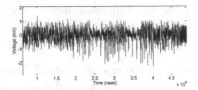

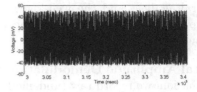

Fig. 6. Demodulated trace (50 kHz - 2 MHz) with analogue filter

Fig. 7. Raw trace with analogue filter (zoomed)

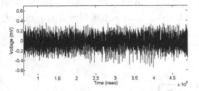

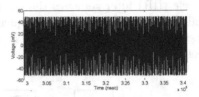

Fig. 8. Demodulated trace (50 kHz - 2 MHz) without analogue filter

Fig. 9. Raw trace without analogue filter (zoomed)

1. Find a suitable trigger point.
2. Align the traces.
3. Locate the DES encryption.
4. Perform the EM analysis.

Data Bus Transfer of Plain- and Ciphertext. As the plaintext for the targeted 3DES operation is known and the ciphertext can be computed in a known-key scenario, we are able to isolate the location of the 3DES encryption by correlating on these values. From the profiling phase with a known key it turns out that the smartcard uses an 8 bit data bus to transfer plain- and ciphertexts. The corresponding values can be clearly identified from 2000 - 5000 traces using a Hamming weight model, as depicted in Fig. 10 and 11.

This first result suggests that the smartcard logic is implemented on a microcontroller which communicates with a separate 3DES hardware engine over

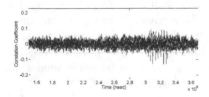

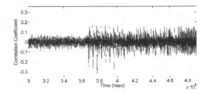

Fig. 10. Correlation coefficients for plaintext bytes (before targeted 3DES encryption) after 5000 traces, Hamming Weight

Fig. 11. Correlation coefficients for ciphertext bytes (after targeted 3DES encryption) after 2000 traces, Hamming Weight

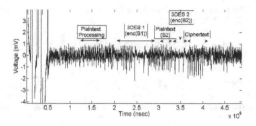

Fig. 12. Overview over operations in amplitude-demodulated trace

a data bus using precharged wires. This assumption is further supported by the
fact that correlation with the plaintext bytes can be observed twice, but with
reversed byte order. The microcontroller probably first receives the plaintext
bytes via the RF module, byte-reverses it and transmits it over the internal bus
to the encryption engine later. The ciphertext is then sent back using the same
byte order as for the second appearance of the plaintext.

From the profiling observations, Fig. 12 was compiled, with the shape of the
3DES operation marked. The first 3DES encryption (3DES 1) results from a
prior protocol step, the correlation with the correct ciphertext appears after the
second 3DES shape only (labeled 3DES 2).

3DES Engine. After having localised the interval of the 3DES operation from
the position of the corresponding plain- and ciphertexts, we now focus on this
part of the trace. Fig. 13 shows a zoomed view of the targeted 3DES operation,
filtered with $f_{lowpass} = 8\,\text{MHz}$ and $f_{highpass} = 50\,\text{kHz}$. The short duration of
the encryption suggests that the 3DES is implemented in a special, separate
hardware module, hence we assume a Hamming distance model[8].

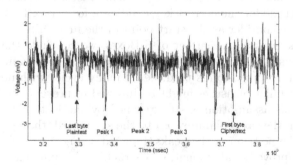

Fig. 13. Part of trace with 3DES encryption, filtered with $f_{lowpass} = 8$ MHz,
$f_{highpass} = 50$ kHz

---

[8] We also considered a Hamming weight model, however, did not reach conclusive
results with it.

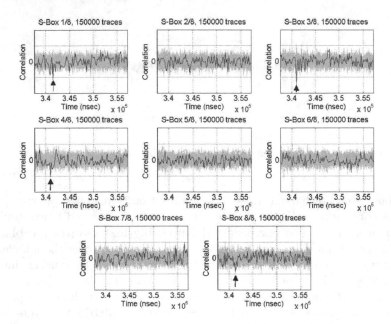

Fig. 14. Correlation coefficients for binwise CPA with peak extraction after 150000 traces, $f_{lowpass} = 8$ MHz, $f_{highpass} = 50$ kHz

The three marked peaks seemingly appear at the end of one complete Single-DES and are thus promising candidates as alignment patterns. Consequently, we conduct a CPA on demodulated traces aligned to each of these peaks, where we consider the Hamming distance between the DES registers (L_0, R_0) and (L_1, R_1), i.e, the state before and after the first round of the first Single-DES. It turns out that for the second peak, results are generally most conclusive. When performing a standard CPA with $L = 150000$ traces, correlation peaks with maximum amplitude for the correct key candidate for S-Box 1 and 3 occur at a position which we consider as the start point of the first DES.

As the attack works for a subset of S-Boxes, we conclude that no masking scheme ([19]) is used to protect the hardware engine. Rather than, we conjecture that hiding in time dimension is used, i.e., dummy cycles with no computation taking place or similar measures might be inserted to prevent correct alignment of the traces. This assumption is strengthened by the fact that even when repeatedly sending the same plaintext B_2 to the smartcard, the shape of the DES operation and the position of the peaks depicted in Fig. 13 vary[9].

In order to improve the alignment, we extract local maxima and minima from the trace part belonging to the first DES operation. The resulting data points (composed of time position and amplitude) are then grouped on the basis of their time coordinate by dividing the time axis into equal intervals or *bins*. Thus, extrema which occur at slightly different points in different traces are assigned

---

[9] This misalignment also hinders improving the SNR by means of averaging.

to the same bin, correcting for timing jitter up to a certain extent. The CPA is performed *binwise*, i.e., the correlation coefficient for each bin is computed from all extrema lying within the corresponding time interval.

The correlation coefficients for this experiment are given in Fig. 14, where the y-axis has been normalised to the theoretical noise level $\frac{4}{\sqrt{L}}$ (cf. [19]), accounting for the different number of data points per bin. It can be seen that using this method, the correct subkey can be identified for S-Box 1, 3, 4 and 8.

4 Future Work

To further improve the attack and to both reduce the number of traces and increase the correlation, we investigate suitable methods for precise alignment within the DES operation and for the detection of dummy operations. For this purpose we are currently evaluating two approaches. On the one hand, we plan to apply CPA in the (short-time) frequency domain ([27], [24]), on the other hand, we optimise our measurement environment to gain more information on the details of the internal operation of the RFID smartcard.

The maximum amplitude of the measurements for our DEMA in the oscilloscope has been approx. 40 mV, while the 8 Bit ADC in the oscilloscope quantises a full scale of 100 mV. Hence, only approx. 100 out of 256 values are currently used for digitising the analogue signal. Accordingly, we expect to carry out an EM analysis with 2.5 times less measurements than before when exploiting the full scale. Besides, the amplitude demodulation that has already has proven its effectiveness when implemented digitally can also be performed in the analogue domain, allowing for a significantly better amplification of the side-channel information contained in the carrier envelope.

It is also promising to further investigate a remote power analysis as described in Sect. 2.2, i.e., whether an EM attack from a distance of several meters is conductable. Since the side-channel signal is contained in the envelope of the carrier wave, it can be expected to be receivable from distant locations in the far field using analogue receiver equipment and suitable antennae.

5 Conclusion

As the main result attained in this paper, we give practical contributions for analysing the security of RFIDs via non-invasive side-channel attacks. We presented a new approach for performing effective EM analyses, realised a corresponding analogue hardware and describe our resulting low-cost measurement environment. We detail on the relevant steps of performing practical real-world EM attacks on commercial contactless smartcards in a black-box scenario and thereby demonstrated the potency of our findings.

This paper pinpoints several weaknesses in the protocol and the actual implementation of widespread cryptographic contactless smartcards, including a vulnerability to DEMA. We investigated the leakage model applicable for the data bus and described a CPA on the 3DES hardware implementation running on the targeted commercial smartcard. We demonstrated the effectiveness of our developed methods, that are generally applicable for analysing all kinds of RFID devices and contactless smartcards, by detailing and performing a full key-recovery attack, leaving no traces, on a black box device.

References

1. Advanced Security Mechanisms for Machine Readable Travel Documents - Extended Access Control (EAC), Password Authenticated Connection Establishment (PACE), and Restricted Identification (RI).,
   http://www.bsi.de/english/publications/techguidelines/tr03110/TR-03110_v200.pdf
2. FIPS 46-3 Data Encryption Standard (DES),
   http://csrc.nist.gov/publications/fips/fips46-3/fips46-3.pdf
3. Agrawal, D., Archambeault, B., Rao, J.R., Rohatgi, P.: The EM Side-Channel(s). In: Kaliski Jr., B.S., Koç, Ç.K., Paar, C. (eds.) CHES 2002. LNCS, vol. 2523, pp. 29–45. Springer, Heidelberg (2003)
4. Brier, E., Clavier, C., Olivier, F.: Correlation Power Analysis with a Leakage Model. In: Joye, M., Quisquater, J.-J. (eds.) CHES 2004. LNCS, vol. 3156, pp. 16–29. Springer, Heidelberg (2004)
5. Carluccio, D.: Electromagnetic Side Channel Analysis for Embedded Crypto Devices. Master's thesis, Ruhr Universität Bochum (2005)
6. Carluccio, D., Lemke, K., Paar, C.: Electromagnetic Side Channel Analysis of a Contactless Smart Card: First Results. In: RFIDSec 2005 Workshop on RFID and Lightweight Crypto (July 2005),
   http://events.iaik.tugraz.at/RFIDandLightweightCrypto05/RFID-SlidesandProceedings/Carluccio-EMSideChannel.pdf
7. Courtois, N.T., Nohl, K., O'Neil, S.: Algebraic Attacks on the Crypto-1 Stream Cipher in MiFare Classic and Oyster Cards. Cryptology ePrint Archive, Report 2008/166 (2008)
8. Eisenbarth, T., Kasper, T., Moradi, A., Paar, C., Salmasizadeh, M., Shalmani, M.T.M.: On the Power of Power Analysis in the Real World: A Complete Break of the KeeLoq Code Hopping Scheme. In: Wagner, D. (ed.) CRYPTO 2008. LNCS, vol. 5157, pp. 203–220. Springer, Heidelberg (2008)
9. Finkenzeller, K.: RFID-Handbuch, 3rd edn. Hanser Fachbuchverlag (October 2002)
10. Garcia, F.D., de Koning Gans, G., Muijrers, R., van Rossum, P., Verdult, R., Schreur, R.W., Jacobs, B.: Dismantling MIFARE Classic. In: Jajodia, S., López, J. (eds.) ESORICS 2008. LNCS, vol. 5283, pp. 97–114. Springer, Heidelberg (2008)
11. Haykin, S.: Communications Systems, 2nd edn., ch. 8. Wiley, Chichester (1983)
12. Hutter, M., Mangard, S., Feldhofer, M.: Power and EM Attacks on Passive 13.56 MHz RFID Devices. In: Paillier, P., Verbauwhede, I. (eds.) CHES 2007. LNCS, vol. 4727, pp. 320–330. Springer, Heidelberg (2007)
13. International Organization for Standardization. ISO/IEC 14443-3: Identification cards - Contactless integrated circuit(s) cards - Proximity cards - Part 3: Initialization and anticollision, 1st edn. (February 2001)
14. International Organization for Standardization. ISO/IEC 14443-4: Identification cards - Contactless integrated circuit(s) cards - Proximity cards - Part 4: Transmission protocol, 1st edn. (February 2001)

15. Kasper, T., Carluccio, D., Paar, C.: An Embedded System for Practical Security Analysis of Contactless Smartcards. In: Sauveron, D., Markantonakis, K., Bilas, A., Quisquater, J.-J. (eds.) WISTP 2007. LNCS, vol. 4462, pp. 150–160. Springer, Heidelberg (2007)
16. Knuth, D.E.: The Art of Computer Programming, 3rd edn., ch. 2. Seminumerical Algorithms. Addison-Wesley, Boston (1998)
17. Kocher, P.C., Jaffe, J., Jun, B.: Differential Power Analysis. In: Wiener, M. (ed.) CRYPTO 1999. LNCS, vol. 1666, pp. 388–397. Springer, Heidelberg (1999)
18. Langer EMV-Technik. Details of Near Field Probe Set RF 2, http://www.langer-emv.de/en/produkte/prod_rf2.htm
19. Mangard, S., Oswald, E., Popp, T.: Power analysis attacks: Revealing the secrets of smart cards. Springer, Secaucus (2007)
20. Microchip. HCS410, KEELOQ Code Hopping Encoder and Transponder Data Sheet, http://ww1.microchip.com/downloads/en/DeviceDoc/40158e.pdf
21. NXP. Data Sheet of Mifare Classic 4k chip MF1ICS70 (2008)
22. Oren, Y., Shamir, A.: Remote Password Extraction from RFID Tags. IEEE Transactions on Computers 56(9), 1292–1296 (2007), http://iss.oy.ne.ro/RemotePowerAnalysisOfRFIDTags
23. Pico Technology. PicoScope 5200 USB PC Oscilloscopes (2008)
24. Plos, T., Hutter, M., Feldhofer, M.: Evaluation of Side-Channel Preprocessing Techniques on Cryptographic-Enabled HF and UHF RFID-Tag Prototypes. In: Dominikus, S. (ed.) Workshop on RFID Security 2008, pp. 114–127 (2008)
25. Plötz, H.: Mifare Classic - Eine Analyse der Implementierung. Master's thesis, Humboldt-Universität zu Berlin (2008)
26. Shanmugam, K.S.: Digital & Analog Communication Systems, ch. 8.3.2. Wiley-India, Chichester (2006)
27. Tiu, C.C.: A New Frequency-Based Side Channel Attack for Embedded Systems. Master's thesis, University of Waterloo (2005)

Identity-Based Identification Scheme Secure against Concurrent-Reset Attacks without Random Oracles*

Pairat Thorncharoensri, Willy Susilo, and Yi Mu

Centre for Computer and Information Security
School of Computer Science & Software Engineering
University of Wollongong, Australia
{pt78,wsusilo,ymu}@uow.edu.au

Abstract. The notion of identity-based cryptography was put forward by Shamir in 1984. This setting has also been considered in identification schemes. Since then, many identity-based identification schemes have been proposed. Nonetheless, most of them only resist against concurrent attacks. In this paper, we consider the most stringent attack in identification schemes, namely the reset attacks. The aim of this paper is to present the *first* identity-based identification scheme that is provably secure against concurrent-reset attacks (CR1) based on the 2-SDH assumption. We shall elaborate the 2-SDH assumption, which is weaker than the well known q-SDH assumption.

Keywords: Identification scheme, Identity-based, Reset Attack, Concurrent Attack, Bilinear Pairings.

1 Introduction

An identification scheme enables one party (*verifier*) to gain assurance that the identity of another (*prover*) is as declared, thereby preventing impersonation [7]. Identification schemes have been considered in a variety of settings. In this paper, we are interested in an asymmetric setting, whereby the prover holds the secret key and the verifier holds the corresponding public key. Traditionally, the prover's public key is generated after the associated secret key is chosen, and therefore it appears to be a random bit string. When we move to a setting where the public key is chosen such that it reflects the identity of the prover, then we achieve an identity-based identification scheme.

One of the primary purposes of identity-based identification is to use only the user's identity to determine whether the prover's access to a resource is granted or denied. Furthermore, the "Internet" model [3] is considered to be an essential resource. In this model, it is essential to consider the security of identity-based identification schemes against active and concurrent attacks. Nevertheless, Bellare, Fischlin, Goldwasser and Micali [1] emphasized that the power of *reset attacks*, where the adversary can reset the prover to the initial state and learn the associated secret key before attempting to impersonate, is over the power of concurrent-active attack. Furthermore, we note that in the "Internet" model based identity-based identification the device used for executing

---

* This work is partially supported by ARC Linkage Project Grant LP0667899.

H.Y. Youm and M. Yung (Eds.): WISA 2009, LNCS 5932, pp. 94–108, 2009.

the interactive protocol is practically a resetable device. More precisely, the computer can be reset into *any* state other than the initial state. It was noted in [1] that a smart card is vulnerable to the reset attack where an adversary can simply disconnect and reconnect its power source to reset its state to the initial state and uses that state for a number of times to gain the private information or some useful information. This case has been discussed independently by Canetti, Goldwasser, Goldreich and Micali in [5] and Bellare, Fischlin, Goldwasser and Micali in [1]. Based on the suggestion in [5] and the example in [1] about the reset attack on identification protocol based on zero-knowledge proof system, it is concluded that *none* of the identification protocol based on zero-knowledge proof system is secure under the reset attack since the knowledge can be extracted by resetting the random tape of the identification protocol for a polynomial number of times. Therefore, the search for a secure identification scheme against reset attack remains an interesting research problem.

In 1984, Shamir introduced the idea of identity-based cryptosystem [16]. In 1986, the first identification scheme was introduced by Fiat, and Shamir [7]. Then, in 1988, the first identity-based identification scheme was introduced by Feige, Fiat, and Shamir [6]. Since then, many standard and identity-based identification schemes have been constructed [9, 13, 14, 11, 10, 12, 15, 17, 8]. Unfortunately, none of these work is secure against reset attacks. Most of them are provable secure against concurrent attacks and the only best known identity-based identification schemes secure under the passive and concurrent-active attacks are Kurosawa and Heng's scheme [11]. In 2004, Bellare, Namprempre and Neven proposed a proof of security for identity-based identification and signature scheme [2] which summarized the security proof of identification and signature schemes. They also formalized the definition of attacks, standard identification schemes, identity-based identification schemes and security properties of those schemes. The first identification protocol secure under the reset attacks was proposed by Canetti, Goldwasser, Goldreich and Micali [5]. However, their scheme in the public key model is inefficient in practice. Bellare, Fischlin, Goldwasser and Micali later gave the other definition for the reset attacks in 2001 [1] and they provided four paradigms for standard identification protocols secure under the reset attacks. The first three paradigms are based on cryptographic primitives, which are stateless signature schemes, encryption schemes and a combination of trapdoor commitment schemes and standard identification protocols secured against non-resetting attacks. The last one is based on the resetable zero knowledge proof of membership which was introduced in [5].

Concurrent-Reset Attacks

In [1], the adversary is given a power to reset the initial state of the honest prover in the concurrent setting. This is known as the strongest adversarial model that is considered for identification protocols. Two types of the reset attacks have been defined [1], namely the concurrent-reset-1 (CR1) attack and the concurrent-reset-2 (CR2) attack. In the CR1 setting, the adversary $\mathcal{A}$ can execute many identification protocols concurrently with the prover $\mathcal{B}$. During the executions $\mathcal{A}$ also can reset $\mathcal{B}$ to the initial state. Simultaneously, $\mathcal{A}$ hopes to gain enough information before attempting to impersonate $\mathcal{B}$ in a future time. Finally, $\mathcal{A}$ will try to impersonate $\mathcal{B}$ to convince a verifier that $\mathcal{A}$ is indeed a valid prover. In the CR2 setting, the setting is defined similarly to CR1, with the addition that during the impersonation stage, $\mathcal{A}$ still can execute many identification protocols concurrently with the honest prover $\mathcal{B}$. $\mathcal{A}$ also can reset $\mathcal{B}$ to the initial state while interleaving the executions. Therefore, the CR1 attack can view as the special case of the CR2 attack. An identification protocol that is secure under the CR2 attack implies that it is also secure under the CR1 attack.

Our Contributions
The contribution of this paper is twofold. Firstly, we improve a definition of CR1, which we call CR1$^+$. Our definition captures a more general case compared to CR1, since we allow the adversary to reset the state of the prover (or its clones) into *any* state that he chooses. In contrast to CR1$^+$, the CR1 attacks only permit the adversary to reset the state of the prover (or its clones) into its initial state. Therefore, we can see that CR1 is a special case of CR1$^+$, where a CR1$^+$ attacker's execution time is less than a CR1 attacker's execution time. Secondly, we present *the first* identity-based identification scheme that is provably secure against impersonation attack under the concurrent-reset attacks, as defined in CR1. Additionally, we also provide an identity-based identification scheme that is secure against impersonation under the passive attack, which is more efficient, in terms of computational ability, than the state-of-the-art scheme due to Kurosawa and Heng [11]. Our scheme secure under CR1$^+$ attack is based on a variant of q-Strong Diffie-Hellman (SDH) assumption, that we call 2-Strong Diffie-Hellman assumption, which is *weaker* than the well known SDH assumption [4, 11].

Paper Organization
The rest of the paper is organized as follows. In Section 2, we review some cryptographic tools and complexity assumptions used throughout this paper. In Section 3, we define the the identity-based identification scheme model, types of attack, the security of identity-based identification scheme under the passive and the concurrent-reset attack settings. A secure identity-based identification scheme under the passive attacks and its security proof is presented in Section 4. In section 5, an identity-based identification scheme secure against impersonation under the concurrent-reset attacks with its proof of security and its experiment is presented. Finally, we present the comparison of our presented schemes with the state-of-the-art of identity-based identification schemes in the literature.

2   Preliminaries

2.1   Bilinear Pairings

Let $\mathbb{G}_1$, $\mathbb{G}_2$ be cyclic groups of prime order p, generated by g_1, g_2, respectively. Let $\mathbb{G}_T$ be a cyclic multiplicative group with the same order p. Let $\hat{e} : \mathbb{G}_1 \times \mathbb{G}_2 \to \mathbb{G}_T$ be a bilinear mapping with the following properties:

1. *Bilinearity:* $\hat{e}(g_1^a, g_2^b) = \hat{e}(g_1, g_2)^{ab}$ for all $g_1 \in \mathbb{G}_1$ and $g_2 \in \mathbb{G}_2$, $a, b \in \mathbb{Z}$.
2. *Non-degeneracy:* There exist $g_1 \in \mathbb{G}_1$ and $g_2 \in \mathbb{G}_2$ such that $\hat{e}(g_1, g_2) \neq 1$.
3. *Computability:* There exists an efficient algorithm to compute $\hat{e}(g_1, g_2)$ for all $g_1 \in \mathbb{G}_2$ and $g_2 \in \mathbb{G}_1$.

A bilinear pairing instance generator is defined as a probabilistic polynomial time algorithm $\mathcal{IG}$ that takes a security parameter 1^k as input and returns an uniformly random tuple $param = (p, \mathbb{G}_1, \mathbb{G}_2, \hat{e}, g_1, g_2)$ of bilinear parameters, including a prime number p of size 1^k, cyclic groups $\mathbb{G}_1$ and $\mathbb{G}_2$ of order p, a multiplicative group $\mathbb{G}_T$ of order p, a bilinear map $\hat{e} : \mathbb{G}_1 \times \mathbb{G}_2 \to \mathbb{G}_T$ and a generator $g_1 \in \mathbb{G}_1$ and a generator $g_2 \in \mathbb{G}_2$. For simplicity, hereafter, we set $\mathbb{G}_1 = \mathbb{G}_2$ and $g_1 = g_2 = g$. We note that our scheme can be easily modified for a general case, when $\mathbb{G}_1 \neq \mathbb{G}_2$. This setting can be viewed as a generic pairing setting which can be changed to other type of pairings (caution needs to be exercised for the the size of $\mathbb{G}_1$ and $\mathbb{G}_2$).

2.2 Complexity Assumptions

Definition 1 (q-Strong Diffie-Hellman (q-SDH) Problem). *Given a $(q+1)$-tuple* $(g, g^x, g^{x^2}, ..., g^{x^q})$ *as input, output a pair* $(c, g^{\frac{1}{x+c}})$ *where* $c \in \mathbb{Z}_p^*$. *An algorithm $\mathcal{A}$ has advantage ϵ in solving q-SDH if* $\mathbf{Pr}\left[\mathcal{A}(g, g^x, g^{x^2}, ..., g^{x^q}) = (c, g^{\frac{1}{x+c}})\right] \geq \epsilon$, *where the probability is over the random choice of $x \in \mathbb{Z}_q^*$ and the random bits consumed by $\mathcal{A}$.*

Assumption 1 ((q, t, ϵ)-Strong Diffie-Hellman Assumption [4]). *We say that the (q, t, ϵ)-SDH assumption holds if no probability polynomial algorithm with time complexity $t(.)$ has advantage at least ϵ in solving the q-SDH problem.*

Definition 2 (2-Strong Diffie-Hellman (2-SDH) Problem). *Given a 3-tuple (g, g^x, g^{x^2}) as input, output a pair $(c, g^{\frac{1}{x+c}})$ where $c \in \mathbb{Z}_p^*$. An algorithm $\mathcal{A}$ has advantage ϵ in solving 2-SDH if* $\mathbf{Pr}\left[\mathcal{A}(g, g^x, g^{x^2}) = (c, g^{\frac{1}{x+c}})\right] \geq \epsilon$, *where the probability is over the random choice of $x \in \mathbb{Z}_q^*$ and the random bits consumed by $\mathcal{A}$.*

Assumption 2 ($(2, t, \epsilon)$-Strong Diffie-Hellman Assumption). *We say that the $(2, t, \epsilon)$-SDH assumption holds if no probability polynomial algorithm with time complexity $t(.)$ has advantage at least ϵ in solving the 2-SDH problem.*

Theorem 1. *q-SDH problem is reducable to 2-SDH problem, and therefore, 2-SDH assumption is a weaker assumption.*

Proof. The proof is trivial, hence, we omitted it. ∎

3 Identity-Based Identification Scheme

In an identity-based identification scheme (IBI-scheme), the algorithms can be classified into two PPT algorithms and one interactive protocol as follows.

1. Setup: It takes a security parameter 1^k as input and generates a pair of public instance P_I and a master private instance S_I. That is, $(P_I, S_I) \leftarrow Setup(1^k)$.
2. Extract: Given an Identity of a prover ID and the master private instance S_I, *Extract* takes ID and S_I as inputs and computes a witness instance W_I and gives it to the prover. That is, $W_I \leftarrow Extract(ID, S_I)$.
3. Identification protocol: A canonical protocol of an identification scheme can be denoted by $CID = (Commit, Challenge, Response, Check)$, where $Commit$, $Challenge$, $Response$ and $Check$ are PPT algorithms used in the following protocol where P is the prover and V is the verifier.

- Step 1. P chooses r at random from a certain domain and computes $x = Commit(r)$. P then sends x to V.
- Step 2. V chooses a challenge c at random from a certain set and sends $Challenge = c$ to P.
- Step 3. P computes a response $y = Response(W_I, r, c)$ and sends y to V.
- Step 4. V checks if $x = Check(P_I, ID, c, y)$. V accepts P if only the prior equation holds.

The above protocol (P, V) in both standard and identity-based identification schemes is often called a canonical protocol. We say that (x, c, y) is a valid transcript for P_I if it satisfies the equation (4) as specified above. Observe that most of identification schemes are transformable from or to digital signature schemes [10, 11]. This is a fast track to construct an identification scheme, but nonetheless these schemes are insecure against concurrent-active or reset attacks.

3.1 Types of Attacks

Passive Attack (PA)
The weakest form of attacks in identification schemes is a passive attack which adversary acts as an eavesdropper attempting to impersonate the prover using only the knowledge of the prover's public key. Furthermore, the adversary is not allowed to interact with the system at all prior to the impersonation stage. Other attacks of an intermediate level such as the honest-verifier attack is considered to be equivalent to the passive attack.

Concurrent Attack (CA) and Concurrent Reset Attack (CR)
The stronger form of passive attacks is the concurrent-active attack. In this scenario, the adversary is allowed to interact with the honest prover several times prior to impersonation, acting as a cheating verifier. Furthermore, he could interact with many different provers (clones) concurrently. All clones have the same secret key. However, they maintain their own independent states. Security against impersonation under the concurrent attack implies security against impersonation under the active attack [2]. As mentioned earlier, in [1], Bellare *et al.* provided a formal definition of the concurrent reset attack and further divided into two different classes, namely CR1 and CR2, depending on the case whether the adversary is still allowed to execute identification protocols with the honest prover during the impersonation stage or not. CR1 can be viewed as a special case of CR2.

CR1$^+$ Attack
We define a stronger type of concurrent reset attack that we call CR1$^+$. This attack is stronger than CR1 in the sense that the adversary is allowed to reset the prover (or clones) to *any* state, instead of just the initial state as defined in CR1. That means, the adversary can still play the role as a cheating verifier prior to impersonation as in the concurrent attack, CR1. However, these states of the prover (clones) can be reset to the initial state or to any other state. Security against impersonation under the CR1$^+$ then implies security against impersonation under the active and concurrent attack, and the CR1 attack.

3.2 Security of Identity-Based Identification Scheme against Impersonation under the Passive Attack

An *imp-pa* adversary $\mathcal{A} = (\hat{U}, \hat{P})$ is a pair of randomized polynomial-time algorithms that consists of the cheating identity-based user and cheating prover, respectively. We consider the following game that comprises of three phases.

Phase 1: KGC runs on input k to produce the master public key and secret key (pk, sk), a random tape is chosen for $\hat{U}$ and it is given an input pk. Then, it interacts with KGC initialized with pk, sk. We can precisely define that $Extract$ is a function of KGC that takes an incoming user's identity ID, pk, sk, and the current state and it returns an user secret key σ associated with pk and identity of user ID. Therefore, the cheating identity-based user $\hat{U}$ can issue a request of the form (ID_i, i) where i is an index of each request form. As a result, the operation $(\sigma_i) \leftarrow Extract(ID_i, i, pk, sk)$ is executed, σ_i is returned to $\hat{U}$ as the secret key associated with user's identity ID_i. These requests of $\hat{U}$ can be arbitrarily interleaved and the next chosen user's identity ID_{i+1} may be relevant with KGC public key and/or the previous user's identities $(ID_1, ..., ID_i)$ and the user secret keys $(\sigma_1, ...\sigma_i)$.

Eventually, $\hat{U}$ outputs user's identity ID_*, whom $\mathcal{A}$ decides to impersonate. This ends the first phase. Note that it is not critical whether $\mathcal{A}$ can output σ_* or not. The most important thing is $\mathcal{A}$ can impersonate the user's identity ID_* so that the game can continue to next phase.

Phase 2: $\mathcal{A}$ makes a request of conversation transcripts between the honest prover (the user's identity ID_*) and the honest verifier. A fresh random tape R is chosen for the honest verifier and the prover ID_*. Let $St_i = (pk, \sigma_*, R, i)$ be the state of information of each request and each transcript is contain $\{Commit_i, Ch_i, Response_i, Check_i\}$ where Ch_i is a random challenge number for each request. These requests of $\mathcal{A}$ can be arbitrarily interleaved and eventually, $\mathcal{A}$ outputs a set of state information St and stops, ending the second phase.

Phase 3: $\mathcal{A}$ now acts as the cheating prover $\hat{P}$, which attempts the impersonation on ID_*. $\mathcal{A}$ is initialized with St, the verifier V is initialized with pk, ID_* and freshly chosen states (or coins) and $\hat{P}$ interacts with V. We say that adversary $\mathcal{A}$ wins if V accepts in this interaction. The *imp-pa* advantage of $\mathcal{A}$, denoted by $\mathrm{Adv}_{IBI,\mathcal{A}}^{imp-pa}(k)$, is the probability that $\mathcal{A}$ wins, taken over the coins of k, the coins of $\hat{U}$, the coins of the prover ID_*, and the coins of V. We say that the IBI is secure against impersonation under the passive-reset attack (IMP-PA secure) if the function $\mathrm{Adv}_{IBI,\mathcal{A}}^{imp-pa}(.)$ is negligible for all *imp-pa* adversaries $\mathcal{A}$ of time complexity polynomial in the security parameter k. Furthermore the time of $\mathcal{A}$ is defined as the execution time of the entire three-phase game, including the time taken by the key generation, extraction, initializations and computation of the entire queries in each phase.

3.3 Security of Identity-Based Identification Scheme against Impersonation under the CR1+Attack

An *imp-cra* adversary $\mathcal{A} = (\hat{U}, \hat{V}, \hat{P})$ is a triple of randomized polynomial-time algorithms, a cheating identity-based user, a cheating verifier and a cheating prover, respectively. The first phase is identical to the imp-pa game, and in this section, we consider the last two phases of the game as follows.

Phase 2: $\mathcal{A}$ now plays the role of a cheating verifier $\hat{V}$. To initialize the setting for this phase, a resetable random tape R_s is first provided to the prover ID_*. The cheating verifier $\hat{V}$ can then issue either a request of the form $(\text{ch}, j, \mathbf{0})$ for the *initial* state or the form $(\text{ch}, j, \mathbf{i})$ for any other state i, where ch is a chosen challenge number and j is the index of prover (or clone) that will be reset. For the form $(\text{ch}, j, 0)$, the initial state $St_{j,0}$ of clone j is set to (pk, σ_*, R_s), the operation $(Commit_{out}, Resp_{out}) \leftarrow P(0, St_{j,0}, \text{ch})$ is executed, $(Commit_{out}, Resp_{out})$ is returned to $\hat{V}$, and remains at the $St_{j,0}$ state. In the other form (ch, j, i), $\hat{V}$ can issue a request of the form (ch, j, i) where the selected challenge number ch is sent to the j-th clone with the chosen current state at i. The j-th clone computes $(Commit_{out}, Resp_{out}) \leftarrow P(i, St_{j,i}, \text{ch})$ and returns $(Commit_{out}, Resp_{out})$ to $\hat{V}$ and it also remains at the $St_{j,i}$ state. These requests of $\hat{V}$ can be arbitrarily interleaved and $\hat{V}$ eventually may output a set of state information St and stops, ending the second phase.

Phase 3: $\mathcal{A}$ now plays the role as a cheating prover $\hat{P}$. $\hat{P}$ is initialized with St (Note that it is not compulsory to give $\hat{P}$ separate coins, or even pk, since these coins can be obtained from $\hat{V}$ from the previous phase via St.), the honest verifier V is initialized with pk, ID_* and a freshly chosen coin, and $\hat{P}$ interacts with V. We say that adversary $\mathcal{A}$ wins if V accepts in this interaction. The imp-cra advantage of $\mathcal{A}$, denoted by $\text{Adv}_{IBI,\mathcal{A}}^{imp\text{-}cra}(k)$ is the probability that $\mathcal{A}$ wins, taken over the coins of k, the coins of $\hat{V}$, the coins of the prover clones, and the coins of V. It is said that the IBI is secure against impersonation under the $CR1^+$ attack (IMP-CRA secure) if the function $\text{Adv}_{IBI,\mathcal{A}}^{imp\text{-}cra}(.)$ is negligible for all imp-cra adversaries $\mathcal{A}$ of time complexity polynomial in the security parameter k. Moreover, the time of $\mathcal{A}$ is defined as in the passive attack model.

4   Identity-Based Identification Scheme against Impersonation under the Passive Attack (IBI-PA)

In this section, firstly we present our scheme that resists against impersonation under the passive attacks. Then, in the next section, we modify our scheme to resist against impersonation under the $CR1^+$ attacks. The scheme is as follows.

1. Setup: Let $(\mathbb{G}_1, \mathbb{G}_T)$ be two multiplicative cyclic groups where $|\mathbb{G}_1| = |\mathbb{G}_T| = p$ for some prime p. g is a generator of $\mathbb{G}_1$ and $\hat{e} : \mathbb{G}_1 \times \mathbb{G}_1 \to \mathbb{G}_T$ is a bilinear pairing function. Given a security parameter 1^k, which is a positive integer, $Setup$ works as follows.
   - Run the public parameters generator to obtain the system parameters $Params = \{\mathbb{G}_1, \mathbb{G}_T, \hat{e}, p, g\}$
   - Select random numbers $\alpha, a, c, \tau \in \mathbb{Z}_p^*$ and compute $P_{Pub} = \hat{e}(g, g)^\alpha$, $A = g^a$, $B = g^\tau$, $C = g^c$, $D = g^{a \cdot c}$, $E = g^{\tau \cdot c}$.

   In this case, a pair of secret and public parameters of KGC is generated where $pk = (\mathbb{G}_1, \mathbb{G}_T, \hat{e}, p, g, P_{pub}, A, B, C, D, E)$ is the set of public parameters and $sk = (\alpha, a, c, \tau)$ is the KGC's master secret key.

2. Extract: Given an identity of a prover ($ID \in \mathbb{Z}_p^*$), the public parameters pk and the master private key sk, $Extract$ takes ID and (α, a, c, τ) as inputs and computes an user I's witness instance W_I as follows:
   - U_i chooses random integers $k, n \in \mathbb{Z}_p^*$ such that $(ID + n \cdot a + \tau) \bmod p \neq 0$ and $k \neq a^{-1}$ and computes $\sigma_1 = g^{\frac{\alpha}{(ID + n \cdot a + \tau)}}$, $\sigma_2 = g^{k \cdot a}$, $\sigma_3 = g^k$, $\sigma_4 = k + ID \cdot c$, $\sigma_5 = g^{k \cdot \tau}$.

   After this point, the user's public key ID and the user's private keys (witness instances) $W_I = (\sigma_1, \sigma_2, \sigma_3, \sigma_4, \sigma_5, n)$ are provided to the prover.

3. Identification protocol: A canonical protocol of identity-based Identification scheme can be denoted by $CID = ($ Commit, Challenge, Response, Check $)$, where Commit, Challenge, Response and Check are PPT algorithms used in the following protocol where P is the prover and V is the verifier.
   - Step 1. P chooses a random integer $r \in \mathbb{Z}_p^*$ and computes $Commit(\sigma_2, \sigma_3, \sigma_5, n, r) = (\sigma_2, \sigma_3, \sigma_5, n, r)$ and then sends $(\sigma_2, \sigma_3, \sigma_5, n, r)$ to V.
   - Step 2. V chooses a random challenge integer $ch \in \mathbb{Z}_p^*$ and sends $Challenge = ch$ to P.
   - Step 3. P computes a response $Response(\sigma_1, \sigma_4, n, r, ch) = z$, which $z = \sigma_1^{\frac{1}{r \cdot \sigma_4 + ch}} \in G_1$, and sends z to V.
   - Step 4. V checks if $P_{pub}(= \hat{e}(g, g)^{\alpha}) \equiv \hat{e}(z, \sigma_2^{n \cdot r} \cdot \sigma_3^{ID \cdot r} \cdot A^{n \cdot ch} \cdot C^{r \cdot ID^2} \cdot D^{n \cdot r \cdot ID} \cdot g^{ID \cdot ch} \cdot \sigma_5^r \cdot E^{r \cdot ID} \cdot B^{ch})$.
   $Check(pk, ID, \sigma_2, \sigma_3, \sigma_5, n, r, ch, z) = 1$ if the above holds.
   - Step 5. V accepts P if only Check holds.

4.1 An Experiment on Identity-Based Identification Scheme against Impersonation under the Passive Attack

Given a random tape R and let $\hat{U}$ denote a cheating user who issues a request form (ID_i, i) to KGC, and KGC returns with a secret key sk_i associated with a public key ID_i. Let $\mathcal{A}$ be an adversary who can break the $IBI\text{-}PA$ identity-based identification scheme with an advantage $\text{Adv}^{imp-pra}$ and $\mathcal{A}$ plays a role in the following game as $\hat{U}$, eavesdropper and $\hat{P}$ in the associated phases as defined earlier. Let an adversary's algorithm $\mathcal{B}$, attempting to solve q-SDH problem with $h, h^x, h^{x^2}, ..., h^{x^q}$ as input and output $(h^{\frac{1}{x+r}}, r)$, be the challenger. $\mathcal{B}$ runs the experiment as follows.

Experiment: Adversary $\mathcal{B}(h, h^x, h^{x^2}, ..., h^{x^q})$

Since $\mathcal{A}$ can make a request of the conversation transcript for at most q_{IP} queries, where $q_S < q$, we may then assume that $\mathcal{A}$ issues exactly $q - 1$ queries. If the actual number of request is less, we can always virtually reduce the value of q so that $q = q_S + 1$. However, $\mathcal{A}$ is required to reveal a number of queries up front. $\mathcal{B}$ randomly chooses integer $l_1, ..., l_{q-1} \in \mathbb{Z}_p^*$. Let $f(y)$ be the polynomial $f(y) = \prod_{i=1}^{q-1}(y + l_i)$. Reform $f(y)$ by expanding to $f(y) = \sum_{i=0}^{q-1} \beta_i y^i$ where $\beta_0, ..., \beta_{q-1} \in \mathbb{Z}_p$ are the coefficients of the polynomial $f(y)$. Let $K_1, ..., K_q$ denote as $h^x, h^{x^2}, ..., h^{x^q}$. Compute as follows:

$$g \leftarrow \prod_{i=0}^{q-1}(K_i)^{\beta_i} = h^{f(x)} \in G_1 \text{ and } g^x \leftarrow \prod_{i=1}^{q}(K_i)^{\beta_{i-1}} = h^{x f(x)} \in G_1.$$

$\mathcal{B}$ randomly chooses an integer $a, c, \alpha, \tau \in \mathbb{Z}_p^*$. Then, $\mathcal{B}$ sets $A = g^a$, $B = g^\tau$, $C = g^c$, $D = g^{a \cdot c}$, $E = g^{\tau \cdot c}$ and $P_{Pub} = \hat{e}(g, g)^\alpha$ be public key $pk = (g, A, B, C, D, E, P_{Pub})$. Initialize with (pk, R) and set i to 0. Let ID-list be the list of user's identities that $\mathcal{A}$ makes a request for a secret key associated with public key (pk, ID) and it is initialized with an empty list.

Phase 1: $\mathcal{B}$ answers $\mathcal{A}$'s requests as follows: $\mathcal{A}$ issues a request of the form (ID_i, i) where ID_i is an adaptively chosen integer in $\mathbb{Z}_p^*$ by $\mathcal{A}$. Without losing generality, assume that $\mathcal{A}$ will never issue a request for the same user's identity again. $\mathcal{A}$ then sends a query to $\mathcal{B}$. Then, $\mathcal{B}$ computes as follows: set $i \leftarrow i + 1$ and then $\mathcal{B}$ randomly selects integers $n_i, k_i \in \mathbb{Z}_p^*$ such that $(ID_i + n_i \cdot a + \tau) \bmod p \neq 0$ and $k_i \neq a^{-1}$ and computes $\sigma_{i,1} = g^{\frac{\alpha}{(ID_i + n_i \cdot a + \tau)}}$, $\sigma_{i,2} = g^{k_i \cdot a}$, $\sigma_{i,3} = g^{k_i}$, $\sigma_{i,4} = k_i + ID_i \cdot c$, $\sigma_{i,5} = g^{k_i \cdot \tau}$ and keeps $(\sigma_{i,1}, \sigma_{i,2}, \sigma_{i,3}, \sigma_{i,4}, \sigma_{i,5}, k_i, n_i, ID_i)$ in ID-list. $\mathcal{B}$ then returns $(\sigma_{i,1}, \sigma_{i,2}, \sigma_{i,3}, \sigma_{i,4}, \sigma_{i,5}, n_i)$ as a secret key associated with the public key ID_i to $\mathcal{A}$.

Phase 2: Eventually after at most $q_{ID} < (p - 2)$ queries, $\mathcal{A}$ outputs ID^* as the public key which $\mathcal{A}$ will attempt to impersonate and then $\mathcal{A}$ now acts as an eavesdropper who can issue a request of the conversation transcript between the prover ID_* and the honest verifier.

If ID_* is in ID-list then $\mathcal{B}$ terminates and returns failure. Otherwise, initialize $\mathcal{A}$ with $(U_{pk}(= (pk, ID_*)), R)$; set i to 0; $\mathcal{B}$ selects a random integer $n \in \mathbb{Z}_p^*$ such that $\tau + n \cdot a + ID_* \neq 0$ and then constructs the partial ID_*'s witness instance as follows: $\sigma_{*,2} = g^{k \cdot a} = g^{a \cdot x}, \sigma_{*,3} = g^k = g^x, \sigma_{*,5} = g^{k \cdot \tau} = g^{\tau \cdot x}$. For the above context, k is x. Let T-list be the list of conversation transcripts that $\mathcal{A}$ requested and it is initialized with an empty list. $\mathcal{B}$ then answers $\mathcal{A}$'s requests as follows: $\mathcal{A}$ issues a request of the conversation transcript between user ID_* and the honest verifier and then $\mathcal{B}$ performs as follows: set $i \leftarrow i + 1$ and then $\mathcal{B}$ randomly chooses $r_i \in \mathbb{Z}_p^*$ such that r_i is not contained in T-list. If $i \geq q$ then $\mathcal{B}$ terminates and returns failure. $\mathcal{B}$ sets $(\sigma_{*,2}, \sigma_{*,3}, \sigma_{*,5}, n, r_i)$ as a commitment. $\mathcal{B}$ computes $ch_i = r_i(l_i - c \cdot ID_*)$ as a challenge number. $\mathcal{B}$ must generate a response with $g^{\frac{1}{x+l_i}}$. To do so, let $f_i(y)$ be the polynomial $f_i(y) = \frac{f(y)}{(y+l_i)} = \prod_{j=1, j \neq i}^{q-1}(y + l_j)$. As before, we reform $f_i(y)$ into $f_i(y) = \sum_{j=0}^{q-2} \delta_j y^j$. Compute $g^{\frac{1}{x+l_i}} \leftarrow \prod_{i=0}^{q-2} K_j^{\delta_i} = h^{\frac{f(x)}{(x+l_i)}} \in G_1$. Let $z_i = g^{\frac{\alpha}{r_i(\tau + n \cdot a + ID_*) \cdot (x+l_i)}}$ be the response. If z_i have been in T-list then $\mathcal{B}$ terminates and returns failure. $\mathcal{B}$ then keeps (r_i, ch_i, z_i) in T-list and returns $(\sigma_{*,2}, \sigma_{*,3}, \sigma_{*,5}, n, r_i, ch_i, z_i)$ as a conversation transcript to $\hat{V}$. Until $\hat{V}$ outputs the state information St on at most $q_{IP} < q$ queries and then stops.

Phase 3: Now, $\mathcal{A}$ changes the status to $\hat{P}$ and cannot longer issue a request of the conversation transcript between user ID_* and the honest verifier to $\mathcal{B}$. $\hat{P}$ starts the impersonation process.

- $\hat{P}$ first randomly chooses an integer $r \in \mathbb{Z}_p^*$ and then sends $(\sigma_{*,2}, \sigma_{*,3}, \sigma_{*,5}, n, r)$ to V.
- $\mathcal{B}$ randomly selects an integer $ch \in \mathbb{Z}_p^*$ such that $l_* = (ch + r \cdot c \cdot ID_*)/r$ and $l_* \notin \{l_1, ..., l_{q-1}\}$ and sends ch to $\hat{P}$.
- $\hat{P}$ returns z_*.

Define $Check(pk, z_*, ch, \sigma_{*,2}, \sigma_{*,3}, \sigma_{*,5}, n, r, ID_*)$ as:

$$P_{pub} \overset{?}{=} \hat{e}(z_*, \sigma_{*,2}^{n \cdot r} \cdot \sigma_{*,3}^{ID_* \cdot r} \cdot A^{n \cdot ch} \cdot C^{r \cdot ID_*^2} \cdot D^{n \cdot r \cdot ID_*} \cdot g^{ID \cdot ch} \cdot \sigma_{*,5}^{r} \cdot E^{r \cdot ID_*} \cdot B^{ch}),$$

where $P_{pub} = \hat{e}(g, g)^{\alpha}$. If the equality holds, then the output of $Check(pk, z_*,$ $ch, \sigma_{*,2}, \sigma_{*,3}, \sigma_{*,5}, n, r, ID_*)$ is 1. Otherwise, it outputs 0. If $Check(pk, z_*, ch,$ $\sigma_{*,2}, \sigma_{*,3}, \sigma_{*,5}, n, r, ID_*) = 1$ then $\mathcal{B}$ computes the output as follows: Since

$l_* = \frac{r \cdot ID_* \cdot c + ch}{r}$, $Z = z_*^{\frac{r(\tau + n \cdot a + ID_*)}{\alpha}} = g^{\left(\frac{\alpha}{(\tau + n \cdot a + ID_*)(r(x + ID_* \cdot c) + ch)}\right)\left(\frac{r(\tau + n \cdot a + ID_*)}{\alpha}\right)}$

$= g^{\frac{1}{x + (ID_* \cdot c + ch/r)}}$ Hence, $Z = g^{\frac{1}{x + l_*}}$. Since $g = h^{f(x)}$, $Z = h^{\frac{f(x)}{x + l_*}}$. Let us

denote by $Z^* = h^{\frac{1}{x + l_*}}$. Let $f_*(y)$ be the polynomial $f_*(y) = \frac{f(y)}{(y + l_i)}$ such that there exists some polynomial $\gamma(y) = \sum_{i=0}^{q-2} \gamma_i y^i$ and some $\gamma_{-1} \in Z_p^*$. Then we reform $f_*(y)$ into $\gamma(y)$ as: $f(y) = \gamma(y)(y + l_*) + \gamma_{-1}$. The exponent of Z, where Z is $h^{\frac{f(x)}{x + l_*}}$, can then be written as $\frac{f(x)}{(x + l_*)} = \frac{\gamma_{-1}}{x + l_*} + \sum_{i=0}^{q-2} \gamma_i x^i$. Since $f(x) = \prod_{i=1}^{q-1}(x + l_i)$ and $l_* \notin \{l_1 \ldots, l_{q-1}\}$, hence, $(x + l_*)$ does not divide $f(x)$ and $\gamma_{-1} \neq 0$. Then $\mathcal{B}$ computes $Z^* \leftarrow \left(Z. \prod_{i=1}^{q-1} K_i^{-\gamma_i}\right)^{1/\gamma_{-1}} =$ $g^{\frac{1}{(x + l_*)}}$. $\mathcal{B}$ wins the game and returns (Z^*, l_*) as the solution of q-SDH. Else, $\mathcal{B}$ returns failure. ∎

4.2 Proof of Security

Theorem 2. *Let $IBI = (Setup, Extract, P, V)$ be the IBI-PA identity-based identification scheme associated to q-SDH assumption. Let $\mathcal{A} = (\hat{U}, \hat{P})$ be an imp-pa adversary of time complexity $t(.)$ attacking IBI. Then there exists a q-SDH adversary $\mathcal{B}$ of time complexity $t'(.)$ solving q-SDH problem such that for every security parameter k*

$$\text{Adv}_{IBI,\mathcal{A}}^{imp-pa}(k) \geq e^2 \cdot \text{Adv}_{\mathcal{B}}^{q-SDH}(k) \text{ or, in a simply term,}$$

$$\epsilon \geq e^2 \cdot \epsilon',$$

where e is the natural logarithm. Moreover, the time complexity t of $\mathcal{A}$ is $t \leq t' - ((6 + 3q_{ID} + q_{IP})C_G + 2C_P)$, where C_G be a computation time of group exponential operation and C_P be a computation time of bilinear group pairing operation, and $\mathcal{A}$ can request queries at most $q_{ID} < p - 1$ and at most $q_{IP} < q$.

Corollary 1. *If q-SDH assumption is (q, t', ϵ')-secure then the IBI-PA identity-based identification scheme associated to q-SDH assumption is $(q_{ID}, q_{IP}, t, \epsilon)$-secure against impersonation under the passive attack.*

Proof. Due to the page limitation, please find the proof for Corollary 1 in the full version of this paper [18].

5 Identity-Based Identification Scheme against Impersonation under the CR1⁺Attack (IBI-CRA)

In this section, we present our identity-based identification scheme that is secure against CR1⁺attack. The $Setup$ phase is the same as the IBI-PA scheme from Section 4, and therefore is omitted. We will describe the rest of the scheme as follows:

1. <u>Extract:</u> Given an identity of the prover ($ID \in \mathbb{Z}_p^*$), the public parameters pk and the master private key sk, $Extract$ takes ID and (α, a, c, τ) as inputs and computes an user I's secret key(or an user I's witness instance) W_I as follows.

   - U_i chooses random integers $v, k, n_1, n_2 \in \mathbb{Z}_p^*$ such that $(ID + n_1 \cdot a + \tau)$ mod $p \neq 0$, $(ID + n_2 \cdot a + \tau)$ mod $p \neq 0$, and $k \neq a^{-1}$ and computes $\sigma_1 = g^{\frac{\alpha - v}{(ID + n_1 \cdot a + \tau)}}$, $\sigma_2 = g^{\frac{v}{(ID + n_2 \cdot a + \tau)}}$, $\sigma_3 = g^{k \cdot a}$, $\sigma_4 = g^k$, $\sigma_5 = k + ID \cdot c$, $\sigma_6 = g^{k \cdot \tau}$.

   Thereafter, the user's public key ID and the user's private keys (witness instances) $W_I = (\sigma_1, \sigma_2, \sigma_3, \sigma_4, \sigma_5, \sigma_6, n_1, n_2)$ are provided to the prover.

2. <u>Identification protocol:</u> A canonical protocol of identity-based Identification scheme can be denoted by $CID = ($ $Commit, Challenge, Response, Check$ $)$, where $Commit, Challenge, Response$ and $Check$ are PPT algorithms used in the following protocol where P is the prover and V is the verifier.

   - Step 1. P chooses a random integer $r \in \mathbb{Z}_p^*$ and computes $Commit(\sigma_3, \sigma_4, \sigma_6, n_1, n_2, r) = (\sigma_3, \sigma_4, \sigma_6, n_1, n_2, r)$ and then sends $(\sigma_3, \sigma_4, \sigma_6, n_1, n_2, r)$ to V.
   - Step 2. V chooses random challenge integers $ch_1, ch_2 \in \mathbb{Z}_p^*$ and sends $Challenge = (ch_1, ch_2)$ to P.
   - Step 3. P computes a response $Response(\sigma_1, \sigma_2, \sigma_5, n, r, ch_1, ch_2) = (z_1, z_2)$, which $z_1 = \sigma_1^{\frac{1}{ch_1 \cdot \sigma_5 + r}}$, $z_2 = \sigma_2^{\frac{1}{ch_2 \cdot \sigma_5 + r}} \in G_1$, and sends (z_1, z_2) to V.
   - Step 4. V checks if $P_{pub}(= \hat{e}(g, g)^\alpha) \equiv \hat{e}(z_1, \sigma_3^{n_1 \cdot ch_1} \cdot \sigma_4^{ID \cdot ch_1} \cdot A^{n_1 \cdot r} \cdot C^{ch_1 \cdot ID^2} \cdot D^{n_1 \cdot ch_1 \cdot ID} \cdot g^{ID \cdot r} \cdot \sigma_6^r \cdot E^{r \cdot ID} \cdot B^{ch_1}) \cdot \hat{e}(z_2, \sigma_3^{n_2 \cdot ch_2} \cdot \sigma_4^{ID \cdot ch_2} \cdot A^{n_2 \cdot r} \cdot C^{ch_2 \cdot ID^2} \cdot D^{n_2 \cdot ch_2 \cdot ID} \cdot g^{ID \cdot r} \cdot \sigma_6^r \cdot E^{r \cdot ID} \cdot B^{ch_2})$. Then, $Check(pk, ID, \sigma_3, \sigma_4, \sigma_6, n_1, n_2, r, ch_1, ch_2, z_1, z_2) = 1$ if the above equation holds.
   - Step 5. V accepts P if only $Check$ is 1.

5.1 An Experiment on Identity-Based Identification Scheme against Impersonation under the CR1$^+$Attack

Given a random tape R and a resetable random tape R_s and let $\hat{U}$ denote a cheating user who issues a request form (ID_i, i) to KGC and KGC returns with a secret key sk_i associated with a public key ID_i. Let $\mathcal{A}$ be an adversary who can break the $IBI\text{-}CRA$ identification scheme with an advantage $\text{Adv}^{imp\text{-}cra}$ and $\mathcal{A}$ plays a role in the following game as $\hat{U}, \hat{V}$ and $\hat{P}$, respectively. Let an adversary's algorithm $\mathcal{B}$, attempting to solve 2-SDH problem with h, h^x, h^{x^2} as input and output $(h^{\frac{1}{x+r}}, r)$, be the challenger. $\mathcal{B}$ executes the experiment as follows.

Experiment: Adversary $\mathcal{B}(h, h^x, h^{x^2})$

$\mathcal{B}$ first randomly chooses an integer $a, c, \alpha, \tau \in \mathbb{Z}_p^*$ and sets $g^x = h^{x^2}$, $g = h^x$, $g^{\frac{1}{x}} = h$. Then, $\mathcal{B}$ sets $A = g^a$, $B = g^\tau$, $C = g^c$, $D = g^{a \cdot c}$, $E = g^{\tau \cdot c}$ and $P_{Pub} = \hat{e}(g, g)^\alpha$ be public key $pk = (g, A, B, C, D, E, P_{Pub})$. Perform the initialization with (pk, R) and set i to 0. Let ID-list be the list of user's identities that $\mathcal{A}$ makes a request for a secret key associated with public key (pk, ID) and it is initialized with an empty list.

Phase 1: $\mathcal{B}$ answers $\mathcal{A}$'s requests as follows. $\mathcal{A}$ issues a request of the form (ID_i, i) where ID_i is an adaptively chosen integer in $\mathbb{Z}_p^*$ by $\mathcal{A}$. Without losing generality, assume that $\mathcal{A}$ will never issue a request for the same user's identity again. Then, $\mathcal{A}$ sends a query to $\mathcal{B}$. $\mathcal{B}$ responds as follows: Set $i \leftarrow i+1$ and then $\mathcal{B}$ randomly selects integers $n_{i,1}, n_{i,2}, k_i, v_i \in \mathbb{Z}_p^*$ such that $(ID_i + n_{i,1} \cdot a + \tau) \bmod p \neq 0$ and $(ID_i + n_{i,2} \cdot a + \tau) \bmod p \neq 0$ and $k_i \neq a^{-1}$ and computes $\sigma_{i,1} = g^{\frac{\alpha - v_i}{(ID_i + n_{i,1} \cdot a + \tau)}}$, $\sigma_{i,2} = g^{\frac{v_i}{(ID_i + n_{i,2} \cdot a + \tau)}}$, $\sigma_{i,3} = g^{k_i \cdot a}$, $\sigma_{i,4} = g^{k_i}$, $\sigma_{i,5} = k_i + ID_i \cdot c$, $\sigma_{i,6} = g^{k_i \cdot \tau}$ and keeps $(\sigma_{i,1}, \sigma_{i,2}, \sigma_{i,3}, \sigma_{i,4}, \sigma_{i,5}, \sigma_{i,6}, k_i, n_{i,1}, n_{i,2}, ID_i)$ in ID-list. $\mathcal{B}$ then returns $(\sigma_{i,1}, \sigma_{i,2}, \sigma_{i,3}, \sigma_{i,4}, \sigma_{i,5}, \sigma_{i,6}, n_{i,1}, n_{i,2})$ as a secret key associated with the public key ID_i to $\mathcal{A}$.

Phase 2: Eventually after at most $q_{ID} < (p-2)$ queries, $\mathcal{A}$ outputs ID^* as the public key which $\mathcal{A}$ will attempt to impersonate and then $\mathcal{A}$ now acts as a cheating verifier $\hat{V}$. $\mathcal{B}$, simultaneously, acts as the honest prover of user ID^*. If ID_* is in ID-list then $\mathcal{B}$ terminates and returns failure, otherwise, initialize $\mathcal{A}$ with $(U_{pk}(= (pk, ID_*)), R)$; set i to 0; $\mathcal{B}$ selects random integers $n_1, n_2 \in \mathbb{Z}_p^*$ such that $\tau + n_1 \cdot a + ID_* \neq 0$ and $\tau + n_2 \cdot a + ID_* \neq 0$. $\mathcal{B}$ then constructs the partial ID_*'s witness instance as follows: $\sigma_{*,3} = g^{k \cdot a} = g^{a \cdot x}, \sigma_{*,4} = g^k = g^x$, $\sigma_{*,6} = g^{k \cdot \tau} = g^{\tau \cdot x}$. For the above context, k is x. Let T-list be the list of conversation transcripts that $\mathcal{A}$ requested and it is initialized with an empty list. $\mathcal{A}$ also sets a counter for a number of provers(clones) $\nu = 0$. $\mathcal{B}$ then answers $\hat{V}$'s requests as follows:

- When $\hat{V}$ issues a request of the form $(\mathrm{ch}, j, 0)$, where $c \xleftarrow{R} \mathbb{Z}_p^*$ and $j = \nu \leftarrow \nu + 1$ if $\hat{V}$ requests for a new clone(prover), else any value such that $j \leq \nu$, and $\mathcal{B}$ computes as follows: set $r_{0,j} \in R_s \leftarrow r$; $i \leftarrow 0$; $j \leftarrow j$. If $i \geq (p-2)$ then $\mathcal{B}$ terminates and returns failure. $\mathcal{B}$ sends $(\sigma_{*,3}, \sigma_{*,4}, n_1, n_2, r_{0,j})$ as a commitment. $(ch_{0,j,1}, ch_{0,j,2}) \xleftarrow{R} \mathrm{ch} \xleftarrow{R} \mathbb{Z}_p^*$. Then, $\hat{V}$ issues challenge numbers $(ch_{0,j,1}, ch_{0,j,2})$ to $\mathcal{B}$. Next, $\mathcal{B}$ computes $k_3 = -\frac{(ch_{0,j,2} \cdot ID_* \cdot c + r_{0,j})(\tau + n_2 \cdot a + ID_*)}{(ch_{0,j,1} \cdot ID_* \cdot c + r_{0,j})(\tau + n_1 \cdot a + ID_*)}$. If $k_3 \cdot ch_{0,j,1}(\tau + n_1 \cdot a + ID_*) + ch_{0,j,2}(\tau + n_2 \cdot a + ID_*) = 0$ or $(ch_{0,j,2} \cdot ID_* \cdot c + r_{0,j})(\tau + n_2 \cdot a + ID_*) = 0$ or $(ch_{0,j,1} \cdot ID_* \cdot c + r_{0,j})(\tau + n_1 \cdot a + ID_*) = 0$ then $\mathcal{B}$ terminates and returns failure. Otherwise, $z_{0,j,1} = g^{x \cdot (k_3 \cdot ch_{0,j,1}(\tau + n_1 \cdot a + ID_*) + ch_{0,j,2}(\tau + n_2 \cdot a + ID_*))}$ and $z_{0,j,2} = g^{x \cdot (k_3 \cdot ch_{0,j,1}(\tau + n_1 \cdot a + ID_*) + ch_{0,j,2}(\tau + n_2 \cdot a + ID_*))}$ be the response. $\mathcal{B}$ then keeps $(r_{0,j}, ch_{0,j,1}, ch_{0,j,2}, z_{0,j,1}, z_{0,j,2})$ in T-list and returns $(z_{0,j,1}, z_{0,j,2})$ to $\hat{V}$.

- When $\hat{V}$ issues a request of the form (ch, j, i), where $c \xleftarrow{R} \mathbb{Z}_p$ and j be any value such that $j \leq \nu$, and $\mathcal{B}$ then does as follows: First, $\mathcal{B}$ sets $r_{i,j} \in R_s \leftarrow r$; $i \leftarrow i+1$. If $i \geq (p-2)$ then $\mathcal{B}$ terminates and returns failure. $\mathcal{B}$ sends $(\sigma_{*,3}, \sigma_{*,4}, n_1, n_2, r_{i,j})$ as a commitment $(ch_{i,j,1}, ch_{i,j,2}) \xleftarrow{R} \mathrm{ch} \xleftarrow{R} \mathbb{Z}_p$. Then, $\hat{V}$ issues challenge numbers $(ch_{i,j,1}, ch_{i,j,2})$ to $\mathcal{B}$. Next, $\mathcal{B}$ computes $k_3 = -\frac{(ch_{i,j,2} \cdot ID_* \cdot c + r_{i,j})(\tau + n_2 \cdot a + ID_*)}{(ch_{i,j,1} \cdot ID_* \cdot c + r_{i,j})(\tau + n_1 \cdot a + ID_*)}$. If $k_3 \cdot ch_{i,j,1}(\tau + n_1 \cdot a + ID_*) + ch_{i,j,2}(\tau + n_2 \cdot a + ID_*) = 0$ or $(ch_{i,j,2} \cdot ID_* \cdot c + r_{i,j})(\tau + n_2 \cdot a + ID_*) = 0$

or $(ch_{i,j,1} \cdot ID_* \cdot c + r_{i,j})(\tau + n_1 \cdot a + ID_*) = 0$ then $\mathcal{B}$ terminates and returns failure. Otherwise, $z_{i,j,1} = g^{x \cdot (k_3 \cdot ch_{i,j,1}(\tau + n_1 \cdot a + ID_*) + ch_{i,j,2}(\tau + n_2 \cdot a + ID_*))}$ and $z_{i,j,2} = g^{x \cdot (k_3 \cdot ch_{i,j,1}(\tau + n_1 \cdot a + ID_*) + ch_{i,j,2}(\tau + n_2 \cdot a + ID_*))}$ be the response. $\mathcal{B}$ then keeps $(r_{i,j}, ch_{i,j,1}, ch_{i,j,2}, z_{i,j,1}, z_{i,j,2})$ in T-list and returns $(z_{i,j,1}, z_{i,j,2})$ to $\hat{V}$.

Until $\hat{V}$ outputs the state information St or makes at most $q_{IP} < (p-1)$ queries and then stops.

Phase 3: Now, A from $\hat{V}$ changes the status to $\hat{P}$. $\hat{P}$ begins the impersonation process.

- $\hat{P}$ first randomly chooses an integer $r \in \mathbb{Z}_p^*$ and then sends $(\sigma_{*,3}, \sigma_{*,4}, \sigma_{*,6}, n_1, n_2, r)$ to V.
- $\mathcal{B}$ first sets $ch_1 = ch_2$ and randomly selects an integer $ch_1 \in \mathbb{Z}_p^*$ such that $ch_1 \neq \frac{-r}{c \cdot ID_*}$ and (r, ch_1, ch_2) is not in T-list and sends ch_1, ch_2 to $\hat{P}$.
- $\hat{P}$ returns z_1, z_2.

Define $Check(pk, ID, z_1, z_2, ch_1, ch_2, \sigma_{*,3}, \sigma_{*,4}, \sigma_{*,6}, n_1, n_2, r, ID_*)$ as:
$$P_{pub} \overset{?}{=} \hat{e}(z_1, \sigma_{*,3}^{n_1 ch_1} \cdot \sigma_{*,4}^{ID_* ch_1} \cdot A^{n_1 r} \cdot C^{ch_1 ID_*^2} \cdot D^{n_1 ch_1 ID_*} \cdot g^{IDr} \cdot \sigma_{*,6}^r \cdot E^{rID_*} \cdot B^{ch_1}) \cdot$$
$$\hat{e}(z_2, \sigma_{*,3}^{n_2 ch_2} \cdot \sigma_{*,4}^{ID_* ch_2} \cdot A^{n_2 r} \cdot C^{ch_2 ID_*^2} \cdot D^{n_2 ch_2 ID_*} \cdot g^{IDr} \cdot \sigma_{*,6}^r \cdot E^{r \cdot ID_*} \cdot B^{ch_2}),$$
where $P_{pub} = \hat{e}(g, g)^\alpha$. If the equality holds, then output 1, otherwise, output 0. If $Check$ outputs 1 then $\mathcal{B}$ checks if $P_{pub} \neq \hat{e}(z_1^{\tau + n_1 a + ID_*}, (g^x \cdot g^{ID_* c})^{ch_1} g^r) \cdot \hat{e}(z_2^{\tau + n_2 a + ID_*}, (g^x \cdot g^{ID_* c})^{ch_2} g^r)$, then $\mathcal{B}$ returns failure. Otherwise, $\mathcal{B}$ computes the solution for 2-SDH problem as follows: let $m = ID_* \cdot c + \frac{r}{ch_1}$. Then compute $Z = (z_1^{ch_1(\tau + n_1 a + ID_*)} \cdot z_2^{ch_2(\tau + n_2 a + ID_*)})^{\frac{1}{\alpha}} = g^{(\frac{(\alpha - v)ch_1(\tau + n_1 a + ID_*)}{\alpha(\tau + n_1 a + ID_*)(ch_1(x + ID_* c) + r)})} \cdot g^{(\frac{v \cdot ch_2(\tau + n_2 \cdot a + ID_*)}{\alpha(\tau + n_2 \cdot a + ID_* \cdot c)(ch_2(x + ID_* \cdot c) + r)})}$. Since $ch_1 = ch_2$ and $g = h^x$, $Z = g^{\frac{1}{x + ID_* \cdot c + r/ch_1}} = g^{\frac{1}{x+m}} = h^{\frac{x}{x+m}} = h^{\frac{-m}{x+m}} \cdot h$. Hence, let $Z^* = (Z^{-1} \cdot h)^{\frac{1}{m}} = h^{\frac{1}{x+m}}$. $\mathcal{B}$ wins the game and returns (Z^*, m) as the solution of 2-SDH. Else, $\mathcal{B}$ returns failure. Note that v is assumed to be a variable that does not need to be known for solving 2-SDH. ∎

5.2   Proof of Security

Theorem 3. *Let* $IBI = (Setup, Extract, P, V)$ *be the* $IBI\text{-}CRA$ *identity-based identification scheme associated to 2-SDH assumption. Let* $\mathcal{A} = (\hat{U}, \hat{V}, \hat{P})$ *be an imp-cra adversary of time complexity* $t(.)$ *attacking* IBI. *Then there exists a 2-SDH adversary* $\mathcal{B}$ *of time complexity* $t'(.)$ *solving 2-SDH problem such that for every security parameter* k

$$Adv_{IBI,\mathcal{A}}^{imp\text{-}cra}(k) \geq (1/(1 - \frac{1}{p-1})^{q_{ID}})(1/(1 - \frac{1}{p-1})^{q_{IP}}) \cdot \frac{(p-1)^2}{p^2 - 2p} \cdot Adv_{\mathcal{B}}^{2-SDH}(k)$$

$$Adv_{IBI,\mathcal{A}}^{imp\text{-}cra}(k) \geq e^2 \cdot \frac{(p-1)^2}{p^2 - 2p} \cdot Adv_{\mathcal{B}}^{2-SDH}(k) \text{ or, in a simply term,}$$

$$\epsilon \geq e^2 \cdot \frac{(p-1)^2}{p^2 - 2p} \cdot \epsilon',$$

where e is the natural logarithm. Moreover, the time complexity t of $\mathcal{A}$ is $t \leq t' - ((8 + 4q_{ID} + 2q_{IP})C_G + 5C_P)$, where C_G be a computation time of group exponential operation and C_P be a computation time of bilinear group pairing operation, and $\mathcal{A}$ can request queries at most $q_{ID} < p - 2$ and at most $q_{IP} < p - 1$. For simplicity, we assumed that the entire computation time of the multiplication in $\mathbb{Z}_p^*$ for the entire phases is one unit time.

Corollary 2. *If 2-SDH assumption is $(2, t', \epsilon')$-secure then the IBI-CRA identity-based identification scheme associated to 2-SDH assumption is $(q_{ID}, q_{IP}, t, \epsilon)$-secure against impersonation under the CR1$^+$ attacks.*

Proof. Due to the page limitation, please find the proof for Corollary 2 in the full version of this paper [18].

6 Efficiency

The performance comparison between our identity-based identification scheme and the state-of-the-art identity-based identification scheme secure against passive attack and concurrent-active attack proposed by Kurosawa and Heng [11] is provided in Table 1. Let C_E be a computation of group exponential operation, C_P be a computation of bilinear group pairing operation and C_M be a computation time of bilinear group multiplicative operation. Note that KH-IBI-PA and KH-IBI-CA are Kurosawa-Heng IBI secured against the passive and concurrent attacks, respectively.

Table 1. Table: Bandwidth and Computation Comparison

Computation	KH-IBI-PA	KH-IBI-CA	IBI-PA	IBI-CRA
Prover	$3C_E+C_P+4C_M$	$6C_E+2C_P+6C_M$	C_E	$2C_E$
Verifier	$3C_E+C_P+4C_M$	$6C_E+2C_P+6C_M$	$9C_E+C_P+8C_M$	$18C_E+2C_P+16C_M$
Total	$6C_E+2C_P+8C_M$	$12C_E+4C_P+12C_M$	$10C_E+1C_P+8C_M$	$20C_E+2C_P+16C_M$
Bandwidth	KH-IBI-PA	KH-IBI-CA	IBI-PA	IBI-CRA
Public(bits)	3403	5451	2210	2210
Secret(bits)	331	662	2721	3052
Communi-cation (bits)	1515	3190	2019	2337

7 Conclusion

We started our paper by providing a stronger definition of CR1$^+$ attack to capture the most stringent attack in id-based identification scheme. The CR1$^+$ attack is a stronger variant than the CR1 attack proposed in [1]. In our definition, we allow the adversary to reset the prover (or it clones) to *any* state instead of its initial state, as defined in [1]. Therefore, the CR1 attack in [1] is a special case of our CR1$^+$ attacks. Then, we provided two id-based identification schemes which are secure under passive attack and secure against the CR1$^+$ attack. The complexity assumption used in our proof is weaker than the state-of-the-art identification scheme by Kurosawa and Heng [11].

References

1. Bellare, M., Fischlin, M., Goldwasser, S., Micali, S.: Identification protocols secure against reset attacks. In: Pfitzmann, B. (ed.) EUROCRYPT 2001. LNCS, vol. 2045, pp. 495–511. Springer, Heidelberg (2001)
2. Bellare, M., Namprempre, C., Neven, G.: Security proofs for identity-based identification and signature schemes. In: Cachin, C., Camenisch, J.L. (eds.) EUROCRYPT 2004. LNCS, vol. 3027, pp. 268–286. Springer, Heidelberg (2004)
3. Bellare, M., Rogaway, P.: Entity authentication and key distribution. In: Stinson, D.R. (ed.) CRYPTO 1993. LNCS, vol. 773, pp. 232–249. Springer, Heidelberg (1994)
4. Boneh, D., Boyen, X.: Short signatures without random oracles. In: Cachin, C., Camenisch, J.L. (eds.) EUROCRYPT 2004. LNCS, vol. 3027, pp. 56–73. Springer, Heidelberg (2004)
5. Canetti, R., Goldreich, O., Goldwasser, S., Micali, S.: Resettable zero-knowledge. In: STOC 2000, pp. 235–244. ACM, New York (2000)
6. Feige, U., Fiat, A., Shamir, A.: Zero-knowledge proofs of identity. J. Cryptology 1(2), 77–94 (1988)
7. Fiat, A., Shamir, A.: How to prove yourself: Practical solutions to identification and signature problems. In: Odlyzko, A.M. (ed.) CRYPTO 1986. LNCS, vol. 263, pp. 186–194. Springer, Heidelberg (1987)
8. Freeman, D.: Pairing-based identification schemes. technical report HPL-2005-154, Hewlett-Packard Laboratories (August 2005)
9. Kim, M., Kim, K.: A new identification scheme based on the bilinear diffie-hellman problem. In: Batten, L.M., Seberry, J. (eds.) ACISP 2002. LNCS, vol. 2384, pp. 362–378. Springer, Heidelberg (2002)
10. Kurosawa, K., Heng, S.-H.: From digital signature to id-based identification/signature. In: Bao, F., Deng, R., Zhou, J. (eds.) PKC 2004. LNCS, vol. 2947, pp. 248–261. Springer, Heidelberg (2004)
11. Kurosawa, K., Heng, S.-H.: Identity-based identification without random oracles. In: Gervasi, O., Gavrilova, M.L., Kumar, V., Laganá, A., Lee, H.P., Mun, Y., Taniar, D., Tan, C.J.K. (eds.) ICCSA 2005. LNCS, vol. 3481, pp. 603–613. Springer, Heidelberg (2005)
12. Kurosawa, K., Heng, S.-H.: The power of identification schemes. In: Yung, M., Dodis, Y., Kiayias, A., Malkin, T.G. (eds.) PKC 2006. LNCS, vol. 3958, pp. 364–377. Springer, Heidelberg (2006)
13. Ohta, K., Okamoto, T.: A modification of the fiat-shamir scheme. In: Goldwasser, S. (ed.) CRYPTO 1988. LNCS, vol. 403, pp. 232–243. Springer, Heidelberg (1990)
14. Okamoto, T.: Provably secure and practical identification schemes and corresponding signature schemes. In: Brickell, E.F. (ed.) CRYPTO 1992. LNCS, vol. 740, pp. 31–53. Springer, Heidelberg (1993)
15. Schnorr, C.-P.: Efficient signature generation by smart cards. J. Cryptology 4(3), 161–174 (1991)
16. Shamir, A.: Identity-based cryptosystems and signature schemes. In: Blakely, G.R., Chaum, D. (eds.) CRYPTO 1984. LNCS, vol. 196, pp. 47–53. Springer, Heidelberg (1985)
17. Shoup, V.: On the security of a practical identification scheme. J. Cryptology 12(4), 247–260 (1999)
18. Thorncharoensri, P., Susilo, W., Mu, Y.: Identity-based identification scheme secure against concurrent-reset attacks without random oracles (full version). Can be obtained from the first author (2009)

Construction of Odd-Variable Boolean Function with Maximum Algebraic Immunity

Shaojing Fu[1], Longjiang Qu[1,2], and Chao Li[1]

[1] Department of Mathematics and System Science,
National University of Defense Technology,
Changsha 410073, China
[2] National Mobile Communications Research Laboratory,
Southeast University, Nanjing 210096, China
shaojing1984@yahoo.cn

Abstract. Because of the algebraic attacks, a high algebraic immunity is now an important criteria for Boolean functions used in stream ciphers. In this paper, Construction of odd-variables Boolean functions with maximum algebraic immunity (AI) was investigated. At first, we study the method of known construction and apply the method to construct odd-variable Boolean functions with maximum AI. Then we present a new construction of odd-variable Boolean functions and we prove that the constructed Boolean functions has maximum AI.

Keywords: Cryptography; Boolean function; Algebraic attack; Algebraic immunity.

1 Introduction

Algebraic attack to LFSR-based stream cipher was proposed by Coutois and Meier in 2003 [5]. Its main idea is to deduce the security of a stream cipher to solve an over-defined system of multivariate nonlinear equations whose unknowns are the bits of the some LFSR-based stream ciphers such as Toyocrypt [9], LILI-128 [5] and SFINKS [6] etc were successfully attacked.

To resist algebraic attack, a new cryptographic property of Boolean functions which is known as algebraic immunity (AI) has been proposed by Meier et al [10]. The AI of a Boolean function expresses its ability to resist standard algebraic attack. Thus the AI of Boolean function used in cryptosystem should be sufficiently high. Courtois and Meier [5,10] showed that, for any n-variable Boolean function, its AI is upper bounded by $\lceil n/2 \rceil$. If the bound is achieved, we say the Boolean function have maximum AI. Obviously, a Boolean function with maximum AI has strongest ability to resist standard algebraic attack. Therefore, the construction of Boolean functions with maximum AI is of great importance, and several constructions of Boolean functions with large algebraic immunity have been investigated [2,4,7,8,10,11,12,13].

All these known constructions can be divided into two main classes. The first one contains functions in even numbers n of variables and is obtained by an

H.Y. Youm and M. Yung (Eds.): WISA 2009, LNCS 5932, pp. 109–117, 2009.
© Springer-Verlag Berlin Heidelberg 2009

iterative construction [3], where it is shown that their algebraic degrees are close to n and their nonlinearity is $2^{n-1} - \binom{n-1}{n/2}$. The second class contains symmetric functions or functions whose values depend on the Hamming weight of the input vectors except for a few inputs [2,4,7,8,10,11,12,13]. The nonlinearities of these functions are often not exceeding $2^{n-1} - \binom{n-1}{\lfloor n/2 \rfloor}$.

In this paper, We will work in the direction of first construction and study the construction of odd-variable Boolean functions with maximum algebraic immunity. We first study the method of [3]'s construction and apply the method to construct Boolean functions, in odd number of variables, has a maximum algebraic immunity(AI). Then a new construction is obtained through a doubly indexed recursive relation.

The paper is organized as follows. Section 2 provides basic definitions and notations. In Section 3, We first study the method of Carlet'construction and apply the method to construct Boolean functions with maximum AI. In Section 4, we construct maximum AI Boolean functions on odd-variable by a doubly indexed recursive relation. Section 5 concludes this paper.

2   Preliminaries

Let $\mathbb{F}_2$ be the binary finite field, the vector space of dimension n over $\mathbb{F}_2$ is denoted by $\mathbb{F}_2^n$. A Boolean function on n variables may be viewed as a mapping from $\mathbb{F}_2^n$ into $\mathbb{F}_2$. The set of all n-variable Boolean function is denoted by B_n. A Boolean function $f(x_1, x_2, \cdots, x_n)$ is also interpreted as the output column of its truth table, that is, a binary string of length 2^n having the form:

$$\{f(0,0,\cdots,0), \quad f(0,0,\cdots,1), \quad \cdots, \quad f(1,1,\cdots,1)\}.$$

The weight of f is the number of ones in its output column, and is denoted by $wt(f)$. The support of f denoted by $supp(f)$ is the set of inputs $X \in F_2^n$ such that $f(X) = 1$.

Definition 1. *An n-variable function f is balanced if and only if $wt(f) = 2^{n-1}$.*

Let us denoted the addition operator over $\mathbb{F}_2$ by $+$. An n-variable function $f(x_1, \cdots, x_n)$ can be seen as a multivariate polynomial over $\mathbb{F}_2$, that is,

$$f(x_1, \cdots, x_n) = a_0 + \sum_{i=1}^{n} a_i x_i + \sum_{1 \leq i < j \leq n} a_{i,j} x_i x_j + \cdots + a_{1,2,\cdots,n} x_1 x_2 \cdots x_n$$

where the coefficients $a_0, a_i, a_{i,j}, a_{1,2,\cdots,n}$ is a constant in $\mathbb{F}_2$. This representation of f is called the algebraic normal form (ANF) of f. The algebraic degree $\deg(f)$ of f is the number of variables in the highest order term with nonzero coefficient. A Boolean function is affine if it has algebraic degree at most 1.

A nonzero n-variable Boolean function g is called an annihilator of an n-variable Boolean function f if $f * g = 0$. We denote the set of all annihilators of f by $AN(f)$. That is,

$$AN(f) = \{g \in B_n | g * f = 0\}$$

Definition 2. *For* $f \in B_n$, *the algebraic immunity(AI) of* f *is the minimum degree of non-zero functions* $g \in B_n$ *such that* $g * f = 0$ *or* $g * (f + 1) = 0$. *Namely,*

$$AI(f) = min\{\deg(g)|0 \neq g \in AN(f) \cup AN(1+f)\}$$

To check that a function f has good algebraic immunity, it is necessary and sufficient to check that f and $f + 1$ do not admit nonzero annihilators of low degrees. Indeed, if f or $f + 1$ has an annihilator g of low degree d, then $f * g$ either is null or equals g and therefore has degree at most d; conversely, if we have $f * g = h$ where $g \neq 0$ and h, g have degrees at most d, then either $g = h$, and then g is an annihilator of $f + 1$, or $g \neq h$, and we have $f * g = f * h$, which proves that $f * (g + h)$ and shows that $g + h$ is a nonzero annihilator of f of degree at most d.

Lemma 1. *[5] Let* f *be an* n-*variable boolean functions, then* $AI(f) \leq \lceil n/2 \rceil$.

From now on, we use a binary string of length 2^n to represent an n-variable Boolean function. We denote by "$\|$" the concatenation of binary strings. For example, let $s, t \in B_1$, and $s = x_1$, $t = x_1 + 1$. In the truth table representation, they are $s = 01$, $t = 10$. Let $u = s\|t = 0110$, then $u = x_1 + x_2$.

3   Construction of Boolean Functions with Maximum AI

In [3], Carlet presented a construction to design a Boolean function of even variables with maximum algebraic immunity. The construction is iterative in nature. At each step, two variables are added and the algebraic immunity is increased by 1. The constructed function is not balanced, but the bias with respect to balancedness tends to zero when tends to infinity. And the constructed functions can be used in the secondary construction to lead to functions satisfying all of the necessary cryptographic criteria.

Construction in [3]: We denoted by $\phi_{2k+2} \in B_{2k+2}$ the function defined by the recursion

$$\phi_{2k+2} = \phi_{2k}\|\phi_{2k}\|\phi_{2k}\|\phi_{2k-1}^1$$

and ϕ_{2k}^1 is defined itself by a doubly indexed recursion

$$\phi_{2k}^i = \phi_{2k-2}^{i-1}\|\phi_{2k-2}^i\|\phi_{2k-2}^i\|\phi_{2k-2}^{i+1} \ (i \geq 1).$$

$\phi_{2k}^0 = \phi_{2k}$, $\phi_0^i = i(mod \ 2)$ for $i \geq 0$.

Theorem 1. *[3] The Boolean functions* ϕ_{2k+2} *constructed above have maximum algebraic immunity.*

Carlet's method can be applied to construct Boolean functions of odd number of variables with algebraic immunity. Now we present the improved construction.

Construction 1: We denoted by $\phi_{2k+1} \in B_{2k+1}$ the function defined by the recursion

$$\phi_{2k+1} = \phi_{2k-1} \| \phi_{2k-1} \| \phi_{2k-1} \| \phi_{2k-1}^1$$

and ϕ_{2k-1}^1 is defined itself by a doubly indexed recursion

$$\phi_{2k-1}^i = \phi_{2k-3}^{i-1} \| \phi_{2k-3}^i \| \phi_{2k-3}^i \| \phi_{2k-3}^{i+1} \ (i \geq 1)$$

where $\phi_{2k+1}^0 = \phi_{2k+1}$, $\phi_1^i = 01$ if i is even, $\phi_1^i = 10$ if i is odd.

Similar as the proof of Theorem 1 in [3], we can get the following results.

Theorem 2. *The Boolean function ϕ_{2k+1} constructed in Construction 1 have maximum algebraic immunity.*

4   New Construction of Odd-Variable Boolean Functions with Maximum AI

In this section, We will give a new construction of Boolean functions with maximum algebraic immunity for any odd number of variables, this construction is based on a doubly indexed recursive relation.

Construction 2: We denoted by $\phi_{2k+1} \in B_{2k+1}$ the function defined by the recursion

$$\phi_{2k+1} = \phi_{2k-1} \| \phi_{2k-1} \| \phi_{2k-1}^1 \| \phi_{2k-1}^2$$

and ϕ_{2k-1}^1 and ϕ_{2k-1}^2 is defined itself by a doubly indexed recursion

$$\begin{cases} \phi_{2k-1}^1 = \phi_{2k-3} \| \phi_{2k-3}^1 \| \phi_{2k-3}^1 \| \phi_{2k-3}^3, \\ \phi_{2k-1}^i = \phi_{2k-3}^{i-2} \| \phi_{2k-3}^i \| \phi_{2k-3}^i \| \phi_{2k-3}^{i+2}, \ (i \geq 2). \end{cases}$$

$\phi_{2k+1}^0 = \phi_{2k+1}$, $\phi_1^i = 01$ if i is even, $\phi_1^i = 10$ if i is odd.

To prove that ϕ_{2k+1} has algebraic immunity k, we need intermediate results. For technical reasons, during our proofs, we will encounter certain situations when the degree of a function is negative. As such functions cannot exist, we will replace those functions by function 0.

Lemma 2. *Assume that the function ϕ_{2i+1} has been generated by Construction 2 for $0 \leq i \leq k$ and that $AI(\phi_{2i+1}) = i + 1$ for $0 \leq i \leq k$. If, for some $1 \leq i \leq k$ and $j \geq 0$, there exist $g \in AN(\phi_{2i-1}^1)$ such that $\deg(g) \leq i - 2$, then $g = 0$.*

Proof. Since the function $\phi_{2i+1} \in B_{2i+1}$ defined by the recursion

$$\phi_{2i+1} = \phi_{2i-1} \| \phi_{2i-1} \| \phi_{2i-1}^1 \| \phi_{2i-1}^2$$

has the ANF as follows,

$$\phi_{2i+1} = (x_{2i} + 1)(x_{2i+1} + 1)\phi_{2i-1} + (x_{2i} + 1)x_{2i+1}\phi_{2i-1} +$$
$$x_{2i}(x_{2i+1} + 1)\phi^1_{2i-1} + x_{2i}x_{2i+1}\phi^2_{2i-1}.$$

then for $g \neq 0 \in B_{2i-1}$,

$$g * \phi^1_{2i-1} = 0 \Rightarrow (x_{2i} + 1)x_{2i+1}g * \phi_{2i+1} = 0$$

That is to say $(x_{2i} + 1)x_{2i+1}g$ is an annihilator of ϕ_{2i+1}.

$$AI(\phi_{2i+1}) = i + 1 \Rightarrow \deg((x_{2i} + 1)x_{2i+1}g) \geq i + 1 \Rightarrow \deg(g) \geq i - 1.$$

Hence $g = 0$.

Lemma 3. *Assume that the function ϕ_{2k+1} has been generated by Construction 2 for $0 \leq i \leq k$ and that $AI(\phi_{2k+1}) = k + 1$ for $0 \leq i \leq k$. If, for some $0 \leq i \leq k$ and $j \geq 0$, there exist $g \in AN(\phi^j_{2i+1})$ and $h \in AN(\phi^{j+1}_{2i+1})$ such that $\deg(g + h) \leq i - 1 - \lfloor j/2 \rfloor$, then $g = h$.*

Proof. We prove Lemma 3 by induction on i.

For the base step $i = 0$, $\deg(g + h) \leq 0 - 1 - \lfloor j/2 \rfloor \leq -1$ implies that such a function cannot exist, i.e., $g + h$ is identically 0, which gives $g = h$.

Now we prove the inductive step. Assume that, for $i < l$, the induction assumption holds (for every $j \geq 0$). We will show it for $i = l$(and for every $j \geq 0$). Suppose that there exist $g \in AN(\phi^j_{2l+1})$ and $h \in AN(\phi^{j+1}_{2l+1})$ with $\deg(g + h) \leq l - 1 - \lfloor j/2 \rfloor$. By construction, if $j = 0$, we have

$$\phi^0_{2l+1} = \phi^0_{2l-1}\|\phi^0_{2l-1}\|\phi^1_{2l-1}\|\phi^2_{2l-1}$$

if $j = 1$, we have

$$\phi^1_{2l+1} = \phi^0_{2l-1}\|\phi^1_{2l-1}\|\phi^1_{2l-1}\|\phi^3_{2l-1}$$

if $j \geq 2$ then we have

$$\phi^j_{2l+1} = \phi^{j-2}_{2l-1}\|\phi^j_{2l-1}\|\phi^j_{2l-1}\|\phi^{j+2}_{2l-1}$$
$$\phi^{j+1}_{2l+1} = \phi^{j-1}_{2l-1}\|\phi^{j+1}_{2l-1}\|\phi^{j+1}_{2l-1}\|\phi^{j+3}_{2l-1}$$

Let us denoted

$$g = g_1\|g_2\|g_3\|g_4$$
$$h = h_1\|h_2\|h_3\|h_4$$

It is obvious that the ANF of $g + h$ is as follows,

$$g + h = (g_1 + h_1) + x_{2l}(g_1 + h_1 + g_2 + h_2) + x_{2l+1}(g_1 + h_1 + g_3 + h_3)$$
$$+ x_{2l}x_{2l+1}(g_1 + h_1 + g_2 + h_2 + g_3 + h_3 + g_4 + h_4)$$

Then we can divide our proof into four cases.

1. It is clear that $\deg(g_1 + h_1) \leq \deg(g + h) = l - 1 - \lfloor j/2 \rfloor$.
   If $j = 0$ *or* 1, then $g_1 \in AN(\phi_{2l-1})$ and $h_1 \in AN(\phi_{2l-1})$, and therefore $g_1 + h_1 \in AN(\phi_{2l-1})$ with $\deg(g_1 + h_1) \leq \deg(g + h) = l - 1 - \lfloor j/2 \rfloor = l - 1$. since $AN(\phi_{2l-1}) = l$, then $g_1 + h_1 = 0$ i.e., $g_1 = h_1$.
   If $j \geq 2$, we have $g_1 \in AN(\phi_{2l-1}^{j-2})$ and $h_1 \in AN(\phi_{2l-1}^{j-1})$, then

$$\deg(g_1 + h_1) \leq l - 1 - \lfloor j/2 \rfloor = (l - 1) - 1 - \lfloor (j - 2)/2 \rfloor$$

   which implies that $g_1 = h_1$, according to the induction assumption.
2. Since $g_1 = h_1$, then

$$g + h = \; x_{2l}(g_2 + h_2) + x_{2l+1}(g_3 + h_3) + $$
$$x_{2l}x_{2l+1}(g_2 + h_2 + g_3 + h_3 + g_4 + h_4)$$

   then $\deg(g_2 + h_2) \leq \deg(g + h) - 1 = (l - 1) - 1 - \lfloor j/2 \rfloor$. We have $g_2 \in AN(\phi_{2l-1}^{j})$ and $h_2 \in AN(\phi_{2l-1}^{j+1})$, which implies that $g_2 = h_2$, according to the induction assumption.
3. If $j = 0$, then $g_3 \in AN(\phi_{2l-1}^1)$ and $h_3 \in AN(\phi_{2l-1}^1)$, and

$$\deg(g_3 + h_3) \leq \deg(g + h) - 1 = l - 2 - \lfloor j/2 \rfloor = l - 2,$$

   By Lemma 2, we have $g_3 = h_3$.
   If $j \geq 1$, we have $g_3 \in AN(\phi_{2l-1}^{j})$ and $h_3 \in AN(\phi_{2l-1}^{j+1})$, and

$$\deg(g_3 + h_3) \leq \deg(g + h) - 1 = (l - 1) - 1 - \lfloor j/2 \rfloor,$$

   which implies that $g_3 = h_3$, according to the induction assumption.
4. Since $g_1 = h_1$, $g_2 = h_2$ and $g_3 = h_3$, then $\deg(g_4 + h_4) \leq \deg(g + h) - 2 = (l - 1) - 1 - \lfloor (j + 2)/2 \rfloor$, we have $g_4 \in AN(\phi_{2l-1}^{j+2})$ and $h_4 \in AN(\phi_{2l-1}^{j+3})$, which implies that $g_4 = h_4$, according to the induction assumption.

Hence, we get $g = h$.

Lemma 4. *Assume that the function ϕ_{2i+1} has been generated by Construction 2 for $0 \leq i \leq k$ and that $AI(\phi_{2i+1}) = i + 1$ for $0 \leq i \leq k$. If, for some $0 \leq i \leq k$ and $j \geq 0$, there exist $g \in AN(\phi_{2i+1}^{j}) \cap AN(\phi_{2i+1}^{j+1})$ such that $\deg(g) \leq i + \lfloor j/2 \rfloor$, then $g = 0$.*

Proof. We prove Lemma 4 by induction on i.
   For the base step $i = 0$, we have from construction $\phi_{2i+1}^{j} = \phi_{2i+1}^{j+1} + 1$(this can easily be checked by induction). Hence,

$$g \in AN(\phi_{2i+1}^{j}) \cap AN(\phi_{2i+1}^{j+1}) \Rightarrow g = 0$$

Now we prove the inductive step. Assume that, for $i < l$, the induction assumption holds (for every $j \geq 0$). We will show it for $i = l$(and for every $j \geq 0$). Suppose that $g \in AN(\phi_{2i+1}^{j}) \cap AN(\phi_{2i+1}^{j+1})$ with $\deg(g) \leq l + \lfloor j/2 \rfloor$. Let us denoted

$$g = g_1 \| g_2 \| g_3 \| g_4$$

It is obvious that the ANF of g is as follows,

$$g = g_1 + x_{2l}(g_1 + g_2) + x_{2l+1}(g_1 + g_3)$$
$$+ x_{2l}x_{2l+1}(g_1 + g_2 + g_3 + g_4)$$

If $j \geq 2$, By construction, we have

$$\phi_{2l+1}^j = \phi_{2l-1}^{j-2} \| \phi_{2l-1}^j \| \phi_{2l-1}^j \| \phi_{2l-1}^{j+2}$$
$$\phi_{2l+1}^{j+1} = \phi_{2l-1}^{j-1} \| \phi_{2l-1}^{j+1} \| \phi_{2l-1}^{j+1} \| \phi_{2l-1}^{j+3}$$

we can finish our proof by three steps.

1. It is clear that $\deg(g_4) \leq \deg(g) \leq (l-1) + \lfloor(j+2)/2\rfloor$. Since $g_4 \in AN(\phi_{2l-1}^{j+2}) \cap AN(\phi_{2l-1}^{j+3})$, we have $g_4 = 0$, according to the induction assumption.
2. Since $g_4 = 0$, then

$$g = g_1 + x_{2l}(g_1 + g_2) + x_{2l+1}(g_1 + g_3) + x_{2l}x_{2l+1}(g_1 + g_2 + g_3)$$

and $\deg(g_1 + g_2), \deg(g_1 + g_3), \deg(g_2 + g_3) \leq l + \lfloor j/2 \rfloor - 1$, which implies that $\deg(g_1), \deg(g_2), \deg(g_3) \leq l + \lfloor j/2 \rfloor - 1$. Since $g_2, g_3 \in AN(\phi_{2l-1}^j) \cap AN(\phi_{2l-1}^{j+1})$, we have $g_2 = g_3 = 0$, according to the induction assumption.
3. Now we have

$$g = g_1(1 + x_{2l} + x_{2l+1} + x_{2l}x_{2l+1})$$

which implies that

$$\deg(g_1) \leq \deg(g) - 2 \leq l + \lfloor j/2 \rfloor - 2 = (l-1) + \lfloor(j-2)/2\rfloor$$

Since $g_1 \in AN(\phi_{2l-1}^{j-2}) \cap AN(\phi_{2l-1}^{j-1})$, we have $g_1 = 0$, according to the induction assumption.

If $j = 1$, we can prove $g_2 = g_3 = g_4 = 0$ similar as the case $j \geq 2$. Now we only show $g_1 = 0$, Since

$$g = g_1(1 + x_{2l} + x_{2l+1} + x_{2l}x_{2l+1})$$

then $\deg(g_1) \leq \deg(g) - 2 \leq l + \lfloor j/2 \rfloor - 2 \leq l - 1$, since $g_1 \in AN(\phi_{2l-1})$, then $g_1 = 0$.

If $j = 0$, we can prove $g_4 = 0$ similar as the case $j \geq 2$, then

$$g = g_1 + x_{2l}(g_1 + g_2) + x_{2l+1}(g_1 + g_3) + x_{2l}x_{2l+1}(g_1 + g_2 + g_3)$$

which implies $\deg(g_1 + g_2) \leq l + \lfloor j/2 \rfloor - 1 = l - 1$, since $g_1, g_2 \in AN(\phi_{2l-1})$, then $g_1 = g_2$. Now we have

$$g = g_1 + x_{2l+1}(g_1 + g_3) + x_{2l}x_{2l+1}g_3$$

which implies $\deg(g_3) \leq \deg(g) - 1 = l - 2$, By Lemma 2 we have $g_3 = 0$. Then

$$g = (1 + x_{2l+1})g_1$$

which implies $\deg(g_1) \leq \deg(g) - 1 = l - 1$, since $g_1 \in AN(\phi_{2l-1})$, then $g_1 = 0$. Hence, we get $g = h$. $\square$

Now we present the main result.

Theorem 3. *The Boolean function ϕ_{2k+1} constructed in Construction 2 have maximum algebraic immunity.*

Proof. We prove Theorem 3 by induction on k,

$$\phi_{2k+1} = \phi_{2k-1}\|\phi_{2k-1}\|\phi_{2k-1}^1\|\phi_{2k-1}^2$$

when $k = 0$, $\phi_{2k+1} = 01$, then $AI(\phi_{2k+1}) = 1$.

In the inductive step, we assume the hypothesis true until $k - 1$ and we have to prove that any $2k+1$-variable nonzero function g such that $g*(\phi_{2k+1}) = 0$ has degree at least $k+1$ (proving that any nonzero function such that $g*(\phi_{2k+1}+1) = 0$ has degree at least $k+1$ is similar). Suppose that such a function g with degree at most k exists. Then, g can be decomposed as

$$g = g_1\|g_2\|g_3\|g_4$$

where $g_1, g_2 \in AN(\phi_{2k-1})$, $g_3 \in AN(\phi_{2k-1}^1)$ and $g_4 \in AN(\phi_{2k-1}^2)$ the ANF of g is as follows,

$$\begin{aligned} g = \ & g_1 + x_{2l}(g_1 + g_2) + x_{2l+1}(g_1 + g_3) \\ & + x_{2l}x_{2l+1}(g_1 + g_2 + g_3 + g_4) \end{aligned}$$

If $\deg(g) \le k$, we have $\deg(g_1+g_2) \le k-1$, then $g_1+g_2 = 0$, which give, $g_1 = g_2$. Therefore,

$$g = \ g_1 + x_{2l+1}(g_1 + g_3) + x_{2l}x_{2l+1}(g_3 + g_4)$$

then $\deg(g_3 + g_4) \le k - 2 = (k - 1) - \lfloor 1/2 \rfloor - 1$, According to Lemma 4 and Lemma 5, we have $g_3 = g_4 = 0$. Therefore,

$$g = \ g_1 + x_{2l+1}g_1$$

then $\deg(g_1) \le \deg(g) - 1 = k - 1$, since $g_1 \in AN(\phi_{2k-1})$, then $g_1 = 0$. Hence $g_1 = g_2 = 0$. This completes the proof. $\square$

5   Conclusion

Possessing a high algebraic immunity is a necessary criteria for Boolean functions used in stream ciphers against algebraic attacks. In this paper, We first study the method of Carlet'construction and then apply the method to construct odd-variable Boolean functions with maximum AI. Furthermore, we construct maximum AI Boolean functions on odd number of variables by a doubly indexed recursive relation. Our constructed functions can be combined with the secondary constructions to lead to functions satisfying all of the necessary cryptographic criteria. However, it is still an open problem to generalize our construction to obtain more Boolean functions with maximum algebraic immunity. Furthermore, there are still some problems need to be studied such as whether the constructed functions can achieve high nonlinearities and be robust against fast algebraic attacks.

Acknowledgments

The work in this paper is supported by the National Natural Science Foundation of China (No:60803156) and the open research of Natural Mobile Communications Research Laboratory, Southeast University(No:W200807).

References

1. Canteaut, A.: Open problems related to algebraic attacks on stream ciphers. In: Ytrehus, Ø. (ed.) WCC 2005. LNCS, vol. 3969, pp. 120–134. Springer, Heidelberg (2006)
2. Carlet, C.: A method of construction of balanced functions with optimum algebraic immunity, http://eprint.iacr.org/2006/149
3. Carlet, C., Dalai, D.K., Gupta, K.C., et al.: Algebraic Immunity for Cryptographically Significant Boolean Functions: Analysis and Construction. IEEE Transactions on Information Theory 52, 3105–3121 (2006)
4. Carlet, C., Feng, K.: An Infinite Class of Balanced Functions with Optimal Algebraic Immunity, Good Immunity to Fast Algebraic Attacks and Good Nonlinearity. In: Pieprzyk, J. (ed.) ASIACRYPT 2008. LNCS, vol. 5350, pp. 425–440. Springer, Heidelberg (2008)
5. Courtois, N., Meier, W.: Algebraic attacks on stream ciphers with linear feedback. In: Biham, E. (ed.) EUROCRYPT 2003. LNCS, vol. 2656, pp. 345–359. Springer, Heidelberg (2003)
6. Courtois, N.: Cryptanalysis of SFINKS. In: Won, D.H., Kim, S. (eds.) ICISC 2005. LNCS, vol. 3935, pp. 261–269. Springer, Heidelberg (2006)
7. Dalai, D.K., Maitra, S., Sarkar, S.: Basic theory in construction of Boolean functions with maximum possible annihilator immunity. Des. Codes, Cryptography 40(1), 41–58 (2006)
8. Li, N., Qi, W.: Construction and analysis of Boolean functions of 2t+1 variables with maximum algebraic immunity. In: Lai, X., Chen, K. (eds.) ASIACRYPT 2006. LNCS, vol. 4284, pp. 84–98. Springer, Heidelberg (2006)
9. Mihaljevic, W., Imai, H.: Cryptanalysis of Toyocrypt-HS1 stream cipher. IEICE Transactions on Fundamentals E85, 66–73 (2002)
10. Meier, W., Pasalic, E., Carlet, C.: Algebraic attacks and decomposition of Boolean functions. In: Cachin, C., Camenisch, J.L. (eds.) EUROCRYPT 2004. LNCS, vol. 3027, pp. 474–491. Springer, Heidelberg (2004)
11. Qu, L.J., Li, C., Feng, K.Q.: A note on symmetric Boolean functions with maximum algebraic immunity in odd number of variables. IEEE Transactions on Information Theory 53, 2908–2910 (2007)
12. Qu, L.J., Li, C.: On the 2^m-variable Symmetric Boolean Functions with Maximum Algebraic Immunity. Science in China Series F-Information Sciences 51, 120–127 (2008)
13. Qu, L.J., Feng, G.Z., Li, C.: On the Boolean Functions with Maximum Possible Algebraic Immunity: Construction and A Lower Bound of the Count, http://eprint.iacr.org/2005/449

Efficient Publicly Verifiable Secret Sharing with Correctness, Soundness and ZK Privacy

Kun Peng and Feng Bao

Institute for Infocomm Research, Singapore
dr.kun.peng@gmail.com

Abstract. A PVSS is a secret sharing scheme with public verification of share validity. A general PVSS must support efficient and immediate secret recovery and have no special requirement on the secret to be shared. No existing general PVSS scheme can achieve correctness, soundness, ZK privacy and practical efficiency simultaneously. A new general PVSS scheme is designed to overcome the existing drawbacks. It is correct, sound and efficient. Moreover, its public verification procedure is strict honest-verifier zero knowledge. In addition, it has an efficient and immediate secret recovery function and has no special requirement on the secret. Another contribution in this paper is that the public verification procedure has independent value.

Keywords: PVSS, honest-verifier ZK, public share verification, long challenge.

1 Introduction

In secret sharing [13], a dealer shares a secret among multiple parties such that only certain groups of the share holders can recover the secret using an efficient recovery function. VSS (verifiable secret sharing) [10,11] is a special secret sharing scheme, where the share holders can verify that they can recover a unique secret. When there is a dispute between the dealer and a share holder, PVSS (publicly verifiable secret sharing) is needed to solve the dispute without revealing the secret or any share. In PVSS, any party can use a public verification procedure to verify that a unique secret is shared among the share holders without knowledge of the secret or any share. The public verification procedure is called public share verification in this paper, which is the most important technique in PVSS. Four properties are required in public share verification in PVSS.

- Correctness: if the dealer honestly shares the secret, public share verification can be passed.
- Soundness: if public share verification is passed, a unique secret is shared among the share holders.
- Privacy: the dealer's proof in public share verification is honest-verifier zero knowledge.
- Practical efficiency: public share verification is efficient enough to be employed in practical applications.

H.Y. Youm and M. Yung (Eds.): WISA 2009, LNCS 5932, pp. 118–132, 2009.
© Springer-Verlag Berlin Heidelberg 2009

We specially emphasize privacy of public share verification. We require that public share verification must be strict honest-verifier zero knowledge. Strict honest-verifier zero knowledge is defined in this paper as follows.

- If the verifier is honest, the proof transcript can be simulated without any difference in distribution and thus no information about any share is revealed in public share verification.
- If Fiat-Shamir heuristic [4] and a hash function is employed, public share verification is a public verification revealing no information about any share.

As we will show later in this section, most existing general PVSS schemes cannot achieve strict honest-verifier zero knowledge and reveal some information about the shares in their public share verification procedure. This problem will be solved in this paper.

In this paper, only general PVSS with efficient recovery function and able to share any practical secret is studied. PVSS with delayed recovery defined in [2] and traced back to [1] is ignored as it does not support immediate efficient secret recovery. Applications of PVSS to sharing of special secret (like sharing of secret factorization in [6]) are not included. Other applications of PVSS like [15] is not included either. The special PVSS scheme in [12] can only share a secret when its discrete logarithm to a given base is known and actually shares the logarithm. Although two special formatting mechanisms are proposed in [12] to extend the message space, it is recognised in the same paper that special limitation to the message space still exists after the extension. So the PVSS in [12] is excluded.

None of the existing general PVSS schemes [14,5,2] can satisfy all the four required properties. Public share verification in [2] is not honest-verifier zero knowledge as the response in the commitment-challenge-response proof to prove validity of a share directly reveals information about the share. Distribution of the response in the proof protocol in [2] is obviously different from the distribution of a random response generated by someone without any knowledge of any share. Moreover, to achieve soundness, the public share verification in [2] requires that the response must be in a special range. This special requirement not only makes the information revealment from the response more serious but also leads to three problems. Firstly, to guarantee that the response in public verification of every share falls in the special range with a non-negligible probability, either the message space must be very small or very large modulus must be used in multiplication and exponentiation. That means either practicality or efficiency must be sacrificed. Secondly, even if the sacrifice has been made, the response may still fall out of the special range, which compromises correctness and leads to rewinding of the proof. Thirdly, the range test is not precise such that either some valid secret or share cannot pass the range test or some invalid secret and share can pass the range test. In [14] there are two PVSS protocols, one based on discrete logarithm and the other based on e^{th} root. The eth root based protocol employs an inappropriate response in its public share verification, which reveals information about the share and compromises privacy of PVSS like in [2]. The discrete logarithm based protocol only supports 1-bit-long challenges and has to be run for many times to guarantee strong soundness. Moreover, double exponentiation operations employed in the discrete logarithm based protocol is more costly than normal exponentiation operations. So the discrete logarithm based protocol lacks efficiency and practicality. The PVSS scheme in [5] has a similar problem with the eth root based PVSS protocol in [14] and does not support strict honest-verifier zero knowledge privacy.

Actually, privacy receives too little attention in existing general PVSS schemes. It is recognised in [14] that it is difficult to prove privacy of its eth root based PVSS protocol, which remains an open question. It is claimed in [2] that its public share verification is "zero-knowledge" without any proof or explanation. If zero-knowledge in [2] means strict honest-verifier zero knowledge, the claim is obviously wrong. In fact, even computational honest-verifier zero knowledge is not guaranteed in the public share verification in [2] as it is unknown how to simulate its proof transcript. In [5], the public share verification protocol is presented in the form of a few proof primitives. Some of the primitives are claimed to be perfect witness hiding, statistical witness hiding or statistical witness indistinguishable. The claims are unsystematic, not proved and some of them are based on inappropriate conditions (e.g. a variable must be in a certain range, which is difficult to precisely verify). There is no formal and convincing guarantee of any kind of zero knowledge of public share verification in [5]. Although we believe public share verification in the discrete logarithm based PVSS protocol in [14] is perfect zero knowledge, as stated before it only supports one-bit challenge and is impractical.

In this paper, a new PVSS scheme is designed to overcome the existing problems. The new scheme employs a discrete-logarithm-based commitment function to commit the secret and Paillier encryption algorithm to encrypt the shares. A novel zero knowledge proof protocol is designed to publicly prove that the encrypted shares can be used to recover a unique secret. The new PVSS scheme has a few advantages over the existing PVSS schemes and related techniques. Firstly, it has an efficient and immediate secret recovery function and has no special requirement on the secret. Secondly, it achieves correctness and soundness with an assumption called logarithm-root assumption defined in Section 3. Thirdly, its public share verification procedure is strict honest-verifier zero knowledge and its zero knowledge property is formally proved. Finally, it employs a long challenge and is efficient. The new PVSS scheme is presented in two steps. In the first step, a basic protocol is designed to satisfy parts of the desired requirements. In the second step, the basic protocol is optimised into an advanced protocol to achieve all the desired requirements. For convenience of demonstrating honest-verifier zero knowledge, the two protocols are described in the form of interactive proof protocols. Obviously they can be transformed into non-interactive protocols using the well known Fiat-Shamir heuristic [4] and a hash function such that they are publicly verifiable.

An additional contribution in this paper is that the public share verification procedure in the new PVSS scheme can be employed in any cryptographic application where it is required to publicly prove and verify that the same secret is committed in a discrete-logarithm-based commitment and encrypted as an exponent in an encryption algorithm (e.g. Paillier encryption). Namely, it is useful when two logarithms must be proved to be equal where the orders of their bases are different. Even when one or two of the two orders are unknown, the new proof and verification mechanism can still work. The new proof and verification mechanism is efficient (as it employs a long challenge and does not need to be repeated) and strict honest-verifier zero knowledge while there exists no efficient strict honest-verifier zero knowledge solution to this kind of proof.

2   Parameters and Symbols

The sharing dealer is D, who shares a secret among a set of share holders $A = \{A_1, A_2, \ldots, A_n\}$. The sharing threshold is t such that any subset of A with size $t + 1$ can reconstruct the secret while any subset of A with a size smaller than $t + 1$ cannot obtain any information about the secret. The following parameters and symbols are used in this paper.

- Large primes p and q of similar size are generated where $p = 2p'$, $q = 2q'$ and p' and q' are primes. $N = pq$ is published but its factorization is concealed. Integer g with order $m = 2p'q'$ modulo N is published. These parameters can be generated using a polynomial algorithm by a trusted party.
- The message space is Z_m and any integer in it can be shared. D can share any integer with practical sense as m is much larger than any practical message (e.g. m is more than 1024 bits long).
- Each A_i sets up Paillier encryption for $i = 1, 2, \ldots, n$. The public key of A_i consists of integers N_i and g_i where $N_i = p_i q_i$, $N_i > (t + 1)mn^t$ and p_i, q_i are secret large primes. The encryption function for A_i is $E_i()$ where $E_i(m) = g_i^m r^{N_i} \bmod N_i^2$ and r is randomly chosen from $Z_{N_i}^*$. More detials of key generation (e.g. choice of g_i) can be found in [8].
- L is a security parameter. In this paper, an L-bit challenge will be used in a three-step zero knowledge proof protocol. It is required that 2^L must be smaller than p, q, any p_i and q_i. On the other hand, L cannot be too small as 2^{-L} must be small enough to guarantee soundness of the ZK proof. In practice, these two requirements can be easily satisfied simultaneously. For example, p, q, p_i and q_i should be at least 1024 bits long for $i = 1, 2, \ldots, n$. When L is 128, $2^L < p$, $2^L < q$, $2^L < p_i$ and $2^L < q_i$ for $i = 1, 2, \ldots, n$ and 2^{-L} is small enough to guarantee very strong soundness of the ZK proof protocol.
- $x \in_u S$ means x is uniformly distributed in set S.

3   Security Fundamentals

It is obvious that order problem defined as follows is hard.

Definition 1. *Order problem: given g and N a polynomial party without knowledge of factorization of N has to find with a non-negligible probability a non-zero integers x such that $g^x = 1 \bmod N$.*

The reason is simple. If a non-zero integer x is found to satisfy $g^x = 1 \bmod N$, then as explained in [3,7] N can be factorized in polynomial time. The following security assumption is used in this paper.

Definition 2. *Logarithm-root assumption: the probability that a polynomial algorithm can calculate integers s, x_1 and x_2 to satisfy $g^s x_1^N / x_2^{N_i} = 1 \bmod N$ without knowledge of factorization of N is negligible unless $s = 0$, $x_1 = 1$ and $x_2 = 1$.*

4   The Basic PVSS Scheme

When the dealer is trusted, the basic PVSS scheme works like all the existing secret sharing schemes and a secret can be shared and reconstructed. When the dealer is not trusted, each share holder can verify validity of its share. When there is dispute about validity of a share between a share holder and the dealer, any observer can publicly verify validity of the share, which is not revealed.

4.1   Secret Sharing and Reconstruction

Secret sharing and reconstruction in the new secret sharing scheme are the same as those in most existing threshold secret sharing schemes (e.g. [13]). A secret s is shared by D among $A_1, A_2, \ldots, A_n$ as follows.

1. Share generation and distribution
   (a) D randomly chooses integers $f_1, f_2, \ldots, f_t$ and generates a polynomial $F(x) = \sum_{j=0}^{t} f_j x^j$ where $f_0 = s$.
   (b) D generates $s_i = F(i)$ as A_i's share for $i = 1, 2, \ldots, n$.
   (c) D sends $c_i = E_i(s_i) = g_i^{s_i} r_i^{N_i} \bmod N_i^2$ to A_i for $i = 1, 2, \ldots, n$ where r_i is randomly chosen from $Z_{N_i}^*$.
2. Secret reconstruction
   (a) Each share holder A_i decrypts c_i and obtains its share s_i.
   (b) Any $t + 1$ sharers can be used to reconstruct the secret: $s = \sum_{i \in S} s_i u_i$ where $u_i = \prod_{j \in S, j \neq i} \frac{j}{j-i}$ and S contains the indices of the $t + 1$ shares.

The secret sharing procedure and secret reconstruction procedure have the following two properties like in the existing threshold secret sharing schemes under the assumption that it is hard to break the employed encryption algorithm.

- A dealer can share its secret among n share holders such that no information about the secret is revealed if the number of cooperating share holders is no more than t.
- If $t + 1$ shares are put together, the secret can be reconstructed.

4.2   Share Verification

The secret sharing protocol in Section 4.1 works only when D is honest and does not deviate from the secret sharing procedure. If D may deviate from the secret sharing procedure, there is no guarantee that a unique secret can be reconstructed from any $t + 1$ shares. So the share holders need to verify validity of their shares when they do not trust the dealer. A share verification procedure is designed as follows.

- D publishes sharing commitments $C_j = g^{f_j} \bmod N$ for $j = 0, 1, \ldots, t$.
- Each A_i verifies $g^{s_i} = \prod_{j=0}^{t} C_j^{i^j} \bmod N$. If the verification succeeds, A_i accepts s_i. Otherwise, it rejects c_i as an invalid encrypted share.

This share verification procedure is the same as that in the VSS scheme in [9] except that Paillier encryption is specified in the new scheme to construct the confidential communication channel between the dealer and the share holders. Soundness of the share verification procedure is illustrated in Theorem 1. The claim in Theorem 1 has been proved in [9] in its parameter setting. The trivial difference between the two schemes does not prevent the theorem from being applied to the parameter setting in this paper. So Theorem 1 is not proved and interested readers are referred to [9]. If information-theoretical privacy is required in the commitment procedure, the information-theoretically hiding commitment function in [10] is supported in the new PVSS scheme as well. The only difference is that with the commitment function in [10], the PVSS scheme will become a little more complex. For simplicity, the unconditional-binding and computational-hiding commitment function presented above is employed when describing the new PVSS scheme.

Theorem 1. *If there are $t + 1$ shares such that each of them s_i satisfies $g^{s_i} = \prod_{j=0}^{t} C_j^{i^j} \bmod N$, then a unique secret can be reconstructed as $\sum_{i \in S} s_i u_i$ where $u_i = \prod_{j \in S, j \neq i} \frac{j}{j-i}$ and S contains the indices of the $t + 1$ shares.*

4.3   Public Share Verification

The main advantage of our new scheme lies in its new public share verification mechanism, by which any observer can publicly verify validity of c_i when there is a dispute about its validity between A_i and D. The new public verification mechanism does not reveal any information about s_i, so that s_i need not to be revealed to solve the dispute. In other words, the new public share verification mechanism enables a public verification that the share encrypted in c_i is a correct share of the secret committed in the commitments. More precisely, two logarithms must be publicly verified to be equal where the orders of their bases are different and the two orders are unknown. As stated in Section 1, the public proof and verification of equality of such two logarithms in [2] cannot be employed due to its breach of privacy and drawbacks in the range test. A new public share verification mechanism is designed as follows. The idea in the new mechanism is to employ a certain modulus when calculating the response

1. D randomly chooses integers $r \in Z$, $t_1 \in Z_N$ and $t_2 \in Z_{N_i}^*$ and publishes

$$a_1 = g^r t_1^{N_i} \bmod N$$
$$a_2 = g_i^r t_2^{N_i} \bmod N_i^2$$

2. A verifier publishes a random L-bit integer

$$c$$

3. D publishes

$$w = r - cs_i \bmod N_i$$
$$v_1 = t_1 g^{(r-cs_i)\%N_i} \bmod N$$
$$v_2 = (t_2/r_i^c) g_i^{(r-cs_i)\%N_i} \bmod N_i^2$$

where % stands for quotient in calculation of division.

Public verification:

$$g^w v_1^{N_i} (\textstyle\prod_{j=0}^{t} C_j^{i^j})^c = a_1 \bmod N \tag{1}$$
$$g_i^w v_2^{N_i} c_i^c = a_2 \bmod N_i^2 \tag{2}$$

Fig. 1. The first public proof and verification Protocol

in the commitment-challenge-response proof protocol such that the response does not reveal any information about the share. As the two orders of the two bases are unknown and thus the employed modulus is neither of them, there is a certain deviation in the response. To counteract the deviation, additional responses are generated. Both responses are used in the verification such that correctness and soundness can be achieved. When this novel proof technique is employed, two parallel three-step proof protocols are needed, which are detailed in Figure 1 and Figure 2 respectively. The proof protocol in Figure 1 guarantees that D can calculate integers s_i, r_i' and r_i in polynomial time such that $\prod_{j=0}^{t} C_j^{i^j} = g^{s_i} r'^{N_i} \bmod N$ and $c_i = g_i^{s_i} r_i^{N_i} \bmod N_i^2$. Obviously, that is not enough to ensure the share encrypted in c_i is valid. So the proof protocol in Figure 2 is employed, which together with the proof protocol in Figure 1 guarantees that the same share s_i is committed in the commitment variables and encrypted in c_i.

4.4 Analysis

We focus on public share verification of the new PVSS scheme and prove the following theorems.

Theorem 2. *The proof protocol in Figure 1 is correct. More precisely, when D is honest and strictly follows the protocol, he can pass the verification in (1) and (2).*

1. D randomly chooses integers $r \in Z$, $t \in Z_N$ and publishes

$$a = g^r t^N \bmod N$$

2. A verifier publishes a random L-bit integer

$$c$$

3. D publishes

$$w = r - cs_i \bmod N$$
$$v = tg^{(r-cs_i)\%N} \bmod N$$

where % stands for quotient in calculation of division.

Public verification:

$$g^w v^N (\textstyle\prod_{j=0}^{t} C_j^{i^j})^c = a \bmod N \qquad (3)$$

Fig. 2. The second public proof and verification Protocol

Proof: When D is honest and strictly follows the protocol,

$$g^w v_1^{N_i} (\textstyle\prod_{j=0}^{t} C_j^{i^j})^c = g^{(r-cs_i) \bmod N_i} (t_1 g^{(r-cs_i)\%N_i})^{N_i} (g^{s_i})^c$$
$$= g^{((r-cs_i) \bmod N_i)+((r-cs_i)\%N_i)N_i} t_1^{N_i} g^{cs_i}$$
$$= g^{r-cs_i} t_1^{N_i} g^{cs_i} = g^r t_1^{N_i} = a_1 \bmod N$$

and

$$g_i^w v_2^{N_i} c_i^c = g_i^{(r-cs_i) \bmod N_i} (t_2 g_i^{(r-cs_i)\%N_i} / r_i^c)^{N_i} g_i^{cs_i} r_i^{cN_i}$$
$$= g_i^{((r-cs_i) \bmod N_i)+((r-cs_i)\%N_i)N_i} t_2^{N_i} r_i^{-cN_i} g_i^{cs_i} r_i^{cN_i}$$
$$= g_i^{r-cs_i} t_2^{N_i} r_i^{-cN_i} g_i^{cs_i} r_i^{cN_i} = g_i^r t_2^{N_i} = a_2 \bmod N_i^2 \qquad \square$$

Theorem 3. *The proof protocol in Figure 2 is correct. More precisely, when D is honest and strictly follows the protocol, he can pass the verification in (3).*

Proof: When D is honest and strictly follows the protocol,

$$g^w v^N (\textstyle\prod_{j=0}^{t} C_j^{i^j})^c = g^{(r-cs_i) \bmod N} (tg^{(r-cs_i)\%N})^N (g^{s_i})^c$$
$$= g^{((r-cs_i) \bmod N)+((r-cs_i)\%N)N} t^N g^{cs_i}$$
$$= g^{r-cs_i} t^N g^{cs_i} = g^r t^N = a \bmod N \qquad \square$$

Theorem 4. *The proof protocol in Figure 1 is strict honest-verifier zero knowledge.*

Proof: Any party without any knowledge about s_i or r_i can simulate the proof transcript a_1, a_2, c, w, v_1, v_2 to satisfy the verification equations in Figure 1 as follows.

1. Randomly chooses c from $\{0, 1, \ldots, 2^L - 1\}$, w from Z_{N_i}, v_1 from Z_N and v_2 from $Z_{N_i^2}^*$.
2. Calculates $a_1 = g^w v_1^{N_i} (\prod_{j=0}^t C_j^{i^j})^c \bmod N$ and $a_2 = g_i^w v_2^{N_i} c_i^c \bmod N_i^2$.

In both the proof transcript in Figure 1 and the simulated transcript, the following and only the following conditions are met.

- a_1 is in the form of $g^{z_1} z_2^{N_i} \bmod N$ where $z_1 \in_u Z$, $z_2 \in_u Z_N$;
- a_2 is uniformly distributed in the ciphertext space of A_i's Paillier encryption system;
- c is uniformly distributed in $\{0, 1, \ldots, 2^L - 1\}$;
- w is uniformly distributed in Z_{N_i};
- v_1 is uniformly distributed in Z_N;
- v_2 is uniformly distributed in $Z_{N_i^2}^*$;
- Equations (1) and (2) are satisfied.

So the simulated transcript and the proof transcript in Figure 1 have the same distribution if the verifier randomly chooses c in the protocol in Figure 1.   □

Theorem 5. *The proof protocol in Figure 2 is strict honest-verifier zero knowledge.*

Proof: Any party without any knowledge about s_i can simulate the proof transcript a, c, w, v to satisfy the verification equation in Figure 2 as follows.

1. Randomly chooses c from $\{0, 1, \ldots, 2^L - 1\}$, w from Z_N and v from Z_N.
2. Calculates $a = g^w v^N (\prod_{j=0}^t C_j^{i^j})^c \bmod N$.

In both the proof transcript in Figure 2 and the simulated transcript, the following and only the following conditions are met.

- a is in the form of $g^{z_1} z_2^N \bmod N$ where $z_1 \in_u Z$, $z_2 \in_u Z_N$;
- c is uniformly distributed in $\{0, 1, \ldots, 2^L - 1\}$;
- w is uniformly distributed in Z_N;
- v is uniformly distributed in Z_N;
- Equation (3) is satisfied.

So the simulated transcript and the proof transcript in Figure 2 have the same distribution if the verifier randomly chooses c in the protocol in Figure 2.   □

Theorem 6. *If D can pass the verification in Figure 1 and Figure 2 with a non-negligible probability, it can calculate in polynomial time s_i and r_i such that $\prod_{j=0}^t C_j^{i^j} = g^{s_i} \bmod N$ and $c_i = g_i^{s_i} r_i^{N_i} \bmod N_i^2$.*

To prove Theorem 6, two lemmas are proved first.

Lemma 1. *If in the proof protocol in Figure 2 given two different challenges c and c' to the same commitment a, D can provide two responses (w, v) and (w', v') to pass the verification respectively, then D can calculate integers $\hat{s}_i$ and $\hat{r}_i$ in polynomial time such that $\prod_{j=0}^t C_j^{i^j} = g^{\hat{s}_i} \hat{r}_i^N \bmod N$.*

Proof: That given two different challenges c and c' to the same commitment a the dealer can provide two responses (w, v) and (w', v') to pass the verification respectively implies

$$g^w v^N (\textstyle\prod_{j=0}^t C_j^{i^j})^c = a \bmod N \tag{4}$$

$$g^{w'} v'^N (\textstyle\prod_{j=0}^t C_j^{i^j})^{c'} = a \bmod N \tag{5}$$

(4) divided by (5) yields

$$g^{w-w'} (v/v')^N = (\textstyle\prod_{j=0}^t C_j^{i^j})^{c'-c} \bmod N \tag{6}$$

D can use the Euclidean algorithm to calculate α and β in polynomial time such that $\beta(c' - c) = \alpha N + GCD(N, c' - c)$. Note that $N = pq$, p and q are primes, $c < 2^L$, $c' < 2^L$, $2^L < p$ and $2^L < q$. So $-2^L < c' - c < 2^L$ and $GCD(N, c' - c) = 1$. Thus,

$$(6) \implies g^{\beta(w-w')}((v/v')^\beta)^N = (\textstyle\prod_{j=0}^t C_j^{i^j})^{\beta(c'-c)} = (\textstyle\prod_{j=0}^t C_j^{i^j})^{\alpha N+1} \bmod N$$

$$\implies g^{\beta(w-w')}((v/v')^\beta)^N / (\textstyle\prod_{j=0}^t C_j^{i^j})^{\alpha N} = \textstyle\prod_{j=0}^t C_j^{i^j} \bmod N$$

$$\implies g^{\beta(w-w')}((v/v')^\beta / (\textstyle\prod_{j=0}^t C_j^{i^j})^\alpha)^N = \textstyle\prod_{j=0}^t C_j^{i^j} \bmod N$$

Therefore, D can calculate integers $\hat{s}_i = \beta(w-w')$ and $\hat{r}_i = (v_1/v')^\beta / (\textstyle\prod_{j=0}^t C_j^{i^j})^\alpha$ $\bmod N$ in polynomial time such that $\textstyle\prod_{j=0}^t C_j^{i^j} = g^{\hat{s}_i} \hat{r}_i^N \bmod N$.   □

Lemma 2. *If in the proof protocol in Figure 1 given two different challenges c and c' to the same commitment (a_1, a_2), D can provide two responses (w, v_1, v_2) and (w', v'_1, v'_2) to pass the verification respectively, then D can calculate integers s_i, r'_i and r_i in polynomial time such that $\textstyle\prod_{j=0}^t C_j^{i^j} = g^{s_i} r'^{N_i}_i \bmod N$ and $c_i = g_i^{s_i} r_i^{N_i} \bmod N_i^2$.*

Proof: That given two different challenges c and c' to the same commitment (a_1, a_2) the dealer can provide two responses (w, v_1, v_2) and (w', v'_1, v'_2) to pass the verification respectively implies

$$g^w v_1^{N_i} (\textstyle\prod_{j=0}^t C_j^{i^j})^c = a_1 \bmod N \tag{7}$$

$$g_i^w v_2^{N_i} c_i^c = a_2 \bmod N_i^2 \tag{8}$$

$$g^{w'} v'^{N_i}_1 (\textstyle\prod_{j=0}^t C_j^{i^j})^{c'} = a_1 \bmod N \tag{9}$$

$$g_i^{w'} v'^{N_i}_2 c_i^{c'} = a_2 \bmod N_i^2 \tag{10}$$

(7) divided by (9) yields

$$g^{w-w'} (v_1/v'_1)^{N_i} = (\textstyle\prod_{j=0}^t C_j^{i^j})^{c'-c} \bmod N \tag{11}$$

(8) divided by (10) yields

$$g_i^{w-w'} (v_2/v'_2)^{N_i} = c_i^{c'-c} \bmod N_i^2 \tag{12}$$

D can use the Euclidean algorithm to calculate α and β in polynomial time such that $\beta(c' - c) = \alpha N_i + GCD(N_i, c' - c)$. Note that $N_i = p_i q_i$, p_i and q_i are primes, $c < 2^L$, $c' < 2^L$, $2^L < p_i$ and $2^L < q_i$. So $-2^L < c' - c < 2^L$ and $GCD(N_i, c' - c) = 1$. Thus,

$$(11) \implies g^{\beta(w - w')}((v_1/v'_1)^\beta)^{N_i} = (\textstyle\prod_{j=0}^t C_j^{ij})^{\beta(c' - c)} = (\textstyle\prod_{j=0}^t C_j^{ij})^{\alpha N_i + 1} \bmod N$$
$$\implies g^{\beta(w - w')}((v_1/v'_1)^\beta)^{N_i}/(\textstyle\prod_{j=0}^t C_j^{ij})^{\alpha N_i} = \textstyle\prod_{j=0}^t C_j^{ij} \bmod N$$
$$\implies g^{\beta(w - w')}((v_1/v'_1)^\beta/(\textstyle\prod_{j=0}^t C_j^{ij})^\alpha)^{N_i} = \textstyle\prod_{j=0}^t C_j^{ij} \bmod N$$

$$(12) \implies g_i^{\beta(w - w')}((v_2/v'_2)^\beta)^{N_i} = c_i^{\beta(c' - c)} = c_i^{\alpha N_i + 1} \bmod N_i^2$$
$$\implies g_i^{\beta(w - w')}((v_2/v'_2)^\beta)^{N_i}/c_i^{\alpha N_i} = c_i \bmod N_i^2$$
$$\implies g_i^{\beta(w - w')}((v_2/v'_2)^\beta/c_i^\alpha)^{N_i} = c_i \bmod N_i^2$$

Therefore, D can calculate integers $s_i = \beta(w - w')$, $r'_i = (v_1/v'_1)^\beta/(\prod_{j=0}^t C_j^{ij})^\alpha \bmod N$ and $r_i = (v_2/v'_2)^\beta/c_i^\alpha \bmod N_i^2$ in polynomial time such that $\prod_{j=0}^t C_j^{ij} = g^{s_i} r'_i^{N_i} \bmod N$ and $c_i = g_i^{s_i} r_i^{N_i} \bmod N_i^2$. □

Proof of Theorem 6: Lemma 1 and Lemma 2 illustrate that if D can pass the public share verification with a non-negligible probability it can calculate integers $\hat{s}_i$, $\hat{r}_i$, s_i, r'_i and r_i in polynomial time such that $\prod_{j=0}^t C_j^{ij} = g^{\hat{s}_i}\hat{r}_i^N \bmod N$, $\prod_{j=0}^t C_j^{ij} = g^{s_i} r'_i^{N_i} \bmod N$ and $c_i = g_i^{s_i} r_i^{N_i} \bmod N_i^2$.
So
$$g^{\hat{s}_i}\hat{r}_i^N = g^{s_i} r'_i^{N_i} \bmod N$$

Namely,

$$g^{\hat{s}_i - s_i}\hat{r}_i^N/r'_i^{N_i} = 1 \bmod N$$

According to logarithm-root assumption, $g^{\hat{s}_i - s_i} = 1 \bmod N$, $\hat{r}_i^N = 1 \bmod N$ and $r'_i^{N_i} = 1 \bmod N$. Therefore, $\prod_{j=0}^t C_j^{ij} = g^{s_i} \bmod N$. □

Theorem 2 and Theorem 3 guarantee that the proposed public share verification mechanism is correct. Theorem 4 and Theorem 5 guarantee that the proposed public share verification mechanism is strict honest-verifier zero knowledge. The proposed public share verification mechanism can be transformed into a non-interactive protocol using Fiat-Shamir heuristic [4] and a hash function such that public verification is implemented. Unfortunately, guarantee of soundness by Theorem 6 is incomplete. When D is honest and strictly follows the proposed PVSS protocol, $s_i \leq (t + 1)mn^t < N_i$ and public share verification is sound as illustrated in Theorem 6. However, if D deviates from the PVSS protocol and $s_i \geq N_i$, the share A_i obtains after decryption of c_i is not s_i and thus inconsistent with the committed secret. This problem will be solved in next section.

5   The Advanced PVSS Scheme

As illustrated in last section, when D is dishonest and generates an s_i larger than N_i, the public share verification mechanism in the basic PVSS scheme cannot guarantee that the share decrypted from c_i is the committed share. To solve this problem, the advanced PVSS scheme is proposed as follows.

- The secret sharing procedure, secret commitment procedure, share verification procedure and secret reconstruction procedure are the same as in the basic PVSS scheme.
- Public share verification procedure is upgraded as follows.
    1. D runs the proof protocols in Figure 1 and Figure 2.
    2. D publishes $D_i = g^{N_i - s_i} \bmod N$ and runs the proof protocols in Figure 3.
    3. Anyone can publicly verify

$$D_i \prod_{j=0}^{t} C_j^{i^j} = g^{N_i} \bmod N \qquad (13)$$

The upgraded public share verification mechanism can be transformed into a non-interactive protocol using Fiat-Shamir heuristic [4] and a hash function such that public verification is implemented. The proof protocol is Figure 3 is very similar to the proof protocol is Figure 2 and the only difference is that the secret s_i is replaced with $N_i - s_i$ and the commitment $\prod_{j=0}^{t} C_j^{i^j}$ is replaced with D_i.

Theorem 7. *An honest D can pass the proof protocol is Figure 3.*

Theorem 8. *The proof protocol in Figure 3 is strict honest-verifier zero knowledge.*

1. D randomly chooses integers $r \in Z$, $t \in Z_N$ and publishes

$$a = g^r t^N \bmod N$$

2. A verifier publishes a random L-bit integer

$$c$$

3. D publishes

$$w = r - c(N_i - s_i) \bmod N$$
$$v = tg^{(r-c(N_i-s_i))\%N} \bmod N$$

where % stands for computation of quotient.

Public verification:

$$g^w v^N D_i^c = a \bmod N$$

Fig. 3. The additional public proof and verification Protocol in the advanced PVSS

Proof of Theorem 7 is similar to that of Theorem 3 and is not repeated. Proof of Theorem 8 is similar to that of Theorem 5 and is not repeated. Thus, the upgraded public share verification is correct and strict honest-verifier zero knowledge.

Theorem 9. *The public share verification mechanism in the advanced PVSS scheme is sound. More precisely, if verification in Figure 1, Figure 2, Figure 3 and (13) is passed with a non-negligible probability, then a unique secret can be reconstructed from the messages in any $t + 1$ encrypted shares.*

To prove Theorem 9, a lemma is proved first.

Lemma 3. *If in the proof protocol in Figure 3 given two different challenges c and c' to the same commitment a, D can provide two responses (w, v) and (w', v') to pass the verification respectively, then D can calculate integers s_i' and d_i in polynomial time such that $D_i = g^{s_i'} d_i^N \bmod N$.*

Proof of Lemma 3 is similar to that of Lemma 1 and is not repeated.

Proof of Theorem 9: According to Theorem 6, D can calculate in polynomial time s_i and r_i such that $\prod_{j=0}^{t} C_j^{i^j} = g^{s_i} \bmod N$ and $c_i = g_i^{s_i} r_i^{N_i} \bmod N_i^2$. According to Lemma 3, D can calculate s_i' and d_i in polynomial time such that $D_i = g^{s_i'} d_i^N \bmod N$. So D can calculate $s_i + s_i'$ and d_i in polynomial time such that

$$D_i \prod_{j=0}^{t} C_j^{i^j} = g^{s_i + s_i'} d_i^N \bmod N \tag{14}$$

(14) and (13) imply that

$$g^{s_i + s_i' - N_i} d_i^N = 1 \bmod N$$

So according to logarithm-root assumption, $s_i + s_i' = N_i \bmod m$. Namely $s_i + s_i' = N_i + km$ where k is an integer.

If $k \neq 0$, then D can calculate in polynomial time a non-zero integer km such that $g^{km} = 1 \bmod N$. This is contradictory to the fact that order problem is hard. So $k = 0$ and $s_i < N_i$. Thus, $D(c_i) = s_i$ where $\prod_{j=0}^{t} C_j^{i^j} = g^{s_i} \bmod N$. So $C_0 = g^{\sum_{i \in S} D(c_i) u_i} \bmod N$ for any S containing the indices of $t + 1$ shares where $u_i = \prod_{j \in S, j \neq i} \frac{j}{j-i}$. So for any two sets S_1 and S_2, each containing the indices of $t + 1$ shares, $\sum_{i \in S_1} D(c_i) u_i = \sum_{i \in S_2} D(c_i) u_i + k'm$ where k' is an integer. If $k' \neq 0$, the dealer can calculate in polynomial time a non-zero integer $k'm = \sum_{i \in S_1} D(c_i) u_i - \sum_{i \in S_2} D(c_i) u_i$ such that $g^{k'm} = 1 \bmod N$ as it knows all the shares. This is contradictory to the fact that order problem is hard. So $k' = 0$ and $\sum_{i \in S_1} D(c_i) u_i = \sum_{i \in S_2} D(c_i) u_i$. Therefore, a unique secret can be reconstructed from the messages in any $t + 1$ encrypted shares.   □

6   Conclusion

A new PVSS scheme is proposed to overcome the existing problems in PVSS. It has an efficient and immediate secret recovery function and has no special requirement on the secret, which is an advantage over the non-general PVSS schemes. Correctness, soundness, strict honest-verifier zero knowledge public verification and practical efficiency are achieved with a reasonable security assumption in the new PVSS scheme, which is an advantage over the existing general PVSS schemes. If information-theoretical privacy is required in the commitment procedure, the information-theoretically hiding commitment function in [10] can be employed and the public share verification procedure can be slightly modified to suit the change. The public share verification procedure in the new PVSS scheme has independent value as it is the only efficient and strict honest-verifier zero knowledge solution to cryptographic applications where it is required to publicly prove and verify that the same secret is committed in a discrete-logarithm-based commitment and encrypted as an exponent in an encryption algorithm (e.g. Paillier encryption).

References

1. Bellare, M., Goldwasser, S.: Verifiable partial key escow. In: ACCCS 1997, pp. 78–91 (1997)
2. Boudot, F., Traore, J.: Efficient public verifiable secret sharing schemes with fast or delayed recovery. In: Varadharajan, V., Mu, Y. (eds.) ICICS 1999. LNCS, vol. 1726, pp. 87–102. Springer, Heidelberg (1999)
3. DeLaurentis, J.: A futher weakness in the common modulus protocol for the rsa cryptoalgorithm. Cryptologia 8, 253–259 (1984)
4. Fiat, A., Shamir, A.: How to prove yourself: practical solutions to identification and signature problems. In: Odlyzko, A.M. (ed.) CRYPTO 1986. LNCS, vol. 263, pp. 186–194. Springer, Heidelberg (1987)
5. Fujisaki, E., Okamoto, T.: A practical and provably secure scheme for publicly verifiable secret sharing and its applications. In: Nyberg, K. (ed.) EUROCRYPT 1998. LNCS, vol. 1403, pp. 32–46. Springer, Heidelberg (1998)
6. Mao, W.: Guaranteed correct sharing of integer factorization with off-line shareholders. In: Imai, H., Zheng, Y. (eds.) PKC 1998. LNCS, vol. 1431, p. 60. Springer, Heidelberg (1998)
7. Miller, G.: Riemann's hypothesis and tests for primality. Journal of Computer and System Science 13, 300–317 (1976)
8. Paillier, P.: Public key cryptosystem based on composite degree residuosity classes. In: Stern, J. (ed.) EUROCRYPT 1999. LNCS, vol. 1592, pp. 223–238. Springer, Heidelberg (1999)
9. Pedersen, T.: Distributed provers with applications to undeniable signatures. In: Davies, D.W. (ed.) EUROCRYPT 1991. LNCS, vol. 547, pp. 221–242. Springer, Heidelberg (1991)
10. Pedersen, T.: Non-interactive and information-theoretic secure verifiable secret sharing. In: Davies, D.W. (ed.) EUROCRYPT 1991. LNCS, vol. 547, pp. 129–140. Springer, Heidelberg (1991)

11. Pedersen, T.: Distributed Provers and Verifiable Secret Sharing Based on the Discrete Logarithm Problem. PhD thesis, Computer Science Department, Aarhus University,Aarhus, Denmark (1992)
12. Schoenmakers, B.: A simple publicly verifiable secret sharing scheme and its application to electronic voting. In: Wiener, M. (ed.) CRYPTO 1999. LNCS, vol. 1666, pp. 149–164. Springer, Heidelberg (1999)
13. Shamir, A.: How to share a secret. Communication of the ACM 22(11), 612–613 (1979)
14. Stadler, M.: Publicly verifiable secret sharing. In: Maurer, U.M. (ed.) EUROCRYPT 1996. LNCS, vol. 1070, pp. 190–199. Springer, Heidelberg (1996)
15. Young, A., Yung, M.: Auto-recoverable auto-certifiable cryptosystems. In: Nyberg, K. (ed.) EUROCRYPT 1998. LNCS, vol. 1403, pp. 17–31. Springer, Heidelberg (1998)

ID-Based Adaptive Oblivious Transfer

Fangguo Zhang[1], Xingwen Zhao[1], and Xiaofeng Chen[2]

[1] School of Information Science and Technology,
Sun Yat-Sen University, Guangzhou 510275, P.R. China
Guangdong Key Laboratory of Information Security Technology
isszhfg@mail.sysu.edu.cn
[2] Key Laboratory of Computer Networks and Information Security,
Ministry of Education, Xidian University, Xi'an 710071, P.R. China

Abstract. The adaptive oblivious transfer is a variant of oblivious transfer. To the best of our knowledge, the existing adaptive oblivious transfer schemes are all constructed in the setting of public key infrastructure. In this paper, we first introduce the notion of ID-based adaptive oblivious transfer. We also show that a generic construction of ID-based adaptive oblivious transfer can be obtained if there exists an ID-based unique blind signature. Another main contribution in this paper is that we propose a provable secure round-optimal ID-based unique blind signature scheme and the resulting ID-based adaptive oblivious transfer scheme.

Keywords: ID-based cryptosystem, Adaptive, Oblivious transfer, Blind signature.

1   Introduction

Oblivious transfer (OT) is a cryptographic primitive for designing secure protocols, which is introduced by Rabin [19], and extended by Even et al. [12], and Brassard et al. [8] respectively. It considers the scenario in which Alice (as a sender) has some information and Bob (as a receiver/chooser) is interested in obtaining part of these information such that Alice cannot figure out which part is given to Bob. Meanwhile, Bob must not learn anything about the rest of the information. The oblivious transfer is so useful that many researchers concentrate on it and propose plenty of schemes [1,2,21,18,16,11,17,15].

The adaptive oblivious transfer means that the sender has N messages and the receiver can choose to receive k of them one-after-the-other. That is to say, the receiver may obtain M_{i-1} before deciding on M_i (i.e., the i-th value may depend on the first $i - 1$ values). The first adaptive oblivious transfer scheme is given by Naor and Pinkas [17]. Recently, Camenisch et al. [9] propose two adaptive oblivious transfer schemes. They also presented a generic solution for constructing adaptive oblivious transfer from unique blind signature schemes (UBSS). Later, Green et al. [14] propose another adaptive oblivious transfer by using the blind version of extracting algorithm from identity-based encryption (IBE) scheme.

H.Y. Youm and M. Yung (Eds.): WISA 2009, LNCS 5932, pp. 133–147, 2009.
© Springer-Verlag Berlin Heidelberg 2009

Identity-based (ID-based) cryptography, introduced by Shamir [20], is a public-key system in which an arbitrary string can be used as the public key, such as email addresses or IP addresses. As a result, identity-based cryptography significantly reduces the system complexity and the cost for establishing and managing the public key authentication framework known as Public Key Infrastructure (PKI).

In this paper, we first introduce the notion of ID-based adaptive oblivious transfer, which is useful in some certain applications. For example, when the ID-based oblivious transfer is used to sell digital goods, a buyer can use the ID of vendor to encrypt the index of the goods. Our generic construction of ID-based oblivious transfer follows the solution of Camenisch et al. [9]. More precisely, we propose a round-optimal ID-based unique blind signature scheme and then construct the adaptive oblivious transfer from it. We argue that Green et al. [14]'s construction from IBE is much different with ours since their OT scheme is not ID-based. Due to the advantage of ID-based cryptosystem, the ID-based adaptive oblivious transfer schemes proposed in this paper can be used to design the digital goods selling protocol[1] and the private information retrieval scheme(PIR) [10] under the ID-based setting.

The rest of this paper is organized as follows: In section 2 we introduce some preliminaries. In section 3, we describe the model and security notions for ID-based adaptive oblivious transfer. In section 4, we present the generic construction of ID-based adaptive oblivious transfer. In section 5, we first present a new ID-based unique blind signature scheme, and then use it to construct an ID-based adaptive oblivious transfer scheme. We consider how to obtain the $IDAOT_{k \times 1}^{N}$ scheme from PKI-based unique signature in section 6. Finally, we conclude in section 7.

2   Preliminaries

2.1   Bilinear Groups with Composite Order

We briefly review the necessary facts about bilinear groups with composite order using the same notation as in [6].

Let $\mathcal{G}$ be an algorithm called a *group generator* that takes as input a security parameter $\lambda \in \mathbb{Z}$ and outputs a tuple $(p, q, \mathbb{G}, \mathbb{G}_T, e)$ where p, q are two distinct primes, $\mathbb{G}$ and $\mathbb{G}_T$ are two cyclic groups of order $n = pq$, and e is a map $e : \mathbb{G} \times \mathbb{G} \to \mathbb{G}_T$, which called a bilinear pairing if this map satisfies the following properties:

1. **Bilinearity:** For all $u, v \in \mathbb{G}$ and $a, b \in \mathbb{Z}_n^*$, we have $e(u^a, v^b) = e(u, v)^{ab}$.
2. **Non-degeneracy:** There exists a $g \in \mathbb{G}$, such that $e(g, g) \neq 1$. In other words, if g is a generator of $\mathbb{G}$, then $e(g, g)$ generates $\mathbb{G}_T$.
3. **Computability:** There is an efficient algorithm to compute $e(u, v)$ for all $u, v \in \mathbb{G}$.

Such groups can be found on supersingular elliptic curves. Here is an example: Let $n = pq$, and L to be smallest prime in $\{n-1, 2n-1, 3n-1, \dots\}$, where $L = 3$

mod 4. An elliptic curves over F_L is the set of geometric solutions $P = (x, y)$ to the equation $y^2 = x^3 + x$. Then $E(F_L)$ has subgroup G of order n. The pairing will be $e : G \times G \to (F_{L^2})^{(|L|^2-1)/N}$. Other examples can be found in [6].

2.2 The One-More Square-Root-Exponents Assumption

In this subsection, we introduce a new hard problem named "One-More Square-Root-Exponents problem". Our new ID-based unique blind signature scheme is based on this assumption.

Definition 1 (One-More-SREP). *One-More Square-Root-Exponents Problem in $(\mathbb{G}, \mathbb{G}_T)$ is given as follows: For an integer $n = pq$ and p, q are big prime numbers, and $x^2 \in_R QR_n$, an one-way hash function $H_2 : \{0, 1\}^* \to \mathbb{G}$, given a BLS signature oracle $(\cdot)^x$, the adversary $\mathcal{A}$ is required to output $t + 1$ tuples $< (m_1, H_2(m_1)^x), \ldots, (m_{t+1}, H_2(m_{t+1})^x) >$, on the condition that the adversary makes at most t oracle queries.*

We denotes $Adv_{\mathcal{A},m}^{OMSREP}(t)$ as the probability that the adversary $\mathcal{A}$ solves the One-More Square-Root-Exponents Problem. Then we say that the One-More-SREP is (t, ϵ)-hard if $Adv_{\mathcal{A},m}^{OMSREP}(t) < \epsilon$, where ϵ is negligible for any adversary $\mathcal{A}$ whose time-complexity is polynomial in the security parameter t.

Solving One-More-SREP is as hard as the forgeability of BLS signature scheme [7], assuming the intractability of the quadratic residuosity problem modulo a composite $n = pq$ where p, q are large primes.

Definition 2 (One-More-SREP Assumption). *We say that the (t, ϵ)-One-More-SREP assumption holds in $(\mathbb{G}, \mathbb{G}_T)$ if no t-time adversary has advantage at least ϵ in solving the One-More-SREP in $(\mathbb{G}, \mathbb{G}_T)$, i.e., One-More-SREP is (t, ϵ)-hard in $(\mathbb{G}, \mathbb{G}_T)$.*

2.3 ID-Based Blind Signature

An ID-based blind signature scheme is considered be the combination of a general blind signature scheme and an ID-based one, *i.e.*, it is a blind signature, but its public key for verification is just the signer's identity. It consists of the following four algorithms, **Setup, Extract, Blind Signature Issuing protocol**, and **Verification**. The security of an ID-based blind signature scheme consists of two requirements: the blindness property and the unforgeability. We say *the blind signature scheme is secure* if it satisfies these two requirements. For detailed description of the definition of ID-based blind signature and the security models, the readers can refer to [22].

As in [9], we say an ID-based blind signature scheme is *unique* (called ID-based unique blind signature) if for each ID, and each message $m \in \{0, 1\}^*$, there exists at most one signature $s \in \{0, 1\}^*$ such that the **Verification** holds.

3 Model and Security Notions

In this section, we present a model and the security notions for the ID-based adaptive oblivious transfer.

3.1 Definitions

An *ID-based Adaptive k-out-of-N Oblivious Transfer (IDAOT$_{k \times 1}^N$)* consists of three participants: a trusted Private Key Generator (PKG), a Sender (S), a Receiver(R) and four algorithms, namely (Setup, Extract, S_I, R_I S_T, R_T) where

1. Setup: A probabilistic polynomial algorithm that takes a security parameter λ, and returns PARAMS (system parameters) and master-key of the PKG.
2. Extract: Takes as input system parameters, master-key, and an arbitrary $ID \in \{0,1\}^*$, and returns a private key sk. Here ID is an arbitrary string that will be used as a public key.
3. S_I, R_I: This is an interactive protocol between the sender and the receiver for the initialization. The sender runs $S_I(M_1, \ldots, M_N)$ (The inputs are the messages $M_1, \ldots, M_N$) to obtain state value S_0, and the receiver runs R_I algorithm without input to obtain state value R_0.
4. S_T, R_T: This is an interactive protocol between the sender and the receiver for the k times transfers. The i-th transfer proceeds as follows: the sender runs $S_T(S_{i-1})$ to obtain state value S_i, and the receiver runs $R_T(R_{i-1}, \sigma_i)$ where $1 \leq \sigma_i \leq N$ is the index of the message to be received. This produces state information R_i and the message M_{σ_i} or $\perp$ indicating failure.

3.2 Security Notions

After the Setup and Extract phases, others are same as the PKI based. So, the security for ID-based oblivious transfer can be defined according to a simulation-based definition similar to the presentation in [9]. The $Real_{\hat{S}, \hat{R}}(N, k, M_1, \ldots, M_N, \Sigma)$ means the experiment for arbitrary sender and receiver algorithms $\hat{S}$ and $\hat{R}$, and the $Ideal_{\hat{S}, \hat{R}}(N, k, M_1, \ldots, M_N, \Sigma)$ means the experiment that trusted party $\mathbb{T}$ handles the data exchanges between the sender $\hat{S}$ and the receiver $\hat{R}$. The experiment details are copied from [9] and listed as follows.

Real experiment. The experiment $Real_{\hat{S}, \hat{R}}(N, k, M_1, \ldots, M_N, \Sigma)$ occurs between the arbitrary sender and receiver algorithms $\hat{S}$ and $\hat{R}$. It proceeds as follows. $\hat{S}$ is given messages $M_1, \ldots, M_N$ as input and interacts with $\hat{R}(\Sigma)$, where Σ is an adaptive selection algorithm that, on input messages $M_{\sigma_1}, \ldots, M_{\sigma_{i-1}}$, outputs the index σ_i of the next message to be queried. In their first run, $\hat{S}$ and $\hat{R}$ produce initial states S_0 and R_0 respectively. Next, the sender and receiver engage in k interactions. In the i-th interaction for $1 \leq i \leq k$, the sender and receiver interact by running $S_i \leftarrow \hat{S}(S_{i-1})$ and $(R_i, M_i^*) \leftarrow \hat{R}(R_{i-1})$, and update their states to S_i and R_i, respectively. Note that M_i^* may be different from M_{σ_i} when either participant cheats. At the end of the k-th interaction, sender and receiver output strings S_k and R_k respectively. The output of the $Real_{\hat{S},\hat{R}}$ experiment is the tuple (S_k, R_k).

Ideal experiment. In experiment $Ideal_{\hat{S}',\hat{R}'}(N, k, M_1, \ldots, M_N, \Sigma)$, the (possibly cheating) sender algorithm $\hat{S}'(M_1, \ldots, M_N)$ generates messages $M_1^*, \ldots, M_N^*$ and hands these to the trusted party $\mathbb{T}$. In each of the k transfer phases, $\mathbb{T}$ receives a bit b_i from the sender $\hat{S}'$ and an index σ_i^* from the (possibly cheating) receiver $\hat{R}'(\Sigma)$. If $b_i = 1$ and $\sigma_i^* \in \{1, \ldots, N\}$, then $\mathbb{T}$ hands $M_{\sigma_i^*}^*$ to the receiver; otherwise, it hands $\perp$ to the receiver. At the end of the k-th transfer, $\hat{S}'$ and $\hat{R}'(\Sigma)$ output a string S_k and R_k; the output of the experiment is the pair (S_k, R_k). Note that the sender's bit b_i models its ability to make the current transfer fail. However, the sender's decision to do so is independent of the index σ_i that is being queried by the receiver.

Then we have the following notions:

Security under ID Attack: Assume that $\mathcal{A}$ is the adversary (he/she can be a sender or receiver or any third party) holding the system parameters and the identity public key ID of the sender. The $\mathcal{A}$ performs the ID attack as follows: $\mathcal{A}$ queries **Extract** q_E ($q_E > 0$) times with (PARAMS, $ID_i \neq ID$) for $i = 1, \cdots, q_E$. **Extract** returns to $\mathcal{A}$ the q_E corresponding secret key sk_i. We assume that q_E is limited by a polynomial in k. If $\mathcal{A}$ can get a (ID_i', sk_i'), such that (ID_i', sk_i') has the same function with (ID, sk), then he/she can play the role of sender. As noted by Naor in [5], any **Extract** phase in ID-based cryptosystem gives a public key signature scheme. When this signature scheme is secure, then $\mathcal{A}$ can not get such (ID_i', sk_i'), this means that $\mathcal{A}$ learns nothing from query results.

Sender-Secure: We say that $OT_{k\times1}^N$ is (t, t', t_D, ϵ)-sender-secure if for any real-world cheating receiver $\hat{R}$ running in time t, there exists an ideal-world receiver $\hat{R}'$ running in time t' such that for any $N \in [1, t]$, any $k \in [0, N]$, any messages $M_1, \ldots, M_N$, and any selection algorithm Σ, no distinguisher D running in time t_D has success probability greater than ϵ to distinguish the distributions $Real_{S,\hat{R}}$ $(N, k, M_1, \ldots, M_N, \Sigma)$ and $Ideal_{S',\hat{R}'}(N, k, M_1, \ldots, M_N, \Sigma)$. The distinguisher D is allowed to query **Extract** oracle except on ID of the sender. With the obtained private keys, the distinguisher D can practice the oblivious transfer at will.

Receiver-Secure: We say that $OT_{k\times1}^N$ is (t, t', t_D, ϵ)-receiver-secure if for any real-world cheating sender $\hat{S}$ running in time t, there exists an ideal-world sender $\hat{S}'$ running in time t' such that for any $N \in [1, t]$, any $k \in [0, N]$, any messages $M_1, \ldots, M_N$, and any selection algorithm Σ, no distinguisher D running in time t_D has success probability greater than ϵ to distinguish the distributions $Real_{\hat{S},R}(N, k, M_1, \ldots, M_N, \Sigma)$ and $Ideal_{\hat{S}',R'}(N, k, M_1, \ldots, M_N, \Sigma)$.

4   Generic Construction of ID-Based Adaptive Oblivious Transfer from ID-Based Unique Blind Signature

In this section, we present a generic construction for ID-based adaptive oblivious transfer $(IDAOT_{k\times1}^N)$. Our construction is identical to the generic construction of [9], except that there is PKG in our construction to issue private keys for the senders and receivers with their IDs as the public keys, and that the unique blind signature scheme is also ID-based. We suppose that the ID-based unique blind signature scheme consists of four algorithms $\mathcal{IDBS}$=(**Setup**, **Extract**, **BlindSign**, **Verify**).

A generic $IDAOT^N_{k\times1}$ is a tuple (Setup, Extract, S_I, R_I S_T, R_T) where:

- Setup Input a security parameter λ, it outputs system parameters *params* and a master secret key *msk* for PKG.
- Extract The same as **Extract** in the ID-based blind signature scheme. Input one PKG's secret key *msk*, and the user's ID, the PKG outputs the corresponding private key for the user. For the sender with identity ID_S, the private key is SK_S. For the receiver with identity ID_R, the private key is SK_R.
- (S_I, R_I): The sender (signer) has N messages $\{M_1, \ldots, M_N\}$. For $i = 1, \ldots, N$, the sender calculates $s_i = \mathbf{Sign}(sk_S, i)$, and $C_i = H(i, s_i) \oplus M_i$, here H is a hash function. $\{ID_S, C_1, \ldots, C_N\}$ are sent to the receiver. At the end of this algorithm, the sender and the receiver reach the initial state, S_0 and R_0, respectively.
- (S_T, R_T): In the i-th transfer ($1 \leq i \leq k$), the receiver selects $\sigma_i \in \{1, \ldots, N\}$ as input, and executes the interactive **BlindSign** protocol with the signer, with the signer's input SK_S. After the protocol, the receiver obtains the blind signature $s_{\sigma_i}=\mathbf{BlindSign}(sk_S, \sigma_i)$ from the signer, with the signer knows neither the σ_i nor the s_{σ_i}. The s_{σ_i} should be the same as the one used to encrypt M_{σ_i} in (S_I, R_I) phase, for the signature scheme is an unique blind signature scheme.
  The receiver verifies the s_{σ_i} by running **Verify**$(ID_S, \sigma_i, s_{\sigma_i})$. If it outputs 0, the receiver outputs $\bot$. Else, the receiver outputs $M_{\sigma_i}=C_{\sigma_i} \oplus H(\sigma_i, s_{\sigma_i})$. If $i = k$ the receiver stops. Else the receiver can decide the next choice by selecting a different σ_{i+1} and running the (S_T, R_T) again. It means that the running states of the sender and the receivers evolve from the states S_i and R_i to S_{i+1} and R_{i+1}.

It is obviously that if the signature scheme educed from Extract phase is secure, then $\mathcal{A}$ learns nothing under the ID attack.

Theorem 1 states that the sender's security is implied by the one-more unforgeability of $\mathcal{IDBS}$, while Theorem 2 states that the receiver's security follows from the selective-failure blindness of $\mathcal{BS}$. The proofs are identical to those in [9] except considering the ID attacks, and provided in Appendix A.

Theorem 1. *If the $\mathcal{IDBS}$ is (t'', q''_S, ϵ'')-unforgeable, then the $IDAOT^N_{k\times1}$ scheme described above is $(t, t', t_D, q_H, \epsilon)$-sender-secure in the random oracle model for any $\epsilon \geq \epsilon''$, $k \leq q''_S$, $t \leq t'' - q_H \cdot t_{Verify}$, and $t' \geq t + t_{Ext} + k \cdot t_{Sign} + q_H \cdot t_{Verify}$, where t_{Ext}, t_{Sign} and t_{Verify} are the time steps required for an execution of the key extraction, blind signing and verification algorithm of $\mathcal{IDBS}$, respectively.*

Theorem 2. *If the $\mathcal{IDBS}$ is (t'', ϵ'') selective-failure blind, then the $IDAOT^N_{k\times 1}$ scheme described above is $(t, t', t_D, q_H, \epsilon)$-receiver-secure in the random oracle model for all $\epsilon \geq 2 \cdot \epsilon'$ and $t_D \leq t'' - t - t_\Sigma - k \cdot t_{User}$, $t' \geq t + t_\Sigma + (q_H + k) \cdot t_{Verify} + k \cdot t_{User}$ where t_{User}, t_{Verify} and t_Σ are the running times of the User, Verify and Σ algorithms, respectively.*

5   A Concrete $IDAOT^N_{k\times 1}$ Scheme

As shown above, the ID-based adaptive oblivious transfer can be constructed if there exists ID-based unique blind signature scheme. In this section, we first propose a round-optimal ID-based unique blind signature scheme and then use it to construct a concrete $IDAOT^N_{k\times 1}$ scheme.

5.1   The Proposed ID-Based Unique Blind Signature Scheme

The round-optimal ID-based unique blind signature scheme is described as follows:

1. **Setup.** The PKG generates the system parameters: $(\mathbb{G}, \mathbb{G}_T, e, n, g, H_1, H_2)$. $H_1 : \{0, 1\}^* \to \mathbb{Z}_n^*$, $H_2 : \{0, 1\}^* \to \mathbb{G}$. The master secret key is (p, q), here $n = pq$ and p, q are big prime numbers.
2. **Extract.** Given a signer's identity ID, PKG computes the private key $s_{id} = H_1(ID\|\mathsf{tag})^{\frac{1}{2}} \bmod n$ associated with ID, here tag is increased from 0 by step 1, untill $H_1(ID\|\mathsf{tag})$ is a quadratic residue modulo n. Then PKG returns s_{id} and tag to the signer.
3. **BlindSign.** The signer has ID and private key s_{id}. The signer interacts with the user to issue blinded signature.
   - **Blind:** For message m, the user picks a random number $r \in \mathbb{Z}_n$, computes $R = H_2(m)^r \in \mathbb{G}$ and sends R to the signer.
   - **Sign:** The signer computes $s = R^{s_{id}} \in \mathbb{G}$, and returns s and tag to the user.
   - **Unblind:** The user computes $s' = s^{r^{-1} \bmod n}$.
4. **Verify:** Given ID, tag, m, s', the the user verifies that

$$e(s', \, s') \stackrel{?}{=} e(H_2(m), \, H_2(m))^{H_1(ID\|\mathsf{tag})}$$

The verification of the signature is justified by the following equations:

$$\begin{aligned}
e(s', \, s') \\
&= e(s^{r^{-1} \bmod n}, \, s^{r^{-1} \bmod n}) \\
&= e((R^{s_{id}})^{r^{-1} \bmod n}, \, (R^{s_{id}})^{r^{-1} \bmod n}) \\
&= e(((H_2(m)^r)^{s_{id}})^{r^{-1} \bmod n}, \, ((H_2(m)^r)^{s_{id}})^{r^{-1} \bmod n}) \\
&= e(H_2(m)^{s_{id}}, \, H_2(m)^{s_{id}}) \\
&= e(H_2(m), \, H_2(m))^{s_{id}^2} \\
&= e(H_2(m), \, H_2(m))^{H_1(ID\|\mathsf{tag})}
\end{aligned}$$

On the blindness of our ID-based blind signature scheme, we can state the following theorem:

Theorem 3. *The proposed scheme is blind.*

Proof. We follow the blindness model presented in [22]. Let $\mathcal{A}$ be the **Signer** or a probabilistic polynomial-time algorithm that controls the **Signer** and has (ID, s_{id}) from **Extract**$(params, ID)$.

If $\mathcal{A}$ gets $\perp$, it is easy to see that $\mathcal{A}$ wins the game with probability exactly the same as a random guessing of b, *i.e.*, with probability $1/2$.

Suppose that $\mathcal{A}$ gets $s'(m_b)$ and $s'(m_{1-b})$, instead of $\perp$. For $i = 0, 1$, let R_i, s_i be the data exchanged during the signature issuing protocol, and (R_0, s_0) and (R_1, s_1) are given to $\mathcal{A}$. Then it is sufficient to show that there exist a random factors d that maps R_i, s_i to R_j, s_j for each $i, j \in \{0, 1\}$. Suppose $H_2(m_i) = g^{y_1}$ and $H_2(m_j) = g^{y_2}$, with unknown y_1 and y_2. Then we have $R_i = H_2(m_i)^{r_i} = g^{y_1 \cdot r_i \mod n}$ and $R_j = H_2(m_j)^{r_j} = g^{y_2 \cdot r_j \mod n}$. There does exist a random factors $d = (y_1 \cdot r_i - y_2 \cdot r_j \mod n)$ that maps R_i to R_j, though we can not calculate it out. Therefore, even an infinitely powerful $\mathcal{A}$ has no advantage in determining which message R_i, s_i belong to, and the probability that $\mathcal{A}$ guesses the right b is $\frac{1}{2}$.

Taking two cases into account, the probability that $\mathcal{A}$ wins is $\frac{1}{2}$. Therefore, the proposed scheme is blind. $\square$

Theorem 4. *The proposed ID-based unique blind signature scheme is unforgeable, under the One-More-SREP assumption.*

Proof. Our proof is in the random oracle model (the hash function is seen as a random oracle, i.e., the output of the hash function is uniformly distributed). Suppose that a forger $\mathcal{F}$ (t, q_E, q_H, ϵ)-break the signature scheme using an adaptive chosen message attack. We will use $\mathcal{F}$ to construct an algorithm $\mathcal{A}$ to solve One-More-SREP. Suppose $\mathcal{A}$ is given a signature oracle $(\cdot)^x$.

Now $\mathcal{A}$ selects hash function $H_1 : \{0, 1\}^* \to \mathbb{Z}_n^*$, plays the role of the signer. He parses the x^2 as $(ID\|tag)$, then the public key for $\mathcal{A}$ is ID. $\mathcal{A}$ will answer hash oracle queries, Extract queries and signing queries. We assume that $\mathcal{F}$ never repeats a hash query or a signature query.

- Step1: $\mathcal{A}$ prepares q_E responses $\{x_1, x_2, \ldots, x_{q_E} \in \mathbb{Z}_n^*\}$ of the Extract oracle queries, where x_i are distributed randomly in this response set. $\mathcal{A}$ maintains a Hash-List for hash and Extract oracle. For each $1 \le i \le q_E$, $\mathcal{A}$ selects a randomly $tag_{i,0} \in \{1, \ldots, k\}$ and sets $H_1(ID_i\|tag_{i,0}) = x_i^2$, with ID_i still unknown for index i. An index counter j is set to 0.
- Step2: When $\mathcal{F}$ makes a $H_1(\cdot)$ oracle query on a new ID ID_{new} and a tag tag_{new}, if ID_{new} does not exist in Hash-List, $\mathcal{A}$ increases j by 1 and sets $ID_j = ID_{new}$, $tag_{j,1} = tag_{new}$. If $tag_{j,1} = tag_{j,0}$, $\mathcal{A}$ returns x_j^2 for $H_1(ID_j\|tag_{j,1})$. Else $\mathcal{A}$ selects randomly unused $r \in \mathbb{Z}_n^*$ sends r to $\mathcal{F}$, and $\mathcal{A}$ also records r in Hash-List as $H_1(ID_j\|tag_{j,1})$. If ID_{new} already exists as ID_j, $\mathcal{A}$ checks if any $tag_{j,l}$ $(0 \le l \le k)$ equals to tag_{new}. If a certain

$tag_{j,l}$ equals to tag_{new}, $\mathcal{A}$ returns $H_1(ID_j||tag_{j,l})$. If no $tag_{j,l}$ ($0 \leq l \leq k$) equals to tag_{new}, an empty array $tag_{j,l'}$ is assigned to contain tag_{new}. Then a randomly unused $r \in \mathbb{Z}_n^*$ is recorded as $H_1(ID_j||tag_{j,l'})$ and sent to $\mathcal{F}$.

- Step3: $\mathcal{F}$ makes a Extract oracle query for a new ID ID_{new}. If there exists an item in Hash-List with $ID_j = ID_{new}$, $\mathcal{A}$ returns $s_{id_j} = x_j$ and $tag_{j,0}$ to $\mathcal{F}$ as the response. Otherwise, $\mathcal{A}$ increases j by 1 and records $ID_j = ID_{new}$. $\mathcal{A}$ sends x_j and $tag_{j,0}$ to $\mathcal{F}$.
- Step4: $\mathcal{F}$ makes a hash oracle query for a l-th message on hash function H_2, $1 \leq l \leq q_H$. $\mathcal{A}$ returns $H_2(m_l)$ to $\mathcal{F}$ as the response, at the same time, $\mathcal{A}$ records the message pair $(m_l, H_2(m_l))$ in its Hash-List. Otherwise, $\mathcal{A}$ reports failure and aborts.
- Step5: $\mathcal{F}$ makes a signature oracle query for a l-th message m_l, $1 \leq l \leq t$. $\mathcal{A}$ forwards the query to the signature oracle $(\cdot)^x$ and forwards the returned $H_2(m_l)^{rx}$ to $\mathcal{F}$ as the response. In this step, $\mathcal{A}$ knows nothing about the message m_l because its hash value is blinded by r.
- Step6: Eventually, $\mathcal{F}$ halts and outputs $t+1$ message-signature tuples $< (m_1, H_2(m_1)^x), \ldots, (m_{t+1}, H_2(m_{t+1})^x) >$. Then $\mathcal{A}$ verifies if each message-signature tuple $(m_l, H_2(m_l)^x)$ satisfies $e(H_2(m_l)^x, H_2(m_l)^x) = e(H_2(m_l), H_2(m_l))^{H_1(ID||\mathsf{tag})}$, where $1 \leq l \leq t+1$. If so, $\mathcal{A}$ outputs these $t+1$ tuples as a solution to One-More-SREP.

We don't need to care about the forgery of combining several message-signature pairs to form a new message-signature pair. Because this kind of forgery requires finding message for the hash function $H_2(\cdot)$, violating the one-wayness of $H_2(\cdot)$.

$\square$

5.2 An $IDAOT_{k \times 1}^N$ Scheme

From the ID-based unique blind signature proposed above, we can easily construct an ID-based adaptive oblivious transfer scheme. We just give out the detailed construction here.

- Setup: Input a security parameter λ, PKG generates the system parameters: $(\mathbb{G}, \mathbb{G}_T, e, n = pq, g, H_1, H_2)$, with $H_1 : \{0,1\}^* \rightarrow \mathbb{Z}_n^*$, $H_2 : \{0,1\}^* \rightarrow \mathbb{G}$. (p, q) is the master secret key of PKG.
- Extract: Input the sender's ID, PKG outputs the corresponding private key $s_{id} = H_1(ID||\mathsf{tag})^{\frac{1}{2}} \bmod n$. Here tag is increased from 0 by step 1, untill $H_1(ID||\mathsf{tag})$ is a quadratic residue modulo n. Then PKG returns s_{id} and tag to the signer. Note that this Extract phase is actually the signing phase of Rabin signature scheme.
- (S_I, R_I): The sender (or signer) has N messages $\{M_1, \ldots, M_N\}$. For $i = 1, \ldots, N$, the signer calculates $s_i = H_2(i)^{s_{id}}$, and $C_i = H_1(i, s_i) \oplus M_i$. $\{ID_S, \mathsf{tag}, C_1, \ldots, C_N\}$ are sent to the receiver. At the end of this algorithm, the signer and the receiver reach the initial state, S_0 and R_0, respectively.
- (S_T, R_T): In the i-th transfer ($1 \leq i \leq k$), the receiver selects $\sigma_i \in \{1, \ldots, N\}$ as input, runs the interactive protocol **BlindSign** with the signer. In the

protocol, the sender's input is s_{id}. After the protocol, the receiver obtains the blind signature $s_{\sigma_i} = H_2(\sigma_i)^{s_{id}}$ from the sender, with the sender knows neither the σ_i nor the s_{σ_i}. The sender also sends tag to the receiver.

The receiver verifies the s_{σ_i} by testing $e(s_{\sigma_i}, s_{\sigma_i}) \stackrel{?}{=} e(H_2(\sigma_i), H_2(\sigma_i))^{H_1(ID\|\mathtt{tag})}$. If it outputs 0, the receiver outputs $\perp$. Else, the receiver outputs $M_{\sigma_i} = C_{\sigma_i} \oplus H_1(\sigma_i, s_{\sigma_i})$. If $i = k$ the receiver stops. Else the receiver can decide the next choice by selecting a different σ_{i+1} and running the (S_T, R_T) again. It means that the running states of the sender and the receivers evolve from the states S_i and R_i to S_{i+1} and R_{i+1}.

6 $IDAOT^N_{k \times 1}$ Scheme from PKI-Based Unique Signature

At Eurocrypt 2004, Bellare, Neven, and Namprempre [3] demonstrated that identity-based signature schemes can be constructed from any PKI-based signature scheme. Galindo et al.[13] extended Bellare et al.'s method, and gave a generic construction of some identity-based signature schemes with additional properties (such as identity-based blind signatures, proxy signatures, verifiably encrypted signatures, ...) from PKI-based signature schemes with the same properties. The ID-based unique blind signature can be obtained using Galindo et al.'s method from PKI-based unique blind signature. So, the $IDAOT^N_{k \times 1}$ scheme can be constructed from PKI-based unique blind signature.

As an example, we give another concrete $IDAOT^N_{k \times 1}$ scheme from BLS signature [7] and blind BLS signature [4] schemes as follows.

- Setup: Input a security parameter λ, PKG generates the system parameters: $(\mathbb{G}, \mathbb{G}_T, e, n, g, H_1, H_2)$, with $H_1 : \{0,1\}^* \to \mathbb{Z}^*_n$, $H_2 : \{0,1\}^* \to \mathbb{G}$. Here n is a prime number. The master public key of PKG is defined as $u = g^x$, whereas the master secret key stored by PKG is x.
- Extract: The sender has public-secret key pair $(pki, ski) = (u_i = g^{x_i}, x_i)$ with identity ID_i. Input the sender's ID and his public key $(ID_i, pki = u_i)$, PKG outputs $\sigma = H_2(ID_i\|u_i)^x$. The resulting secret key of the identity ID_i is $s_{idi} = (ski = x_i, pki = u_i, \sigma)$.
- (S_I, R_I): The sender (or signer) has N messages $\{M_1, \ldots, M_N\}$. For $i = 1, \ldots, N$, the signer calculates $s_i = H_2(i)^{x_i}$, and $C_i = H_1(i, s_i) \oplus M_i$. $\{ID_i, C_1, \ldots, C_N\}$ are sent to the receiver. At the end of this algorithm, the signer and the receiver reach the initial state, S_0 and R_0, respectively.
- (S_T, R_T): In the i-th transfer $(1 \leq i \leq k)$, the receiver selects $\sigma_i \in \{1, \ldots, N\}$ as input, runs the interactive protocol with the signer as follows:
  - The receiver picks up a random number $r \in \mathbb{Z}^*_n$, sends $g^r H_2(\sigma_i)$ to the sender.
  - The sender computes $s'_{\sigma_i} = (g^r H_2(\sigma_i))^{x_i}$, and sends $(s'_{\sigma_i}, u_i, \sigma)$ to the receiver.
  - The receiver computes $s_{\sigma_i} = s'_{\sigma_i}/u^r_i = H_2(\sigma_i)^{x_i}$, verifies that $e(s_{\sigma_i}, g) \stackrel{?}{=} e(H_2(\sigma_i), u_i)$ and $e(\sigma, g) \stackrel{?}{=} e(H_2(ID_i\|u_i), u)$.

After the protocol, the receiver obtains the blind signature $s_{\sigma_i} = H_2(\sigma_i)^{x_i}$ from the sender, with the sender knows neither the σ_i nor the s_{σ_i}.

The receiver verifies the s_{σ_i} and σ if they are valid. If it outputs 0, the receiver outputs $\perp$. Else, the receiver outputs $M_{\sigma_i} = C_{\sigma_i} \oplus H_1(\sigma_i, s_{\sigma_i})$. If $i = k$ the receiver stops. Else the receiver can decide the next choice by selecting a different σ_{i+1} and running the (S_T, R_T) again. It means that the running states of the sender and the receivers evolve from the states S_i and R_i to S_{i+1} and R_{i+1}.

From the previous scheme [17,9,14], as well as our scheme, we can see that the key pair of the sender can not be reused for the new set of N messages $\{M'_1, \ldots, M'_N\}$. Because they all use the unique values to encrypt the message. For instance, in [9] the ciphertext for M_i is $C_i = H(i, s_i) \oplus M_i$, and the value $H(i, s_i)$ is used to encrypt M_i. Because s_i is the unique blind signature, if the key pair and the index i are not changed, s_i hence $H(i, s_i)$ will be the same for every new set of N messages $\{M'_1, \ldots, M'_N\}$. And the receiver can run the protocol once and obtain the decryption key forever.

Due to the advantages of the identity-based cryptography, the sender can set his identity as $ID_S \| date$, for every new set of N messages, he will ask PKG to compute his new secret key. So, the ID information can be reused.

7 Conclusion

In this paper, we introduce the notion of ID-based adaptive oblivious transfer. The primitive is useful in some applications since it enjoys the inherent advantages of the identity-based cryptography. We also show that a generic construction of ID-based adaptive oblivious transfer can be given if there exists ID-based unique blind signature. Moreover, we propose a round-optimal ID-based unique blind signature scheme and prove its security. To the best of our knowledge, the resulting oblivious transfer scheme is the first ID-based adaptive oblivious transfer scheme.

Acknowledgements

We would like to thank the anonymous reviewers for their helpful comments and suggestions. This work has been supported by the National Natural Science Foundation of China (No. 60773202, 60633030), 973 Program(No. 2006CB303104) and Program of the Science and Technology of Guangzhou, China (No. 2008J1-C231-2).

References

1. Aiello, W., Ishai, Y., Reingold, O.: Priced oblivious transfer: How to sell digital goods. In: Pfitzmann, B. (ed.) EUROCRYPT 2001. LNCS, vol. 2045, pp. 119–135. Springer, Heidelberg (2001)

2. Abe, M., Ohkubo, M., Suzuki, K.: 1-out-of-n signatures from a variety of keys. In: Zheng, Y. (ed.) ASIACRYPT 2002. LNCS, vol. 2501, pp. 415–432. Springer, Heidelberg (2002)
3. Bellare, M., Namprempre, C., Neven, G.: Security proofs for identity-based identification and signature schemes. In: Cachin, C., Camenisch, J.L. (eds.) EUROCRYPT 2004. LNCS, vol. 3027, pp. 268–286. Springer, Heidelberg (2004)
4. Boldyreva, A.: Efficient threshold signature, multisignature and blind signature schemes based on the Gap-Diffie-Hellman -group signature scheme. In: Desmedt, Y.G. (ed.) PKC 2003. LNCS, vol. 2567, pp. 31–46. Springer, Heidelberg (2002)
5. Boneh, D., Franklin, M.: Identity-based encryption from the Weil pairing. In: Kilian, J. (ed.) CRYPTO 2001. LNCS, vol. 2139, pp. 213–229. Springer, Heidelberg (2001)
6. Boneh, D., Goh, E., Nissim, K.: Evaluating 2-DNF formulas on ciphertexts. In: Kilian, J. (ed.) TCC 2005. LNCS, vol. 3378, pp. 325–341. Springer, Heidelberg (2005)
7. Boneh, D., Lynn, B., Shacham, H.: Short signatures from the Weil pairing. In: Boyd, C. (ed.) ASIACRYPT 2001. LNCS, vol. 2248, pp. 514–532. Springer, Heidelberg (2001)
8. Brassard, G., Crépeau, C., Robert, J.M.: All-or-nothing disclosure of secrets. In: Odlyzko, A.M. (ed.) CRYPTO 1986. LNCS, vol. 263, pp. 234–238. Springer, Heidelberg (1987)
9. Camenisch, J., Neven, G., Shelat, A.: Simulatable adaptive oblivious transfer. In: Naor, M. (ed.) EUROCRYPT 2007. LNCS, vol. 4515, pp. 573–590. Springer, Heidelberg (2007)
10. Chor, B., Goldreich, O., Kushilevitz, E., Sudan, M.: Private Information Retrieval. In: 36th Annual Symposium on Foundations of Computer Science, pp. 41–50. IEEE, Los Alamitos (1995)
11. Chu, C.K., Tzeng, W.G.: Efficient k-out-of-n oblivious transfer schemes with adaptive and non-adaptive queries. In: Vaudenay, S. (ed.) PKC 2005. LNCS, vol. 3386, pp. 172–183. Springer, Heidelberg (2005)
12. Even, S., Goldreich, O., Lempel, A.: A randomized protocol for signing contracts. Communications of the Association for Computing Machinery 28(6), 637–647 (1985)
13. Galindo, D., Herranz, J., Kiltz, E.: On the Generic Construction of Identity-Based Signatures with Additional Properties. In: Lai, X., Chen, K. (eds.) ASIACRYPT 2006. LNCS, vol. 4284, pp. 178–193. Springer, Heidelberg (2006)
14. Green, M., Hohenberger, S.: Blind Identity-based encryption and simulatable oblivious transfer. In: Kurosawa, K. (ed.) ASIACRYPT 2007. LNCS, vol. 4833, pp. 265–282. Springer, Heidelberg (2007)
15. Lindell, Y.: Efficient fully-simulatable oblivious transfer. In: Malkin, T.G. (ed.) CT-RSA 2008. LNCS, vol. 4964, pp. 52–70. Springer, Heidelberg (2008)
16. Kalai, Y.T.: Smooth projective hashing and two-message oblivious transfer. In: Cramer, R. (ed.) EUROCRYPT 2005. LNCS, vol. 3494, pp. 78–95. Springer, Heidelberg (2005)
17. Naor, M., Pinkas, B.: Computationally secure oblivious transfer. Journal of Cryptology 18(1), 1–35 (2005)
18. Ogata, W., Kurosawa, K.: Oblivious keyword search. Journal of Complexity 20(2-3), 356–371 (2004)
19. Rabin, M.O.: How to exchange secrets by oblivious transfer, Technical Report TR-81, Harvard Aiken Computation Laboratory (1981)

20. Shamir, A.: Identity-based cryptosystems and signature schemes. In: Blakely, G.R., Chaum, D. (eds.) CRYPTO 1984. LNCS, vol. 196, pp. 47–53. Springer, Heidelberg (1985)
21. Tzeng, W.G.: Efficient 1-Out-n Oblivious Transfer Schemes. In: Naccache, D., Paillier, P. (eds.) PKC 2002. LNCS, vol. 2274, pp. 159–171. Springer, Heidelberg (2002)
22. Zhang, F., Kim, K.: ID-based blind signature and ring signature from pairings. In: Zheng, Y. (ed.) ASIACRYPT 2002. LNCS, vol. 2501, pp. 533–547. Springer, Heidelberg (2002)

Appendix A

Proof of Theorem 1

Proof. For any real-world cheating receiver $\hat{R}$, consider the ideal-world receiver $\hat{R}'$ that works as follows. $\hat{R}'$ generates a fresh pair (ID'_S, sk_{id}) from **Extract** algorithm for the blind signature scheme $\mathcal{BS}$ and chooses random strings $C_1, \dots, C_N \leftarrow \{0, 1\}^l$. It then feeds the string $(ID'_S, C_1, \dots, C_N)$ as input to $\hat{R}$ to obtain initial state R_0. During the transfer phase, when $\hat{R}$ engages in a transfer protocol, $\hat{R}'$ simulates the honest sender by executing the blind signature protocol as prescribed by **BlindSign**(sk_{id}, σ_i). To answer random oracle queries, $\hat{R}'$ maintains an initially empty associative array HT[·] and a counter ctr. When $\hat{R}$ performs a random oracle query H(x), $\hat{R}'$ responds with HT[x], or proceeds as follows if this entry is undefined.

> If $x = (i, s)$ and Verify$(ID'_S, i, s)=1$ and $i \in [1, N]$ then
> $\quad ctr \leftarrow ctr + 1$; If $ctr > k$ then halt with output $\bot$
> $\quad$ Obtain M_i from the ideal functionality
> $\quad$ HT[x]$\leftarrow M_i \oplus C_i$
> else HT[x]$\leftarrow \{0, 1\}^l$.

When eventually $\hat{R}$ outputs a string τ, $\hat{R}'$ halts with the same output τ. The running time t' of $\hat{R}'$ is that of $\hat{R}$ plus the time of a key extraction, k signing interactions and up to q_H signature verifications, i.e. $t' = t + t_{Ext} + k \cdot t_{Sign} + q_H \cdot t_{Verify}$.

The only differences between the two environments occur in the initialization message and the simulation of the random oracle. In both environments, the initialization message contains a fresh public key ID'_S and N random l-bit strings. The responses to $\hat{R}$'s random oracle queries are random l-bit strings, except for queries of the form H(i, s) where s is the unique signature such that Verify$(ID'_S, i, s) = 1$. For these queries, the unique value $M_i \oplus C_i$ is returned. Since C_i is a randomly chosen string, the distribution of $M_i \oplus C_i$ will also be a random l-bit string. So if $\hat{R}'$ does not abort, $\hat{R}'$ perfectly simulates the environment of the $Real_{S,\hat{R}}(N, k, M_1, \dots, M_N, \Sigma)$ experiment.

$\hat{R}'$ just relays messages between its signing oracle and $\hat{R}$ to simulate $\hat{R}$'s transfer queries. So $\hat{R}'$ aborts with probability not greater than $\hat{R}$, which is ϵ''.

From the above, we know that as long as $\hat{R}'$ does not abort, the distributions $Real_{S,\hat{R}}(N, k, M_1, \dots, M_N, \Sigma)$ and $Ideal_{S',\hat{R}'}(N, k, M_1, \dots, M_N, \Sigma)$ are identical, so any unbounded distinguisher $\mathcal{D}$ can not have any advantage in telling them apart, even if he is allowed to query **Extract** oracle and practice the oblivious transfer as described in subsection 3.2. The probability that $\hat{R}'$ aborts however introduces a statistical difference ϵ'' between the distributions. Hence, even an unbounded distinguisher $\mathcal{D}$ has advantage at most $\epsilon \leq \epsilon''$.   □

Proof of Theorem 2

Proof. For any real-world cheating sender $\hat{S}$, consider the ideal-world sender $\hat{S}'$ that works as follows. On input $(M_1, \ldots, M_N)$, it runs $\hat{S}(M_1, \ldots, M_N)$. It records each of the random oracle queries performed by the $\hat{S}$ algorithm and responds by providing consistent random values (i.e., if the adversary queries some value x twice, return the same result on each query). Let $(ID_S, C_1, \ldots, C_N)$ be the output of the sender's initialization process. At this point, $\hat{S}'$ reviews each of the random oracle queries $H(h_i)$ made by $\hat{S}$. If it succeeds at parsing h_i as (σ_i, s_i) such that **Verify**$(ID_S, \sigma_i, s_i) = 1$ and $\sigma_i \in \{1, \ldots, N\}$, then it sets $M_i^* \leftarrow C_i \oplus H(\sigma_i, s_i)$. For all $1 \leq j \leq N$ such that M_j^* has not been set by this process, it chooses a random value for M_j^*. It sends $(M_1^*, \ldots, M_N^*)$ to the trusted party. To handle the next k interactive queries, $\hat{S}'$ runs the honest receiver algorithm to obtain the initial state R_0. At each transfer, $\hat{S}'$ set $\sigma_i = 1$, that's to say, it simulates an honest receiver that queries the first message M_1 k times. Remember that $\hat{S}'$ is not given the selection algorithm Σ as input, so it cannot simulate an honest receiver on the real values. If algorithm $\hat{S}$ completes the protocol and the output of simulation is a valid signature, then $\hat{S}'$ sends 1 to the trusted party. Otherwise, it sends 0, indicating an abort for this query. (Note that it is here that we need the selective-failure property of $\mathcal{BS}$. Namely, it implies that the probability that the user algorithm fails is independent of the message being queried, and hence is the same for index 1 and the real index that would be generated through Σ.) At the end of the k-th query, $\hat{S}$ outputs a string S_k, and the ideal sender $\hat{S}'$ also outputs the same string S_k.

Given the simulation that $\hat{S}'$ provides to $\hat{S}$ above, one can see that $Ideal_{\hat{S}',R'}$ $(N, k, M_1, \ldots, M_N, \Sigma)$ and $Real_{\hat{S},R}(N, k, M_1, \ldots, M_N, \Omega)$ are identically distributed, where Ω is the selection algorithm that always outputs $\sigma_i = 1$. Suppose we are given an algorithm $\mathcal{D}$ that distinguishes outputs of $Ideal_{\hat{S}',R'}(N, k, M_1, \ldots, M_N, \Sigma) = Real_{\hat{S},R}(N, k, M_1, \ldots, M_N, \Omega)$ from outputs of $Real_{\hat{S},R}(N, k, M_1, \ldots, M_N, \Sigma)$. Then consider the following adversary $\mathcal{A}$ against the selective-failure blindness of $\mathcal{BS}$. Algorithm $\mathcal{A}$ first chooses a random bit $b' \leftarrow \{0, 1\}$ and runs $\hat{S}$ to obtain public key ID_S and ciphertexts $C_1, \ldots, C_N$. If $b' = 0$, then $\mathcal{A}$ runs $\hat{S}$ in interaction with normal receiver on indices generated by Σ; if $b' = 1$, it uses all ones as indices. Let i be the first index σ_i generated by Σ such that $\sigma_i = 1$. (If such index doesn't exist, then $\mathcal{D}$ is trying to distinguish between two identical distributions, so obviously it has advantage zero.) Algorithm $\mathcal{A}$ outputs ID_S as the public key and $M_0 = \sigma_i$, $M_1 = 1$ as the messages on which it wishes to be challenged. It simulates the i-th interaction with $\hat{S}$ by relaying messages between $\hat{S}$ and the first instantiation of the

User algorithm that $\mathcal{A}$ is faced with (which implements either User(ID_S, σ_i) or User(ID_S, 1)). It then continues running $\hat{S}$, simulating interactions as before. Eventually, $\hat{S}$ outputs a string S_k. $\mathcal{A}$ runs the distinguisher $\mathcal{D}$ on input S_k to obtain a bit b_D. Assume that $b_D = 0$ indicates that $\mathcal{D}$ guesses S_k was drawn from $Real_{\hat{S},R}(N, k, M_1, \ldots, M_N, \Sigma)$, and that $b_D = 1$ indicating that it was drawn from $Real_{\hat{S},R}(N, k, M_1, \ldots, M_N, \Omega)$. If $b_D = b' = 0$, then algorithm $\mathcal{A}$ outputs $b_A = 0$; if $b_D = b' = 1$, algorithm $\mathcal{A}$ outputs $b_A = 1$; otherwise, $\mathcal{A}$ outputs a random bit $b_A \leftarrow \{0, 1\}$. Let b be the hidden bit chosen by the blindness game that $\mathcal{A}$ has to guess. It is clear from the simulation that if $b = b' = 0$, then S_k follows the distribution $Real_{\hat{S},R}(N, k, M_1, \ldots, M_N, \Sigma)$. Likewise, if $b = b' = 1$, then S_k follows $Real_{\hat{S},R}(N, k, M_1, \ldots, M_N, \Omega)$. Using a standard probability analysis, one can see that the advantage of $\mathcal{A}$ in breaking the blindness of $\mathcal{BS}$ is $\epsilon' = \epsilon/2$. The running time t' of $\hat{S}'$ is at most the running time t of $\hat{S}$ plus that of $q_H + k$ signature verifications, k executions of receiver simulation and one execution of Σ. The running time t'' of $\mathcal{A}$ is at most the running time t_D of $\mathcal{D}$ plus the running time t of $\hat{S}$ plus the running time of Σ plus the time required for k executions of receiver simulation. □

Unknown Plaintext Template Attacks

Neil Hanley[1], Michael Tunstall[2], and William P. Marnane[1]

[1] Department of Electrical and Electronic Engineering,
University College Cork, Ireland
neilh@eleceng.ucc.ie, l.marnane@ucc.ie
[2] Department of Computer Science, University of Bristol,
Merchant Venturers Building, Woodland Road,
Bristol BS8 1UB, United Kingdom
tunstall@cs.bris.ac.uk

Abstract. In this paper we present a variation of the template attack classification process that can be applied to block ciphers when the plaintext and ciphertext used are unknown. In a naïve implementation this attack can be applied to any round of a block cipher. We also show that when a block cipher is implemented with the masking countermeasure a similar attack can be applied to the first round of the cipher. We demonstrate that the attack works in practice by applying it to implementations of AES on 8051 and ARM7 microprocessors. We also demonstrate that the attack can be applied to implementations of block ciphers that use the masking countermeasure when three points are selected from which templates are constructed, or two points if the plaintext can be guessed.

Keywords: block ciphers, side channel attack, power analysis, template attack.

1 Introduction

Side-Channel Analysis (SCA) focuses on how a cipher is implemented in hardware or software, rather than focusing on the cryptographic strength of a cipher. An implementation of a cryptographic algorithm can leak some side channel information, which can, potentially, lead to information on the key being extracted. Information from a multitude of sources can be exploited such as the time required to compute an algorithm [15], the power consumption [16] or electromagnetic emanations [11] during the computation, or a combination of these sources [1].

There are a number of different methods for using the power consumption to attack an implementation of a block cipher, such as Simple Power Analysis (SPA) [16] and Differential Power Analysis (DPA) [7,16]. SPA makes use of visible differences in a power trace to deduce which operations have occurred, and key bits can be extracted if conditional paths can be identified in an implementation of a cryptographic algorithm. In DPA, hypotheses on intermediate states of an algorithm are correlated with the instantaneous power consumption (termed traces) acquired while computing an algorithm to derive information on

H.Y. Youm and M. Yung (Eds.): WISA 2009, LNCS 5932, pp. 148–162, 2009.

the key being used. Template Attacks (TA), first introduced in [8], take a different approach to key extraction. They require a profiling stage of an open device to create power consumption *templates*, followed by a classification stage of the device under attack. The key can then be determined by matching the power consumption of the device under attack to the previously computed templates.

In this paper we introduce a classification process for a template based attack that allows for key extraction without knowledge of the plaintext (or ciphertext). Only power traces taken from the cryptographic device while an encryption is being performed are required. As in a known-plaintext attack, full control over the identical profiling device is still required. We show that when a block cipher is implemented naïvely, information from any round can be extracted to derive the secret key being used. We also show that when an implementation is protected with the masking countermeasure that this is no longer the case. However, by combining our classification process with the attack described in [20], the key can still be recovered from the *first round* using only the acquired power traces in this scenario.

This paper is organised as follows. Related work is outlined in Section 2, and a brief overview of known plaintext template attacks is given in Section 3. Section 4 then explains the theory behind the unknown plaintext attack. This is followed up with experimental results on two different platforms in Section 5. The attack in relation to masking is dealt with in Section 6. Finally, conclusions are drawn from the work in Section 7.

2   Related Work

The idea of a two stage attack, a profiling and classification stage, was first considered in [10]. Template attacks were then introduced in [8], and expanded in [21]. In [2] template attacks were used to recover single bits of the key at a time, as well as introducing an attack where the adversery has access to a profiling device with a slightly biased random number generator. Subspace-based template attacks were investigated in [5,22].

When implementing block ciphers on embedded devices it is common to use a masking scheme, where a random value is generated for each instance of the block cipher, and every operation is implemented such that the input and output are correct when XORed with this random value [17]. It was suggested in [3,4] that it is only necessary to apply this countermeasure to the first and last few rounds of a block cipher. In [12] it was demonstrated that techniques used in differential cryptanalysis can be used to apply side channel analysis to internal rounds of block ciphers. In this paper we show that a template attack can be applied to any round of a block cipher, and affirm that countermeasures need to be included in every round of an implementation of a block cipher (as shown in [12]).

A method of applying Differential Power Analysis to a block cipher in counter mode [9] is described in [13], where the secret key is derived by guessing the initial state of portions of the counter. We show that when a block cipher is implemented using a masking scheme as described above, the techniques presented in [13] can be modified to attack an implementation of a block cipher.

In [20] a template attack on an implementation of a block cipher that uses a masking scheme is presented that retrieves the secret key. We show that a similar attack can be used to derive the secret key used in a masked implementation of a block cipher where the plaintext and ciphertext are unknown.

3   Overview of Template Attacks

Template attacks are one of the strongest forms of side channel attack; however, they rely on the assumption that an attacker possesses an identical device to the one under attack, and has full control over its inputs. Template attacks work by building up a set of templates for an intermediate value using a large number of acquired traces, where a trace is a recording of the power consumption of the device being attacked while it is executing an algorithm. The classification stage then matches traces to a particular template using a multivariate Gaussian probability distribution. The correct key value should be returned with a higher probability than the incorrect values. The computationally intensive and time-consuming template building stage need only be completed once for a particular device. The same templates can then be used to mount multiple attacks on identical devices. An overview of how templates are constructed and used to attack an implementation of a cryptographic algorithm are given in Sections 3.1 and 3.2 respectively.

3.1   Template Construction

As described in Equations (1) and (2), templates consist of a pair of estimates for the mean vector $\mathbf{m}_i$ and the noise covariance matrix $\mathbf{C}_i$, for $i \in \{1, \ldots, n\}$, where n is the number of different possible values that can be present at the point which an attacker wishes to analyse. These values are constructed from a large number of traces, $\mathbf{t}_j$ where $j \in \{1, \ldots, k\}$, and k is typically in the region of 1000. However, the actual value of k will vary from one device to another.

$$\mathbf{m}_i = \frac{1}{k} \sum_{j=1}^{k} \mathbf{t}_{i,j} \qquad (1)$$

where $\mathbf{t}_{i,j}$ represents the j-th acquisition of the i-th possible value.

$$\mathbf{C}_i = \frac{1}{k-1} \sum_{j=1}^{k} (\mathbf{t}_{i,j} - \mathbf{m}_i)(\mathbf{t}_{i,j} - \mathbf{m}_i)^T \qquad (2)$$

If, for example, the target operation of the attack is a MOV instruction for the output of the SubBytes (SBOX) operation in the AES encryption algorithm, a template must be built for each of the 256 valid outputs of the SBOX. In [7], it is shown that the power consumption of a microprocessor can correspond to the Hamming weight of the value being examined (this holds for both our 8051 and our ARM7 microprocessors). Therefore, it is reasonable to instead make templates for all possible Hamming weights of the SBOX output byte, i.e. $n = 9$. Then for each Hamming weight possibility, there exists a template consisting of (m_i, C_i).

When recording power traces, the sampling rate is often high, to capture small fluctuations in power consumption. This leads to the length of a trace being very large (e.g. for the ARM traces used in Section 5 a sampling rate of 125Ms/s led to a trace length of 120,000 points for the first two rounds of the AES algorithm). It is computationally unfeasible to construct templates including all of these points. Methods are available to reduce redundant information in a trace, such as integration within a clock cycle or extracting the maximum value per clock cycle [17]. These methods significantly reduce processing time; however, for the construction of templates further reduction is required to extract the features that the templates will be based on. One option is to sum the absolute differences of the mean traces and select the required number of highest points, ensuring that only one point per clock cycle is retained, as described in [21]. Note that if certain trace reduction methods have already been employed, only one point per clock cycle may be present to begin with. Another option is to use Principal Component Analysis (PCA) [14] as demonstrated in [5]. PCA is a data dimensionality reduction technique which projects the traces into a subspace where only the required principal directions of maximum variability are retained for template construction.

3.2   Template Classification

To extract key information, an attack trace t is required, which is a power trace of the operation that is being targeted. The trace must first be reduced in size and features selected, in the same manner that the template generation traces were reduced. For each of the n templates, the probability of the trace corresponding to a given template can be calculated using Equation (3):

$$\Pr\left(\mathbf{t} \mid \mathbf{m}_i, \mathbf{C}_i\right) = \frac{1}{\sqrt{(2\pi)^n \left|\mathbf{C}_i\right|}} \cdot e^{-\frac{1}{2}(\mathbf{m}_i - \mathbf{t})\mathbf{C}_i^{-1}(\mathbf{m}_i - \mathbf{t})^T} \tag{3}$$

Using the known plaintext (or ciphertext), the intermediate target operation that the templates were generated for is hypothesised for each possible key value in the key set K. Therefore, each key value maps to one of the templates. As the Hamming weight of an intermediate value is being targeted, each key value does not map to a unique template. The probability of each of the l possible key values, k_j, can then be calculated using Bayes' theorem, which is given in Equation (4):

$$\Pr\left(k_j \mid \mathbf{t}\right) = \frac{p\left(\mathbf{t} \mid k_j\right) \cdot p\left(k_j\right)}{\sum_{l=1}^{K}\left(p\left(\mathbf{t} \mid k_l\right) \cdot (k_l)\right)} \tag{4}$$

The success of the attack is increased if a set of D power traces, $\mathbf{T}$, for a constant secret key are available. In this scenario, either Bayes' theorem applied iteratively or Equation (5) can be used, thereby increasing the power of the attack, as shown in [20].

$$\Pr(k_j \mid \mathbf{T}) = \frac{\left(\prod_{x=1}^{D} p\left(\mathbf{t}_x \mid k_j\right)\right) \cdot \Pr\left(k_j\right)}{\sum_{l=1}^{K} \left(\left(\prod_{x=1}^{D} p\left(\mathbf{t}_x \mid k_l\right)\right) \cdot p\left(k_l\right)\right)} \tag{5}$$

4   Key Extraction with Unknown Plaintext

If we consider a naïve implementation of a block cipher, a template attack can be used to extract key information without using any known plaintext (or ciphertext) values by targeting two separate intermediate states in the encryption algorithm.

When attacking a block cipher a natural choice would be to build templates on α and β, where $\beta = S(\alpha \oplus k)$ and S is a substitution table, and k is some portion of the secret key or subkey. For AES, α, β and k would each be eight bit values. Assuming that the device under attack follows the Hamming weight model, the attacker would be interested in constructing templates to identify the Hamming weight of α and β and using this to derive k. If the second template is built on β, where β occurs before the substitution table (i.e. $\beta = \alpha \oplus k$), the key cannot be extracted as there is a linear relationship between the two template sets. We note that in some architectures an attacker would be obliged to model some previous state that would affect the power consumption of α and β, referred to as the Hamming distance model [17]. In this paper we just consider the Hamming weight, as this corresponds to our experimental results.

The advantage for an attacker is that the attack can be applied to any round of a block cipher, allowing countermeasures that are implemented on certain rounds to be circumvented. In this section we informally discuss the expected number of observations an attacker would be required to make to derive k if applying this attack in general, and, more specifically, to the SubBytes function of AES [19].

4.1   Approximating the Number of Key Hypotheses

The amount of possible values that α and β can take can be computed from the Hamming weight h as $\binom{8}{h}$. Trivially, we can say that if either α and β have a Hamming weight of zero or eight, then the number of hypotheses returned for the key will be dictated by the Hamming weight of the other variable.

For further analysis, we assume that the function S is a one-to-one transformation that allows β to be modelled as a random variable. However, it should be noted that for a given observation of $H(\beta)$, there will be $\binom{8}{H(\beta)}$ distinct possible values for β. Given these assumptions, we can view the expected number of valid key hypotheses as a variant of the Classical Occupancy Problem [18].

For a given observation of the Hamming weight of α and β, each of the $\binom{8}{H(\beta)}$ possible values for β will return $\binom{8}{H(\alpha)}$ distinct key values. However, given our assumptions on S, this means that we have a repeated generation of $\binom{8}{H(\alpha)}$ distinct key hypotheses, so some key values will occur more than once.

If we take the example of $H(\alpha) = 1$ and $H(\beta) = 1$, the first evaluation for a given value of β will give $\binom{8}{\alpha}\binom{8}{1} = 8$ possible key hypotheses. In order to evaluate the effect of considering a second possible value of β, one needs to compute the number of new key hypotheses one would expect to see after generating the next eight distinct key hypotheses.

The statistically expected increase in the number of key hypotheses in this case can be computed as:

$$E(\# \text{ key hypotheses } (X)) = \Sigma_{i=0}^{8}(n + i) \cdot \Pr(X = i) \qquad (6)$$

where X is the expected increase, and that n is the number of keys seen before, and that between zero and eight new key hypotheses are possible.

Computing the probability that the increase in the number of key hypotheses is i, $\Pr(X = i)$, is somewhat problematic, as when listing the combinations of events where $X = 2$ the probability of a new or previously observed key hypothesis is dependent on how many key values have already been observed. For example, if we consider the probability of $X = 2$, there are $\binom{8}{2} = 28$ possible combinations of observations that have to be evaluated, but each possibility occurs with the same probability.

This can be general case of $\Pr(X = i)$ can be computed as:

$$\Pr(X = i) = \binom{v}{i} \cdot \frac{n^{(v-i)} \cdot (m - n)^{(i)}}{m^{(v)}} \qquad (7)$$

where n is number of keys seen before, m is the total number of possible key hypotheses, and v is the number of key hypotheses being generated. The function $x^{(y)}$ is defined in Equation (8):

$$x^{(y)} = x \cdot (x - 1) \cdot (x - 2) \cdots (x - y + 1) \qquad (8)$$

However, this assumes that the $\binom{8}{H(\beta)}$ values are random and uniformly distributed. It is therefore necessary to apply a correction to the above formula to take into account key hypotheses that will be impossible because of previous observations. If, for example, we consider the case where $H(\alpha) = 0$ and $H(\beta) = 1$. The first value possible for β will return one key hypothesis, as there is only one possible value for α. The second possible value for β must produce a different hypothesis to the first value, as S is a one-to-one transformation. If we return to the initial example of $H(\alpha) = 1$ and $H(\beta) = 1$, then eight key hypotheses will be returned by analysing the first value possible for β. When the second value is analysed, only seven of these values can be reproduced. A third possible value of β will have two different key hypotheses that it cannot generate as it is a different value to the previous two etc.

This means that when computing the expected number of key hypotheses, the number of values for β that have already been analysed need to be taken into account. When making observations with the ℓ-th possible value of β, the probability in Equation (7) is adjusted in Equation (9):

$$\begin{aligned}\Pr(X = i) &= \binom{v}{i} \cdot \frac{(n - (\ell - 1))^{(v-i)} \cdot (m - (\ell - 1) - (n - (\ell - 1)))^{(i)}}{(m - (\ell - 1))^{(v)}} \\ &= \binom{v}{i} \cdot \frac{(n - \ell + 1)^{(v-i)} \cdot (m - n)^{(i)}}{(m - \ell + 1)^{(v)}}\end{aligned} \quad (9)$$

From this, we can compute the expected number of hypotheses (Z) for a given pair of observations $H(\alpha)$ and $H(\beta)$ by computing

$$E(Z) = \sum_{i=1}^{\binom{\log_2 m}{H(\beta)}} \sum_{j=0}^{\binom{\log_2 m}{H(\alpha)}} (v + j) \cdot \binom{v}{j} \cdot \frac{(n - i + 1)^{(v-j)} \cdot (m - n)^{(i)}}{(m - j + 1)^{(v)}} \quad (10)$$

This can be used to compute the expected number of key hypotheses returned for a pair of observations $H(\alpha)$ and $H(\beta)$, where we are looking at attacking the function $\beta = S(\alpha \oplus k)$, where each variable is one byte, and we know the Hamming weight of α and β. The expectations are shown in Table 1.

Table 1. The approximate expected number of key hypotheses given observations $H(\alpha)$ and $H(\beta)$

		$H(\alpha)$								
		0	1	2	3	4	5	6	7	8
	0	1	8	28	56	70	56	28	8	1
	1	8	58.134	156.036	221.568	236.940	221.568	156.036	58.134	8
	2	28	156.036	247.789	255.838	255.982	255.838	247.789	156.036	28
	3	56	221.568	255.838	256	256	256	255.838	221.568	56
$H(\beta)$	4	70	236.940	255.982	256	256	256	255.982	236.940	70
	5	56	221.568	255.838	256	256	256	255.838	221.568	56
	6	28	156.036	247.789	255.838	255.982	255.838	247.789	156.036	28
	7	8	58.134	156.036	221.568	236.940	221.568	156.036	58.134	8
	8	1	8	28	56	70	56	28	8	1

The overall expectation can be computed from this by multiplying each table entry by the probability of it occurring, i.e. $\frac{\binom{8}{H(\alpha)} \cdot \binom{8}{H(\beta)}}{2^{16}}$, and computing the sum of the result. This produces an expected number of key hypotheses for a single observed pair, $H(\alpha)$ and $H(\beta)$, to be 246.3. This means that one would expect to acquire $\log_2(256/246.334) = 0.055527$ bits of information from one observed $H(\alpha)$ and $H(\beta)$, and to derive all eight bits would therefore be expected to require $8/0.055527 = 144.074$ observations.

4.2 Computing the Actual Number of Key Hypotheses for AES

The above description will only give an approximation of the expected number of hypotheses, as a random one-to-one function is used as a model. The number

Table 2. The expected number of key hypotheses given observations $H(\alpha)$ and $H(\beta)$

		$H(\alpha)$							
		0	1	2	3	4	5	6	7	8
	0	1	8	28	56	70	56	28	8	1
	1	8	54	163	219	236	218	153	58	8
	2	28	146	251	256	256	256	249	153	28
	3	56	222	256	256	256	256	256	226	56
$H(\beta)$	4	70	244	256	256	256	256	256	246	70
	5	56	222	256	256	256	256	256	226	56
	6	28	146	251	256	256	256	249	153	28
	7	8	54	163	219	236	218	153	58	8
	8	1	8	28	56	70	56	28	8	1

of hypotheses returned can be computed for attacking an implementation of AES by counting all the valid key hypotheses for a given pair of observations $H(\alpha)$ and $H(\beta)$ using the actual SubBytes function. The expected number of hypotheses returned is shown in Table 2.

As described above, the overall expectation can be computed by multiplying each table entry by the probability of it occurring, i.e. $\frac{\binom{8}{H(\alpha)} \cdot \binom{8}{H(\beta)}}{2^{16}}$, and computing the sum of the result. This produces an expected number of key hypotheses for an observed $H(\alpha)$ and $H(\beta)$ to be 246.5. This means that one would expect to acquire $\log_2(256/246.507) = 0.0545172$ bits of information from one observed $H(\alpha)$ and $H(\beta)$, and to derive all eight bits would therefore be expected to require $8/0.0545172 = 146.743$ observations. This is assuming that the Hamming weight is correctly identified every time.

5   Experimental Results

An attack was carried out on an unprotected software implementation of AES on a 8051 microprocessor. Two sets of templates were built for an unknown plaintext attack. The targeted operations, i.e. the α and β from Section 4, were the Hamming weights of the first plaintext byte and the first byte of the output of the SubBytes function in the first round. For the template generation, 9000 traces were used, each with a pseudorandomly generated plaintext and a constant key. The sum of the absolute difference method (Section 3) was used to select the points of interest in the trace. A seperate set of 1000 traces were used to conduct the attack. For comparison, the results of a known plaintext attack, using the same templates for the output of the SubBytes function, are given in Figure 1. As can be seen, after only five traces, the correct key is returned with a probability of 1. Figure 2 shows the results of the unknown plaintext attack. After approximately 85 traces, the correct key value is returned with a probability close to 1. This is consistent with the expected number of values calculated in Section 4.2, which indicates that the Hamming weight of the target value is identified correctly each time.

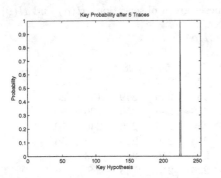

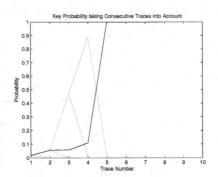

(a) Each possible key probability after 5 traces

(b) Key probabilites after each trace. Correct key hypothesis is given in black

Fig. 1. Known plaintext template attack

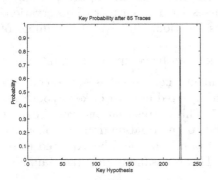

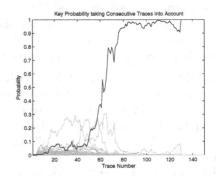

(a) Each possible key probability after 85 traces

(b) Key probabilites after each trace. Correct key hypothesis is given in black

Fig. 2. 8-bit microprocessor unknown plaintext template attack

As the traces are statistically independent, each key probability can be multiplied on a trace by trace basis. In an ideal scenario when a key hypothesis is classified as incorrect, it has a probability of 0 so will subsequently be eliminated by multiplication. However, in a practical template attack, the correct Hamming weight is not classified with a probability of 1, with incorrect values consequently 0. Therefore, keys are not ruled out on a trace by trace basis, but instead are multiplied by a smaller probability. The correct key value then emerges with the highest probability as can be seen in Figure 2(b).

5.1 Trace Signal-to-Noise Ratio

The number of required traces for an unknown plaintext attack on the microprocessor is consistent with the expected number of traces, as for the 85 traces used the Hamming weight value is identified correctly for the plaintext in 97.65% of

Table 3. Comparison of Identified Hamming Weights (HW) for 1000 traces

HW error	8-bit Microprocessor		ARM	
	plaintext	sbox	plaintext	sbox
0	99.5	96.8	47.0	45.9
1	0.5	3.2	46.0	45.1
2	0.0	0.0	6.2	7.8
3	0.0	0.0	0.8	1.1
> 3	0.0	0.0	0.0	0.1

cases, and for the SubBytes function it is classified correctly in 92.94% of cases. As incorrect classifications occur infrequently, they only have a small impact on the overall result.

This can be compared to the result of an attack on an unprotected AES implementation on a 32-bit ARM7TDMI [6] microprocessor, shown in Figure 3. A comparison of the percentage of correct Hamming weight classifications between targeted platforms is given in Table 3. The percentage of incorrect classifications is also given for the different Hamming weight errors. The AES functions on the 32-bit ARM were implemented using the same algorithms used on the 8-bit microprocessor implementation, i.e. no optimisations that take advantage of the 32-bit architecture were included, to allow for a fair comparison with the results given above. The templates were built using 10000 traces, using the sum of the absolute difference method to select the points of interest as before, and a further 1000 traces were used to conduct the attack. As the Hamming weight was identified correctly less frequently on the ARM chip than on the 8-bit microprocessor, considerably more traces were required to determine the correct key to allow for the incorrect classifications reducing the correct key probability.

In other similar cases using the same templates and a different set of attack data, the unknown plaintext classification did not return the correct answer. The

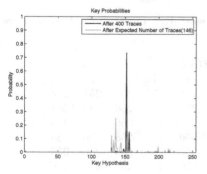

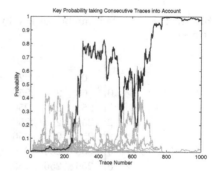

(a) Each possible key probability after 146 & 400 traces

(b) Key probabilites after each trace. Correct key hypothesis is given in black

Fig. 3. ARM unknown plaintext template attack

incorrect Hamming weight classifications not leading to any, or an incorrect, key value being returned. When attacking such noisy platforms the attack is helped by higher instances of low and/or high Hamming weights in the initial stages, as fewer key values are assigned high probabilities.

6   Application to Masked Implementations

A common side channel countermeasure is to use Boolean masking, where every intermediate state is stored in memory XORed with some random value that varies for each execution [17]. This random value will be chosen to be a convenient size for a given block cipher, e.g. for AES this would typically be an 8-bit value because of the SubBytes function. The target equation would then become

$$\beta \oplus \tau = S'(\alpha \oplus k \oplus \tau) = S(\alpha \oplus k) \oplus \tau \tag{11}$$

where τ is a random value generated for each execution of the implementation of AES.

To attack a masked implementation of AES our observations then become $H(\alpha \oplus \tau)$ and $H(\beta \oplus \tau)$. There will be some values of τ that will not be possible given a pair of observations (dictated by the structure of the function S), the simplest method of evaluating the expected number of hypotheses returned by evaluating a pair of observations is to count all the possibilities.

As described in Section 4.2, the overall expectation can be computed from this by multiplying each table entry by the probability of it occurring, i.e. $\frac{\binom{8}{H(\alpha \oplus \tau)} \cdot \binom{8}{H(\beta \oplus \tau)}}{2^{16}}$, and computing the sum of the result. This produces an expected number of key hypotheses for an observed $H(\alpha \oplus \tau)$ and $H(\beta \oplus \tau)$ to be 255.9. This means that one would expect to acquire $\log_2(256/255.992) = 0.00004299$ bits of information from one observed $H(\alpha \oplus \tau)$ and $H(\beta \oplus \tau)$, and to derive all eight bits would therefore be expected to require $8/0.00004299 = 186063$ observations, once again assuming the hamming weight is correctly identified each time.

Table 4. The expected number of key hypotheses given observations $H(\alpha \oplus \tau)$ and $H(\beta \oplus \tau)$

		0	1	2	3	4	5	6	7	8
	0	163	254	256	256	256	256	256	254	163
	1	254	256	256	256	256	256	256	256	254
	2	256	256	256	256	256	256	256	256	256
	3	256	256	256	256	256	256	256	256	256
$H(\beta \oplus \tau)$	4	256	256	256	256	256	256	256	256	256
	5	256	256	256	256	256	256	256	256	256
	6	256	256	256	256	256	256	256	256	256
	7	254	256	256	256	256	256	256	256	254
	8	163	254	256	256	256	256	256	254	163

The column headers 0–8 correspond to $H(\alpha \oplus \tau)$.

6.1   Attacking Masked Implementations

Eliminating the Mask: To successfully attack a masked implementation an attacker would be required to construct a third set of templates on τ; a known plaintext version of this attack is described in [20]. In this scenario, the attack is restricted to the first round only, as one of the templates must be built on α before it is masked, not $\alpha \oplus \tau$ which would allow any round to be targeted. In the case of building the templates for $\alpha \oplus \tau$ instead of just α, removing the mask from both $\alpha \oplus \tau$ and $\beta \oplus \tau$ by XORing both sets of values with the third template τ limits the amount the valid keyspace can be reduced, preventing key extraction. On the other hand, using the templates built on τ, the mask on β can be removed by XORing all possible combinations together. Then the templates based on α can be used to extract the secret data k in $\beta = S(\alpha \oplus k)$ as before. The attack reverts as the previously explained method without masking, at the cost of extra traces being required to deal with the extra unknown value τ. Hence, similar to the result presented in [20], Boolean masking does not prevent key extraction in the context of an unknown plaintext template attack.

Predicting the Plaintext: It has also been shown in [13] that a differential power analysis [7,16] is possible against an implementation of a block cipher where a counter is part of the plaintext, e.g. counter mode used for random number generation [9]. The initial value of the counter does not need to be known for an attacker to mount an attack against an implementation where no countermeasures are present. The same technique could be used to guess the values of a counter present in the plaintext to guess values of α such that information can be derived in a similar manner, by evaluating all the possible values of τ and $\beta \oplus \tau$, from templates constructed on $\alpha \oplus \tau$ and $\beta \oplus \tau$. The advantage of this attack over eliminating the mask is that templates only need to be constructed on two points.

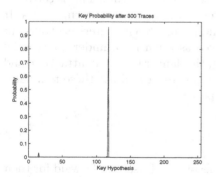 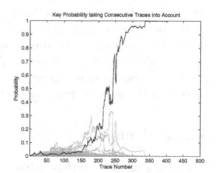

(a) Each possible key probability after 300 traces

(b) Key probabilites after each trace. Correct key hypothesis is given in black

Fig. 4. Simulated unknown plaintext template attack

6.2   Simulation Results

Figure 4 demonstrates a simulation of an unknown plaintext attack on a masked implementation of AES using three templates as described in the first part Section 6.1. This simulation assumes that the attacker is able to use templates built for the plaintext, mask and masked output of the first SubBytes operation to correctly identify the Hamming weight at the desired points in time. As can be seen from Figure 4(b), a greater number of traces than before are required to extract the correct secret key value with a high probability because of the extra variation that the third template introduces.

7   Conclusion

In this paper we describe a template attack where an attacker can derive the secret key used in an instantiation of a block cipher without knowledge of the plaintexts or ciphertexts used during the classification stage, and an implementation of this attack applied to AES is demonstrated.

We also show that this attack can be extended to block ciphers that are implemented using the Boolean masking countermeasure, using similar techniques to those described in [13] and [20]. A simulation of a template attack is described that assumes that an attacker would be able to identify the Hamming weight at three different points of the first round of the AES. In our attack we assume that the function S' is implemented such that both the input and output are masked with the same random value τ. In a real implementation, the input and output may be masked with two different values and our attack would no longer apply. However, the same random values would appear in all rounds, as it would not be practical to construct multiple examples of the SubBytes function in memory. An attack could then be constructed where the same point in two adjacent rounds would be analysed, rather than two points in the same round.

In the attacks implemented it is assumed that an attacker can characterise a device to build templates on intermediate states of an algorithm. This is certainly possible when block ciphers are implemented without any countermeasures. It would not be expected that an attacker would be able to characterise a transient random value, such as the random value used as a mask to hinder Differential Power Analysis. However, as observed in [20], the danger is that an attacker could characterise certain commands required by a processor and use these templates to implement the attacks described in this paper.

Acknowledgements

The authors would like to thank Robert McEvoy and Elisabeth Oswald for their helpful suggestions regarding the work presented in this paper. In particular, the authors would like to thank Elisabeth Oswald for providing the power traces for one of the attacks described in Section 5. The work described in this paper has been supported in part by the Informatics Commercialisation Initiative of

Enterprise Ireland, the European Commission IST Programme under Contract IST-2002-507932 ECRYPT and EPSRC grant EP/F039638/1 "Investigation of Power Analysis Attacks".

References

1. Agrawal, D., Rao, J.R., Rohatgi, P.: Multi-channel attacks. In: Walter, C.D., Koç, Ç.K., Paar, C. (eds.) CHES 2003. LNCS, vol. 2779, pp. 2–16. Springer, Heidelberg (2003)
2. Agrawal, D., Rao, J.R., Rohatgi, P., Schramm, K.: Templates as master keys. In: Rao, J.R., Sunar, B. (eds.) CHES 2005. LNCS, vol. 3659, pp. 15–29. Springer, Heidelberg (2005)
3. Akkar, M.-L., Bévan, R., Goubin, L.: Two power analysis attacks against one-mask methods. In: Roy, B., Meier, W. (eds.) FSE 2004. LNCS, vol. 3017, pp. 332–347. Springer, Heidelberg (2004)
4. Akkar, M.-L., Goubin, L.: A generic protection against high-order differential power analysis. In: Johansson, T. (ed.) FSE 2003. LNCS, vol. 2887, pp. 192–205. Springer, Heidelberg (2003)
5. Archambeau, C., Peeters, E., Standaert, F.-X., Quisquater, J.-J.: Template attacks in principal subspaces. In: Goubin, L., Matsui, M. (eds.) CHES 2006. LNCS, vol. 4249, pp. 1–14. Springer, Heidelberg (2006)
6. ARM Limited. ARM7TDMI technical reference manual (revision r4p1). ARM Doc No. DDI 0210, Issue C (2004), http://infocenter.arm.com/help/topic/com.arm.doc.ddi0210c/DDI0210B.pdf
7. Brier, E., Clavier, C., Olivier, F.: Correlation power analysis with a leakage model. In: Joye, M., Quisquater, J.-J. (eds.) CHES 2004. LNCS, vol. 3156, pp. 16–29. Springer, Heidelberg (2004)
8. Chari, S., Rao, J.R., Rohatgi, P.: Template attacks. In: Kaliski Jr., B.S., Koç, Ç.K., Paar, C. (eds.) CHES 2002. LNCS, vol. 2523, pp. 13–28. Springer, Heidelberg (2003)
9. Dworkin, M.: Recommendation for Block Cipher Modes of Operation: Methods and Techniques. National Institute of Standards and Technology (December 2001)
10. Fahn, P.N., Pearson, P.K.: Ipa: A new class of power attacks. In: Koç, Ç.K., Paar, C. (eds.) CHES 1999. LNCS, vol. 1717, pp. 725–726. Springer, Heidelberg (1999)
11. Gandolfi, K., Mourtel, C., Olivier, F.: Electromagnetic analysis: Concrete results. In: Koç, Ç.K., Naccache, D., Paar, C. (eds.) CHES 2001. LNCS, vol. 2162, pp. 251–261. Springer, Heidelberg (2001)
12. Handschuh, H., Preneel, B.: Blind differential cryptanalysis for enhanced power analysis. In: Biham, E., Youssef, A.M. (eds.) SAC 2006. LNCS, vol. 4356, pp. 163–173. Springer, Heidelberg (2007)
13. Jaffe, J.: A first-order DPA attack against AES in counter mode with unknown initial counter. In: Paillier, P., Verbauwhede, I. (eds.) CHES 2007. LNCS, vol. 4727, pp. 1–13. Springer, Heidelberg (2007)
14. Jolliffe, I.T.: Principal Component Analysis, 2nd edn. Springer Series in Statistics. Springer, Heidelberg (2002)
15. Kocher, P.: Timing attacks on implementations of Diffie-Hellman, RSA, DSS, and other systems. In: Koblitz, N. (ed.) CRYPTO 1996. LNCS, vol. 1109, pp. 104–113. Springer, Heidelberg (1996)

16. Kocher, P., Jaffe, J., Jun, B.: Differential power analysis. In: Wiener, M.J. (ed.) CRYPTO 1999. LNCS, vol. 1666, pp. 388–397. Springer, Heidelberg (1999)
17. Mangard, S., Oswald, E., Popp, T.: Power Analysis Attacks — Revealing the Secrets of Smart Cards. Springer, Heidelberg (2007)
18. Menezes, A.J., van Oorschot, P.C., Vanstone, S.A.: Handbook of Applied Cryptography. CRC Press, Boca Raton (1997)
19. National Institute of Standards and Technology. Advanced Encryption Standard (AES), FIPS–197 (2001)
20. Oswald, E., Mangard, S.: Template attacks on masking — resistance is futile. In: Abe, M. (ed.) CT-RSA 2007. LNCS, vol. 4377, pp. 243–256. Springer, Heidelberg (2007)
21. Rechberger, C., Oswald, E.: Practical template attacks. In: Lim, C.H., Yung, M. (eds.) WISA 2004. LNCS, vol. 3325, pp. 440–456. Springer, Heidelberg (2005)
22. Standaert, F.-X., Archambeau, C.: Using subspace-based template attacks to compare and combine power and electromagnetic information leakages. In: Oswald, E., Rohatgi, P. (eds.) CHES 2008. LNCS, vol. 5154, pp. 411–425. Springer, Heidelberg (2008)

On Comparing Side-Channel Preprocessing Techniques for Attacking RFID Devices

Thomas Plos, Michael Hutter, and Martin Feldhofer

Institute for Applied Information Processing and Communications (IAIK),
Graz University of Technology, Inffeldgasse 16a, 8010 Graz, Austria
{Thomas.Plos,Michael.Hutter,Martin.Feldhofer}@iaik.tugraz.at

Abstract. Security-enabled RFID tags become more and more important and integrated in our daily life. While the tags implement cryptographic algorithms that are secure in a mathematical sense, their implementation is susceptible to attacks. Physical side channels leak information about the processed secrets. This article focuses on practical analysis of electromagnetic (EM) side channels and evaluates different preprocessing techniques to increase the attacking performance. In particular, we have applied filtering and EM trace-integration techniques as well as Differential Frequency Analysis (DFA) to extract the secret key. We have investigated HF and UHF tag prototypes that implement a randomized AES implementation in software. Our experiments prove the applicability of different preprocessing techniques in a practical case study and demonstrate their efficiency on RFID devices. The results clarify that randomization as a countermeasure against side-channel attacks might be an insufficient protection for RFID tags and has to be combined with other proven countermeasure approaches.

Keywords: RFID, Differential Frequency Analysis, Side-Channel Analysis, Electromagnetic Attacks.

1 Introduction

During the last few years, Radio-Frequency Identification (RFID) has emerged from a simple identification technique to the enabler technology for buzzwords like "ambient intelligence" or the "Internet of things". Additional features like sensors and actuators allow applications in many different fields apart from supply-chain management and inventory control. Sarma *et al.* [19] have been the first who addressed the importance of security for passive RFID tags. The introduction of security allows tags to prove their identity by means of cryptographic authentication. Furthermore, privacy issues could be solved and a protected access to the tag's memory becomes possible.

In 2003, it was stated e.g. by Weis *et al.* [20] that strong cryptography is unfeasible on passive tags due to the fierce constraints concerning power consumption and chip area. Since then, many attempts have been made to implement standardized cryptographic algorithms in hardware complying with the

H.Y. Youm and M. Yung (Eds.): WISA 2009, LNCS 5932, pp. 163–177, 2009.

requirements of passive RFID tags. Among the most popular publications on that are realizations of the Advanced Encryption Standard (AES) [6], Elliptic-Curve Cryptography (ECC) [3,8], and GPS [9,15].

Unfortunately, having a crypto module of a secure algorithm in hardware on the tag is not sufficient for a secure RFID system. Due to the fact that an adversary always tries to break the weakest link in a system (and this is the RFID tag that is easily available for attacks), further attacks have to be considered. Side-channel attacks target at the implementation of a cryptographic device. They are very powerful in retrieving the secret key by measuring some physical property like power consumption, electromagnetic emanation, or timing behavior *etc.* Differential power analysis (DPA) [13] attacks and differential electromagnetic analysis (DEMA) [18,1] attacks gained a lot of attention during the last ten years.

In the findings of Hutter *et al.* [11] for HF tags as well as in the work of Oren *et al.* [16] and Plos [17] for the UHF frequency range, it has been shown that passive RFID tags are also susceptible to side-channel attacks. Even in the presence of the strong electromagnetic field of the reader DEMA attacks are possible. Hence, as far as a cryptographic algorithm is implemented on a tag, appropriate countermeasures have to be implemented. According to [14], countermeasures can principally be divided in either hiding or masking.

A very efficient way of implementing hiding, especially for low-resource devices like RFID tags, is to randomize the execution of the algorithm. This means that the performed operations of the algorithm occur at different moments in time in each execution. Randomization can be done by shuffling and by randomly inserting dummy cycles [14]. The reason why randomization is very cost efficient in terms of hardware resources is that the implementation is mainly done in the control logic. Moreover, spending additional clock cycles for randomizing the execution of the algorithm is convenient since the data rates used in RFID systems are rather low.

Differential Frequency Analysis (DFA)—not to confuse with differential fault analysis, which uses the same acronym—has been first mentioned by Gebotys *et al.* [7] in 2005. There, the authors successfully applied DFA to attack cryptographic algorithms running on a Personal Digital Assistant (PDA) device. The principle idea of DFA is to transform measured side-channel traces from the time domain to the frequency domain. The Fast Fourier Transform (FFT) is an operation that can be used for this transformation. Since the FFT is time-shift invariant, the time delays introduced by the side-channel analysis countermeasures are removed in the frequency domain. Further advantage of DFA especially for attacking RFID tags is that misaligned traces are of no concern. Misalignments do often occur due to the interfering reader field and difficulties in triggering appropriate events on the tag. Another approach that uses the frequency domain for handling misaligned traces has been presented by Homma *et al.* [10] in 2006. They have been able to diminish the displacement between traces by using a so-called phase-only correlation after transformation to the frequency domain.

In this work we show how DFA can be used to extract the secret key out of RFID devices that implement randomization countermeasures. We compare DFA with preprocessing techniques such as filtering and trace integration. Commercially-available RFID tags today do not contain cryptographic algorithms with randomization countermeasures implemented. In order to perform and analyze the proposed attacks, we used semi-passive RFID-tag prototypes for HF and UHF frequency as target of evaluation. In these prototypes it is possible to implement e.g. the AES algorithm with randomization countermeasures in software. Our results show that DFA is a powerful technique, especially when analyzing the electromagnetic emanation of RFID devices.

This article is structured as follows. In Section 2, we describe the HF and the UHF RFID-tag prototypes used throughout the analysis. Section 3 provides insights about different hiding techniques that are applied in practice followed by attacking techniques on them in Section 4. The design of the randomized AES implementation used during the experiments is presented in Section 5 and the measurement setups for the attacks are shown in Section 6. Results are given in Section 7. The article closes with conclusions in Section 8.

2   RFID-Tag Prototype Implementation

In order to perform side-channel analysis on RFID devices, we have developed two different RFID-tag prototypes. Using prototypes provides many advantages. They can be used to demonstrate new applications and protocols, making an invention more informative and imaginable. Prototypes are also suitable for identifying weaknesses more easily by modifying and testing the device in real terms. In cryptographic systems, prototypes allow the analysis of side channels by measuring, for example, the electromagnetic emanation. This article focuses on such analyses by using prototypes that implement security mechanisms.

Two RFID-tag prototypes have been designed and developed. One prototype operates in the HF frequency band at 13.56 MHz and one prototype works in the UHF frequency band at 868 MHz. Both devices have been assembled using discrete components. In Figure 1, a picture of the two prototypes is shown. They principally consist of an antenna, an analog front-end, a microcontroller, a clock oscillator, a serial interface, a JTAG interface, and a power-supply connector. Both devices differ in their antenna design, the analog front-end, the clock source, and the software that runs on the microcontroller. The remaining components are the same. As a microcontroller, the ATmega128 [2] has been used, which manages reader requests and tag responses by following the specification of the used RF communication protocol. The microcontroller is able to communicate with a PC over a serial interface. It furthermore supports In-System Programming (ISP) and has a JTAG interface for debug control and system programming. Both devices are semi-passive where the microcontroller is powered by an external power source, typically a battery, while the RF communication is done passively without any signal amplification.

In the following sections, the design of the HF and the UHF tag prototype is described in a more detail.

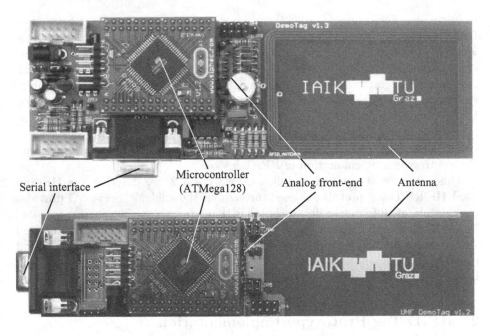

Fig. 1. Picture of the HF (top) and the UHF (bottom) tag prototype

2.1   HF-Tag Prototype

The HF-tag prototype uses a self-made antenna according to ISO 7810. It consists of a coil with four windings that allows the communication with a reader over the air interface. The antenna is tuned to resonate at a carrier frequency of 13.56 MHz, which is realized by a matching RLC circuit. This circuit narrows the frequency range and can also be considered as a band-pass filter that passes the carrier frequency but attenuates unwanted and spurious frequencies. The matched signals are then preprocessed by an analog front-end that is used to transform the analog signals into the digital world. First, the signals are rectified using a bridge rectifier. Small-signal Shottky diodes have been assembled that provide low voltage drops and low leakage currents. Second, the voltage is regulated by a Zener diode. At the third stage, a comparator is used to identify reader modulations. The output of the comparator is then connected to the microcontroller that rises an interrupt and starts the receiving process. The microcontroller is clocked by a 13.56 MHz quartz crystal that has been assembled on board. For sending data from the tag to the reader, a load modulation circuit is available that consists of a shunt and a transistor. The microcontroller triggers the transistor that switches the shunt and thus modulates the tag response.

The tag prototype can communicate using several protocol standards. It implements ISO 15693, ISO 14443 (type A and B), ISO 14443-4 and ISO 18092. The software is written in C while parts have been implemented in assembly language due to timing constraints. Moreover, it implements a user-command interface that allows easy administration over the serial interface. For our experiments, we have used the ISO 14443-A protocol standard [12] and have included some proprietary commands that implement a simple challenge-response protocol. First, the reader sends 16 bytes of plaintext to the prototype. The prototype encrypts the plaintext using the Advanced Encryption Standard (AES). Second, the reader retrieves the ciphertext and verifies the encrypted result. Furthermore, we have implemented a command to set different randomization parameters for the AES encryption. These parameters are used to randomize the encryption process that is commonly used as a countermeasure for side-channel analysis.

2.2 UHF-Tag Prototype

The second tag prototype operates in the UHF frequency band. In contrast to the HF-tag prototype it uses a half-wave dipole antenna consisting of two wires directly integrated to the layout of the printed circuit board (PCB). The antenna, whose length is about 150 mm, is optimized for a frequency of 868 MHz and it is connected to the analog front-end. Like for the HF-tag prototype, an adjustable capacitor is placed in parallel to its antenna. This capacitor is used for matching the antenna to the input impedance of the analog front-end. Signals that are received by the antenna are first rectified by a charge-pump rectifier. This rectifier performs demodulation and voltage multiplication all at once. Special detector diodes, which have a low voltage drop and are constructed to operate up to some GHz, are used in the rectifier circuit. Subsequently, signals are filtered and passed to a comparator before feeding them to the microcontroller. For tag-to-reader communication, a backscatter-modulation circuit is provided within the analog front-end. This circuit, which works similar to the one used by the HF-tag prototype, uses a transistor to switch an impedance (shunt and capacitor) in parallel to the tag antenna. A 16 MHz quartz crystal is assembled on board in order to generate the system clock for the microcontroller.

The UHF-tag prototype supports the ISO 18000-6C standard (EPC Gen2 [5]) which is the most widespread protocol in the UHF frequency range. Implementation of the protocol is done in software on the microcontroller. The software for the UHF-tag prototype is also mainly written in C while time-critical routines are directly realized via assembly language. Also the same challenge-response protocol has been implemented that allows encryption of received data, as well as a dedicated command to adjust the parameters for the AES randomization.

3 Hiding as a Countermeasure against Side-Channel Analysis

Hiding data-dependent information of a cryptographic device can be achieved by two different approaches. The first approach blurs the data-dependent information by varying the power-consumption characteristic in its amplitude. The second approach randomizes the execution of operations in the time dimension. However, hiding can also occur in an *unintended* manner. There, misaligned traces in the amplitude and also in the time make the analysis of side channels largely infeasible. Measurements on contactlessly-powered devices like RFID tags are a typical example for this scenario where the acquired EM traces have to be aligned or preprocessed before the analysis in order to perform a successful attack. In the following, a short description of hiding in the amplitude and hiding in the time dimension is given. A more detailed description can be found in [14].

3.1   Hiding in the Amplitude Dimension

In fact, the measurement of side channels that leak from RFID devices is a challenging task. RFID readers emit a very strong field in order to allow a certain reading range. This field is necessary to power the tags, to allow a communication, and in most cases also to provide a clock signal to the tags. However, the field interferes and perturbs the measurement of the weak side-channel emissions. In addition, if the reader field and the clock signal of the tag differ in their frequency, a superposition of signals can be perceived. This results in periodic rises and falls in the amplitude of the measured EM traces. Measurements on HF RFID-tag prototypes, whose clock frequency differ from 13.56 MHz, are a typical example where the reader field interferes the measurement of interesting side-channel emissions. Measurements on UHF tags are another example where they often include their own oscillators. The internal clock allows the communication with multiple reader frequencies such as used in different countries (868 MHz in Europe, 915 MHz in USA, or 950 MHz in Japan).

In contrast to these *unintended* interferences, variations in the amplitude are also often generated purposely. This has its reason in the fact, that variations in the amplitude dimension essentially lower the signal-to-noise (SNR) ratio and thus make the measurement of side channels harder to perform. This kind of hiding is commonly-used as a countermeasure against such attacks. Devices often integrate noise generators or perform several operations in parallel to increase the overall noise [21].

3.2   Hiding in the Time Dimension

Hiding can also emerge in the time dimension where traces are misaligned either in an unintended or in an intended manner. Unintended time variations often occur due to the absence of adequate trigger signals for measurement. Especially in RFID environments, triggering is often performed on the communication instead of the measured emanation. For example, the end of the last reader command before executing the targeted algorithm can be used to trigger the measurement. This trigger signal does not always appear at the same position in time which leads to misaligned traces and thus to unintended hiding.

Intended time variations are referred to hiding through randomization. There are two possibilities on how the execution of an algorithm can be randomized. The first possibility is to insert dummy operations such as additional rounds (or only parts of it). These dummy operations can be processed before or after

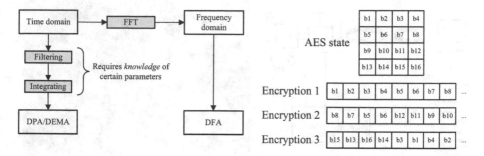

Fig. 2. Overview of the preprocessing steps necessary for DEMA and DPA as well as DFA attacks

Fig. 3. Principle of shuffling used in the randomized AES implementation

the execution of the actual algorithm. The second possibility is to shuffle the sequence of operations [4]. In respect of AES, several operations can be randomized such as AddRoundKey, SubBytes, ShiftRows, or MixColumns.

4   Attacking Techniques on Hiding

There exist techniques that increase the performance of attacks on hiding through trace preprocessing. The most obvious and commonly-used preprocessing technique is filtering. By applying different filters, it is possible to reduce noise that originates from narrow-band interferers such as RFID readers. Filtering of these perturbing signals helps to evade hiding in the amplitude dimension. Though this requires knowledge of the appropriate filter parameters to preserve data-dependent information in the traces. In contrast, hiding in the time dimension can be obviated by integration of power or EM traces. Specific points in time are summed up before performing the attack. In practice, only points are chosen that exhibit a high side-channel leakage. These points form a kind of comb or window that can be swept through the trace in order to obtain the highest correlation. This technique is often referred to as *windowing*. However, it is evident that this technique implies the knowledge of certain points in time where the leakage of information is high. If no knowledge of this leakage is available, it shows that the performance of this attack is rather low due to the integration of unimportant points.

Another related technique uses FFT to transform the traces into the frequency domain. Instead of performing differential analysis in the time domain (as done in standard DPA and DEMA attacks), the analysis is performed in the frequency domain. This allows a time-invariant analysis of side-channel leakages across the overall signal spectrum. Such an analysis is also referred as Differential Frequency Analysis (DFA) [7]. Figure 2 illustrates the necessary preprocessing steps for conducting DEMA and DPA attacks as well as DFA attacks in presence of hiding. In this article, all three discussed types of preprocessing techniques are analyzed

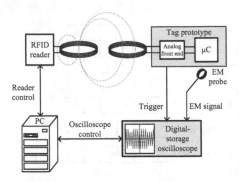 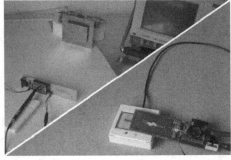

Fig. 4. Schematic view of the general measurement setup used to gather the EM emissions of the tag prototypes

Fig. 5. Picture of the measurement setup using UHF (upper left) and HF (lower right) RFID-tag prototypes

in terms of their efficiency. These preprocessing techniques are applied on EM measurements having increased noise in both amplitude and time dimension. This noise is caused by an interfering RFID reader and by a randomized AES implementation that is described in the following.

5   Description of the Randomized AES Implementation

In our experiments, a 128-bit AES implementation has been used that offers hiding in the time dimension. First, the implementation allows to choose additional rounds that are randomly executed either at the beginning or the end of the actual algorithm. Second, it allows the shuffling of bytes $b1$ to $b16$ within the AES state. There, the sequence of the columns and the sequence of the rows can be randomized as shown in Figure 3. In order to set specific randomization parameters during our experiments, we have implemented a custom command that can be sent over the air interface. These parameters define the number of dummy rounds and the number of shuffling operations. In particular, it is possible to define the sequence of the columns as well as the sequence of the rows within the AES state. If no dummy rounds are inserted and all bytes of the state are shuffled, 16 different positions can be taken over time for one state operation. Regarding side-channel analysis, the correlation coefficient through randomization is then reduced linearly by a factor of 16. The number of necessary traces to succeed an attack increases by a factor of $16^2 = 256$. However, the quadratic influence is only correct when no preprocessing method like windowing or DFA is applied [14].

6   Measurement Setups

The measurement setup used for our experiments is shown in Figure 4. It comprises different devices such as a PC, a standard RFID reader, a digital-storage oscilloscope, the tag prototype, and a measurement probe. The RFID reader and the digital-storage oscilloscope are directly connected to the PC that controls the overall measurement process. Matlab is running on the PC and is used to apply the preprocessing techniques and to conduct the side-channel analysis. The RFID reader communicates with the tag prototype via the air interface. For the HF-tag prototype, the ISO 14443-A protocol has been used while the ISO 18000-6C protocol has been used for the UHF-tag prototype. The HF-tag prototype has been placed directly upon the reader antenna. The UHF-tag prototype has been placed 30 cm in front of the UHF reader. Two channels of the digital-storage oscilloscope (*LeCroy LC584AM*) are used in our experiments. One channel is connected to the trigger pin of the tag prototype, the other channel is connected to the measurement probe. Signals have been sampled with 2 GS/s. Both tag prototypes have been programmed to release a trigger event whenever a new AES encryption is started. This trigger event causes the oscilloscope to record the EM emissions of the tag prototype using magnetic near-field probes. We have used two probes from *EMV Langer* which is the RF R 400 for the HF measurements and the RF B 3-2 for the UHF measurements. Figure 5 shows a picture of the measurement setup for the HF and one for the UHF-tag prototype.

7   Results

In this section, the results of the performed side-channel attacks on our RFID-tag prototypes are presented. Attacks have been performed on the electromagnetic emissions of the HF and the UHF-tag prototype. The target of all attacks has been the first byte of the first round of AES. As a power model, the Hamming weight has been used.

First, we have analyzed the impact of misaligned traces in the amplitude dimension. For this, we have measured EM emissions of our prototypes that are interfered by unsynchronized reader signals. Note that no randomization of the AES state is enabled in this experiment. In order to perform attacks on such kind of hiding, we investigated two different preprocessing approaches. The first approach applies filtering techniques to suppress the interfering noise of the reader. The second approach applies an FFT before performing the DFA attack. We compare both techniques in their practical efficiency and performance. Second, we show results of attacks that have been performed on misaligned traces in both amplitude and in the time dimension. For this experiment, we have enabled the randomization of the AES implementation which is commonly-used in practice to counteract against side-channel attacks. For this scenario, we have applied trace-integration techniques by windowing. We also compare the results with the results obtained by DFA.

7.1   Attacks on Hiding in the Amplitude Dimension

At first, we focus on typical measurements in RFID environments. The misalignment of EM traces is often caused by readers that interfere the EM measurement

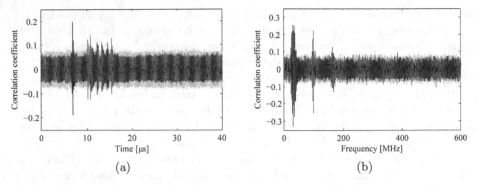

Fig. 6. Result of the filtered DEMA attack (a) and DFA attack (b) on the HF-tag prototype when using hiding in the amplitude domain

of RFID devices. In our experiment, we consider the scenario where the clock signal of our prototype and the reader carrier are desynchronized. This is already the case for our UHF-tag prototype which operates at 16 MHz and which communicates with an 868 MHz reader. For the HF prototype, we have used a 13.56 MHz quartz crystal that is assembled on board. This quartz crystal is also unsynchronized with the communicating 13.56 MHz reader. Both devices have been placed inside the reader field, which interfered the measurement due to additional noise. After the acquisition of 2 000 traces, we have performed filtering techniques to circumvent the interferer and to decrease the noise at this juncture. For the HF prototype, a bandstop filter has been designed using Matlab that filters the 13.56 MHz carrier. For the UHF prototype, a low-pass filter has been used that passes all frequencies below 200 MHz. We have performed a filtered DEMA attack and a DFA attack using FFT.

In Figure 6(a), the result of the filtered DEMA attack for the HF-tag prototype is shown. The correct key hypothesis is plotted in black while all other key hypotheses are plotted in gray. The correct key hypothesis leads to a correlation coefficient of 0.20. Figure 6(b) shows the result of the DFA attack. Three peaks in the electromagnetic spectrum are clearly discernable, which represent high data-dependent frequency emissions. The highest absolute correlation coefficient has been 0.33 and occurred at a frequency of around 33 MHz.

In Figure 7(a), the result of the filtered DEMA attack is presented that has been performed on the UHF-tag prototype. A maximum absolute correlation coefficient of 0.63 has been obtained for the correct key hypothesis. Figure 7(b) shows the result of the DFA attack. As opposed to the results of the HF-tag prototype, many peaks occurred up to a frequency of about 600 MHz. The highest correlation that has been obtained is 0.28.

For the UHF-tag prototype, the results show a higher correlation coefficient compared to the results of the HF prototype. This is explained by the fact that our UHF measurement setup provides a higher SNR. On the one hand, a different EM probe has been used for the measurement that allows the probe to be drawn nearer to the surface of the chip. On the other hand, our experiments have shown

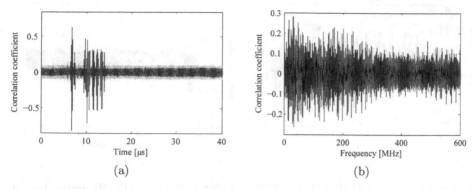

Fig. 7. Result of the filtered DEMA attack (a) and DFA attack (b) on the UHF-tag prototype when using hiding in the amplitude domain

that the UHF reader produces lower noise compared to the HF reader. However, when the result of the filtering technique and the result of the DFA are compared to each other, it shows that the DFA attack leads to a higher correlation in noisier environments while it is less effective in measurements where a low noise source is present.

7.2   Attacks on Hiding in the Amplitude and Time Dimension

Next, we consider the scenario where a side-channel countermeasure is enabled on the tag side. In addition to the noise of the reader, we have activated the hiding mechanism using AES randomization. As stated in Section 5, we are able to shuffle all bytes within the AES state. This leads to 16 different positions in time where a byte may be processed during one round. Nonetheless, the results of our experiments have shown that for the HF tag no significant correlation has been obtained for the case where we have preprocessed the traces using the trace integration (windowing) technique. By performing the attack in the frequency domain using DFA, we successfully revealed the correct key byte. However, we decided to reduce the number of shuffling bytes to 8 for the HF-measurement scenario in order to succeed the attack in both cases. The attacks for the UHF-measurements scenario, in contrast, have been successful when randomizing all 16 bytes of the AES state. We performed software filtering as described in the section above to reduce the noise of the RFID reader for the DEMA attacks in the time domain. For each experiment, 10 000 traces have been captured.

The attack using windowing as a preprocessing technique has been performed as follows. We summed up 100 points in time which showed the highest correlation in a previously performed standard DEMA attack. This defines an integration window that involves points with high data-dependent information. For a better visualization of the window matching, we have further implemented an automatic sweep that slides the window from the beginning of the trace to its end. At each position in time, all points of the window are summed up and a DEMA attack is performed afterwards. This results in a correlation trace where a peak

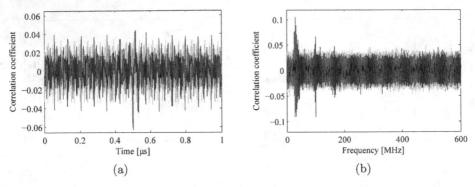

Fig. 8. Result of the windowing attack (a) and DFA attack (b) on the HF-tag prototype when using hiding in the amplitude domain and shuffling 8 bytes of the AES state

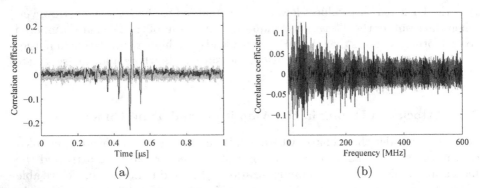

Fig. 9. Result of the windowing attack (a) and DFA attack (b) on the UHF-tag prototype when using hiding in the amplitude domain and shuffling 16 bytes of the AES state

occurs in time when the window fits best the specified data-dependent locations. In Figure 8(a), the result of the attack on the HF-tag prototype is shown where we have zoomed only into the interesting region in time. A peak is observable which has a maximum absolute correlation coefficient of 0.06. In Figure 8(b), the result of the performed DFA attack is given. Note that neither filtering nor other trace-alignment techniques have been applied before. Two peaks are discernable that arise at about 30 MHz and 100 MHz. These data-dependent frequencies are the same as those we have already obtained in the previous experiment (see Figure 6(b)). The highest correlation coefficient is 0.10.

After that, we have focused on our UHF-tag prototype. We have applied the same integration technique as used for the HF-tag prototype. In contrast to the attack on the HF-tag prototype where 8 bytes have been randomized, now 16 bytes have been shuffled within the AES state. Figure 9(a) shows the trace-integration result of the UHF-tag prototype. A maximum absolute correlation coefficient of 0.23 has been obtained. In Figure 9(b), the result of the performed

DFA attack is shown again without using any filtering or trace-alignment techniques. There, a maximum correlation coefficient of 0.14 is obtained.

By taking a closer look at our results, it becomes clear that DFA poses a powerful and easy preprocessing technique that is able to reveal the secret key of our RFID-tag prototypes. DFA provides not only high correlation even in noisy environments but can also be successfully applied against randomization countermeasures without having any knowledge of either interfering frequencies nor data-dependent locations.

8 Conclusions

In this article, we present results of performed DEMA and DFA attacks on HF and UHF RFID-tag prototypes. We addressed the issue of misaligned traces that are captured during EM measurements. These traces are interfered by the reader field, which results in a lower SNR within the amplitude dimension. In addition to that, we have investigated a randomized AES implementation in software that hides the leakage of side channels in the time dimension. We performed several attacks by applying filtering, trace integration, and DFA preprocessing techniques. Our experiments prove that DFA is a powerful technique that allows a fast and time-invariant analysis even in environments where traces are misaligned due to noise and randomization. Filtering techniques, in contrast, need the knowledge of the noise-source frequency and might also suppress interesting leakages. Applying integration techniques is a time-consuming task that requires the knowledge of data-dependent locations to design an appropriate integration window. Moreover, if the degree of randomization is increased, the number of windowing points has to be increased as well. We conclude that DFA offers many advantages especially when neither knowledge of the device nor possibilities of noise reduction are given. All side-channel attacks performed on the RFID-tag prototypes with the randomized AES implemented in software have been successful by applying DFA. This also clarifies that RFID devices that are using randomization as a countermeasure suffer from this kind of attack. The effort for attacking commercially-available RFID tags is assumed to be higher, since they will have their cryptographic algorithm and the countermeasure realized in dedicated hardware. Nevertheless, combining randomization with other countermeasure approaches as proposed in [14] might be a good solution to provide a higher degree of security.

Acknowledgements

This work has been supported by the Austrian Government through the research program FIT-IT Trust in IT Systems (Project POWER-TRUST under the Project Number 816151 and Project CRYPTA under the Project Number 820843).

References

1. Agrawal, D., Archambeault, B., Rao, J.R., Rohatgi, P.: The EM Side-Channel(s). In: Kaliski Jr., B.S., Koç, Ç.K., Paar, C. (eds.) CHES 2002. LNCS, vol. 2523, pp. 29–45. Springer, Heidelberg (2003)
2. Atmel Corporation. 8-bit AVR Microcontroller with 128K Bytes In-System Programmable Flash (August 2007),
   http://www.atmel.com/dyn/resources/prod_documents/doc2467.pdf
3. Batina, L., Guajardo, J., Kerins, T., Mentens, N., Tuyls, P., Verbauwhede, I.: Public-Key Cryptography for RFID-Tags. In: Workshop on RFID Security 2006 (RFIDSec 2006), Graz, Austria, July 12-14 (2006)
4. Clavier, C., Coron, J.-S., Dabbous, N.: Differential Power Analysis in the Presence of Hardware Countermeasures. In: Paar, C., Koç, Ç.K. (eds.) CHES 2000. LNCS, vol. 1965, pp. 252–263. Springer, Heidelberg (2000)
5. EPCglobal. EPC Radio-Frequency Identity Protocols Class-1 Generation-2 UHF RFID Protocol for Communications at 860 MHz - 960 MHz Version 1.0.9 (January 2005), http://www.epcglobalinc.org/
6. Feldhofer, M., Dominikus, S., Wolkerstorfer, J.: Strong Authentication for RFID Systems using the AES Algorithm. In: Joye, M., Quisquater, J.-J. (eds.) CHES 2004. LNCS, vol. 3156, pp. 357–370. Springer, Heidelberg (2004)
7. Gebotys, C.H., Ho, S., Tiu, C.C.: EM Analysis of Rijndael and ECC on a Wireless Java-Based PDA. In: Rao, J.R., Sunar, B. (eds.) CHES 2005. LNCS, vol. 3659, pp. 250–264. Springer, Heidelberg (2005)
8. Hein, D., Wolkerstorfer, J., Felber, N.: ECC is Ready for RFID – A Proof in Silicon. In: Avanzi, R., Keliher, L., Sica, F. (eds.) SAC 2008. LNCS, vol. 5381, pp. 401–413. Springer, Heidelberg (2008)
9. Hofferek, G., Wolkerstorfer, J.: Coupon Recalculation for the GPS Authentication Scheme. In: Grimaud, G., Standaert, F.-X. (eds.) CARDIS 2008. LNCS, vol. 5189, pp. 162–175. Springer, Heidelberg (2008)
10. Homma, N., Nagashima, S., Imai, Y., Aoki, T., Satoh, A.: High-Resolution Side-Channel Attack Using Phase-Based Waveform Matching. In: Goubin, L., Matsui, M. (eds.) CHES 2006. LNCS, vol. 4249, pp. 187–200. Springer, Heidelberg (2006)
11. Hutter, M., Mangard, S., Feldhofer, M.: Power and EM Attacks on Passive 13.56 MHz RFID Devices. In: Paillier, P., Verbauwhede, I. (eds.) CHES 2007. LNCS, vol. 4727, pp. 320–333. Springer, Heidelberg (2007)
12. International Organization for Standardization (ISO). ISO/IEC 14443: Identification Cards - Contactless Integrated Circuit(s) Cards - Proximity Cards (2000)
13. Kocher, P.C., Jaffe, J., Jun, B.: Differential Power Analysis. In: Wiener, M. (ed.) CRYPTO 1999. LNCS, vol. 1666, pp. 388–397. Springer, Heidelberg (1999)
14. Mangard, S., Oswald, E., Popp, T.: Power Analysis Attacks – Revealing the Secrets of Smart Cards. Springer, Heidelberg (2007)
15. McLoone, M., Robshaw, M.J.B.: New Architectures for Low-Cost Public Key Cryptography on RFID Tags. In: Proceedings of IEEE International Symposium on Circuits and Systems (ISCAS 2007), New Orleans, USA, May 27-30, pp. 1827–1830. IEEE, Los Alamitos (2007)
16. Oren, Y., Shamir, A.: Remote Password Extraction from RFID Tags. IEEE Transactions on Computers 56(9), 1292–1296 (2007)
17. Plos, T.: Susceptibility of UHF RFID Tags to Electromagnetic Analysis. In: Malkin, T.G. (ed.) CT-RSA 2008. LNCS, vol. 4964, pp. 288–300. Springer, Heidelberg (2008)

18. Quisquater, J.-J., Samyde, D.: ElectroMagnetic Analysis (EMA): Measures and Counter-Measures for Smart Cards. In: Attali, S., Jensen, T.P. (eds.) E-smart 2001. LNCS, vol. 2140, pp. 200–210. Springer, Heidelberg (2001)
19. Sarma, S.E., Weis, S.A., Engels, D.W.: RFID Systems and Security and Privacy Implications. In: Kaliski Jr., B.S., Koç, Ç.K., Paar, C. (eds.) CHES 2002. LNCS, vol. 2523, pp. 454–469. Springer, Heidelberg (2003)
20. Weis, S.A., Sarma, S.E., Rivest, R.L., Engels, D.W.: Security and Privacy Aspects of Low-Cost Radio Frequency Identification Systems. In: Hutter, D., Müller, G., Stephan, W., Ullmann, M. (eds.) Security in Pervasive Computing. LNCS, vol. 2802, pp. 201–212. Springer, Heidelberg (2004)
21. Witteman, M.: Advances in Smartcard Security. Information Security Bulletin (7), 11–22 (2002)

You Cannot Hide behind the Mask: Power Analysis on a Provably Secure S-Box Implementation*

J. Pan, J.I. den Hartog, and Jiqiang Lu

Eindhoven University of Technology,
Den Dolech 2, 5612 AZ, Eindhoven, The Netherlands

Abstract. Power analysis has shown to be successful in breaking symmetric cryptographic algorithms implemented on low resource devices. Prompted by the breaking of many protected implementations in practice, researchers saw the need of validating security of implementations with formal methods. Three generic S-box implementation methods have been proposed by Prouff el al., together with formal proofs of their security against 1st or 2nd-order side-channel analysis. These methods use a similar combination of masking and hiding countermeasures. In this paper, we show that although proven resistant to standard power analysis, these implementation methods are vulnerable to a more sophisticated form of power analysis that combines Differential Power Analysis (DPA) and pattern matching techniques. This new form of power analysis is possible under the same assumptions about power leakage as standard DPA attacks and the added complexity is limited: our experiments show that 900 traces are sufficient to break these algorithms on a device where 150 traces are typically needed for standard DPA. We conclude that the defense strategies—hiding by repeating operations for each possible value, and masking and hiding using the same random number—can create new vulnerabilities.

Keywords: Power analysis, side-channel analysis, provable security, block cipher S-box.

1 Introduction

With the expansion of electronic data-processing systems, small cryptographic devices like smart card and key tokens have tremendous possibilities for devising solutions for crucial applications, such as financial transaction and user identification. Based on cryptography, these devices could safely store secret keys and execute cryptographic algorithms. The security of cryptographic devices is therefore of vital importance and is the prime concern during design and development of such a device and the software running on it.

Power analysis has shown to be a practical threat to the security of cryptographic devices. Using the fact that the power consumption of the device is

* This work is funded by the Dutch Sentinels-project PinpasJC TIF.6687.

H.Y. Youm and M. Yung (Eds.): WISA 2009, LNCS 5932, pp. 178–192, 2009.

dependent on the data that is being processed within the device, power analysis extracts confidential information—such as the secret key used in a cryptographic algorithm—and leads to powerful and easy to conduct attacks. Ensuring the security of cryptographic devices against power analysis is an ongoing arms race at the forefront of scientific research, with new attacks being discovered, new countermeasures being introduced (see [1,2,7,14] for example).

As many implementations which are supposedly secured with effective countermeasures turn out to be broken, be it by incorrect use of the countermeasure or use of inadequate countermeasures, the need for a formal proof of the security of an implementation arises. Prouff et al. [9,10] proposed three generic S-box implementation methods that were demonstrated secure against 1st/2nd-order DPA attacks within the proof-of-security framework presented in [12]. In their methods, a combination of masking and hiding is used. On the one hand, the input and output of the S-box are masked by random values; on the other hand, the S-box look-up is executed for all possible values in a loop, during which the expected result, the masked output of S-box for the given input, is produced at a random moment in time for each execution. As will be explained in detail in Section 2, these implementations can thwart many powerful power analysis attacks.

In this paper, we introduce a new form of power analysis that can break the S-box implementation methods in [9,10], and any implementation methods that are of the same fashion. The attack makes use of the facts that the S-box look-up is executed for all possible values in an order that is predictable, and that the random value used for masking the intermediate result is also the value used for shuffling the look-ups. The new attack does not require a leakage of information that is more than required for standard DPA and the added complexity is limited. Using a device where 150 power traces are typically needed to break an unprotected implementation of AES S-box, practical experiments show that our attack on the protected implementation of AES S-box requires only 900 traces for success. Our theoretical analysis indicates that even when the noise is relatively high (e.g. SNR=0.015), our attack is still less than two orders of magnitude (i.e. 100 times) harder than a standard DPA attack. Moreover, the cost of our attack is independent from the number of random values used for the protection, as long as they are handeled in the same fashion as in the implementation methods presented in [9,10]. As a result, this new form of power analysis will reasonably be applicable in any setting where standard DPA is possible. Thus, though the implementation methods in [9,10] are formally proven to resist (1st/2nd-order) DPA attacks, they are not secure in any setting where such attacks are possible.

The remainder of the paper is organized as follows. Section 2 summarizes some popular power analysis techniques and countermeasures, followed by brief descriptions of the attacked S-box implementation methods proposed in [9,10]. Section 3 describes our attack strategy in detail. The practical and the theoretical aspects of the attack are discussed in Section 4 and Section 5, respectively. Finally, Section 6 provides some conclusions.

2 Preliminary

2.1 Power Analysis and Countermeasures

Many forms of power analysis have been discussed in literature in the past decade. Among them, the most common ones are Simple Power Analysis (SPA) [6] attacks and Differential Power Analysis (DPA) [4] attacks. With detailed knowledge of the implementation of the cryptographic algorithm under attack, SPA attacks can derive confidential information directly from a small number of power traces measured from the device. DPA attacks, on the other hand, combine a large number of power traces to extract secrets and thus remain applicable even when there is a lot of noise in the measurements. Moreover, they do not require detailed knowledge of the attacked device. Other popular forms of power analysis are template attacks [3] and collision attacks [11]. Template attacks can extract information from a single power trace by matching it against some pre-built templates. Collision attacks detect equal intermediate values that the device manipulates by comparing the power consumptions that correspond to different executions of the cryptographic algorithm under attack or to different moments in time in the same execution.

To defend against power analysis attacks, two general countermeasures exist: Masking and hiding [5]. Masking conceals the intermediate results of a cryptographic algorithm with random values. In the most common form, an intermediate result v is split into n shares (random masks $r_1, \ldots, r_{n-1}$ and $v \oplus r_1 \oplus \ldots \oplus r_{n-1}$) which all need to be combined to find any information on v. At each point in time only a single share is used ensuring that no information is leaked about the masked sensitive intermediate result. Hiding uses techniques such as random insertion of dummy operations and shuffling of (independent) instructions to ensure that in different executions the same operation does not (or only with a small probability) happen at the same moment in time. As a result, power traces are misaligned and the power consumption in any point in time will not (or at most very weakly) be correlated to any sensitive data.

In turn, attacks exist that are efficient despite these countermeasures (see e.g. [5,13]). An n-order DPA attack examines n locations of a power trace that correspond to the secret shares of an intermediate result. Biased mask attacks force bias into the masks, for instance by using templates, enabling DPA attacks examining a sensitive intermediate result that is protected by the no longer uniformly distributed masks. In windowing, the power consumption for a complete region, the 'window', is used rather than a single point in time. This prevents misalignment as long as the targeted intermediate result is computed within the window. As the number of traces needed for a successful attack grows with the size of the window, windowing is typically combined with trace alignment techniques such as pattern matching. In pattern matching, a pattern chosen from one trace is matched to the other traces to detect the same executed operation or equal processed data, allowing realignment of a trace.

Table 1. The provably secure S-box implementation algorithms proposed in [9,10]. INPUT: $\tilde{x} = x \oplus r$ (algorithm 1) or $\tilde{x} = x \oplus r_1 \oplus r_2$ (algorithms 2, 3); OUTPUT: $S(x) \oplus s$ (algorithm 1) or $S(x) \oplus s_1 \oplus s_2$ (algorithms 2, 3).

Algorithm 1	Algorithm 2	Algorithm 3
$R_0 \leftarrow s$	$r_3 \leftarrow rand(n)$	
$R_1 \leftarrow s$	$r' \leftarrow (r_1 \oplus r_3) \oplus r_2$	$b \leftarrow rand(1)$
for $a = 0$ to $2^n - 1$	**for** $a = 0$ to $2^n - 1$	**for** $a = 0$ to $2^n - 1$
$\quad cmp \leftarrow compare(a, r)$	$\quad a' \leftarrow a \oplus r'$	$\quad cmp \leftarrow compare_b(r_1 \oplus a, r_2)$
$\quad R_{cmp} \leftarrow R_{cmp} \oplus S(\tilde{x} \oplus a)$	$\quad T[a'] \leftarrow (S(\tilde{x} \oplus a) \oplus s_1) \oplus s_2$	$\quad R_{cmp} \leftarrow (S(\tilde{x} \oplus a) \oplus s_1) \oplus s_2$
end	**end**	**end**
$cmp \leftarrow compare(R_0, R_1)$	**return** $T[r_3]$	**return** R_b
return $R_0 \oplus (cmp \times R_1)$		

2.2   Provably Secure S-Box Implementations

Table 1 depicts the algorithms of the three provably secure S-box implementation proposed in [9,10]. In these algorithms, an S-box look-up maps an input in $\mathbb{F}_2^n$ to an output in $\mathbb{F}_2^m$ through a table S. Sensitive intermediate result x, which is often a function of the plaintext and the key, is concealed by an input mask r (resp. $r_1 \oplus r_2$). The masked sensitive intermediate result $\tilde{x}$ is taken as the input of the algorithms and a masked S-box output $S(x) \oplus s$ (resp. $S(x) \oplus s_1 \oplus s_2$) is returned at the end. The input masks and the output masks are independent variables. The core idea of the algorithms is to compute $S(\tilde{x} \oplus a) \oplus s$ (resp. $S(\tilde{x} \oplus a) \oplus s_1 \oplus s_2$) for every $a \in \mathbb{F}_2^n$ and return only the *expected result* $S(x) \oplus s$ (resp. $S(x) \oplus s_1 \oplus s_2$) at the end.

It was formally proven in [9,10] that algorithm 1 is secure against 1st-order DPA attacks and that algorithms 2 and 3 are also secure against 2nd-order DPA attacks. The security of the implementations lies in countermeasures including masking, shuffling and insertion of dummy operations. During each execution $2^n - 1$ dummy S-box look-ups are performed, between which the look-up for the expected result is executed. The 2^n look-ups are shuffled according to the value of the input mask (i.e. r or $r_1 \oplus r_2$), randomizing the moment in time the expected result is computed.

An S-box implemented in this fashion can thwart most of the pratical power analysis reported till now [5,13]. Standard 1st/2nd-order DPA attacks do not work for these implementations because the intermediate results are masked and randomized. Biased mask attacks and windowing are no longer practical, because there are too many possible points in time where the expected result could be computed, and the attacks cannot succeed without an extreme increase of the number of power measurements (about 2^n times as many as needed for an unprotected implementation). Pattern matching can detect equal (masked) intermediate values, however, without further information about the masks this is not enough to reveal the secret key.

Table 2. A model of the algorithms shown in Table 1. INPUT: $\tilde{x} = x \oplus r$; OUTPUT: $S(x) \oplus s$.

Model	Instantiations	
	Algorithm 1	Algorithms 2, 3
for $a = 0$ to $2^n - 1$		
$\quad \leftarrow \tilde{x} \oplus a$		
$\quad \leftarrow S(\tilde{x} \oplus a)$	$R =$	
$\quad \leftarrow S(\tilde{x} \oplus a) \oplus R$		$R = s$
end	$\begin{cases} s & \text{if } a = r, \\ \bigoplus_{\substack{j=0 \\ j \neq r}}^{a-1} S(\tilde{x} \oplus j) \oplus s & \text{if } a \neq r. \end{cases}$	$r = r_1 \oplus r_2$
$\quad \leftarrow S(x) \oplus s$		$s = s_1 \oplus s_2$
$S(\tilde{x} \oplus a) \oplus R = S(x) \oplus s$ if $a = r$		

3   Description of the New Power Analysis Attack

The algorithms depicted in Table 1 all use the same approach to secure an S-box implementation. Even though there are some differences in the usages of masks, registers and memory, these algorithms produce intermediate results in a similar fashion which can be summarized by an implementation model that is shown in Table 2. In this model, masks used for the same intermediate result are summed up and treated as one. Note that although it can be instantiated differently depending on the algorithm, variable R is randomized from execution to execution. The expected result is always computed during loop iteration $a = r$.

The model in Table 2 exposes three potential vulnerabilities to power analysis. First, the 2^n S-box look-ups are executed depending on the same unknown value of a few bits (e.g. $\tilde{x}$) so that guessing the unknown allows the prediction of the inputs and outputs of all the S-box during an execution. Second, the S-box look-up is performed for a large number (i.e. 2^n) of times in each execution with different inputs, opening the possibility for a 1st-order DPA attack even within a single trace. Third, the same random variable (i.e. r) is used for both masking and hiding, hence defeating one countermeasure immediately leads to the destruction of the other.

We introduce an attack that combines these vulnerabilities and can break a protected S-box implementation that is in line with the model in Table 2. This attack is illustrated in Figure 1 and consists of five steps. To simplify the notations, in the rest of this section we assume that the sensitive intermediate result x is the XOR of the plaintext and the key of the attacked device, and the variable R is always equal to the output mask s in the attacked implementation (as in Algorithms 2, 3 in Table 2). As discussed in Sections 4 and 5, the attack functions equally for implementations where a different R or a different x is used.

Step 1: Acquiring Power Traces. First, we randomly generate N plaintexts $p_1, \ldots, p_N$ and let the attacked device process these plaintexts based on its secret key K. During the execution of p_i, an input mask r_i and an output mask s_i are randomly generated on the device and the masked sensitive data $\tilde{x}_i = p_i \oplus (K \oplus r_i)$ is computed. The sensitive data x_i is thereby $p_i \oplus K$. Next, the S-box implementation is used to execute on $\tilde{x}_i$ to produce $S(x_i) \oplus s_i$.

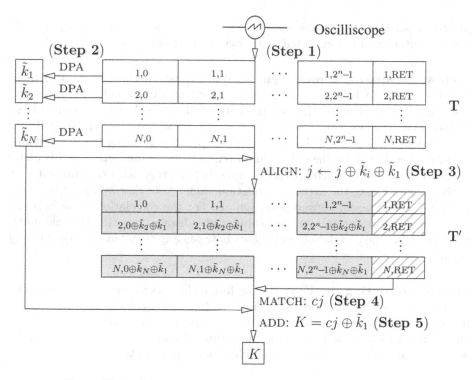

Fig. 1. Block diagram illustrating the new power analysis attack

During this execution a power trace is recorded: $\mathbf{t}_i = (t_{i,0}, \ldots, t_{i,2^n-1}, t_{i,\mathrm{RET}})$, where *subtrace* $t_{i,j}$, $0 \leq j \leq 2^n - 1$, corresponds to loop iteration $a = j$ and subtrace $t_{i,\mathrm{RET}}$ corresponds to the return of the expected result $S(x_i) \oplus s_i$. The power traces obtained for all the plaintexts can be written as matrix $\mathbf{T} = (\mathbf{t}_1, \ldots, \mathbf{t}_N)'$.

Step 2: Recovering Masked Keys. Note that each trace $\mathbf{t}_i$ has 2^n subtraces $t_{i,0}$ trough $t_{i,2^n-1}$ which all process intermediate values depending on some (unknown) masked key $\tilde{k}_i = K \oplus r_i$, (known) plaintext p_i and (known) loop iteration indices $a = 0..2^n - 1$. This allows us to recover $\tilde{k}_i$ by performing a standard 1st-order DPA attack using the subtraces. (See e.g. [5] for details of such an attack.) Note that we perform an attack for each trace, i.e. N attacks in total.

Step 3: Aligning Power Traces. The effect of the hiding countermeasure can be removed by aligning the power traces. Having found the value of the masked keys we rearrange the traces to ensure that the computation of the expected result always happens at a fixed (but unknown) location for every power trace. We use cj to denote (the index of the subtrace containing) this location. We rearrange trace $\mathbf{t}_i$ by moving subtrace $t_{i,j\oplus\tilde{k}_i\oplus\tilde{k}_1}$ to position j leaving $t_{i,\mathrm{RET}}$ where it is: $\mathbf{t}_i' = (t_{i,0\oplus\tilde{k}_i\oplus\tilde{k}_1}, \ldots, t_{i,2^n-1\oplus\tilde{k}_i\oplus\tilde{k}_1}, t_{i,\mathrm{RET}})$. Note that the computation of the expected result happens in loop iteration r_i. Since trace $\mathbf{t}_1' = \mathbf{t}_1$ remains unchanged it holds that $cj = r_1$. For an arbitrary trace $\mathbf{t}_i'$, subtrace $t_{i,cj\oplus\tilde{k}_i\oplus\tilde{k}_1}$ is placed at position cj. Because $cj \oplus \tilde{k}_i \oplus \tilde{k}_1 = r_1 \oplus \tilde{k}_i \oplus \tilde{k}_1 = r_i$, in every trace $\mathbf{t}'_i$ the computation of the expected result occurs at position cj.

Step 4: Comparing Power Signals. Within matrix $\mathbf{T}' = (\mathbf{t}'_1, \ldots, \mathbf{t}'_N)$ there are two columns which work with the expected result: the column cj and the last column RET. Due to the non-linearity of the S-box and the randomness of the output masks, the expected results $S(x_i) \oplus s_i$, $i = 1..N$, are statistically independent from other intermediate values processed during the loop. Thus, column cj of $\mathbf{T}'$ will be the only column related with column RET, and because of this, we can find cj by comparing column RET to every other column of $\mathbf{T}'$—it is the column giving the highest correlation value. Note that every subtrace $t_{i,j}$ contains several power consumption signals. For the correlation value between a column of subtraces and another column of subtraces, we take the maximum value amongst the correlation coefficients between any signal in the first and any signal in the second subtrace.

Step 5: Recovering the Key. Having found the masked key $\tilde{k}_1$ and the mask $r_1 (= cj)$ we can recover now the secret key $K = \tilde{k}_1 \oplus cj$. Note that if we make an error in determining $\tilde{k}_1$ we will make the same error in realigning the traces; the column cj will be shifted by the same amount and $K = \tilde{k}_1 \oplus cj$ still gives the correct result. We provide a more detailed analysis in Section 5.

4   Practical Experiments

We validate our attack on a protected S-box implementation described in Section 3 by experiments with the AES S-box. To obtain the power traces for these experiments the power signals of an 8-bit microcontroller clocked at 3.57 Mhz (1 clock cycle per 280 ns) are sampled at rate 1 GHz (1 sample per ns). In the experiments we focus on steps 2 and 4 of the attack described in Section 3. These steps contain the statistical analysis of the traces which mostly determines the effectiveness of the attack. The steps 1, 3 and 5 include trace processing and data computations which are well known to be feasible in practice and do not need to be repeated. E.g. we create the subtraces $t_{i,j}$ (step 1) by using already measured and extracted S-box computations rather than generating them anew.

Experimental Validation of Step 2. In effect, step 2 performs, for each trace, a 1st-order DPA attack on an unprotected S-box using all 2^n possible plaintexts. We take the 256 power traces measured while the microcontroller executes AES S-box look-ups using plaintexts $0, 1, \ldots, 255$ and a randomly selected key. Figure 2 shows the results of the attack for the correct key hypothesis (160 in this case). There exist multiple peaks in this graph because an S-box look-up takes more than one nanoseconds to be executed on the microcontroller. The highest peak $\rho = 0.59$ occurs at 9514 ns, denoting the point in time the output of the S-box is processed. Figure 3 plots the results of the attack for 9514 ns and all

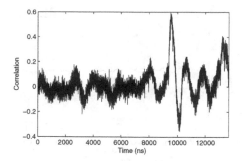

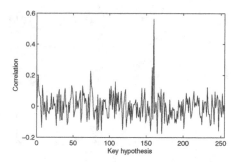

Fig. 2. The results of the DPA attack for the correct key hypothesis

Fig. 3. The results of the DPA attack for all the key hypotheses at 9514 ns

the key hypotheses. As expected, the highest peak occur at the correct hypothesis 160 in this graph. The significance of the peaks in both figures indicates that the DPA attack has successfully revealed the key.

Note that a DPA attack may already succeed with fewer power traces. Figure 4 shows the evolution of the results for all the key hypotheses with an increasing number of traces used. The result for the correct hypothesis is plotted in black and the results for the incorrect hypotheses are plotted in light gray. The outer dark gray curves mark the confidence interval for correlation coefficient that is equal to zero (see Section 5). Within the interval is the expected region for the incorrect key hypotheses. The point where the black curve leaves this region gives an estimation for the number of traces required for a successful attack. Figure 4 shows that approximately 150 power traces are already sufficient to find the key. Using all available 256 traces the attack will almost always succeed; for nearly all traces we will find the correct masked key. Therefore, in step 3 nearly all traces can be correctly aligned.

Experimental Validation of Step 4. In step 4 of the attack the column of subtraces RET is compared to each of the other columns of subtraces. As the expected result is statistically independent from other intermediate results of the algorithm and every location of the traces corresponds to independent intermediate results, in step 4 we compare the **return** of the expected result with the computation of the same number or a random number.

For this purpose, we take the measurements of the power consumption for the transference to memory of $N = 10000$ expected results to obtain the column $(\mathbf{t}'_{i,\mathrm{RET}})_{i=1..N}$. Then, we compare this column to the measurements during the computation of the expected results in $(\mathbf{t}'_{i,cj})_{i=1..N}$. We also compare this column to a column of computation of random results $((\mathbf{t}'_{i,j})_{i=1..N}$ for $j \neq cj)$.

Figure 5 shows the results of the comparison for different numbers of power traces used (up to $N = 10000$). The results for column cj are plotted in black and the results for the other 255 columns are plotted in light gray. Again, the outer dark gray curves indicate the expected region for the traces that are uncorrelated to cj. The point where the black curve leaves this region suggests that the number

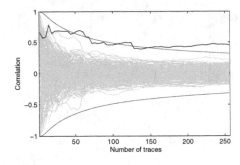
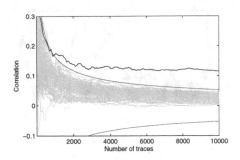

Fig. 4. The results of the DPA attack for different number of power traces

Fig. 5. The results of power signal comparison for different number of power traces

of traces required for a distinctive match between the columns cj and RET is approximately 900. A simple XOR in step 5 completes the attack.

5   Analysis of the Attack Practicality

The attack in Section 4 has succeeded with a relatively small number of power traces. However, more traces could be necessary when the measured power consumption signals have a rather low signal-to-noise ratio (SNR). In this section, we will discuss the practicality of our attack in a more general setting and provide a theoretical assessment of the effectiveness of the attack based on the SNR of the power measurements. Methods that can be used to improve the effectiveness of the attack are also suggested in this section.

Needless to say, due to the insertion of the dummy look-ups, the time required for the measurement of power traces and the hardware capability needed for the storage of the measured traces are, compared with a standard attack on an unprotected implementation, increased by a factor of approximately 2^n for our attack.

In order to mount the DPA attacks in step 2, 2^n subtraces must be extracted from each power trace that correspond to the 2^n loop iterations respectively. Such subtraces can often be spotted by visually inspecting the power trace, in that the S-box look-up is repeated for so many times in a row and that every look-up takes nearly the same amount of time (despite different processed data). Besides, the fact that an S-box look-up is usually captured by a number of points on a trace (see Section 4) can also facilitate the detection of subtraces. Note that this trace division needs to be performed only once for all traces, as in spite of different processed data the executed operations are already aligned in the raw traces for all plaintexts, hence the segmentation of one trace is also valid for other traces.

Next, N standard DPA attacks are performed (separately) using the extracted subtraces to find the masked keys. Although the total number of attacks in this step seems large, one of such attacks can usually be mounted with little effort. Please refer to e.g. [5,8] for a detailed description of a standard DPA attack.

Let ρ_{DPA} denote the expected correlation peak resulted from such a DPA attack for the correct hypothesis of the masked key and the correct moment in time. Let sr be the success rate of such a DPA attack and snr be the SNR of the power signal at the correct moment in time. Based on the rule of thumb introduced by Mangard et al. in [5], the relations between ρ_{DPA}, sr and snr can be roughly defined as in Eq. (1). The detailed deductions of ρ_{DPA} and sr can be found in [5] and Appendix A, respectively.

$$\rho_{\text{DPA}} = \frac{1}{\sqrt{1 + snr^{-1}}}, \qquad sr = \text{cdf}\left(\sqrt{\frac{2^n - 3}{8}} \ln^2 \frac{1 + \rho_{\text{DPA}}}{1 - \rho_{\text{DPA}}}\right). \qquad (1)$$

Having a success rate of sr in step 2 implies that there are on average $sr \cdot N$ correctly aligned power traces resulted from step 3 leaving $(1 - sr) \cdot N$ traces misaligned. Thanks to the non-linearity of an S-box, the misaligned traces must contain only randomly shuffled power consumption signals.

Based on these partially aligned traces, in step 4 column RET of $\mathbf{T}'$ is matched against other columns to find cj. Let ρ_{CMP} denote the expected correlation value given by columns RET and cj. Let snr_1 and snr_2 be the SNRs of the power signals (in RET and cj respectively) that actually result in ρ_{CMP}. Eq. (2) shows some rules of thumb for the relations between ρ_{CMP}, snr_1, snr_2 and N—the number of traces needed for a successful attack, where $z_{0.9999}$ ($= 3.719$) is the quantile of the standard normal distribution $\mathcal{N}(0, 1)$ for probability 0.9999. The derivations of ρ_{CMP} and N are explained in detail in Appendix A and [5] respectively. Please note that the N obtained by Eq. (2) is the number of traces needed for an entire attack.

$$\rho_{\text{CMP}} = \frac{1}{\sqrt{1 + snr_1^{-1}}} \cdot \frac{1}{\sqrt{1 + snr_2^{-1}}} \cdot sr, \qquad N = 3 + 8 \frac{z_{0.9999}^2}{\ln^2 \frac{1 + \rho_{\text{CMP}}}{1 - \rho_{\text{CMP}}}}. \qquad (2)$$

Eqs. (1) and (2) indicate that the number of traces needed for a successful attack grows about quadratically on the success rate of the trace alignment. Increasing the correctness of trace alignment, especially in case of heavy noise, will enormously reduce the number of traces required. Hence, we provide an error-correction method that can be applied at the end of step 3 to detect misaligned traces. Since the S-box look-up is executed for all possible values in a run of the implementation, there must exists one (and only one) subtrace in each trace such that they all correspond to S-box look-ups with equal data (i.e. equal inputs and equal outputs). These subtraces can be derived based on correct masked keys obtained from step 2, while incorrect masked keys will only result in subtraces that correspond to random intermediate data thanks to the non-linearity of the S-box. For example, let us assume two different traces $\mathbf{t}_{i_0}$ and $\mathbf{t}_{i_1}$ and define relation $j_1 = j_0 \oplus \tilde{k}_{i_1} \oplus \tilde{k}_{i_0} \oplus p_{i_1} \oplus p_{i_0}$; subtraces t_{i_0, j_0} and t_{i_1, j_1} correspond to S-box look-ups with equal data if and only if $\mathbf{t}_{i_0}$ and $\mathbf{t}_{i_1}$ are correctly aligned with each other, since $p_{i_0} \oplus \tilde{k}_{i_0} \oplus j_0 = p_{i_1} \oplus \tilde{k}_{i_1} \oplus j_1$. Therefore, by demonstrating for all subtraces of all traces whether or not the relevant subtraces indeed correspond to equal processed data, we can verify if the obtained masked keys are correct and hence identify misaligned traces. To determine equal intermediate results, one can use highest-correlation or least-variance method to the corresponding power signals. If a misaligned trace is found, one can choose either to exclude it from the rest of the attack, or to make another attempt to align this trace by using the 2nd (or the 3rd, etc.) best hypothesis of the masked key resulted from step 2.

However, it is not always necessary in practice to improve the success rate of step 3; the improvement is only worthwhile when the SNR of the power measurements is relatively low. As shown later in this section, in case of the AES S-box, one percent increase of the success rate in step 3 can reduce the number of traces needed by 1686 if $snr = 1/50$ but can save only 66 traces if $snr = 1/10$.

Effectiveness of Our Attack for the AES S-Box. Using the methods presented in this section, we have assessed the effectiveness of our attack for such a protected implementation of the AES S-box. Figure 6 depicts the evolutions of the correlations ρ_{DPA} and ρ_{CMP}, the success rate sr and the number of traces needed N with a decreasing SNR, as well as the evolution of N with a decreasing sr. Note that in order to simplify the demonstration, we have replaced snr_1^{-1} and snr_2^{-1} in Eq. (2) with their mean snr^{-1}. Compared with using two separate SNRs, using the mean leads to a lower estimation of ρ_{CMP} and a higher estimation of N. Therefore, the results presented in Figure 6 show a worse case scenario for the attacker.

Figure 6 shows that the success rate of step 2 is extremely high even when the power measurement is relatively noisy. This is because the AES S-box is especially vulnerable to DPA attacks. The number of traces required for the entire attack, on the other hand, grows rapidly when the noise goes up, implying that the attack may encounter practical difficulties in case of very low SNRs. Fortunately, many commercial cryptographic devices used in practice can give relatively high SNRs (e.g. $\geq 1/20$). In fact, we believe that if a device requires such an elaborate protection as the attacked implementations in this paper, this device must leak a considerable amount of information once without protections; in this case our attack can be highly effective.

Table 3 shows the effectiveness of our attack for some SNRs that are higher than $1/20$. The number of traces need for our attack for the protected AES S-box implementation and the number of traces needed for a standard DPA attack for an unprotected AES S-box implementation are listed shoulder to shoulder in Table 3. It is indicated by our results that while a standard DPA attack requires about 150 traces to break an unprotected implementation, our attack needs about 980 traces to break the protected implementation. This estimation accords almost perfectly with the practical results presented in Section 4, which shows that our theoretical results given in Table 3 can be very close to the reality.

Important Notes. Our attack is possible for the proven secure implementations *not* because of incorrect proofs of Prouff et al.'s or inadequate security metrics of Standaert et al.'s. When directly analyzing the raw power traces that are measured during the executions of Prouff et al.'s algorithms, the security metric of Standaert et al.'s can be perfectly satisfied; therefore the algorithms were believed secure against side-channel analysis in general. However, in fact

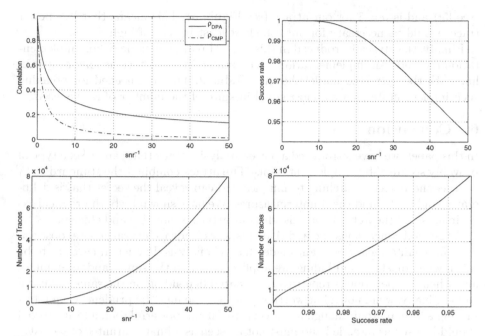

Fig. 6. The estimated results of our attack on the provably secure implementations with the AES S-box. Left-top: ρ_{DPA} and ρ_{CMP} with a decreasing SNR; right-top: sr with a decreasing SNR; left-bottom: N with a decreasing SNR; right-bottom: N with a decreasing sr.

Table 3. The number of traces needed for some high SNRs: N is for our attack on a protected AES S-box; $N(\text{DPA})$ is for a standard 1st-order DPA attack on an unprotected AES S-box

snr^{-1}	1	2	3	4	5	6	7	8	9	10	12	14	16	18	20
N	95	233	427	676	980	1340	1755	2225	2751	3333	4666	6227	8022	10057	12339
$N(\text{DPA})$	39	67	95	122	150	178	206	233	261	289	344	399	455	510	565

the proofs have only shown the resistances against *straightforward* side-channel analysis. With a little preprocessing, such as the steps 2 and 3 in our attack, power signals that contain information about the same sensitive data can be relocated to the same position for all traces; the security of the implementation is thereby no longer provable by the metric for the aligned power traces.

We would also like to point out that our attack is *not* a high-order attack. In a high-order DPA attack, all the unknown secret shares have to be tested resulting in a total testing space that is exponential to the testing space for one share. Whereas in our attack, the required testing space is only linear to the testing space for one share because only the masked key needs to be tested. Moreover, in a noise-free scenario the resulted correlation peak of our attack is greater than that of a high-order DPA attack, e.g. it is $\rho = 0.24$ for a 2nd-order DPA attack (see [5]) and is $\rho = 1$ for our attack (see Figure 6). This means that less power traces would be necessary for our attack at least in case of high SNRs.

Finally, the attack introduced in this paper breaks a generic S-box implementation method. An implementation is vulnerable to our attack as long as it is in line with the extracted model shown in Table 2. The cost needed for breaking such implementations are almost equal in spite of the number of masks used.

6   Conclusion

In this paper, we have introduced a power analysis attack that can break a type of provably secure S-box implementations. The attack combines the standard DPA attacks and pattern matching techniques and can reveal the secret that is deliberately hidden behind the countermeasures. We have shown by both practical experiments and theoretical analysis that our attack is effective and efficient.

As general conclusions, we find that this work leads to several general observations for countermeasures against side-channel analysis. Particular care has to be taken when masking and hiding are applied based on the same random number, and when an operation is repetitively executed depending on the same unknown value. Finally, well known but worth repeating, one must be prudent when interpreting a formal proof; though it is a very useful and powerful tool, a formal proof should be used as intended and must not be seen as a final guarantee of security.

References

1. Brier, É., Clavier, C., Olivier, F.: Correlation power analysis with a leakage model. In: Joye, M., Quisquater, J.-J. (eds.) CHES 2004. LNCS, vol. 3156, pp. 16–29. Springer, Heidelberg (2004)
2. Chari, S., Jutla, C.S., Rao, J.R., Rohatgi, P.: Towards sound approaches to counteract power-analysis attacks. In: Wiener, M.J. (ed.) CRYPTO 1999. LNCS, vol. 1666, pp. 398–412. Springer, Heidelberg (1999)
3. Chari, S., Rao, J.R., Rohatgi, P.: Template attacks. In: Kaliski Jr., B.S., Koç, Ç.K., Paar, C. (eds.) CHES 2002. LNCS, vol. 2523, pp. 13–28. Springer, Heidelberg (2003)
4. Kocher, P., Jaffe, J., Jun, B.: Differential power analysis. In: Wiener, M. (ed.) CRYPTO 1999. LNCS, vol. 1666, pp. 388–397. Springer, Heidelberg (1999)
5. Mangard, S., Oswald, E., Popp, T.: Power Analysis Attacks: Revealing the Secrets of Smart Cards. In: Advances in Information Security. Springer, Heidelberg (2007)
6. Mayer-Sommer, R.: Smartly analyzing the simplicity and the power of simple power analysis on smartcards. In: Paar, C., Koç, Ç.K. (eds.) CHES 2000. LNCS, vol. 1965, pp. 78–92. Springer, Heidelberg (2000)
7. Oswald, E., Mangard, S., Herbst, C., Tillich, S.: Practical second-order DPA attacks for masked smart card implementations of block ciphers. In: Pointcheval, D. (ed.) CT-RSA 2006. LNCS, vol. 3860, pp. 192–207. Springer, Heidelberg (2006)
8. Pan, J., den Hartog, J., de Vink, E.: An operation-based metric on cpa resistance. In: Jajodia, S., Samarati, P., Cimato, S. (eds.) SEC, International Federation for Information Processing, pp. 429–443. Springer, Boston (2008)
9. Prouff, E., Rivain, M.: A generic method for secure sbox implementation. In: Kim, S., Yung, M., Lee, H.-W. (eds.) WISA 2007. LNCS, vol. 4867, pp. 227–244. Springer, Heidelberg (2008)

10. Rivain, M., Dottax, E., Prouff, E.: Block ciphers implementations provably secure against second order side channel analysis. In: Nyberg, K. (ed.) FSE 2008. LNCS, vol. 5086, pp. 127–143. Springer, Heidelberg (2008)
11. Schramm, K., Wollinger, T.J., Paar, C.: A new class of collision attacks and its application to DES. In: Johansson, T. (ed.) FSE 2003. LNCS, vol. 2887, pp. 206–222. Springer, Heidelberg (2003)
12. Standaert, F.-X., Malkin, T.G., Yung, M.: A unified framework for the analysis of side-channel key recovery attacks. Cryptology ePrint Archive, Report 2006/139 (2006), http://eprint.iacr.org/2006/139
13. Tillich, S., Herbst, C.: Attacking state-of-the-art software countermeasures-a case study for AES. In: Oswald, E., Rohatgi, P. (eds.) CHES 2008. LNCS, vol. 5154, pp. 228–243. Springer, Heidelberg (2008)
14. Tillich, S., Herbst, C., Mangard, S.: Protecting AES software implementations on 32-bit processors against power analysis. In: Katz, J., Yung, M. (eds.) ACNS 2007. LNCS, vol. 4521, pp. 141–157. Springer, Heidelberg (2007)

A   The Derivation of Eqs. (1) and (2)

Mangard et al. [5] introduced a rule of thumb to assess the number of power traces needed for a DPA attack. According to them, the number of traces N that are necessary to mount a DPA attack with a confidence α can be calculated by Eq. (3), where $z_\alpha = \text{cdf}^{-1}(\alpha)$ is the quantile of $\mathcal{N}(0, 1)$ for probability α. They also suggested that the number of traces needed for a successful attack can be calculated by Eq. (3) with $\alpha = 0.9999$. Inversely, the success rate α of a DPA attack using N power traces can be derived as in Eq. (4). Letting $N = 2^n$ we obtain the success rate of the DPA attack in step 2 of our attack as the sr in Eq (1).

$$N = 3 + 8\,\frac{z_\alpha^2}{\ln^2 \frac{1+\rho_{\text{DPA}}}{1-\rho_{\text{DPA}}}}\,, \quad (3) \qquad \alpha = \text{cdf}\left(\sqrt{\frac{N-3}{8}}\ln^2\frac{1+\rho_{\text{DPA}}}{1-\rho_{\text{DPA}}}\right). \quad (4)$$

To deduct ρ_{CMP} in (2) we first consider the following general cases. Assume that M_1 and M_2 are two measurements for the same random data variable, of which the Hamming-weight is H. Let P_1 and P_2 denote the noise in M_1 and M_2, respectively. Therefore, the SNR of M_1 is $snr_1 = \sigma^2(H)/\sigma^2(P_1)$ and the SNR of M_2 is $snr_2 = \sigma^2(H)/\sigma^2(P_2)$. The correlation coefficient $\rho(M_1, M_2)$ can be calculated by Eq. (5), where the simplification is made based on the fact that P_1 and P_2 are statistically independent from H.

$$\rho(M_1, M_2) = \rho(H + P_1, H + P_2) \tag{5}$$

$$= \frac{E((H+P_1)\cdot(H+P_2)) - E(H+P_1)\cdot E(H+P_2)}{\sigma(H+P_1)\cdot\sigma(H+P_2)}$$

$$= \frac{E(H^2) - E^2(H)}{\sigma(H+P_1)\cdot\sigma(H+P_2)} = \frac{\sigma^2(H)}{\sqrt{\sigma^2(H)+\sigma^2(P_1)}\cdot\sqrt{\sigma^2(H)+\sigma^2(P_2)}}$$

$$= \frac{1}{\sqrt{1 + snr_1^{-1}}}\cdot\frac{1}{\sqrt{1 + snr_2^{-1}}}.$$

Now, let us consider the case where error exists in one of the measurement. Let I denote a random variable that has the same distribution as M_1 but is statistically independent from M_1. Let X_1 be an erroneous measurement of the processed data such that $Pr(X_1 = M_1) = sr$ and $Pr(X_1 = I) = 1 - sr$. The correlation coefficient $\rho(X_1, M_2)$ then corresponds to the expected correlation coefficient ρ_{CMP} in step 4 of our attack for the correct loop iteration index (see Eq. (2)), which can therefore by developed as in Eq. (6) based on Eq.(5).

$$\rho_{\text{CMP}} = \rho(X_1, M_2) = \frac{E(X_1 \cdot M_2) - E(X_1) \cdot E(M_2)}{\sigma(M_2) \cdot \sqrt{E(X_1^2) - E^2(X_1)}} \tag{6}$$

$$= \frac{E(M_1 \cdot M_2) \cdot sr + E(I \cdot M_2) \cdot (1 - sr) - (E(M_1) \cdot E(M_2) \cdot sr + E(I) \cdot E(M_2) \cdot (1 - sr))}{\sigma(M_2) \cdot \sqrt{E(M_1^2) \cdot sr + E(I^2) \cdot (1 - sr) - (E(M_1) \cdot sr + E(I) \cdot (1 - sr))^2}}$$

$$= \frac{(E(M_1 \cdot M_2) - E(M_1) \cdot E(M_2)) \cdot sr}{\sigma(M_2) \cdot \sigma(M_1)} = \rho(M_1, M_2) \cdot sr$$

$$= \frac{1}{\sqrt{1 + snr_1^{-1}}} \cdot \frac{1}{\sqrt{1 + snr_2^{-1}}} \cdot sr .$$

A Comparative Study of Mutual Information Analysis under a Gaussian Assumption

Amir Moradi[1,*], Nima Mousavi[2], Christof Paar[1], and Mahmoud Salmasizadeh[2]

[1] Horst Görtz Institute for IT Security, Ruhr University Bochum, Germany
[2] Electronics Research Center, Sharif University of Technology, Tehran, Iran
{moradi,cpaar}@crypto.rub.de, nm@ee.sharif.edu, salmasi@sharif.edu

Abstract. In CHES 2008 a generic side-channel distinguisher, Mutual Information, has been introduced to be independent of the relation between measurements and leakages as well as between leakages and data processed. Assuming a Gaussian model for the side-channel leakages, correlation power analysis (CPA) is capable of revealing the secrets efficiently. The goal of this paper is to compare mutual information analysis (MIA) and CPA when leakage of the target device fits into a Gaussian assumption. We first theoretically examine why MIA can reveal the correct key guess amongst other hypotheses, and then compare it with CPA proofs. As our theoretical comparison confirms and shown recently in ACNS 2009 and CHES 2009, the MIA is less effective than the CPA when there is a linear relation between leakages and predictions. Later, we show detailed practical comparison results of MIA and CPA, by means of several alternative parameters, under the same condition using leakage of a smart card as well as of an FPGA.

1 Introduction

1.1 History

In 2000s side-channel attacks made a challenge on cryptographers' point of view that not only mathematical security of a cryptographic algorithm but also physical security of its implementation should be justified to call a system "secure". Since a side-channel adversary needs physical access to the target device, pervasive devices such as smart cards and RFIDs which can operate in an uncontrolled environment are at risk of such powerful implementation attacks, e.g., differential power analysis (DPA) [8] and electromagnetic analysis (EMA) [6,14] which extract key materials by monitoring respectively the power consumption and electromagnetic emanation of the attacked device.

In a DPA attack, measurements are categorized into usually two (but can more, e.g., [9]) sets using a partition function which relies on one (respectively can on more) bit of intermediate value depending on the secret and a known input/output. Then, clear peaks in difference of means of two sets indicate the

---

* Amir Moradi performed most of the work described in this contribution when he has been with Electronics Research Center of Sharif University of Technology.

H.Y. Youm and M. Yung (Eds.): WISA 2009, LNCS 5932, pp. 193–205, 2009.
© Springer-Verlag Berlin Heidelberg 2009

correct key guess. Afterwards, a more general scheme namely correlation power analysis (CPA) [2] has been introduced that uses a hypothetical power consumption model ideally close to the actual leakage function of the target device. Depending on the hypothetical power model, known input/output, and each key guess, hypothetical power values are constructed and compared with measurements by means of Pearson correlation coefficient. Similarly to DPA, clear peaks in correlation coefficients identify the correct key hypothesis. It should be noted that improved side-channel key recovery attacks such as template attacks [4] and higher order attacks [12] have been designed to make the key recovery process more efficient or to defeat a range of countermeasures.

On the other hand, recently Gierlichs et al. introduced Mutual Information Analysis (MIA) [7] in which Mutual Information of guessed intermediate values and measurements is used as a side-channel distinguisher. In contrary to CPA, MIA has been designed to be effective without any knowledge about the particular dependencies between the processed data and leakages as well as between leakages and measurements. Also, CPA works efficiently under a Gaussian assumption when means and variances are estimated. However, MIA still can reveal the secrets if this Gaussian assumption does not hold.

Recently two articles [13] and [15] on mutual information analysis and its application have been published which shows that the topic is of highly interest for the relevant research community. In [13] the authors exposed theoretical foundation of the side-channel distinguishers, including MIA, and assessed their limitations and assets. They have answered a number of questions regarding the efficiency of MIA and the condition where it is better than the other distinguishers. Moreover, they generalized MIA to higher orders. As a short result, they showed that the MIA is less efficient than the CPA when the leakage is a linear function of the predictions, e.g., Hamming weight model. Further, it has been shown that a proposed extension of MIA is more efficient than existing higher-order attacks on masked implementations. Both articles argued that in addition to histograms, which has been used in the original description of MIA [7] to estimate the probability density functions, using other methods such as Kernel ones and parametric ones allows to improve the efficiency of the MIA attacks. The authors of [15] also discussed on the conditions in which the MIA can be more effective than the CPA, and proposed to use alternative probability-distance measure tools (which allows deciding which subkey is the most likely to be the correct one) with different impacts on the attack efficiency.

1.2 Motivation

Although it has been showed that the CPA is more effective than the MIA under a certain assumption, they have not been compared precisely with each other to make it clear how much the CPA is better than the other. What exactly we want to discuss in this paper is to compare the CPA and the MIA when the target device is a so-called general-purpose microcontroller or an FPGA whose leakages fit into a Gaussian assumption. In fact, we are going to answer several questions that arise by comparing MIA and CPA. These questions can be categorized as follows. Note that in all of our comparisons, the histogram method which has been proposed in the original description of MIA, is used.

- How much CPA can reveal the secrets better than the MIA, i.e., the revealed secret reported by CPA is more distinguishable amongst others than that reported by MIA?
- How is the threat of MIA in the presence of noise? Does it still work when CPA can not reveal the secrets, or vice versa?
- As mentioned in [7], it is not essential to use a hypothetical leakage model perfectly matching with the actual leakage function of the attacked device, and it is sufficient to use a leakage function proportional to the particular actual one. What is the result if this inaccurate leakage function is used in a CPA? Does CPA works when MIA does not?
- Does CPA need less number of traces than a corresponding MIA attack, under the same condition?
- How much the hypothetical leakage model is independent of the actual leakage of the attacked device? Does it work without any knowledge about the particular dependencies? Is it the same for CPA? or MIA works more efficiently?
- To the best of our knowledge, success rate of a CPA targeting leakage of a function having linear relation with secrets is not perfectly 100%, so does MIA work better in this situation?

1.3 Organization

This paper is organized as follows. In Section 2 we give an overview of the side-channel model used in our theoretical comparison. Section 3 and Section 4 recall the basic notations of CPA and MIA respectively, and compare the cases where CPA and MIA are not capable of revealing the secret. Later, in Section 5 we present the practical results of MIA and CPA under the same condition for several different situations to give an insight on answer of the aforementioned questions. Finally, conclusions are given by Section 6.

2 Side-Channel Model

In this section, we restate the theoretical model for side-channel leakage of a cryptographic device which has been introduced in [7]. Later, we will use the notations given here to theoretically discuss on CPA and MIA.

As shown by Fig. 1, a cryptographic device which performs a cryptographic algorithm, E, using a secret key, k, on input values, x, and generates outputs y as $E_k(x)$ is taken into account. During the computation of the device, intermediate values depending on the input x and unknown key k are changed and lead to some bit flips modeled by w. Changing the internal states of the target device leaks through a side-channel leakage function $L(w) = l$. However, what a side-channel adversary can observe is a noisy measurement, o, (usually acquired by

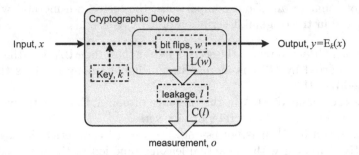

Fig. 1. Side-channel leakage model

a digital oscilloscope) of the leakage l, $o = \mathrm{C}\,(l)$. In short, a noisy function of bit flips of internal state would be observed by a side-channel adversary,

$$o = \mathrm{C}\,(\mathrm{L}\,(w))\,.$$

To mount a side-channel key recovery attack, an adversary collects q queries of measurements, o_i, $0 < i \leq q$, knowing corresponding inputs x_i and/or outputs y_i. Supposing that instants of time when the cryptographic device operates on the target secret are known causes each measurement o_i to contain single value[1]. During the attack, the adversary using "divide and conquer" scheme guesses parts of the secret key k and estimates the bit flips w, then using the measurements o and a side-channel distinguisher tries to find the most probable hypothesis close to the correct one.

3   Correlation Power Analysis

If we model the leakage for each key hypothesis $k \in$ key space K as l_k and similarly to [2] suppose a linear relation between adversary's observation o and leakages, we can write the following equation for the correct key guess, k_0:

$$o = a \cdot l_{k_0} + b, \tag{1}$$

where a is a constant scaling factor of the channel C, and b is a random variable concerning the noise model (Gaussian assumption). We also assume that the random variable b is independent of l_{k_0}. An example for the leakage function satisfying the linear relation of Eq. (1) would be Hamming weight (HW) of the intermediate values as well as HW of a part of the intermediate values since similar to Eq. (1) there is a linear relation between HW and partial HW of a value with an additive and independent random variable, i.e.,

$$\mathrm{HW}(x) = \mathrm{HW}(x_L) + \mathrm{HW}(x_R)\;;\quad x = x_L \| x_R.$$

---

[1] Otherwise the adversary needs to acquire longer measurements and takes each instant of time separately into account.

Considering linear relation of Eq. (1), correlation coefficient between two random variables of adversary's observation, o, and leakage for a wrong key, $l_{k'}$; $k' \in K$, $k' \neq k_0$ can be written as follows:

$$\rho_{ol_{k'}} = \frac{\text{Cov}(o, l_{k'})}{\delta_o \cdot \delta_{l_{k'}}} = \frac{\text{Cov}(a \cdot l_{k_0} + b, l_{k'})}{\delta_o \cdot \delta_{l_{k'}}} = \frac{a \cdot \text{Cov}(l_{k_0}, l_{k'})}{\delta_o \cdot \delta_{l_{k'}}}$$

$$= \frac{a \cdot \text{Cov}(l_{k_0}, l_{k_0})}{\delta_o \cdot \delta_{l_{k_0}}} \cdot \frac{\text{Cov}(l_{k_0}, l_{k'})}{\delta_{l_{k_0}} \cdot \delta_{l_{k'}}} \qquad (2)$$

$$= \rho_{ol_{k_0}} \cdot \rho_{l_{k_0} l_{k'}}.$$

Since $|\rho_{l_{k_0} l_{k'}}| \leq 1$, the leakage random variable has the maximum correlation with observations for the correct key guess. Thus, a CPA reveals the correct key as long as $|\rho_{l_{k_0} l_{k'}}| < 1$ for $\forall k' \in K$; $k' \neq k_0$. In other words, CPA is not capable of revealing the secret when $\exists k' \in K$, $k' \neq k_0$ for which $l_{k'}$ has a linear relation with l_{k_0}. Note that we discuss on this issue more in Section 4.

4   Mutual Information Analysis

By classifying a hypothetical leakage l_k for a key hypothesis $k \in K$ and making histograms of the measurements o, we can estimate $P(l_k)$ and $P(o)$ as the probability distributions of the random variables hypothetical leakage and measurements respectively. Further, we can estimate conditional probability distribution of measurements given a hypothetical leakage as $P(o|l_k)$. As illustrated in [7], we can now estimate conditional entropy $H(o|l_k)$ and hence have an estimation for mutual information of measurements and each key hypothesis as

$$I(o; l_k) = H(o) - H(o|l_k). \qquad (3)$$

First we should point out a markov chain as $l_{k'} \rightarrow l_{k_0} \rightarrow o$ for $\forall k' \neq k_0$ and equivalently the equality of $P(o|l_{k_0} l_{k'}) = P(o|l_{k_0})$. It can be easily verified by considering Eq. (1), the probability distribution of o given l_{k_0} is similar to the distribution of b (with a constant difference) which is also independent of $l_{k'}$. Note that o is not generally independent of $l_{k'}$. Now, writing $I(o; l_{k_0} l_{k'})$ in two ways gives:

$$I(o; l_{k_0} l_{k'}) = I(o; l_{k_0}) + I(o; l_{k'} | l_{k_0}) \overset{(a)}{=} I(o; L_{k_0}) \qquad (4)$$

$$I(o; l_{k_0} l_{k'}) = I(o; l_{k'}) + I(o; l_{k_0} | l_{k'}), \qquad (5)$$

where (a) follows from the aforementioned markov chain. Finally, we can write

$$I(o; l_{k'}) = I(o; l_{k_0}) - I(o; l_{k_0} | l_{k'}). \qquad (6)$$

Since mutual information is a positive function, $I(o; l_{k'}) \leq I(o; l_{k_0})$. Therefore, the leakage random variable for the correct key, l_{k_0}, has the maximum mutual information with random variable observations, and hence MIA gives the correct key as long as $I(o; l_{k_0}|l_{k'}) \neq 0$. In other words, MIA is not capable of revealing the secret when $\exists k' \neq k_0$ for which mutual information of o and l_{k_0} given $l_{k'}$ is zero. Further, a special case is when $\exists k' \neq k_0$ for which there is a one-to-one relation between l_{k_0} and $l_{k'}$. In such a case, $H(l_{k_0}|l_{k'}) = H(l_{k_0}|l_{k'}o) = 0$, so $I(o; l_{k_0}|l_{k'}) = H(l_{k_0}|l_{k'}) - H(l_{k_0}|l_{k'}o) = 0$.

Comparison with CPA. As mentioned in Section 3, CPA does not work if $\exists l_{k'}$, $k' \neq k_0$ for which $\rho_{l_{k_0}l_{k'}} = 1$. This implies a linear relation between l_{k_0} and $l_{k'}$. Since a linear relation is also a one-to-one relation, MIA does not work in this case too. In general, MIA is unsuccessful if $\exists l_{k'}$, $k' \neq k_0$ for which $I(o; l_{k_0}|l_{k'}) = 0$ whereas CPA is successful in this case unless for the aforementioned linear relation. As a consequence, the situations in which CPA is unsuccessful are fewer than that of MIA.

Generally we can say that correlation coefficient captures the **linear** relation between two random variables while mutual information captures **any** relation between two random variables. Therefore, CPA is not successful when there is a **linear** relation between observations and leakage variable for a wrong key guess as much as a **linear** relation between observations and leakage variable for the correct key. However, MIA is not successful when there is **any** relation between observations and leakage variable for a wrong key guess as much as a **linear** relation between observations and leakage variable for the correct key. It seems that in the later case, capturing **any** relation between leakage variable for a wrong key and observations, when a **linear** model holds for the correct key, leads to a weaker attack.

5   Practical Comparison

5.1   Target Devices and Measurement Setup

We developed two experimental platforms. One is a programmable smart card embedded by an Atmel AVR ATmage163 microcontroller [1] in which we have developed an implementation of the AES Rijnael encryption algorithm. Since the microcontroller has an 8-bit architecture, each subbyte transformation is executed separately using pre-computed look-up tables. Further, to obtain the power consumption traces the voltage drop over a 100Ω resistor placed in GND pin of the microcontroller has been measured by a digital oscilloscope with the sampling rate of $500\,\text{MS/s}$. Note that a standard smart card reader, e.g., [5], has been used to communicate with a PC. The second one is a XC2S150 Spartan-II FPGA [16] running again the AES-128 encryption function. Only one combinational Sbox based on the netlist presented in [3] is implemented in the device and is shared in subbytes operation. Side-channel observations are collected by measuring the differential voltage of a 3.3Ω resistor in the V_{CCINT} line with the same sampling rate of the smart card case. The only difference between the two devices is due to the clock signal which is supplied by the smart card reader in the first one and an external signal generator with a frequency of $1\,\text{MHz}$ in the later one.

5.2 Results

Unmatched Power Models - Because of a pre-charged (or pre-discharged) architecture of microcontrollers' data bus usually HW of data transfered by the bus leaks through power traces. Thus, an efficient CPA attack using HW model can be mounted on microcontroller-based devices. However, in MIA it is not needed to apply a power model matching with the particular leakage function, e.g., HW [7]. For instance we can use a part, e.g., 7 bits, of intermediate values instead of their HW. The first question here is due to what happens if this model is used in a CPA. To examine this, we have performed MIA and CPA on 256 traces measured from our microcontroller due to 256 chosen plaintexts to cover all possible plaintexts when the target secret is a key byte and the leakage of the Sbox output is the target leakage. Note that in MIA we need to set a parameter due to the number of bins in which measurements are classified to build the histograms. In the rest of this paper, we denote this parameter as "number of bins". In contrary to [7] we have used an automated way to make histograms, i.e, we have just divided the range of measurement values (of course independently for each instant of time) by the number of bins to equal intervals. Also, we have used the same power model, 7 bits of Sbox output, in a CPA attack. Fig. 2 shows the maximum mutual information and correlation coefficient for each key hypothesis when two bins used for measurement classifications in the MIA. We have repeated this task for all possible power models as a part of the intermediate value on the same power traces, i.e., $\sum_{i=1}^{7} \binom{8}{i} = 254$ power models for MIA and CPA. Interestingly, in all cases both MIA and CPA are capable of revealing clearly the correct key byte amongst other hypotheses[2]. It is because of a high correlation between decimal values and their HWs. Suppose a random variable R containing decimal values of a n-bit binary data. Correlation coefficient of R and $HW(R)$ is $\{1, 0.95, 0.88, \ldots, 0.65, 0.61\}$ for $n = 1, 2, 3, \ldots, 7, 8$. Thus, still there is a high correlation between the actual leakage of the target device, i.e., HW, and the decimal values as the power model in a CPA.

Further, we have done this procedure for all possible HW of a part of the target leakage, i.e., $\sum_{i=1}^{8} \binom{8}{i} = 255$ power models, which led to the same result as expected. As a result, in this case when the target device is a microcontroller with HW leakage, CPA like MIA can recover the secret for all possible leakage functions.

Distinguishability - One important criterion in comparison of MIA and CPA is how much a secret revealed by MIA is more distinguishable amongst other hypotheses than that by CPA (or vice versa). In order to have a parameter to compare the distinguishabilities we have computed the normalized difference between two most probable hypotheses. Suppose that $V = \{v_1, v_2, \ldots, v_n\}$ is a set of correlation coefficient or mutual information values reported by a CPA or

---

[2] Note that we have computed absolute of correlation coefficient in CPAs; for this reason all coefficients seen in Fig. 2 are positive.

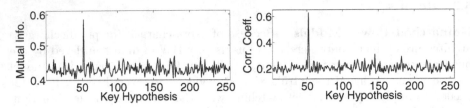

Fig. 2. Mutual information (left) and correlation coefficient (right) values over a key byte hypotheses using 7 bits of an Sbox output as the power model

MIA respectively, and suppose that v_i values are sorted from the biggest to the smallest, i.e., $v_i \geq v_j$; $i > j$. We have used $(v_1 - v_2)/v_1$ as the normalized difference to compare the distinguishabilities. Note that according to the previous results we know that always v_1 relates to the correct hypothesis.

Since the number of bins is a new parameter which does not exist in CPA, we examined its effect on distinguishability too. Although it is suggested in [7] "to use as many bins as there are distinct values in the domain covered by the sample set to ensure that no information is lost", we could not obtain good results by choosing big number of bins, we hence limited the number of bins to $\{2, 3, \ldots, 16\}$. Indeed, by selecting more number of bins the correct hypothesis was not distinguishable clearly amongst other hypotheses (of course using a particular number of traces). First we have considered all possible HW models (similarly to the previous experiments) and all possible number of bins, i.e., 255 power models for CPA and 255×15 power models for MIA. Then, we got average of distinguishability values over number of bits contributed in the power model leading to the diagram shown by Fig. 3(a). Further, we have repeated this scenario for a part of the target leakage, i.e., some bits of the Sbox output, as the power model, i.e., 254 power models for CPA and 254×15 power models for MIA. The comparative diagram is shown by Fig. 3(b). Obviously when a HW model is used, there is not a big difference between the normalized differences in MIA and CPA. Also, by increasing the number of bits contributed in HW model, the correct secret would be more distinguishable. However, better results are achieved by CPA when a part of the target leakage is used as the power model. For instance, when a high number of bits, e.g., 6 and 7, construct the power model, CPAs reveal the secret more distinguishably than MIA especially than those with high number of bins.

Minimum Number of Traces - Another parameter investigated is the minimum number of traces needed to have a successful attack in MIA and CPA. Since the role of thumb illustrated in [11] still is not examined for MIA, we had to repeat the previous attacks using different number of traces. To do so, because of the high computation overload, we have limited our examinations to eight power models for each number of bits contributed in the power model, i.e., $8 \times 7 = 56$ models when a part of the target leakage constructs the power model and $56 + 1$ models when HW power model is used. Moreover, we have limited

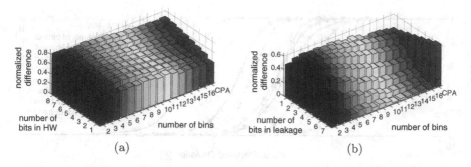

Fig. 3. Average of the normalized difference of two most probable hypotheses in CPA and MIA over the number of bins and over the number of bits contributed in (a) HW model and (b) a part of the target leakage

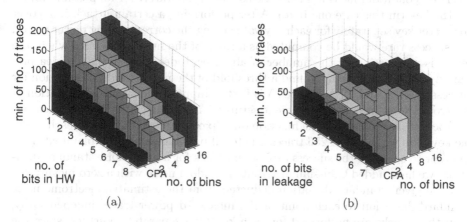

Fig. 4. Average of the minimum number of traces in CPA and MIA over the number of bins and over the number of bits contributed in (a) HW model and (b) a part of the target leakage

the number of bins to $\{2, 4, 8, 16\}$. Getting the average on the minimum number of traces over the number of bits in power models led to diagrams shown by Fig. 4. As expected when a HW model is used in CPA as well as in MIA, the more bits contributed in the model, the less traces are needed. However, an unpredicted treatment from MIA is seen when a part of the target leakage is used as the power model. For instance, when the number of bins is 4, contributing 1 or 7 bits of the Sbox output leads to higher number of traces than other cases. More interestingly, in almost all cases CPA clearly needs less traces to reveal the secret.

Noise Effect - The next criterion by which MIA and CPA are compared is their capability of revealing the secret in the presence of noise. A parameter which we have used to compare them is the success rate. To compute the success rate for

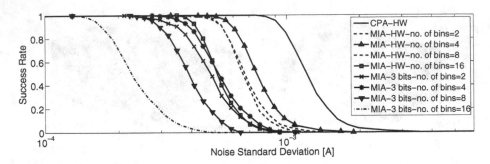

Fig. 5. Success rate of CPA and MIA for different conditions over noise standard deviation

a certain condition, measurements are collected separately for all possible values for the key (in this case one byte). After performing a certain attack to recover the secret key separately for each key value using the corresponding power traces, the success rate would be computed as a ratio of the number of cases where the secret is recovered over the number of all cases. Indeed, based on the idea presented in [10] we are going to find a threshold of the noise standard deviation for successful MIA and CPA attacks. We first computed the standard deviation of electronic noise in our measurement setup by measuring 2 000 traces when our microcontroller runs the same function on a fixed plaintext. Then, for each possible key byte we collected 256 traces due to all possible values of a plaintext byte. In order to compute the success rate of an attack for a given noise standard deviation, we have added Gaussian distributed random noise with a zero mean value and the given standard deviation deducted from the estimated electronic noise standard deviation to each point of the measured power traces independently, then the attacks are performed for each key byte separately. Further, since the noise values are chosen randomly from a distribution, we have repeated each of the above procedure ten times to improve the accuracy of our estimation about the success rate. For instance to obtain the aforementioned threshold for a CPA using HW of 8 bits of the Sbox output, we have performed the attack 256×10 for each of 25 given noise standard deviations, i.e., in sum 64 000 times. This procedure has been repeated for MIA with HW model as well as a model which uses a part of the target leakage. However, we limited ourselves to HW model of 8 bits, a model using 3 bits of the Sbox output, and four different number of bins as $\{2, 4, 8, 16\}$. As shown by Fig. 5, the CPA which uses HW of all 8 bits has the highest threshold, and the noise standard deviation threshold decreases by increasing the number of bins in MIAs. Further, for each power model the best result is achieved by using 4 bins in measurements classification. Note that since the variance of signal is the same for all attacks, $0.095 \, (\text{mA})^2$, one can compare the SNR thresholds for the plotted diagrams in Fig. 5.

Linear Functions - It is clear that the success rate of an attack targeting the leakage of a linear function, e.g., an XOR, is not perfect if the leakage of the

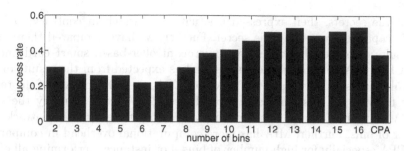

Fig. 6. Success rate of CPA and MIA for different number of bins targeting the leakage of an XOR operation using a HW model

attacked device fits into a HW or HD model. In order to examine the success rate practically we have performed CPAs as well as MIAs on the traces we had collected for the noise effect, i.e., 256 traces for each possible key byte. Fig. 6 shows the success rate of CPA and MIA for different number of bins when the HW of an AddRoundKey output byte is considered as the target leakage. Moreover, we have repeated this task with different leakage models as a part of the AddRoundKey output byte, but neither CPA nor MIA (for any number of bins) led to a success rate over 0.05. From our point of view, in this case MIA does not have an advantage in comparison with CPA.

Independency on the Particular Leakage Function - MIA has been introduced to be independent of the particular relation between measurements and processed data [7]. In order to evaluate this we have measured a set of traces of our FPGA board and performed MIAs and CPAs using a set of different leakage models. Considering a HD model for an Sbox output, of course, MIA as well as CPA are successful. However, when a HW or a part of the Sbox output is used as the leakage model, neither MIA nor CPA can reveal the secret. We applied all power models used before on 10 000 traces of a random plaintext byte that measured separately for each possible key byte, but none of the attacks reported a success rate more than 0. In contrast, using a part of the XOR result of two consecutive Sbox outputs, i.e., bit flips, both MIA and CPA work similarly to the results presented so far. To our knowledge we can not say that MIA or any power analysis attack can work efficiently without any knowledge about the leakage function of the target device. At least we should know how intermediate values affect the side-channel leakage. For instance, we use HW model (or a part of the Sbox output) when attacking a microcontroller, because we know there is a pre-charged bus inside. Generally, one can not perform a universal attack on some traces measured from an unknown unprotected device, of course knowing algorithm and inputs/outputs.

6 Conclusions

We studied MIA in comparison with CPA from theoretical as well as practical points of view. First we have examined the theoretical reason of why MIA can recover the secrets, then expressed a situation in which in contrary to MIA, CPA is capable of revealing the secret. Further, we have compared them using the results of practical attacks on a microcontroller-based smart card and an FPGA board. According to the results, and as expected from the results of [13] and [15] there is no advantage for MIA over CPA when the leakage of the target device fits into a Gaussian assumption. Indeed, MIA works roughly the same as CPA, but with more parameters that affect the efficiency of the attack, i.e., number of bins. Further, MIA has more computational overhead in comparison with CPA especially for high number of bins. For instance, performing all of the attacks illustrated in Section 5.2 took weeks of computation on a 3GHz Intel Core2 PC.

In short, in comparison with CPA, MIA works worse in the presence of noise; it distinguishes the correct guess amongst other hypotheses weaker than CPA considering a part of the target leakage as the power model; it is not perfectly independent of the particular leakage function of the attacked device (similar to CPA); number of the traces which is needed to have a successful MIA attack is more than a corresponding CPA attack; and similarly to CPA, it is not 100% successful when the leakage of a function that has a linear relation with the secret is the target leakage.

Note that the results expressed here are only correct when the target device is an unprotected implementation which has side-channel leakage like HW or HD model. However, the MIA especially its multi-dimensional extension is much more effective than the CPA and respectively than classical higher-order CPA attacks when the leakage of the target device is not linearly proportional to the predictions, e.g., masked implementations.

References

1. Atmel. AVR ATmega163 Data Sheet,
   http://www.atmel.com/dyn/resources/prod_documents/doc1142.pdf
2. Brier, E., Clavier, C., Olivier, F.: Correlation Power Analysis with a Leakage Model. In: Joye, M., Quisquater, J.-J. (eds.) CHES 2004. LNCS, vol. 3156, pp. 16–29. Springer, Heidelberg (2004)
3. Canright, D.: A Very Compact S-Box for AES. In: Rao, J.R., Sunar, B. (eds.) CHES 2005. LNCS, vol. 3659, pp. 441–455. Springer, Heidelberg (2005)
4. Chari, S., Rao, J., Rohatgi, P.: Template Attacks. In: Kaliski Jr., B.S., Koç, Ç.K., Paar, C. (eds.) CHES 2002. LNCS, vol. 2523, pp. 13–28. Springer, Heidelberg (2003)
5. CHIPDRIVE. Smart Card Reader,
   http://www.chipdrive.de/cgi-bin/edcstore.cgi?category=Einkaufen;01
   _Chipkartenleser&user_action=detail&catalogno=P208199.
6. Gandolfi, K., Mourtel, C., Olivier, F.: Electromagnetic Analysis: Concrete Results. In: Koç, Ç.K., Naccache, D., Paar, C. (eds.) CHES 2001, vol. 2162, pp. 251–261. Springer, Heidelberg (2001)
7. Gierlichs, B., Batina, L., Tuyls, P., Preneel, B.: Mutual Information Analysis. In: Oswald, E., Rohatgi, P. (eds.) CHES 2008. LNCS, vol. 5154, pp. 426–442. Springer, Heidelberg (2008)

8. Kocher, P.C., Jaffe, J., Jun, B.: Differential Power Analysis. In: Wiener, M. (ed.) CRYPTO 1999. LNCS, vol. 1666, pp. 388–397. Springer, Heidelberg (1999)
9. Le, T.-H., Clédière, J., Canovas, C., Robisson, B., Servière, C., Lacoume, J.-L.: A Proposition for Correlation Power Analysis Enhancement. In: Goubin, L., Matsui, M. (eds.) CHES 2006. LNCS, vol. 4249, pp. 174–186. Springer, Heidelberg (2006)
10. Macé, F., Standaert, F.-X., Quisquater, J.-J.: Information Theoretic Evaluation of Side-Channel Resistant Logic Styles. In: Paillier, P., Verbauwhede, I. (eds.) CHES 2007. LNCS, vol. 4727, pp. 427–442. Springer, Heidelberg (2007)
11. Mangard, S., Oswald, E., Popp, T.: Power Analysis Attacks: Revealing the Secrets of Smart Cards. Springer, Heidelberg (2007)
12. Messerges, T.S.: Using Second-Order Power Analysis to Attack DPA Resistant Software. In: Paar, C., Koç, Ç.K. (eds.) CHES 2000. LNCS, vol. 1965, pp. 238–251. Springer, Heidelberg (2000)
13. Prouff, E., Rivain, M.: Theoretical and Practical Aspects of Mutual Information Based Side Channel Analysis. In: Abdalla, M., Pointcheval, D., Fouque, P.-A., Vergnaud, D. (eds.) ACNS 2009. LNCS, vol. 5536, pp. 499–518. Springer, Heidelberg (2009)
14. Quisquater, J.-J., Samyde, D.: ElectroMagnetic Analysis (EMA): Measures and Counter-Measures for Smart Cards. In: Attali, S., Jensen, T. (eds.) E-smart 2001. LNCS, vol. 2140, pp. 200–210. Springer, Heidelberg (2001)
15. Veyrat-Charvillon, N., Standaert, F.-X.: Mutual Information Analysis: How, When and Why. In: Clavier, C., Gaj, K. (eds.) CHES 2009. LNCS, vol. 5747, pp. 429–443. Springer, Heidelberg (2009),
http://www.dice.ucl.ac.be/~fstandae/PUBLIS/67.pdf
16. XILINX. Spartan-II FPGA Family Data Sheet,
http://www.xilinx.com/support/documentation/data_sheets/ds001.pdf

Finding Collisions for a 45-Step Simplified HAS-V*

Nicky Mouha[1,2,**], Christophe De Cannière[1,2,***], Sebastiaan Indesteege[1,2,†], and Bart Preneel[1,2]

[1] Department of Electrical Engineering ESAT/SCD-COSIC,
Katholieke Universiteit Leuven. Kasteelpark Arenberg 10, B-3001 Heverlee, Belgium
[2] Interdisciplinary Institute for BroadBand Technology (IBBT), Belgium
{nicky.mouha,christophe.decanniere,sebastiaan.indesteege,
bart.preneel}@esat.kuleuven.be

Abstract. Recent attacks on hash functions start by constructing a differential characteristic. By finding message pairs that satisfy this characteristic, a collision can be found. This paper describes the method of De Cannière and Rechberger to construct generalized characteristics for SHA-1 in more detail. This method is further generalized and applied to a simplified variant of the HAS-V hash function. Using these techniques, a characteristic for 45 steps is found, requiring an effort of about 2^{46} compression function evaluations to find a colliding message pair. A lot of the message bits can still be freely chosen when using this characteristic, greatly increasing its usefulness.

Keywords: Cryptanalysis, hash function, HAS-V, collision.

1 Introduction

Hash functions are an important building block in cryptography. As described in [1], these are functions h that convert an input m of arbitrary length in a deterministic way to a fixed-length output $h(m)$. It is crucial that a number of security properties are satisfied, one of which is the infeasibility of finding collisions (collision resistance). A collision consists of two input values m, m' where $m \neq m'$ for which $h(m) = h(m')$.

Recent attacks by Wang et al. on the widely used hash functions MD4 [2], MD5 [3], RIPEMD [2] and SHA-1 [4], as well as other hash functions, show that it is possible to find collisions for these hash functions much faster than expected by the birthday paradox [1]. In response to these attacks, NIST has

---

* This work was supported in part by the IAP Program P6/26 BCRYPT of the Belgian State (Belgian Science Policy), and in part by the European Commission through the ICT program under contract ICT-2007-216676 ECRYPT II.
** This author is funded by a research grant of the Institute for the Promotion of Innovation through Science and Technology in Flanders (IWT-Vlaanderen).
*** Postdoctoral Fellow of the Research Foundation – Flanders (FWO)
† F.W.O. Research Assistant, Fund for Scientific Research – Flanders (Belgium).

H.Y. Youm and M. Yung (Eds.): WISA 2009, LNCS 5932, pp. 206–225, 2009.
© Springer-Verlag Berlin Heidelberg 2009

launched a competition to find a new hash function standard [5]. Although a lot of cryptanalysis effort is directed to these submissions, we feel that it is still very important to analyze existing hash function standards.

These recent collisions have lead to several practical attacks on network applications. For POP3 [6], it is is possible to mount a password recovery attack. Together with IMAP [7], POP3 [8] is one of the most used protocols for retrieving e-mail. Very recently, a rogue CA certificate was created using a collision for MD5 [9,10]. This certificate allows an attacker to impersonate any website on the Internet secured by HTTPS [11], including websites for banking and e-commerce.

The hash function HAS-V [12] is similar in structure to these hash functions. HAS-V fulfills the need of the KCDSA [13], the Korea Certificate-based Digital Signature Algorithm, to use a hash function with a variable digest size. The only cryptanalytic results on HAS-V known to us are described in [14]. Results using the recent attacks on hash functions have not been published before. The cryptanalysis of a simplified variant of HAS-V is the subject of this paper.

Recent attacks on hash functions focus on the construction of a differential characteristic, that allows collisions to be found with a good probability by finding messages m, m' that satisfy this characteristic. Characteristics are often constructed in an ad hoc way, which does not give any insight into the application of these attacks to other hash functions. This emphasizes the need for automated methods.

One such method, introduced in [15], is further generalized and applied to the simplified HAS-V. Using this method, we found a characteristic for a 45-step collision with an expected work factor of $2^{75.84}$ step function evaluations, which is given in Table 12. Further improvements lead to the better characteristic shown in Table 13, which has a work factor of $2^{51.53}$, making a collision finding attack feasible. If the cost of one step function evaluation is about 2^{-5} compression function evaluations, these work factors are equivalent to about 2^{71} and 2^{46} compression function evaluations, respectively. Note that a lot of bits in the message words can still be freely chosen.

Notation is defined in Table 1. In Sect. 2, a description of a simplified variant of HAS-V is given. An alternative, cyclic description of this hash function is provided as well. The technique for finding NL-characteristics of [15] is further explained, generalized and applied to HAS-V in Sect. 3. Techniques for improving NL-characteristics are laid out in Sect. 4, where good NL-characteristics for a 45-step simplified HAS-V are obtained as well. A conclusion and suggestions for future work are given in Sect. 5.

Appendix A lists the NL-characteristics we obtained. To assist the reader in understanding the more abstract explanation of the graph method in this paper, a simple example is given in Appendix B. Although this method is extensively used to attack SHA-1 in [15], this paper is the first to fully explain it.

2   A Simplified HAS-V

The hash function HAS-V [12] splits a 1024-bit message block into two 512-bit message blocks, which are then processed in two streams. The rounds of each

Table 1. Notation

notation	description
$x \parallel y$	concatenation of the binary strings x and y
$x \wedge y$	bitwise AND of x and y
$x \vee y$	bitwise OR of x and y
$x \oplus y$	bitwise XOR of x and y
$\neg x$	bitwise NOT of x
$x \lll s$	rotation of x to the left by s positions
$x \ggg s$	rotation of x to the right by s positions
$x + y$	addition of x and y modulo 2^{32} (in text)
$x \boxplus y$	addition of x and y modulo 2^{32} (in figures)
$x[i]$	bit selection: 0 if $(x \wedge 2^i) \equiv 0$, 1 otherwise

Table 2. The IV values for the simplified HAS-V

	A	B	C	D	E
IV	0x67452301	0xEFCDAB89	0x98BADCFE	0x10325476	0xC3D2E1F0

stream alternately use message words of the first and the second 512-bit block. In our simplified variant of HAS-V, the right stream is omitted, as well as rounds in the left stream that depend on message words of the second 512-bit message block. As recent collision finding attacks are applied to hash functions with only one stream, simplifying the hash function in this way allows us to focus more easily on the main concepts of these recent attacks. For the same reason, the optional output tailoring is not applied. All other properties of HAS-V are left intact. A description of this simplified HAS-V is now given.

2.1 Description

The input message is padded and split into 512-bit message blocks. A 3-round compression function with 20 steps per round is applied to each of these 512-bit message blocks. This compression function $g(m, h)$ uses a 160-bit chaining input h and a 512-bit message block m as its inputs. The chaining input h_{n+1} of the next call of the compression function is calculated as $h_n + g(m, h_n)$. Here, the addition is done in blocks of 32-bit words, using a total of five adders modulo 2^{32}. The chaining variables for the first compression function call are set to fixed values, referred to as the IV. They are shown in Table 2. The last chaining input h represents the hash value.

Given a 512-bit message block m, consisting of 16 32-bit message words m_i, four extra message words, referred to as XOR-words, are derived from these message words for every round, as specified in Table 3. The extended message words w_i consist of the message words m_i followed by the four XOR-words.

Table 4 shows how the expanded message words W_t are derived as a reordering of the extended message words w_i for every round.

Table 3. Calculation of the XOR-words for the simplified HAS-V

	w_{16}	w_{17}	w_{18}	w_{19}
Round 1	$w_0 \oplus w_1 \oplus w_2 \oplus w_3$	$w_4 \oplus w_5 \oplus w_6 \oplus w_7$	$w_8 \oplus w_9 \oplus w_{10} \oplus w_{11}$	$w_{12} \oplus w_{13} \oplus w_{14} \oplus w_{15}$
Round 2	$w_3 \oplus w_6 \oplus w_9 \oplus w_{12}$	$w_{15} \oplus w_2 \oplus w_5 \oplus w_8$	$w_{11} \oplus w_{14} \oplus w_1 \oplus w_4$	$w_7 \oplus w_{10} \oplus w_{13} \oplus w_0$
Round 3	$w_{12} \oplus w_5 \oplus w_{14} \oplus w_7$	$w_0 \oplus w_9 \oplus w_2 \oplus w_{11}$	$w_4 \oplus w_{13} \oplus w_6 \oplus w_{15}$	$w_8 \oplus w_1 \oplus w_{10} \oplus w_3$
Round 4	$w_7 \oplus w_2 \oplus w_{13} \oplus w_8$	$w_3 \oplus w_{14} \oplus w_9 \oplus w_4$	$w_{15} \oplus w_{10} \oplus w_5 \oplus w_0$	$w_{11} \oplus w_6 \oplus w_1 \oplus w_{12}$
Round 5	$w_{15} \oplus w_9 \oplus w_5 \oplus w_3$	$w_{12} \oplus w_8 \oplus w_6 \oplus w_2$	$w_{13} \oplus w_{11} \oplus w_7 \oplus w_1$	$w_{14} \oplus w_{10} \oplus w_4 \oplus w_0$

Table 4. The message expansion for the simplified HAS-V

t	0	1	2	3	4	5	6	7	8	9	10	11	12	13	14	15	16	17	18	19
Round 1	18	0	1	2	3	19	4	5	6	7	16	8	9	10	11	17	12	13	14	15
Round 2	18	12	5	14	7	19	0	9	2	11	16	4	13	6	15	17	8	1	10	3
Round 3	18	15	9	5	3	19	12	8	6	2	16	13	11	7	1	17	14	10	4	0

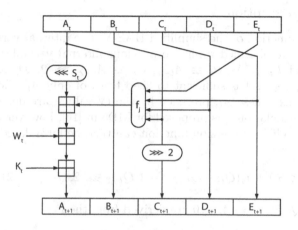

Fig. 1. The HAS-V step function

Figure 1 gives a schematic representation of the HAS-V step function, which is also described by

$$
\begin{cases}
A_{t+1} \leftarrow (A_t \lll S_t) + f_j(B_t, C_t, D_t, E_t) + W_t + K_t \ , \\
B_{t+1} \leftarrow A_t \ , \\
C_{t+1} \leftarrow B_t \ggg 2 \ , \\
D_{t+1} \leftarrow C_t \ , \\
E_{t+1} \leftarrow D_t \ .
\end{cases}
\tag{1}
$$

Here, f_j represents a Boolean function, different for every round j:

$$
\begin{aligned}
f_1(B, C, D, E) &\triangleq (B \wedge C) \oplus (\neg B \wedge D) \oplus (C \wedge E) \oplus (D \wedge E) \ , \\
f_2(B, C, D, E) &\triangleq (B \wedge C) \oplus (\neg B \wedge E) \oplus D \ , \\
f_3(B, C, D, E) &\triangleq (\neg B \wedge C) \oplus (B \wedge D) \oplus (C \wedge E) \oplus (D \wedge E) \ .
\end{aligned}
\tag{2}
$$

Table 5. Constant K_t for the simplified HAS-V

	Round 1	Round 2	Round 3
K_t	0x00000000	0x6ED9EBA1	0xA953FD4E

Table 6. Rotation value S_t for the simplified HAS-V

t	0	1	2	3	4	5	6	7	8	9	10	11	12	13	14	15	16	17	18	19
S_t	5	11	7	13	15	6	13	9	5	11	7	12	8	15	13	8	15	6	7	14

In every step a constant K_t, different for every round, is added. These are listed in Table 5. The rotation value S_t is different for every step of a round. They are given in Table 6.

2.2 Cyclic Description

Using the step function of the simplified HAS-V, five internal variables A_t, B_t, C_t, D_t and E_t are obtained from five previous internal variables, A_{t-1}, B_{t-1}, C_{t-1}, D_{t-1} and E_{t-1}. As $B_t \equiv A_{t-1}$, $C_t \equiv A_{t-2} \ggg 2$, $D_t \equiv A_{t-3} \ggg 2$ and $E_t \equiv A_{t-4} \ggg 2$, it is sufficient to keep track of only A_t when calculating the step functions. These A_t, preceded by the IV values, are denoted by Q_t. A similar cyclic formulation was proposed for MD5 in [16], however there Q_t refer to values of B_t. The compression function can then be formulated alternatively for $t = 0, \ldots, 59$ as:

$$Q_{t+1} \leftarrow (Q_t \lll S_i) + f_j(Q_{t-1}, Q_{t-2} \ggg 2, Q_{t-3} \ggg 2, Q_{t-4} \ggg 2) + W_t + K_t .$$
(3)

The values of Q_t for $t = -4, \ldots 0$ are derived from the IV:

$$(Q_{-4}, Q_{-3}, Q_{-2}, Q_{-1}, Q_0) \leftarrow (E \lll 2, D \lll 2, C \lll 2, B, A) .$$
(4)

It is this cyclic formulation of the simplified HAS-V that will be used from now on in this text.

3   NL-characteristics

For collision attacks, non-linear characteristics (NL-characteristics) start with chaining input and output difference zero. Differences are introduced via the message input, which then cancel themselves out with a sufficiently high probability by following the characteristic. High-probability characteristics are crucial for building fast collision-finding attacks, yet not much is known about their construction. They are often generated manually, using a great deal of intuition and experience. This paper further improves the results from [15], where an automated method is described for constructing these NL-characteristics.

In this paper, NL-characteristics are applied to the simplified HAS-V. An NL-characteristic is a set of conditions $\nabla Q_{-4}, \ldots \nabla Q_T$ and $\nabla W_0 \ldots \nabla W_{T-1}$. Each

Table 7. All possible conditions for $(X[i], X'[i])$

	(0,0)	(1,0)	(0,1)	(1,1)		(0,0)	(1,0)	(0,1)	(1,1)
?	✓	✓	✓	✓	3	✓	✓	-	-
-	✓	-	-	✓	5	✓	-	✓	-
x	-	✓	✓	-	7	✓	✓	✓	-
0	✓	-	-	-	A	-	✓	-	✓
u	-	✓	-	-	B	✓	✓	-	✓
n	-	-	✓	-	C	-	-	✓	✓
1	-	-	-	✓	D	✓	-	✓	✓
#	-	-	-	-	E	-	✓	✓	✓

$\nabla X[i]$ represents a set of possible combinations for $(X[i], X[i]')$, as shown in Table 7. The number of steps T is left variable to be able to study step-reduced versions of this simplified HAS-V. In this paper, results will be obtained for $T = 45$.

3.1 Representation of Conditions on One Bit $\nabla Q_{t+1}[i]$

The step function (3) can be written as follows for every bit, for $0 \leq t < T$ and $0 \leq i < 32$. Indices i are calculated modulo 32. The carry input of the addition is denoted by $C_{t,i}$, the carry output by $C_{t+1,i+1}$.

$$C_{t+1,i+1} \parallel Q_{t+1}[i] \leftarrow Q_t[i - S_t] + f_j(Q_{t-1}[i], Q_{t-2}[i+2], Q_{t-3}[i+2], \\ Q_{t-4}[i+2]) + W_t[i] + K_t[i] + C_{t,i} . \tag{5}$$

To calculate $Q_{t+1}[i]$, the bit positions of the previous state words Q_{t-k} ($0 \leq k < 5$) are schematically represented in Fig. 2.

In the step function for every bit (5), a single, large addition is used with a carry input and output. The resulting carry states $(C_{t+1,i+1}, C'_{t+1,i+1})$ are then the only way in which adjacent bits of the same message word pair (W_t, W'_t) or internal states (Q_{t+1}, Q'_{t+1}) interact. If a particular carry state $(C_{t+1,i+1}, C'_{t+1,i+1})$ cannot occur as the output for the calculation of this bit $(Q_{t+1,i}, Q'_{t+1,i})$, nor as the input of the calculation of the next bit $(Q_{t+2,i+1}, Q'_{t+2,i+1})$, this combination of carries is said to be invalid.

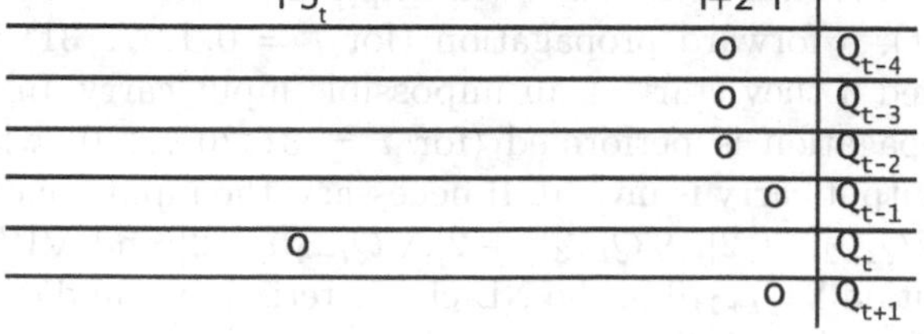

Fig. 2. Calculation of $Q_{t+1}[i]$ from $Q_t[i - S_t]$, $Q_{t-1}[i]$, $Q_{t-1}[i+2]$, $Q_{t-1}[i+2]$ and $Q_{t-1}[i+2]$

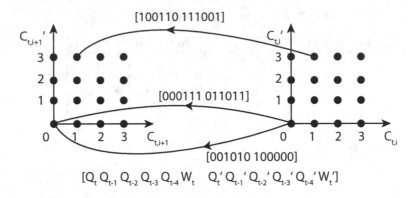

Fig. 3. Explanation of the edges in the graph. When adding four bits (and the carry input), the carry output $C_{t,i}$ can be 0, 1, 2 or 3. The addition of the corresponding four bits of the second message of the collision pair results in the carry $C'_{t,i}$.

For the calculation of every $\nabla Q_{t+1}[i]$, all valid combinations of $(C_{t,i}, C'_{t,i})$, (Q_{t-k}, Q'_{t-k}) for $0 \le k < 5$ and (W_t, W'_t) are represented by an edge in Fig. 3. Imposing new conditions on $\nabla Q_{t+1}[i]$ will lead to the elimination of some of these edges.

The Boolean function f_j and the constant bit $K_t[i]$ are fixed for one bit position. Therefore, they are not included on the edges in Fig. 3. Each of the 2^{16} input bits of the edges then completely determines the six output bits $(Q_{t+1}[i], Q'_{t+1}[i])$ and $(C_{t,i+1}, C'_{t,i+1})$.

3.2   Propagation of Conditions for Every Word ∇Q_{t+1}

Initially, all input conditions are allowed for all bits. As this implies that all outputs are allowed for every bit, this is a self-consistent state. However, as soon as some restrictions are imposed on a bit, this may affect other bits. We refer to this mechanism as the propagation of conditions.

To calculate the possible conditions for every word ∇Q_{t+1} for $0 \le t < T$, it is necessary to do both a forward and a backward propagation over all conditions $\nabla Q_{t+1}[i]$ for $0 \le i < 32$. In Fig. 3, every possible input combination is shown as an edge connecting the input carries $(C_{t,i}, C'_{t,i})$ to the output carries $(C_{t,i+1}, C'_{t,i+1})$. Note that there can be multiple edges between two nodes.

In Fig. 4 (left), a forward propagation (for $i = 0, 1, \ldots, 31$) is done where edges are removed if they start at an impossible input carry. In Fig. 4 (right), a backward propagation is performed (for $i = 31, 30, \ldots, 0$) where edges are removed if the output carry is invalid. If necessary, the input conditions $\nabla Q_t[i - S_t]$, $\nabla Q_{t-1}[i]$, $\nabla Q_{t-2}[i + 2]$, $\nabla Q_{t-3}[i + 2]$, $\nabla Q_{t-4}[i + 2]$ and $\nabla W_t[i]$, as well as the output condition $\nabla Q_{t+1}[i]$ in the NL-characteristic are updated. In this way, one word can affect the conditions of another word.

This step is repeated for every word ∇Q_{t+1} for $0 \le t < T$, until further propagation would not remove additional edges or until at least one condition

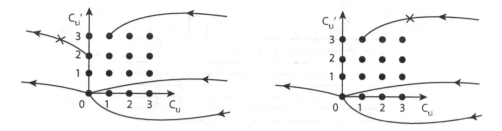

Fig. 4. Removing edges through forward propagation (left) and backward propagation (right)

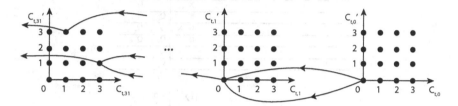

Fig. 5. Remaining valid paths for one word ∇Q_{t+1}

is inconsistent. Every time a message bit $W_t[i]$ is assigned a new value, message bits $W_{t'}[i]$ that are related by the message expansion, are updated as well if necessary. The remaining valid paths for one word are then shown in Fig. 5.

3.3 Double Conditions

Conditions that do not involve one pair of bits, but two pairs of bits, are referred to as "double conditions". The use of these is new to this paper. They are similar to Table 7, except that double conditions apply to four bits instead of two. Thus, there are 2^{16} possible double conditions, instead of 2^4.

For the simplified HAS-V, double conditions can be used in three locations for the calculation of one bit of ∇Q_{t+1}. These are shown in Fig. 6, as the result of the only possibilities of creating an overlap of at least two bits of Fig. 2 with a translated version of this pattern.

The use of the first double condition is explained, the other two cases are analogous. During the calculation of $\nabla Q_{t'+1}[2]$, a double condition is used to represent the possibilities of the joint occurrence of $\nabla Q_{t'+1}[2]$ and $\nabla Q_{t'-1}[2]$. When $\nabla Q_{t+1}[0]$ is calculated, it can be seen that the same double condition now also applies to the joint occurrence of $\nabla Q_{t-2}[2]$ and $\nabla Q_{t-4}[2]$. It is possible that this information leads to the removal of additional edges. If this is the case, the number of iterations needed to construct an NL-characteristic is lowered, and inconsistencies can be found sooner. In our implementation of the search for NL-characteristics, double conditions can be implemented with minimal overhead.

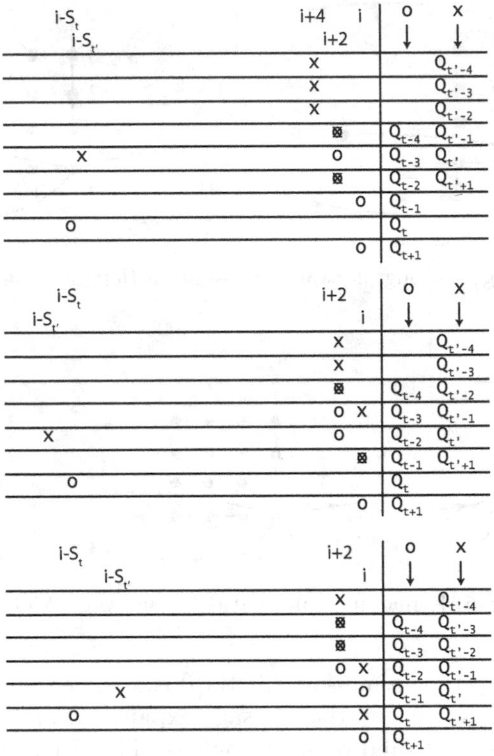

Fig. 6. Double conditions for the HAS-V step function, obtained as the only possible overlaps of at least two bits in Fig. 2 with a translated version of this pattern

3.4   Work Factor

The work factor N_w of an NL-characteristic indicates the expected number of step function evaluations required to find a collision using this characteristic. When building NL-characteristics, the collision search is optimized by lowering the work factor. This concept was introduced in [15].

Message freedom $F_W(t)$. "Single-message modification" [3] (also known as "single-step modification" [2]) can be used during the search process, as there is still freedom left in the choice of several expanded message words W_t. Due to the constraints imposed by the XOR-words, this is not possible for each of the 20 message word pairs (W_t, W_t') of the first round. Of the five message word pairs involved in the calculation of each XOR-word, only the first four can be chosen. The last message word pair cannot be freely chosen, but must equal the XOR of the four others.

The message freedom $F_W(t)$ of a characteristic at step t is the number of ways to choose (W_t, W_t'), without violating any (linear) condition imposed by the message expansion, given fixed values of (W_j, W_j') for $0 \leq j < t$.

The description of the simplified HAS-V indicates that $F_W(t)$ is always 1 for $t = 10, 14, 15, 19$ and $t \geq 20$. For the other values of t, $F_W(t)$ is the product of the number of possibilities for conditions $\nabla W_t[i]$ for $0 \leq i < 32$. This number of possibilities equals the number of checkmarks ($\checkmark$) for the respective conditions in Table 7.

Uncontrolled probability $P_u(t)$. The uncontrolled probability $P_u(t)$ of a characteristic at step t is the probability that the output (Q_{t+1}, Q'_{t+1}) follows the characteristic, given that all input pairs (Q_{t-k}, Q'_{t-k}) for $0 \leq k < 5$ and message word pairs (W_t, W'_t) follow this characteristic as well:

$$P_u(t) = P\left((Q_{t+1}, Q'_{t+1}) \in \nabla Q_{t+1} \mid (Q_{t-k}, Q'_{t-k}) \in \nabla Q_{t-k} \right. \\ \left. \text{for } 0 \leq k < 5, \text{ and } (W_t, W'_t) \in \nabla W_t\right) . \tag{6}$$

This probability can be calculated as the number of remaining paths of Fig. 5, divided by the number of paths for which only the input pairs (Q_{t-k}, Q'_{t-k}) for $0 \leq k < 5$ and the message word pairs (W_t, W'_t) follow the characteristic, but not necessarily the output pair (Q_{t+1}, Q'_{t+1}).

Controlled probability $P_c(t)$. The controlled probability $P_c(t)$ of a characteristic at step t is the probability that there exists at least one pair of message words (W_t, W'_t) following the characteristic, such that the output (Q_{t+1}, Q'_{t+1}) follows the characteristic, given that all input pairs (Q_{t-k}, Q'_{t-k}) for $0 \leq k < 5$ do as well:

$$P_c(t) = P\left(\exists(W_t, W'_t) \in \nabla W_t : (Q_{t+1}, Q'_{t+1}) \in \nabla Q_{t+1} \right. \\ \left. \mid (Q_{t-k}, Q'_{t-k}) \in \nabla Q_{t-k} \text{ for } 0 \leq k < 5\right) . \tag{7}$$

A graph is made for every bit i for the calculation of $(Q_{t+1}[i], Q'_{t+1}[i])$ to determine this probability. Each node of the graph is a carry mask, indicating which of the 16 possible values of $(C_{t,i}, C'_{t,i})$ can occur. Thus, a carry mask can have 2^{16} possible values. Note the analogy with Table 7, where the possible combinations of $(X[i], X[i]')$ are shown.

Let n be the number of possibilities for $(Q_{t-k}[i], Q'_{t-k}[i]) \in \nabla Q_{t-k}[i]$ for $0 \leq k < 5$. For each possibility, we run through all carries $(C_{t,i}, C'_{t,i})$ and all message bit pairs $(W_t[i], W'_t[i])$. A binary 16×16 matrix indicates which transition possibilities from $(C_{t,i}, C'_{t,i})$ to $(C_{t,i+1}, C'_{t,i+1})$ can occur.

Using this 16×16 transition matrix, we can calculate the possible carry masks for bit $i + 1$ using the carry mask of bit i. For the least significant bit ($i = 0$), only one carry mask is possible: the carry is $(0, 0)$ with probability 1. Each of the edges in the graph has probability $1/n$. Unlike in Fig. 5, there is never more than one edge between two nodes. This step is repeated for every $(Q_{t-k}[i], Q'_{t-k}[i]) \in \nabla Q_{t-k}[i]$ for $0 \leq k < 5$.

This calculation is performed for bits $i = 0 \ldots 31$. We now consider the most significant bit ($i = 31$). One carry mask indicates that none of the carries $(C_{t,31}, C'_{t,31})$ are valid. $P_c(t)$ then equals the sum of all the other carry masks.

Table 8. Lowest Hamming weights found for L-characteristics, not taking the weight of ∇Q_{t+1} for $0 \leq t < 20$ into account

	collision	near-collision	pseudo-collision
40 steps	30	26	27
45 steps	75	68	65

Total work factor N_{w}. In the collision search tree, the average number of children of a node at step t is $F_{\mathrm{W}}(t) \cdot P_{\mathrm{u}}(t)$. Only a fraction $P_{\mathrm{c}}(t)$ of the nodes at step t have children at all. The search stops as the last step $T - 1$ of the compression function is reached. We can thus obtain the following recursive relation for the expected number of nodes $N_{\mathrm{s}}(t)$ at every step of the compression function:

$$N_{\mathrm{s}}(t) = \begin{cases} 1 & \text{for } t = T - 1 \ , \\ \max\left(N_{\mathrm{s}}(t+1) \cdot F_{\mathrm{W}}^{-1}(t) \cdot P_{\mathrm{u}}^{-1}(t), P_{\mathrm{c}}^{-1}(t)\right) & \text{for } 0 \leq t < T - 1 \ . \end{cases} \tag{8}$$

The total work factor is then given by

$$N_{\mathrm{w}} = \sum_{t=0}^{T-1} N_{\mathrm{s}}(t) \ . \tag{9}$$

In tables, the base 2 logarithms of $F_{\mathrm{W}}(t)$, $P_{\mathrm{u}}(t)$, $P_{\mathrm{c}}(t)$, $N_{\mathrm{s}}(t)$ and $N_{\mathrm{w}}(t)$ are shown.

A difference with [15], is that in this work, the double conditions of Sect. 3.3 are also taken into account in the calculation of the work factor. These are assumed to be included in the definitions of $P_{\mathrm{u}}(t)$ and $P_{\mathrm{c}}(t)$. This is because double conditions are used in the actual collision search as well. Implementing this is possible with minimal overhead, and can only improve N_{w}. Experimental results of using these double conditions will be given in Sect. 4.

4   Finding NL-characteristics for 45 Steps

To obtain a good NL-characteristic, Stage 1 of [15] consists of obtaining a sparse L-characteristic to use as a starting point. As can be seen in Table 8, no suitable L-characteristic could be found for 45 steps of the simplified HAS-V. The weight of the ∇Q_{t+1} for the first round is not taken into account, assuming for simplicity that these can all be satisfied by single-message modification.

To overcome this problem, we looked for message differences that are localized at a small number of steps of the internal states ∇Q_{t+1}. The Boolean functions f_j are particularly well suited to allow for NL-characteristics consisting of very short collision regions. It can be seen that both f_1 and f_3 allow any input difference to be either passed on or canceled out at the output. For $f_2(B, C, D, E)$, this is also the case for every input difference, except for an input difference at D, which will always lead to an output difference. The HAS-V specification [12]

Table 9. The work factor N_w (in base 2 logarithm) after each of the four stages

	Stage 1	Stage 2	Stage 3	Stage 4
without double conditions	143.87	89.30	81.84	59.92
with double conditions	143.87	81.28	75.84	51.53

reveals that this is by design, in an attempt to satisfy the "Strict Avalanche Criterion (SAC)" [17]. As the attacker can choose both messages m, m' of a collision pair, he can control the output differences of the f-function at certain positions (either probabilistically, or by single- or multi-message modification). This allows for more freedom in the construction of NL-characteristics, while still keeping the probability of the characteristic high.

Differences in the message words m_i are only introduced in $m_{12}[0]$ and $m_{14}[0]$. Due to the message expansion, these differences can be found in $W_{16}[0]$, $W_{18}[0]$, $W_{21}[0]$, $W_{23}[0]$. Before and after this collision region, equality is imposed on the internal state words.

In the short collision region, all conditions for $\nabla Q_{t+1}[i]$ are still unrestricted ("?") at Stage 1.

Stages 2 and 3 are the same as in [15]. In Stage 2, unrestricted conditions ("?") are randomly chosen, and the requirement that they are equal ("-") is imposed. This stage is repeated several times, until a characteristic with a sufficiently low work factor is obtained. Further in Stage 2, conditions ("x") start to appear, which are replaced by either ("u") or ("n") when selected. In Stage 3, local optimizations are performed by going over all "-" conditions, and replacing them by "0" or "1" if this improves the work factor. By repeating Stage 3 several times, the work factor gradually decreases. The end result after Stage 3 is shown in Table 12, with corresponding work factor $N_w = 2^{75.84}$.

After Stage 3, adding a single extra condition will never decrease the work factor. It is possible, however, to reduce the work factor even further. This is done in an additional stage, Stage 4, not described in [15]. In Stage 4, not one, but several conditions are added locally, as long as they do not worsen the work factor. If adding multiple conditions improves the work factor, a minimal set of conditions is derived from these, that still lowers the work factor. This set is obtained by relaxing the additional conditions again, one by one, to see if they had any impact on the global work factor. Only the conditions of this minimal set are kept. Experiments show that it is even possible, that relaxing conditions decreases the work factor of the NL-characteristic. This fourth stage is also repeated several times. The end result is shown in Table 13, where a work factor N_w of $2^{51.53}$ is obtained. After the Stage 4, it is not possible to decrease the work factor by adding or relaxing a single condition.

Note that the characteristics obtained after every stage are not necessarily the best possible. Every stage can thus be performed several times, until a characteristic is found that is good enough.

Experimental results indicating the impact of these double conditions on N_w after each of the four stages, are shown in Table 9.

Although time limits did not allow us to find a colliding message pair, we have verified for reduced versions that the complexity estimates accurately reflect the actual search cost, both with and without the inclusion of double conditions.

5   Conclusion and Future Work

This paper shows how techniques developed for SHA-1 in [15] can be further improved and generalized for a simplified variant of the hash function HAS-V. This simplified variant consists of only a single stream.

For 45 steps of this simplified HAS-V, an NL-characteristic is constructed, requiring about $2^{51.53}$ step function evaluations, or about 2^{46} compression function evaluations, to find a collision. A lot of the message bits can still be freely chosen when using this characteristic.

Stage 1 of method of De Cannière and Rechberger [15], the search for a good L-characteristic, is replaced by the requirement that collisions occur in a very short region. As the method described in this paper can be applied without finding good L-characteristics first, it might be used for hash functions such as RIPEMD-160 [18], for which also no good L-characteristics were found [19].

"Double conditions" are introduced as conditions for two pair of bits. They can be used to speed up the actual collision search.

An extra stage, Stage 4, is introduced to further improve the work factor for finding a collision. It is shown how this additional stage can reduce the work factor from $2^{75.84}$ step function evaluations, or about 2^{71} compression function evaluations, in Table 12, to $2^{51.53}$, or about 2^{46} compression function evaluations, in Table 13.

Acknowledgments. The authors would like to thank their colleagues at COSIC, and the symmetric cryptography subgroup in particular, for their useful comments and suggestions. Special thanks go to Vesselin Velichkov, who greatly helped in improving the clarity of this paper.

Several techniques in this paper were already used in the cryptanalysis of SHA-1 by Christophe De Cannière and Christian Rechberger [15], but had not been explained before. The authors are greatly indebted to Christian Rechberger, not only for his useful comments and suggestions, but also for allowing us build upon his previous work for SHA-1.

References

1. Preneel, B.: Analysis and design of cryptographic hash functions. PhD thesis, Katholieke Universiteit Leuven (1993)
2. Wang, X., Lai, X., Feng, D., Chen, H., Yu, X.: Cryptanalysis of the Hash Functions MD4 and RIPEMD. In: Cramer, R. (ed.) EUROCRYPT 2005. LNCS, vol. 3494, pp. 1–18. Springer, Heidelberg (2005)
3. Wang, X., Yu, H.: How to Break MD5 and Other Hash Functions. In: Cramer, R. (ed.) EUROCRYPT 2005. LNCS, vol. 3494, pp. 19–35. Springer, Heidelberg (2005)

4. Wang, X., Yin, Y.L., Yu, H.: Finding Collisions in the Full SHA-1. In: Shoup, V. (ed.) CRYPTO 2005. LNCS, vol. 3621, pp. 17–36. Springer, Heidelberg (2005)
5. National Institute of Standards and Technology: Announcing Request for Candidate Algorithm Nominations for a New Cryptographic Hash Algorithm (SHA-3) Family. Federal Register 27(212), 62212–62220 (2007),
http://csrc.nist.gov/groups/ST/hash/documents/FR_Notice_Nov07.pdf (2008/10/17)
6. Leurent, G.: Message Freedom in MD4 and MD5 Collisions: Application to APOP. In: Biryukov, A. (ed.) FSE 2007. LNCS, vol. 4593, pp. 309–328. Springer, Heidelberg (2007)
7. Crispin, M.: Internet Message Access Protocol - Version 4rev1. RFC 3501 (Proposed Standard), Updated by RFCs 4466, 4469, 4551, 5032, 5182 (March 2003)
8. Myers, J., Rose, M.: Post Office Protocol - Version 3. RFC (Standard), Updated by RFCs 1939, 2449 (1939)
9. Stevens, M., Lenstra, A.K., de Weger, B.: Chosen-Prefix Collisions for MD5 and Colliding X.509 Certificates for Different Identities. In: Naor, M. (ed.) EUROCRYPT 2007. LNCS, vol. 4515, pp. 1–22. Springer, Heidelberg (2007)
10. Sotirov, A., Stevens, M., Appelbaum, J., Lenstra, A., Molnar, D.A., Osvik, D.A., de Weger, B.: MD5 considered harmful today: Creating a rogue CA certificate. In: 25th Chaos Communications Congress, Berlin, Germany (2008)
11. Rescorla, E.: HTTP Over TLS. RFC 2818, Informational (May 2000)
12. Park, N.K., Hwang, J.H., Lee, P.J.: HAS-V: A New Hash Function with Variable Output Length. In: Stinson, D.R., Tavares, S. (eds.) SAC 2000. LNCS, vol. 2012, pp. 202–216. Springer, Heidelberg (2001)
13. Lim, C.H., Lee, P.J.: A Study on the Proposed Korean Digital Signature Algorithm. In: Ohta, K., Pei, D. (eds.) ASIACRYPT 1998. LNCS, vol. 1514, pp. 175–186. Springer, Heidelberg (1998)
14. Mendel, F., Rijmen, V.: Weaknesses in the HAS-V Compression Function. In: Nam, K.-H., Rhee, G. (eds.) ICISC 2007. LNCS, vol. 4817, pp. 335–345. Springer, Heidelberg (2007)
15. De Cannière, C., Rechberger, C.: Finding SHA-1 Characteristics: General Results and Applications. In: Lai, X., Chen, K. (eds.) ASIACRYPT 2006. LNCS, vol. 4284, pp. 1–20. Springer, Heidelberg (2006)
16. Hawkes, P., Paddon, M., Rose, G.G.: Musings on the Wang et al. MD5 Collision. Cryptology ePrint Archive, Report 2004/264 (2004), http://eprint.iacr.org/
17. Webster, A.F., Tavares, S.E.: On the Design of S-Boxes. In: Williams, H.C. (ed.) CRYPTO 1985. LNCS, vol. 218, pp. 523–534. Springer, Heidelberg (1986)
18. Dobbertin, H., Bosselaers, A., Preneel, B.: RIPEMD-160: A Strengthened Version of RIPEMD. In: Gollmann, D. (ed.) FSE 1996. LNCS, vol. 1039, pp. 71–82. Springer, Heidelberg (1996)
19. Mendel, F., Pramstaller, N., Rechberger, C., Rijmen, V.: On the Collision Resistance of RIPEMD-160. In: Katsikas, S.K., López, J., Backes, M., Gritzalis, S., Preneel, B. (eds.) ISC 2006. LNCS, vol. 4176, pp. 101–116. Springer, Heidelberg (2006)
20. Lipmaa, H., Moriai, S.: Efficient Algorithms for Computing Differential Properties of Addition. In: Matsui, M. (ed.) FSE 2001. LNCS, vol. 2355, pp. 336–350. Springer, Heidelberg (2002)

21. Lipmaa, H., Wallén, J., Dumas, P.: On the Additive Differential Probability of Exclusive-Or. In: Roy, B.K., Meier, W. (eds.) FSE 2004. LNCS, vol. 3017, pp. 317–331. Springer, Heidelberg (2004)
22. Indesteege, S., Preneel, B.: Practical Collisions for EnRUPT. In: Dunkelman, O. (ed.) FSE 2009. LNCS, vol. 5665, pp. 246–259. Springer, Heidelberg (2009)

A   NL-characteristics

The NL-characteristics obtained after Stage 3 of Sect. 4 are shown in Table 12. After Stage 4, Table 13 is obtained. The work factor N_w improves from $2^{75.84}$ to $2^{51.53}$.

B   A Two-Bit Example

B.1   Introduction

Let n denote the word size in bits. We will write the differential probability of addition modulo 2^n as $\text{xdp}^+(\alpha, \beta \to \gamma)$, where α, β and γ are bitstrings, most significant bit first. The best known method to find xdp^+ was an exponential-in-n calculation, before Lipmaa and Moriai introduced their algorithm in [20]. In [21], it was shown how xdp^+ can be calculated as a series of matrix multiplications in linear time in n.

In this section, we will calculate the $\text{xdp}^+(11, 01 \to 10)$ by representing the addition as a graph and applying dynamic programming. We then show the relation of this graph method with [21], as both algorithms can be implemented in $\mathcal{O}(n)$ by using matrix multiplications. Afterwards, we mention several improvements and extensions to the graph method. Although this two-bit example may seem contrived, we found a fully worked-out example to be very useful to help understand the more abstract explanation of Fig. 3-5 in Sect. 3. There, an extension of the graph method is used to represent the step update function of HAS-V.

B.2   Visualizing $\text{xdp}^+(11, 01 \to 10)$ in a Graph

For $\text{xdp}^+(\alpha_1 \parallel \alpha_0, \beta_1 \parallel \beta_0 \to \gamma_1 \parallel \gamma_0) \triangleq \text{xdp}^+(11, 01 \to 10)$, we consider two additions, $z = x + y$ and $z' = x' + y'$, as shown in Fig. 7. For this particular xdp^+, we define the input differences for the least significant bits ($\alpha_0 = x_0 \oplus x_0' = 1$ and $\beta_0 = y_0 \oplus y_0' = 1$), and for the most significant bits ($\alpha_1 = x_1 \oplus x_1' = 1$ and $\beta_1 = y_1 \oplus y_1' = 0$). We assume that all valid inputs x, x' and y, y' are uniformly distributed. We then find $\text{xdp}^+(11, 01 \to 10)$ as the probability that the output has difference $\gamma_0 = z_0 \oplus z_0' = 0$ and $\gamma_1 = z_1 \oplus z_1' = 1$.

The calculation for the least significant bits is shown in Table 10. As there is no carry input for the least significant bits, we only consider $C_0 = C_0' = 0$. We list all values that satisfy the input conditions (α_0 and β_0) for the least significant bit. Note that the output condition ($\gamma_0 = z_0 \oplus z_0' = 0$) is satisfied as well for all valid inputs ($C_0, C_0', x_0, y_0, x_0', y_0'$).

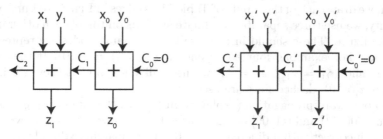

Fig. 7. Calculating $z = x + y$ and $z' = x' + y'$. All variables with subscripts represent one bit.

Table 10. The summation for the least significant bits (z_0, z_0'), where $\alpha_0 = x_0 \oplus x_0' = 1$ and $\beta_0 = y_0 \oplus y_0' = 1$

C_0	C_0'	x_0	y_0	x_0'	y_0'	C_1	C_1'	z_0	z_0'	α_0	β_0	γ_0
0	0	0	0	1	1	0	1	0	0	1	1	0
0	0	0	1	1	0	0	0	1	1	1	1	0
0	0	1	0	0	1	0	0	1	1	1	1	0
0	0	1	1	0	0	1	0	0	0	1	1	0

Table 11. The summation for the most significant bits (z_1, z_1'), where $\alpha_1 = x_1 \oplus x_1' = 1$ and $\beta_1 = y_1 \oplus y_1' = 0$

C_1	C_1'	x_1	y_1	x_1'	y_1'	C_2	C_2'	z_1	z_1'	α_1	β_1	γ_1
0	0	0	0	1	0	0	0	0	1	1	0	1
0	0	0	1	1	1	0	1	1	0	1	0	1
0	0	1	0	0	0	0	0	1	0	1	0	1
0	0	1	1	0	1	1	0	0	1	1	0	1
1	0	0	0	1	0	0	0	1	1	1	0	0
1	0	0	1	1	1	1	1	0	0	1	0	0
1	0	1	0	0	0	1	0	0	0	1	0	0
1	0	1	1	0	1	1	0	1	1	1	0	0
0	1	0	0	1	0	0	1	0	0	1	0	0
0	1	0	1	1	1	0	1	1	1	1	0	0
0	1	1	0	0	0	0	0	1	1	1	0	0
0	1	1	1	0	1	1	1	0	0	1	0	0

We then draw each of these input values as the four rightmost edges in the graph of Fig. 8. Every edge is labeled with the input conditions $[x_0 \; y_0 \; x_0' \; y_0']$, and starts at (C_0, C_0'). Together, these uniquely determine (z_0, z_0') and (C_1, C_1'). For now, the reader can ignore that some lines are dashed.

Next, we do the calculation for the most significant bits, as shown in Table 11. We again list all values that satisfy the input conditions (α_1 and β_1). Note that now, several carry inputs (C_1, C_1') are possible. The output condition ($\gamma_1 = z_1 \oplus z_1' = 1$) is not always satisfied, implying that $\mathrm{xdp}^+(11, 01 \to 10) < 1$.

Again, we draw each of the inputs of Table 11 as edges in Fig. 8. To improve the readability, we use three separate coordinate systems on top of each other on the left of the figure. These should in fact overlap: the same nodes are represented three times. For example, four edges end in $(C_2, C_2') = (0, 0)$. If the output pairs are valid ($\gamma_1 = z_1 \oplus z_1' = 1$), we use full lines, and if they are invalid ($\gamma_1 = z_1 \oplus z_1' = 0$), dashed lines are used.

Due to backward propagation (explained in Fig. 4), the inputs $[x_0 \ y_0 \ \ x_0' \ y_0']$ with values $[00 \ 11]$ and $[11 \ 00]$ become dashed lines as well: they will eventually result in an incorrect output difference $\gamma_1 = 0$. The probability $\text{xdp}^+(11, 01 \to 10)$ is then equal to the number of paths in the graph with valid inputs (x, x', y, y') and outputs z, z' (full lines), divided by the number of paths that have valid inputs $(x, x'y, y')$ (full or dashed lines). This ratio is equal to $8/16$, or $1/2$.

Note that storing this graph does not require a lot of memory. For every bit in the n-bit addition, we need to store 2^6 bits. Each of these 2^6 bits is either set to 1 if the input $(C_0, C_0', x_0, y_0, x_0', y_0')$ is valid, and 0 otherwise. For the entire n-bit addition, we thus need to store only $2^6 n$ bits; the memory requirement is $\mathcal{O}(n)$.

B.3   Calculating $\text{xdp}^+(11, 01 \to 10)$ Using Matrix Multiplications

Similar to [21], we can calculate xdp^+ as a series of matrix multiplications. The graph of Fig. 8 can be seen as a first-order Markov chain. We have $[1 \ 0 \ 0 \ 0]^T$ as the initial distribution, as the input carry of the addition is $C_0 = C_0' = 0$ with probability 1. All three other input carries (C_0, C_0') have probability 0.

As the input conditions for $[x_0 \ y_0 \ \ x_0' \ y_0']$ are given, these specify the transition matrix of the Markov chain. Every column contains the transition probabilities for one carry input (C_0, C_0') to every carry output (C_1, C_1'). We left-multiply the initial distribution by this transition matrix. We do the same for every subsequent bit of the n-bit addition.

Lastly, we sum all probabilities (by left-multiplying by $[1 \ 1 \ 1 \ 1]$): the carry outputs (C_n, C_n') are not used, so all of them are valid. This gives us the total probability of xdp^+.

B.4   Extending the Graph Method

As the reader may have noticed, the matrices of the previous section are larger than those in [21]. This is because we do not take the symmetry into account: although the value of $C_i \oplus C_i'$ would be sufficient, we keep track of the values of (C_i, C_i'). This symmetry exists because we restrict the input differences α, β and the output differences γ to exclusive-or differences.

The graph based method of the previous section, however, can also support the signed differences that were used for the cryptanalysis of MD5 [3], and as well as all the other generalized conditions of Table 7.

It is straightforward to generalize the graph method to the addition of three or more words. In this case, we extend each of the n adders of Fig. 7 to three or more input bits. This will increase the maximal values of the carry (C_i, C_i'):

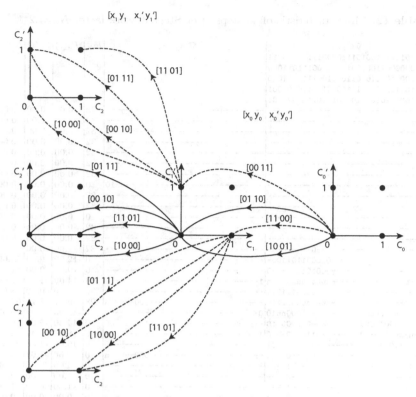

Fig. 8. Graph representation to calculate $\text{xdp}^+(11, 01 \to 10)$. Only valid input pairs are shown. Full lines are used for the eight paths that have valid output pairs ($\gamma_1 = z_1 \oplus z_1' = 1$), and dashed lines are used for paths with invalid output pairs ($\gamma_1 = z_1 \oplus z_1' = 0$). As there are eight of each, the ratio gives $\text{xdp}^+(11, 01 \to 10) = 8/16 = 1/2$. The three coordinate systems on top of each other on the left represent the same nodes three times. This makes the drawing more readable, however note that, for example, four edges end in $(C_2, C_2') = (0, 0)$.

for example, the addition of four bits (and the carry input) can have a maximal carry output of 3.

In this case, value of the carry (C_i, C_i') is equal to all output bits of the adder at position i, except the least significant bit. This can be seen as a variant of Fig. 7, where three or more bits are input to every adder.

In fact, the method can be generalized for any combination of additions, exclusive-ors and Boolean functions, as long as no rotations are present (except at the input or output). It is this calculation that was used for every step of SHA-1 in [15], and is also used for every step of HAS-V in this paper. By constructing higher-order Markov chains, the graph method was used in [22] to efficiently calculate the differential probability of a multiplication by 9, given by $\text{xdp}^+(x, x \ll 3)$.

Table 12. NL-characteristic of 45 steps after Stage 3, work factor $N_w = 2^{75.84}$

t	∇Q_{t+1}	∇W_t	F_W	$P_u(t)$	$P_c(t)$	$N_s(t)$
-5	00001111010010111000011111000011					
-4	01000000110010010101000111011000					
-3	01100010111010110111001111111010					
-2	11101111110011011010101110001001					
-1	01100111010001010010001100000001					
0	--------------------------------	--------------------------------	32	0.00	0.00	0.00
1	--------------------------------	--------------------------------	32	0.00	0.00	0.00
2	--------------------------------	--------------------------------	32	0.00	0.00	0.00
3	--------------------------------	--------------------------------	32	0.00	0.00	0.00
4	--------------------------------	--------------------------------	32	0.00	0.00	0.00
5	--------------------------------	--------------------------------	32	0.00	0.00	0.00
6	--------------------------------	--------------------------------	32	0.00	0.00	0.00
7	--------------------------------	------------------------------10	30	0.00	0.00	0.00
8	--------------------------------	--------------------------------	32	0.00	0.00	0.00
9	--------------------------------	--------------------------------	32	0.00	0.00	0.00
10	--------------------------------	--------------------------------	0	0.00	0.00	0.00
11	--------------------------------	--------------------------------	32	0.00	0.00	0.00
12	--------------------------------	--------------------------------	32	0.00	0.00	0.00
13	--------------------0-----------	--------------------------------	32	-1.00	0.00	22.60
14	----------------010001101100-	--------------------------------	0	-12.00	0.00	53.60
15	----------------001011---1-0	--------------------------------	0	-9.00	0.00	41.60
16	---0----------------unnnnnnnnnnnn	1-----------------------------u	30	-14.00	-1.00	32.60
17	0--1---------------0100u111010	--------------------------------	32	-13.00	0.00	48.60
18	0--1------1--------uu-uu0001110	----------------------000-----u	28	-19.77	-7.77	67.60
19	01-unn--nnu-----------11000nn10	0------------------------------	0	-19.00	-1.97	75.83
20	n-u0uu001-000----------0-0000-10	----------------------1-101--00	0	-14.30	-2.30	56.83
21	u-u-0n11----0---------11-010--10	1-----------------------------u	0	-18.98	-6.42	42.53
22	--1-110---011----------0-n----11	------------------------------10	0	-10.00	-1.00	23.56
23	--0-0-10----------------n-0---0	----------------------000-----u	0	-7.56	-1.61	13.56
24	--1-1--1----------------1---	--------------------------------	0	-4.00	0.00	6.00
25	------------------------0----	--------------------------------	0	-1.00	0.00	2.00
26	------------------------1----	--------------------------------	0	-1.00	0.00	1.00
27	--------------------------------	--------------------------------	0	0.00	0.00	0.00
28	--------------------------------	--------------------------------	0	0.00	0.00	0.00
29	--------------------------------	--------------------------------	0	0.00	0.00	0.00
30	--------------------------------	--------------------------------	0	0.00	0.00	0.00
31	--------------------------------	--------------------------------	0	0.00	0.00	0.00
32	--------------------------------	--------------------------------	0	0.00	0.00	0.00
33	--------------------------------	--------------------------------	0	0.00	0.00	0.00
34	--------------------------------	0-------------------------------	0	0.00	0.00	0.00
35	--------------------------------	--------------------------------	0	0.00	0.00	0.00
36	--------------------------------	--------------------------------	0	0.00	0.00	0.00
37	--------------------------------	--------------------------------	0	0.00	0.00	0.00
38	--------------------------------	--------------------------------	0	0.00	0.00	0.00
39	--------------------------------	--------------------------------	0	0.00	0.00	0.00
40	--------------------------------	--------------------------------	0	0.00	0.00	0.00
41	--------------------------------	0-------------------------------	0	0.00	0.00	0.00
42	--------------------------------	--------------------------------	0	0.00	0.00	0.00
43	--------------------------------	------------------------------10	0	0.00	0.00	0.00
44	--------------------------------	--------------------------------	0	0.00	0.00	0.00

Table 13. NL-characteristic of 45 steps after Stage 4, work factor $N_w = 2^{51.53}$

t	∇Q_{t+1}	∇W_t	F_W	$P_u(t)$	$P_c(t)$	$N_s(t)$
-5	00001111010010111000011111000011					
-4	01000000110010010101000111011000					
-3	01100010111010110111001111111010					
-2	11101111110011011010101110001001					
-1	01100111010001010010001100000001					
0	--------------------------------	--------------------------------	32	0.00	0.00	0.00
1	--------------------------------	--------------------------------	32	0.00	0.00	0.00
2	--------------------------------	--------------------------------	32	0.00	0.00	0.00
3	--------------------------------	--------------------------------	32	0.00	0.00	0.00
4	--------------------------------	--------------------------------	32	0.00	0.00	0.00
5	--------------------------------	--------------------------------	32	0.00	0.00	0.00
6	--------------------------------	--------------------------------	32	0.00	0.00	0.00
7	--------------------------------	---00------------------------10	28	0.00	0.00	0.00
8	--------------------------------	--------------------------------	32	0.00	0.00	0.00
9	--------------------------------	--------------------------------	32	0.00	0.00	0.00
10	--------------------------------	--------------------------------	0	0.00	0.00	0.00
11	--------------------0-----------	--------------------------------	32	-1.00	0.00	0.00
12	--------------------0-0---------	--------------------------------	32	-2.00	0.00	0.00
13	-----------------1-00-----0-----	--------------------------------	32	-4.00	0.00	23.48
14	-----------------0100011011001	--------------------------------	0	-13.00	0.00	51.48
15	------00-----------0001011011110	--------------------------------	0	-17.00	0.00	38.48
16	-1-0--110----------unnnnnnnnnnn	10-------------------1-00---0u	25	-17.83	-4.24	21.48
17	01-1--1------------110100u111010	-------------------------0------	31	-18.00	0.00	28.65
18	00-1-1110011111-----uu1uu0001110	-------------------01000--1--u	25	-20.00	-1.00	41.65
19	01-unn--nnu1111------1011000nn10	000-00000100000------1-100000000	0	-11.58	-8.59	46.65
20	n-u0uu0010000--------0000000-10	000001011110---1------0101011000	0	-6.10	-4.62	35.07
21	u-u-0n11--110---------11-01---10	10-------------------1-00---0u	0	-10.03	-1.00	28.96
22	--1-110---011----------0-n----11	----00-----------------------10	0	-7.46	-0.12	18.93
23	--0-0-10-----------------n-0---0	-------------------01000--1--u	0	-5.48	0.00	11.48
24	--1-1--1----------------1----	--------------------------------	0	-4.00	0.00	6.00
25	-------------------------0----	--------------------------------	0	-1.00	0.00	2.00
26	-------------------------1----	--------------------------------	0	-1.00	0.00	1.00
27	--------------------------------	--------------------------------	0	0.00	0.00	0.00
28	--------------------------------	--------------------------------	0	0.00	0.00	0.00
29	--------------------------------	--------------------------------	0	0.00	0.00	0.00
30	--------------------------------	--------------------------------	0	0.00	0.00	0.00
31	--------------------------------	--------------------------------	0	0.00	0.00	0.00
32	--------------------------------	------------------------0------	0	0.00	0.00	0.00
33	--------------------------------	--------------------------------	0	0.00	0.00	0.00
34	--------------------------	000-00000100000------1-100000000	0	0.00	0.00	0.00
35	--------------------------------	--------------------------------	0	0.00	0.00	0.00
36	--------------------------------	--------------------------------	0	0.00	0.00	0.00
37	--------------------------------	--------------------------------	0	0.00	0.00	0.00
38	--------------------------------	--------------------------------	0	0.00	0.00	0.00
39	--------------------------------	--------------------------------	0	0.00	0.00	0.00
40	--------------------------------	--------------------------------	0	0.00	0.00	0.00
41	--------------------------	000-00000100000------1-100000000	0	0.00	0.00	0.00
42	--------------------------------	--------------------------------	0	0.00	0.00	0.00
43	--------------------------	----00-----------------------10	0	0.00	0.00	0.00
44	--------------------------------	--------------------------------	0	0.00	0.00	0.00

Non-linear Error Detection for Finite State Machines*

Kahraman D. Akdemir, Ghaith Hammouri, and Berk Sunar

CRIS Lab, Worcester Polytechnic Institute
100 Institute Road, Worcester, MA 01609-2280
{kahraman,hammouri,sunar}@wpi.edu

Abstract. We propose the use of systematic nonlinear error detection codes to secure the next-state logic of finite state machines (FSMs). We consider attacks under an adversarial model which assumes an advanced attacker with high temporal and spatial fault injection capability. Due to the non-uniform characteristics of FSMs, simple application of the systematic non-linear codes will not provide sufficient protection. As a solution to this problem, we use randomized masking. Furthermore, we show that our proposal detects injected faults with probability exponentially close to 1.

Keywords: Fault-resilience, state-machines, adversarial-faults.

1 Introduction

Active fault injection attacks are proven to be effective on many cryptosystems. The idea is to reveal secret information using erroneous executions of a device as a result of intentional fault injections. The seminal paper of Boneh et al. [5] demonstrated that it is relatively easy to break the Chinese Remainder Theorem (CRT) based RSA algorithm in the presence of faults. Following this work, Biham et al. [4] introduced differential fault attacks (DFA) and showed that secret key cryptosystems such as DES can also be compromised using active fault injection attacks.

Following the attacks outlined above, various fault injection techniques and attacks have been proposed. Bit-flips using optical laser beams [25], electro-magnetically induced faults [24], variations in the external voltage lines, external clock transients, and temperature variations are some of the active fault attack techniques available in the literature. Survey papers by Bar-El et al. [1] and Naccache [23] can provide more information on side channel attacks (SCAs).

Since active fault attacks pose a serious threat for many cryptosystems, various countermeasures have been proposed. In the Double-Data-Rate (DDR) computation technique [20,21], both edges of the clock are used to make the same computation twice and check for errors by comparing the two results. Another

---

* This material is based upon work supported by the National Science Foundation under Grant No. CNS-0831416.

H.Y. Youm and M. Yung (Eds.): WISA 2009, LNCS 5932, pp. 226–238, 2009.

proposed solution is dual-rail encoding [6,16]. In this technique, one bit of information is represented by two bits, hence making it more difficult to inject a precise fault. Finally, concurrent error detection (CED) [3,14,7,22] is the most common countermeasure against active attacks. For example, triple modular redundancy (TMR) and quadruple modular redundancy (QMR) replicate the original design three and four times, respectively. Next, the result is determined after a majority voting by comparing the outcomes of each individual copy. Non-linear robust error detection codes [13], a subset of CED, provides strong protection against active fault attacks. These codes mainly minimize the maxima of error masking probability under a strong adversarial model. They achieve robustness with reasonable efficiency. The robustness property states that the attacker cannot find an error vector which will be missed for all data values that can be observed at the output of the device. In other words, any error pattern that will be masked is data dependent. Another benefit of this coding scheme is the uniformity of its error detection capability over all possible error vectors. This means that no error vector has an error detection probability smaller than the error detection probabilities of other possible error vectors. For the details of this coding scheme and its possible applications such as the protection of the Advanced Encryption Standard (AES) hardware, the reader is referred to [19,18,11,12,17]. More details about systematic non-linear codes and their robustness characteristics will be provided in Section 3.

The remainder of this paper is organized as follows: Section 2 lays out our main motivation in protecting FSMs against active fault attacks and summarizes our contributions. In Section 3, we introduce background information on nonlinear robust codes. Section 4 discusses the characteristics of the advanced attacker and describes our fault model. Next, the proposed error detection technique is described in Section 5. The security and robustness measure of the proposed scheme are presented in Section 6. In Section 7, hardware scaling results are provided. Finally, in Section 8, the proposed scheme is compared with other FSM protection schemes from a security perspective.

2   Motivation

Finite state machine (FSM) security is an important aspect of cryptographic hardware design. The datapath which usually includes arithmetic units (such as adders, multipliers, division and inversion units) will work on operands with sizes on the order of hundreds to a thousand bits and therefore will take much more chip area when compared to control units. As a result, it is a reasonable design decision to arm these units with strong error detection techniques against active fault injection attacks. However, the security level of a cryptographic device can be reduced to the security level of its most vulnerable and non-secure portion. Hence, secure FSM design should also be one of the crucial design goals, even though the size of the control unit is relatively small.

Most of the CED based error detection techniques, mainly aim to secure the datapath of the crypto-system while leaving the FSM as a possible target point.

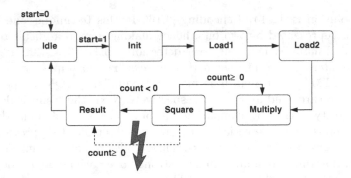

Fig. 1. Fault injection example on the control unit of the Montgomery Ladder Algorithm with Point of Attack indicated by the dashed transition [8]

Also note that the state values in an FSM are kept in either current-state or next-state registers. Consequently, any attack that is shown to be successful on a register level (bit-flips or stuck-at faults, etc.) can be used to attack FSMs with mild effort. Depending on the characteristics of the FSM, various effective attack scenarios can be produced. An example attack scenario from [8] on the FSM of the Montgomery Ladder Algorithm is shown in Figure 1. The figure shows that an attacker can recover the secret exponent using this attack with mild effort. A similar attack can be applied to all exponentiation algorithms that are implemented in a similar fashion. Elliptic curve add-always point multiplication algorithm will potentially suffer from a similar attack [10]. In summary, FSM security is an important problem that needs to be handled carefully.

The first solution to this problem was proposed by [15,2]. This solution simply uses single-bit error detection/correction schemes. However, these schemes mainly target single event upsets and naturally occurring faults such as the ones caused by radioactive radiation.

Another well-known solution which could be proposed is the usage of TMR and QMR based schemes. Both of these schemes will add an acceptable level of security against a weak attacker. However, when an advanced attacker -as will be defined in Section 4- is considered, both QMR and TMR will offer **no** security at all. This point will be further explained in Section 8.

As a solution to the same problem in cryptographic settings against an intelligent adversary, Gaubatz et al. [8] proposed to apply linear error detection codes. Linear codes provide security in an adversarial setting, yet only against weak attackers with limited fault injection capabilities. The reason is that any error pattern, which is also a valid codeword in the utilized coding scheme, will be missed in this detection scheme. Consequently, a more advanced attacker can still easily bypass this error detection mechanism.

In [9], physically unclonable functions (PUFs) are used in order to secure know-path state machines. In this class of FSMs, state transitions do not depend on the inputs, and hence the name known-path state machines. This solution used PUFs to create a fingerprint of the expected and fault free state transitions.

This fingerprint is then used to check if the FSM transitioned correctly while the device is in operation. This is an interesting and efficient error detection technique, yet it requires the golden fingerprint to be kept in a secure location in the device and assumes that this part of the circuit cannot be attacked.

Our Contribution: In this paper, we will use systematic nonlinear error detection codes to secure the next-state logics of FSMs against an advanced attacker. We assert that protecting FSMs with a high error detection probability is a difficult problem due to the non-uniform characteristics of FSMs. In this case, the systematic non-linear codes discussed in Section 3 can not be directly applied to FSMs. As a solution to this problem, we propose using randomized masking which solves this non-uniformity problem. This makes it possible to apply the nonlinear codes to FSMs and use their high error detection capabilities. In our security proof, we show that any injected fault will be detected with probability exponentially close to 1.

3   Background on Robust Codes

In [13], Karpovsky and Taubin proposed a new class of systematic non-linear error detection codes with the following definition.

Definition 1. *[13] Let V be a binary linear (n, k) code with $n \le 2k$ and check matrix $H = [P|I]$ with $rank(P) = n - k$. Then $C_V = \{(x, w)|x \in GF(2^k), w = (Px)^3 \in GF(2^r)\}$.*

To quantify the performance of C_V, they also defined the following error masking probability metric:

$$Q(e) = \frac{|\{x|(x + e_x, w + e_w) \in C_V\}|}{|C_V|}. \tag{1}$$

where the error vector is represented by $e = (e_x, e_w)$. This metric quantifies the number of the information codewords x for which the error pattern e will be masked, and computes the error masking probability by normalizing this number by the code size.

Note that the code C_V is defined to be **robust** if and only if the error masking probability $Q(e) < 1$ for all possible error vectors e (i.e. $k=r$), and **partially robust** if for some e, $Q(e)=1$ (i.e. $k > r$).

The following theorem from [13] quantifies the error detection performance of the proposed nonlinear code C_V.

Theorem 1. *[13] For C_V the set $E = \{e|Q(e) = 1\}$ of undetected errors is a $(k - r)$-dimensional subspace of V, from the remaining $2^n - 2^{k-r}$ errors, $2^{n-1} + 2^{k-1} - 2^{k-r}$ are detected with probability 1 and $2^{n-1} - 2^{k-1}$ errors are detected with probability $1 - 2^{-r+1}$.*

This theorem is constructed and proved under the following uniformity assumption:

Assumption 1: *All the codewords $x \in GF(2^k)$ in C_V are uniformly distributed and have the same probability of being observed.*

In this coding scheme, contrary to linear codes, masked errors are data-dependent. This means that an active adversary trying to inject an undetected error in the data would need to know the value of the data beforehand in order to find an undetectable error pattern. In a linear scheme, the success of a specific error vector is independent from the data because any error vector which is also a valid codeword is masked in this case. Also note that the error detection probability provided by the nonlinear codes has a uniform lower bound. In other words, error detection probability does not dramatically decrease for any specific error pattern.

4   Adversarial Fault Model

Our adversarial model assumes that the details of the particular structure or the function of the device are known to the attacker. Also, we do not limit the attacker's particular fault injection methodology. This model assumes an advanced attacker with high temporal and spatial fault injection capability. In other words, the attacker has high resolution both in time and space. In addition, we model the errors observed at the output of the device in additive nature in which case the error will be the difference between the expected output x and the erroneous output $\tilde{x} = x \oplus e$, where $\oplus$ is the logical XOR operation. This implies that an attacker cannot conduct his attack by overwriting the output values.

Furthermore, the attacker cannot observe any existing data on the circuit in the same clock cycle as the fault injection takes place. This means that the attacker will not be able to adaptively attack the circuit by first reading the existing data and then choosing the appropriate error vector. We also assume that the attacker can pick any specific error vector he desires to reflect to the output, i.e. he can choose the value of e. For a comprehensive analysis, we also allow every error vector (all multiplicities) to be observed at the output of the device.[1]

In our detection model, the device is disabled or the secret information is reseted after an injected fault is detected. As a result, the attacker will have only one chance to successfully inject a fault into the circuit.

5   The Error Detection Technique

Systematic nonlinear codes described in Section 3 provide robustness and error detection uniformity which are crucial when error detection is concerned. As a result, the proposed error detection technique in this paper inherits the main

---

[1] Depending on the specifics of the design, not all the error vectors can be observed at the output of the device. However, we will conduct our error detection study assuming that all the errors can be observed at the output of the proposed design with the same frequency for sake of completeness.

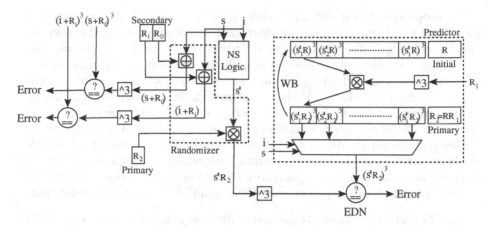

Fig. 2. Proposed error detection technique

structure of these codes to protect next-state logics of FSMs. Note that the security level provided by the nonlinear codes is sufficient for arithmetic circuits even against strong adversaries such as the one discussed in Section 4. However, the structure of these codes needs to be modified to some extent to make them applicable to FSMs.

Assume that we want to apply a nonlinear (n, k, r)-code to protect a specific FSM. When FSMs are concerned, the number of valid codewords is limited by the number of states in the FSM, which is usually a relatively small number. In addition to this, some states of the FSMs can be visited more than others. As a result, FSMs show a non-uniform behavior over a larger domain that is defined by the values of k and r of the applied (n, k, r)-code. Due to this non-uniform characteristic of the FSMs, applying the non-linear (n, k, r)-code becomes a difficult problem. In this case, the security level provided by this method cannot be quantified as in Theorem 1 because FSMs do not satisfy Assumption 1. Consequently, to inherit the useful error detection characteristics of these codes, we need to guarantee that the state register value will have a distribution that is close-to-uniform.

As a solution to this problem, we now formally define a randomized robust code by merging a randomized masking process with the robust code definition introduced by Karpovsky et al. [13] in Definition 1 as follows.

Definition 2. *Let V be a binary linear (n, k) code with $n = 2k$. We define the robust code with randomized masking as $C_V = \{(x, R, w) | x \in GF(2^k), R \in GF(2^k), w = (x \times R)^3 \in GF(2^k)\}$.*

In this code, the masking is achieved by the random string R. The effect of masking is to essentially remove the non-uniform characteristic of the FSM. In this case, the error detection probability can be quantified as in Theorem 1 because Assumption 1 is satisfied with the utilization of randomized masking. Note that because we are working in a finite field the above randomization can be realized by a multiplication as in the definition or by an addition.

The proposed solution built on the defined robust code with randomized masking is shown in Figure 2. There are four main building blocks in this solution, these blocks will be described in more detail shortly. Note that all the variables in this figure are elements of $GF(2^k)$ and all the operations are conducted over $GF(2^k)$. The essential idea here is to use the input values i and s in order to generate the next-state value s'. At the same time, the inputs and their check-sums are used to predict the check-sum of the next-state value. This operation is costly in general. Therefore, we use a multiplexer structure which will help improve the efficiency of the circuit. Also note that when multiplexers are concerned, attacks on the select lines could lead to potential vulnerabilities. As a result, we will carry out separate comparisons to ensure that no errors were injected into the i and s values which are the select lines of the multiplexer in the predictor unit.

- **NS Logic:** This is basically the non-redundant next-state logic of any FSM. Using the input (i) and current-state (s), this block computes the appropriate next-state value (s'). Note that i, s, and s' are padded with zeros so that they will be elements of $GF(2^k)$. In other words, i, s, and $s' \in GF(2^k)$. However, this does not affect the logic that implements the next-state function. This block uses the well-known next-state logic computation methods.
- **Randomizer:** This block applies the randomized masking on the next-state logic (s'). It takes the current-state value (s), next-state value(s'), the primary randomizer (R_2), and the secondary randomizer (R_0, R_i) as its inputs where R_2, $R_0, R_i \in GF(2^k)$. The primary randomizer (R_2) will be coming from the predictor block and is the main randomizer that is used in the current cycle. The secondary randomizer (R_0) is calculated in the previous cycle and is stored to be used in the current cycle. The other secondary randomizer R_i is generated with every input i.

  This block computes $(s'R_2)$. However, the randomizer needs to be passed to the next cycle. As multiplications are more expensive than addition and this multiplication has been already performed, the next state is passed a different version of the randomizer, namely R'_0, which will be the secondary randomizes used in the next cycle. More specifically, the output of this block will be

  $$s'R_2 = s' + s'(R_2 - 1), \tag{2}$$

  where $R'_0 = s'(R_2 - 1)$. Observe that the output of this block is randomized using R_2. This proves that the non-uniform behavior of the FSMs is removed at the error-check (comparison) level. Even though the non-redundant next-state logic block still provides non-uniform behavior, this does not affect the error detection probability because cubing and hence the error-checking is done on the randomly masked version of the next-state value.

  Furthermore, to compute the secondary randomizer for the following cycle, s' is subtracted from the output $s'R_2$ of the randomizer. Keep in mind that all these addition and multiplication operations are conducted over $GF(2^k)$.
- **Predictor:** The function of the predictor block is to predict the check-sum of the randomized next-state logic value (output of the randomizer). In order to achieve this, the predictor takes the input (i) and current-state (s).

In this block, there is a register that holds the $(s'_j R)^3$ values with the initial randomizer $R \in GF(2^k)$ (this is actually the primary randomizer from the previous cycle) where j indexes all possible next-state values. With every clock-cycle, the values in this register are randomized with R_1 that is generated using a true-random number generator (TRNG) and the results are also written back (WB) into the original register. Also note that the primary randomizer $R_2 = R \times R_1$ is stored and sent to the randomizer block. The write-back (WB) is necessary to prevent the attacker from reading the existing values of the original register, calculating the appropriate error vector that will be missed, and injecting the faults that will result with this error vector. In other words, we want to update these register values at every clock cycle because the advanced attacker we are modeling cannot read and write within the same clock cycle. After the computation of the new $(s'_j R_2)^3$ for all j, the input i and the current-state s select the appropriate $(s' R_2)^3$ using a multiplexer circuit. The multiplexer circuit is inspired from the efficient FSM implementations. This is the predicted check-sum of the randomized next-state logic.

As with the randomizer block, the multiplication and cubing operations are conducted over $GF(2^k)$.

- **EDN:** The error detection network (EDN) is mainly a comparator. It compares the cube of the randomized next-state logic with the predicted check-sum. As we mentioned earlier, because s and i are used in the multiplexer structure they have to be randomized with their previous state randomizer and compared to their randomized check-sums $(s+R_0)^3, (i+R_0)^3 \in GF(2^k)$. If the results match in all comparisons, this means that the operation is fault free, yet if there is a mismatch, then an error signal is triggered. The error signal can either reset the secret information or can stop the operation. A PUF based EDN mechanism is proposed in [9]. This provides a fault free EDN block and prevents the attacker from simply attacking and disabling the error signal.

6 Security Analysis

The main idea of the proposed protection scheme is to prevent an attacker from forcing the state machine into an arbitrary state. The predictor's job is focused on using the checksum of the state and the input in order to generate the checksum of the next state. Building a circuit which directly performs this operation can be quite costly. Therefore, we have used a multiplexer structure as explained in the previous section. The multiplexer uses the current input and the current state to compute the check sum of the next state. This operation does not preserve the isolation between the predictor branch and the FSM which could have an effect on the security of the scheme. To solve this problem we perform an overall comparison between the current state and its checksum. We also perform separate comparisons between the actual values and the checksums of both the current state and the input. In this section we will show that the error detection probability will be exponentially close to 1. We start with the following Corollary to Theorem 1.

Corollary 1. *Given a uniformly random $x \in \{0,1\}^k$ and $w = x^3 \in \{0,1\}^k$ the probability that any chosen pair $(e_x, e_w) \neq 0$ satisfies $(x + e_x)^3 = w + e_w$ is lower bounded by $1 - 2^{-k+1}$.*

Proof. Use Theorem 1 with parameters $n = 2k$, $k = r$ and $H = I$.

We will also need the following Lemma.

Lemma 1. *Given a uniformly random $x \in \{0,1\}^k$ and any $w \in \{0,1\}^k$ the probability that any chosen pair (e_x, e_w) such that $e_x \neq 0$ satisfies $(x + e_x)^3 = w + e_w$ is upper bounded by $3 \cdot 2^{-k}$.*

Proof. The equation $(x + e_x)^3 = w + e_w$ is a cubic equation over $GF(2^k)$ and hence will have at most 3 solutions for x for any given w, e_x, e_w. As x is chosen uniformly at random, the probability that x will be the correct solution for a specific w, e_x, e_w will be at most $\frac{3}{2^k}$.

Now we can prove the main theorem.

Theorem 2. *The error detection probability of the scheme depicted in Figure 2 is lower bounded by $1 - 9 \cdot 2^{-k}$.*

Proof. Recall that under the adversarial fault model we are assuming, the attacker will not know R_0, R_i, R and R_1 in the same clock cycle in which he will inject his error. The scheme of Figure 2 will perform a comparison between $s_w = (s + R_0)^3$ and the cube of $(s + R_0)$, and similarly for $i_w = (i + R_i)^3$ and the cube of $(i + R_i)$. As R_i and R_0 are unknown and uniformly random to the attacker, using Corollary 1 we can assume that except with probability at most $2 \cdot 2^{-k+1}$ no error will be injected on s or i. Now we proceed by assuming error injections in every possible location of the circuit. Let the error injected into the FSM reflect to s' as a shift of $e_{s'}$. Similarly, let the error injected into the registers holding $R, R_1, R_2, (S'R)^3$ and $(S'R_2)^3$ be $e_R, e_{R1}, e_{R2}, e_{SR}$ and e_w respectively. Also, let the error injected into the multiplication between s' and R_2 be e_x. Finally, let the error injected into the cubing of R_1 be e_{R3}. The output of the main branch will be

$$(s' + e_{s'}) [(R_1 + e_{R1})(R + e_R) + e_{R2}] + e_x.$$

The output of the predictor side will be

$$((s'R)^3 + e_{SR})[(R_1 + e_{R1})^3 + e_{R3}] + e_w.$$

As the main goal of an attacker is to change s' to another valid state we can assume $e_{s'} \neq 0$. We will now work our way backwards. The EDN will cube the main branch and compare it to the predictor branch. Now the term to the left of e_x in the main branch equation, which we label x, is a uniformly random string in $\{0,1\}^k$ and similarly, the term to left of e_w, which we label w is some string in $\{0,1\}^k$. Therefore, using Lemma 1 if $e_x \neq 0$ the probability of finding (e_x, e_w) to satisfy the comparison will be upper bounded by $3 \cdot 2^{-k}$. If this is not the case we can assume that $e_x = 0$. Next, we assume that $(RR_1) \neq 0$ which will happen with probability at least $1 - 2 \cdot 2^{-k}$. We can now factor RR_1 from x and $(RR_1)^3$ from w. For the left side we get,

$$(RR_1)[(s' + e_{s'})[(1 + e_{R1}R_1^{-1})(1 + e_R R^{-1}) + e_{R2}(RR_1)^{-1}]] .$$

The right side becomes.

$$(RR_1)^3[((s')^3 + e_{SR}R^{-3})[(1 + e_{R1}R_1^{-1})^3 + e_{R3}R_1^{-3}]] + e_w .$$

We can now write the comparison as

$$[(RR_1)(\hat{x} + e_{s'})]^3 \stackrel{?}{=} (RR_1)^3(\hat{w} + e_w) .$$

Where

$$\hat{x} = s'[(1 + e_{R1}R_1^{-1})(1 + e_R R^{-1}) + e_{R2}(RR_1)^{-1}] + e_R e_{R1} e_{s'}(RR_1)^{-1}$$
$$+ e_R e_{s'} R^{-1} + e_R e_{s'} R_1^{-1} + e_{R2} e_{s'}(RR_1)^{-1}$$

which can be seen as a uniformly random string in $\{0,1\}^k$. Similarly,

$$\hat{w} = ((s')^3 + e_{SR}R^{-3})[(1 + e_{R1}R_1^{-1})^3 + e_{R3}R_1^{-3}] + e_2(RR_1)^{-3}]$$

which is some string in $\{0,1\}^k$ which can be dependent on $\hat{x}$. Recall, that $(RR_1) \neq 0$ therefore the comparison above will hold iff

$$(\hat{x} + e_{s'})^3 = (\hat{w} + e_w) .$$

However, $e_{s'} \neq 0$. Now using Lemma 1 the above will hold with probability $1 - 3 \cdot 2^{-k}$. This will make the overall probability of an error not being detected by the scheme of Figure 2 lower bounded by

$$1 - (2 \cdot 2^{-k+1} + 3 \cdot 2^{-k} + 2 \cdot 2^{-k}) = 1 - 9 \cdot 2^{-k}$$

The theorem above insures the security of the proposed scheme against fault attacks. We note here that this scheme needs to carry three separate comparisons. A simpler scheme would be for the main branch to compute $(i + R_i) \times (s + R_0) \times (sR_2)$ then cube this term and compare it to $(i + R_i)^3 \times (s + R_0)^3 \times (sR_2)^3$ coming from the predictor branch. In fact, we conjecture that the error detection probability of such a scheme will still be exponentially close to 1. However, in this paper we do not present a formal proof of this conjecture.

7   Hardware Scalability

As can be observed from Figure 2, the scheme requires 4 k-bit finite field cubings (1 k-bit squaring + 1 k-bit multiplication), 2 k-bit finite field multiplications, and 2 k-bit finite field additions. Since addition over $GF(2)$ is just XORing, field adders can be implemented very efficiently. Squaring over $GF(2)$ can also be implemented quite efficiently. As a result, the effect of the field adders and squarings in the scheme will be minimal. However, multiplication over $GF(2)$ is an expensive operation and hence the multipliers will dominate most of the area of the whole scheme. Since this is the case, the total area of the proposed FSM will scale in parallel with the area of the field multipliers. It is a reasonable assumption to state that the area of a field multiplier will scale as $O(k^2)$. Hence the area of the whole scheme will scale as $O(k^2)$.

The scaling factor $O(k^2)$ essentially determines a trade-off between area overhead and security. When the FSM is running a very sensitive application with a high security risk, one cannot tolerate any errors. Therefore, only an exponentially small probability of failure can be accepted. Of course it does not make sense to pay for such an overhead when the underlying application is not sensitive.

An interesting perspective is to consider the overhead from a complexity point of view. With this perspective one can see that the proposed scheme requires a circuit of size polynomial in k while providing an exponentially small (in k) fault injection probability.

8   Comparison with Other FSM Security Schemes

At this point, it is also important to compare the error detection capability of our scheme with other error detection schemes. Triple modular redundancy (TMR) and quadruple modular redundancy (QMR) are two of the most common error detection techniques against active fault attacks. In TMR, the non-redundant FSM is replicated three times and a majority voting circuit determines the correct result. QMR works in the exact same way, but the non-redundant FSM is replicated four times. Applying linear codes for error detection is another proposed method [8]. Under a weak attacker model, linear codes, TMR, and QMR may provide limited security with minimum error detection probability greater than zero. However, in the advanced attacker model considered in this paper, an attacker can with %100 probability cause invalid state transitions on an FSM protected by these schemes. For example, in TMR and QMR, the attacker can inject the exact same error to all replicas of the original design. This attack will clearly go undetected as all replicas of the circuits will behave in an identical fashion. Similarly, in the linear scheme, the attacker will choose error vectors which are also valid codewords in the utilized code and hence the injected error will be undetected. It should be clear that the strength of the scheme proposed in this paper stems from the exponentially small error detection probability even against an advanced attacker.

Another interesting FSM security scheme based of physically unclonable functions (PUF) was proposed in [9]. This PUF-based scheme is quite efficient with an exponentially small error detection probability even against an advanced attacker. However, it is only applicable to known-path FSMs, which is a specific class of FSMs. The approach we present in this paper is a generic one that can be applied to any FSM.

9   Conclusion

We presented a fault detection scheme in FSMs based on systematic nonlinear error detection codes. Our scheme detects any injected fault with probability exponentially close to 1 under an adversarial model which assumes an advanced attacker with high temporal and spatial fault injection capability. Furthermore, the work here presents a new approach to handling the non-uniform output of a state machine in a way which enables the usage of some classical error detection techniques.

References

1. Bar-El, H., Choukri, H., Naccache, D., Tunstall, M., Whelan, C.: The sorcerer's apprentice guide to fault attacks. Proceedings of the IEEE 94, 370–382 (2006)
2. Berg, M.: Fault tolerant design techniques for asynchronous single event upsets within synchronous finite state machine architectures. In: 7th International Military and Aerospace Programmable Logic Devices (MAPLD) Conference. NASA (September 2004)
3. Bertoni, G., Breveglieri, L., Koren, I., Maistri, P., Piuri, V.: Error analysis and detection procedures for a hardware implementation of the advanced encryption standard. IEEE Transactions on Computers 52(4), 492–505 (2003)
4. Biham, E., Shamir, A.: Differential fault analysis of secret key cryptosystems. In: Kaliski Jr., B.S. (ed.) CRYPTO 1997. LNCS, vol. 1294, pp. 513–525. Springer, Heidelberg (1997)
5. Boneh, D., DeMillo, R., Lipton, R.: On the importance of checking cryptographic protocols for faults. In: Fumy, W. (ed.) EUROCRYPT 1997. LNCS, vol. 1233, pp. 37–51. Springer, Heidelberg (1997)
6. Cunningham, P., Anderson, R., Mullins, R., Taylor, G., Moore, S.: Improving Smart Card Security Using Self-Timed Circuits. In: Proceedings of the 8th International Symposium on Asynchronus Circuits and Systems. IEEE Computer Society, Washington (2002)
7. Gaubatz, G., Sunar, B.: Robust finite field arithmetic for fault-tolerant public-key cryptography. In: Breveglieri, L., Koren, I. (eds.) 2nd Workshop on Fault Diagnosis and Tolerance in Cryptography - FDTC 2005 (September 2005)
8. Gaubatz, G., Sunar, B., Savas, E.: Sequential circuit design for embedded cryptographic applications resilient to adversarial faults. IEEE Transactions on Computers 57(1), 126–138 (2008)
9. Hammouri, G., Akdemir, K., Sunar, B.: Novel puf-based error detection methods in finite state machines. In: Lee, P.J., Cheon, J.H. (eds.) ICISC 2008. LNCS, vol. 5461, pp. 235–252. Springer, Heidelberg (2008)
10. Joye, M.: Highly regular right-to-left algorithms for scalar multiplication. In: Paillier, P., Verbauwhede, I. (eds.) CHES 2007. LNCS, vol. 4727, p. 135. Springer, Heidelberg (2007)
11. Karpovsky, M., Kulikowski, K.J., Taubin, A.: Differential fault analysis attack resistant architectures for the advanced encryption standard. In: Proc. World Computing Congress (2004)

12. Karpovsky, M., Kulikowski, K.J., Taubin, A.: Robust protection against fault-injection attacks on smart cards implementing the advanced encryption standard. In: DSN 2004: Proceedings of the 2004 International Conference on Dependable Systems and Networks (DSN 2004), Washington, DC, USA, p. 93. IEEE Computer Society, Los Alamitos (2004)

13. Karpovsky, M., Taubin, A.: A new class of nonlinear systematic error detecting codes. IEEE Trans. Info. Theory 50(8), 1818–1820 (2004)

14. Karri, R., Wu, K., Mishra, P., Kim, Y.: Concurrent error detection schemes for fault-based side-channel cryptanalysis of symmetric block ciphers. IEEE Transactions on computer-aided design of integrated circuits and systems 21(12), 1509–1517 (2002)

15. Krasniewski, A.: Concurrent error detection in sequential circuits implemented using fpgas with embedded memory blocks. In: Proceedings of the 10th IEEE International On-Line Testing Symposium (IOLTS 2004) (2004)

16. Kulikowski, K., Venkataraman, V., Wang, Z., Taubin, A., Karpovsky, M.: Asynchronous balanced gates tolerant to interconnect variability. In: IEEE International Symposium on Circuits and Systems, ISCAS 2008, pp. 3190–3193 (2008)

17. Kulikowski, K., Wang, Z., Karpovsky, M.: Comparative Analysis of Robust Fault Attack Resistant Architectures for Public and Private Cryptosystems. In: Proceedings of the 2008 5th Workshop on Fault Diagnosis and Tolerance in Cryptography, pp. 41–50. IEEE Computer Society, Washington (2008)

18. Kulikowski, K.J., Karpovsky, M., Taubin, A.: Robust codes for fault attack resistant cryptographic hardware. In: Workshop on Fault Diagnosis and Tolerance in Cryptography (FTDC 2005) (2005)

19. Kulikowski, K.J., Karpovsky, M., Taubin, A.: Fault attack resistant cryptographic hardware with uniform error detection. In: Breveglieri, L., Koren, I., Naccache, D., Seifert, J.-P. (eds.) FDTC 2006. LNCS, vol. 4236, pp. 185–195. Springer, Heidelberg (2006)

20. Maistri, P., Leveugle, R.: Double-Data-Rate Computation as a Countermeasure against Fault Analysis. IEEE Transactions on Computers 57(11), 1528–1539 (2008)

21. Maistri, P., Vanhauwaert, P., Leveugle, R.: Evaluation of Register-Level Protection Techniques for the Advanced Encryption Standard by Multi-Level Fault Injections. In: 22nd IEEE International Symposium on Defect and Fault-Tolerance in VLSI System (DFT 2007), pp. 499–507 (2007)

22. Mitra, S., McCluskey, E.: Which concurrent error detection scheme to choose? In: Proc. of Int. Test Conference (ITC), pp. 985–994. IEEE Press, Los Alamitos (2000)

23. Naccache, D.: Finding faults. IEEE Security and Privacy 3(5), 61–65 (2005)

24. Schmidt, J., Hutter, M.: Optical and em fault-attacks on crt-based rsa: Concrete results. In: Austrochip 2007: Proceedings of the 15th Austrian Workshop on Microelectronics (2007)

25. Skorobogatov, S., Anderson, R.: Optical Fault Induction Attacks. In: Cryptographic Hardware and Embedded Systems-Ches 2002: 4th International Workshop, Redwood Shores, CA, USA, August 13-15 (2002) (Revised Papers)

Quadratic Equations from a Kind of S-boxes

Jia Xie, Weiwei Cao, and TianZe Wang

The state key laboratory of information securtiy, Institute of software,
Chinese Academy of Science, Beijing, 100790, China
The state key laboratory of information securtiy,
Graduate School of Chinese Academy of Sciences
{xiejia,wtz}@is.iscas.ac.cn, gowiwin@gmail.com

Abstract. Algebraic attack studies ciphers from the point of view of solving equations. It is important to measure the security of block ciphers how many linearly independent bi-affine or quadratic equations they satisfy. As the S-box is the main nonlinear part of block ciphers, it really makes sense to get the number of linearly independent bi-affine and quadratic equations that an S-box satisfies to analyse the security of block ciphers. The article answers this question for two S-boxes based on APN power functions, and shows how to find out the equations by two toy examples. The techniques can be generalized to other S-boxes constructed by power functions. According to these conclusions, we can estimate the safety of such kind of block ciphers.

Keywords: algebraic attack, Trace form representation, APN functions, power functions.

1 Introduction

In algebraic attack, a cryptosystem is broken by solving a set of multivariate equations over a finite field(such as F_2) which describes the whole system. The complexity of the attack depends on the number of such equations, their type and their algebraic degree. The main idea goes back to Shannon's work in [5]. Here, multivariate equations are said to be linearly independent when each distinct monomial is considered as a new variable.

Nicolas T.Courtois and Wili.Meier proposed several kinds of ways about algebraic attack on stream ciphers in 2003[1]. After that, It has been a hot topic in cryptography, and great improvement has been made such as in [2,3,4]. All the results above are about stream ciphers. At the same time, algebraic attack does not look so powerful in block ciphers.

Algebraic attack was also applied to block ciphers such as AES[11]. The S-box of AES can be expressed as an overdefined system of algebraic equations. If we can obtain an algebraic system with the number of equations exceeds that of the monomials from an S-box, a block cipher with the S-box can be represented by many equations with smaller number of variables. By solving these multivariate equations by so called the XSL algorithm, we may find the key of the block cipher.

H.Y. Youm and M. Yung (Eds.): WISA 2009, LNCS 5932, pp. 239–253, 2009.

In [11], Γ was defined as $\Gamma = ((t-r)/n)^{\lceil (t-r)/n \rceil}$ to measure the complexity to solve the algebraic system using XSL algorithm, where r is the number of equations, and t is the number of monomials in them. It was proposed that if $\Gamma \geq 2^{32}$, the S-box is safe. That is to say, $(t-r)/n$ should not be less than 10. In [11], it was shown that $\Gamma = 2^{22.9}$ for AES S-box.

Note that an improvement of algorithms on solving multivariate equations may result in different ways to solve the problem. However, it is true that Γ reflects a difficulty of solving multivariate equations in some sense, and we will use it to estimate the resistance to algebraic attack in this paper.

There exists a kind of functions, APN functions, which have the best resistance to affine attack and differential analysis, so they are good candidates to construct S-boxes. Algebraic attack may be a good method to analyse them.

J. Cheon and D.H. Lee estimated the number of quadratic equations we can get from APN functions[7,8]. However, the results were proved not to be correct by simulations[10].

In [10], Courtois et al. gave some results about the S-box of AES cipher. But they did not generalize the results to other power S-boxes and only provided experimental results for other APN functions.

In this article, we will develop techniques to find out the number of linearly independent bi-affine and quadratic equations obtained from APN power functions, which can be used to construct S-boxes. Furthmore, we can find out the form of the equations. This method can be generalized to other S-boxes constructed by power functions. In the discussion, we can see that neither of the conclusions in [8,10] are perfect.

2 Preliminary Knowledge

2.1 Notations

The number of 1s in the binary representation of n is called the weight of n with the symbol $wt(n)$, and we will use wt for short in the context.

Let F be the set of functions from $GF(2^n)$ to $GF(2)$, and define B to be the set of boolean functions in n variables.

A function from F_{2^n} to F_{2^m} is called a vector boolean function. When m=1, it is merely called a boolean function.

Let $\{\omega_0, ..., \omega_{n-1}\}$ be a basis of F_{2^n}, then there exists an isomorphism ϕ : $F_{2^n} \to F_2^n$ given by

$$\phi(\textstyle\sum a_i \omega_i) = (a_0, ..., a_{n-1}).$$

Let $(f_0, ..., f_{n-1}) = \phi \circ f$, and define the algebraic degree of f by the degree of its component functions.

Lemma. Let $f(x) = x^d \in F_{2^n}$, $\phi \circ f = (f_0, ... f_{n-1})$, then $\deg(f) = wt(d)$.

Proof. Let $x = \sum_{i=0}^{n-1} x_i \omega_i$, $d = \sum_{j=0}^{n-1} d_j 2^j$,and

$$x^d = (\sum_{i=0}^{n-1} x_i \omega_i)^{\sum_{j=0}^{n-1} d_j 2^j} = \prod_{d_j=1} (\sum_{i=0}^{n-1} x_i \omega_i^{2^j}) = \prod_{d_j=1} (\sum_{i=0}^{n-1} x_i (\sum_{k=0}^{n-1} a_{ik} \omega_k)) \ (a_{ik} \in F_2)$$

$$= \prod_{d_j=1} (\sum_{k=0}^{n-1} (\sum_{i=0}^{n-1} a_{ik} x_i) \omega_k) = \sum_{k=0}^{n-1} f_k(x_0, ..., x_{n-1}) \omega_k .$$

Since $\deg(f_j) = wt(d)(j = 0, ..., n-1)$, the degree of x^d is $wt(d)$ too.

On F_{2^n} , a cyclotomic coset is defined by $C_S = \{s, s*2, ..., s*2^{n_s-1}\}$, where n_s is the smallest positive integer such that $s \equiv s*2^{n_s} (\mod 2^n - 1)$. The subscript s is chosen as the smallest integer in C_s , and s is called the coset leader of C_s . For example, if n=4, the cyclotomic cosets modulo 15 are:

$$C_0 = \{0\}, C_1 = \{1, 2, 4, 8\}, C_3 = \{3, 6, 12, 9\}, C_5 = \{5, 10\}, C_7 = \{7, 14, 13, 11\} ,$$

where $\{0, 1, 3, 5, 7\}$ are coset leaders modulo 15.

It is easy to see that the elements in one coset are of the same weight. And if b and c belong to the same coset, we can get c from b by cyclic shift in binary.

2.2 Almost Perfect Nonlinear Power Functions

Definition1. Let $X = \{x \mid f(x) \neq g(x), f \in B, g \in B, x \in F_2^n\}$, $d(f, g) = \mid X \mid$ is the hamming distance of f and g .

Let $f(x)$ be a boolean function with n variables, $\Phi = \{\varphi(x)$ $= wx + a : x \in F_2^n, w \in F_2^n, a \in F_2\}$,

$N_f = \min\{(d(f, \varphi)) \mid \varphi \in \Phi\}$ is called the nonlinearity of $f(x)$. The functions with maximal nonlinearity is called the maximal nonlinear functions.

The nonliearity of a boolean function is used to measure its resistance to linear analysis.

Definition2. A function $F : F_{2^n} \rightarrow F_2$ is called almost perfect nonlinear(APN) if each equation: $F(x+a) + F(x) = b$, $a \in F_{2^n}^*$, $b \in F_{2^n}$ has not more than two solutions $x \in F_{2^n}$.

The known APN power functions: $y = x^d, x \in F_{2^n}$ [14]

 I. Gold: d= $2^k + 1$, $\gcd(k, n) = 1 (1 \leq k \leq n)$

 II. Kasami: d= $2^{2k} - 2^k + 1, \gcd(k, n) = 1 (2 \leq k \leq \lfloor n / 2 \rfloor)$

 III. Dobbertin: $d = 2^{4s} + 2^{3s} + 2^{2s} + 2^s - 1 (n = 5s)$

 IV. Welch: d= $2^m + 3$ (n=2m+1)

V. Niho: $d = \begin{cases} 2^m + 2^{m/2} - 1, m = 2t \\ 2^m + 2^{(3m+1)/2} - 1, m = 2t+1 \end{cases}$ $(n=2m+1)$

VI. Inverse: d=-1(n=2m+1)

The S-box of AES cipher is constructed by the inverse exponent.

From the definition, we can see APN functions have the best resistance to differential analysis. Also it is proved that maximally nonlinear function is APN if n is odd[14]. Especially, Gold and Kasami exponents are proved to be maximally nonlinear, while Welch and Niho exponent are deemed to be maximally nonlinear. Therefore, these APN functions are considered to be good candidates of S-boxes of block ciphers.

3 Quadratic Equations from S-boxes Based on APN Functions

Power functions are often used to construct S-boxes. Since APN power functions have the best resistance to diffential analysis and linear attack, we try to analyse them by algebraic attack.

Niho and Kasami exponent are proved or considered to be maximally nonlinear. Since the number of equations with low degree has great influence on the complexity of attack, we will find out the number of equations with low degree with the two exponents. From the analysis, we can see that the method can be generalized to other power functions.

Firstly, we will give several theorems. The conclusions are used in[11]. We will amend them and give the whole proof.

Theorem1. $\{\alpha_0, ..., \alpha_{n-1}\}$ is a basis of F_{2^n}. $y \in F_{2^n}$, $y = \sum_i \alpha_i y_i$. $f(y)$ is a boolean function, and $f(y) = \sum_{i=0}^{n-1} b_i y_i$. Then $f(y)$ has a unique polynomial representation in $F_{2^n} : f(y) = Tr_1^n(cy)$, $c \in F_{2^n}$.

Proof. Suppose $\{\beta_0, ..., \beta_{n-1}\}$ to be the corresponding dual basis of $\{\alpha_0, ..., \alpha_{n-1}\}$ over F_{2^n}, and $y = \sum_{i=0}^{n-1} y_i \alpha_i$. So $y_i = Tr_1^n(\beta_i y)$.

$$f(y) = \sum_i b_i y_i = \sum_i b_i Tr_1^n(\beta_i y) = Tr_1^n(\sum_i b_i \beta_i y) = Tr(cy),$$

where $c = \sum_i b_i \beta_i$, $c \in F_{2^n}$. c is determined by b_i, β_i, and it is unique.

This ends the proof.

Theorem2. $\{\alpha_0, ..., \alpha_{n-1}\}$ is a basis of F_{2^n}, $x = \sum_i \alpha_i x_i$, $y = \sum_i \alpha_i y_i$. $g(x, y)$ is a biaffine boolean function, and $g(x, y) = \sum_{i,j} a_{ij} x_i y_j$. Then $g(x, y)$ has a unique polynomial representation in F_{2^n}: $\sum_{i,j} a_{i,j} x_i y_j = \sum_{k=0}^{n-1} Tr_1^n(c_k x^{2^k} y)$, $c_k \in F_{2^n}$.

Proof. Suppose $\{\beta_0,...,\beta_{n-1}\}$ to be the corresponding dual basis of $\{\alpha_0,...,\alpha_{n-1}\}$ over F_{2^n}.

Let $x = \sum\limits_{i=0}^{n-1} x_i\alpha_i$, $y = \sum\limits_{i=0}^{n-1} y_i\alpha_i$. So $x_i = Tr_1^n(\beta_i x)$, $y_i = Tr_1^n(\beta_i y)$.

$$\sum_{i,j} a_{i,j} x_i y_j = \sum_{i,j} a_{i,j} Tr_1^n(\beta_i x) Tr_1^n(\beta_j y)$$

$$= Tr_1^n(\sum_{i,j} a_{i,j} Tr_1^n(\beta_i x)\beta_j y)$$

$$= Tr_1^n(\sum_{i,j} a_{i,j} \sum_{k=0}^{n-1}(\beta_i x)^{2^k} \beta_j y)$$

$$= \sum_{k=0}^{n-1} Tr_1^n(\sum_{i,j} a_{i,j}\beta_i^{2^k} \beta_j x^{2^k} y).$$

Let $c_k = \sum\limits_{i,j} a_{i,j}\beta_i^{2^k} \beta_j$, $c_k \in F_{2^n}$, where $\sum\limits_{i,j} a_{i,j} x_i y_j = \sum\limits_{k=0}^{n-1} Tr_1^n(c_k x^{2^k} y)$.

c_k ($k = 0,1,...,n-1$) are determined by $a_{i,j}, \beta_i, \beta_j, k$, so the polynomial is unique.

This ends the proof.

Theorem3. $\{\alpha_0,...,\alpha_{n-1}\}$ is a basis of F_{2^n}, $y = \sum\limits_i \alpha_i y_i$. $h(y)$ is a boolean function,

and $h(y) = \sum\limits_{i\leq j} a_{ij} y_i y_j$. Then $h(y)$ has a unique polynomial representation in F_{2^n}:

$$\sum_{i\leq j} b_{i,j} y_i y_j = Tr_1^n(b_0 y)+ \sum_{1\leq k\leq m} Tr_1^n(b_k y^{2^k+1}), \ b_i \in F_{2^n}, 0\leq i \leq m, \text{ where } n=2m+1;$$

$$\sum_{i\leq j} b_{i,j} y_i y_j = Tr_1^n(b_0 y)+ \sum_{1\leq k\leq m-1} Tr(b_k y^{2^k+1})+Tr_1^m(b_m y^{2^m+1}), \ b_i \in F_{2^n}, 0\leq i \leq m,$$

where n=2m.

Proof. Suppose $\{\beta_0,...,\beta_{n-1}\}$ to be the corresponding dual basis of $\{\alpha_0,...,\alpha_{n-1}\}$ over F_{2^n}, and $y = \sum\limits_{i=0}^{n-1} y_i\alpha_i$. Then $y_i = Tr_1^n(\beta_i y)$.

Use the same method as above, and we can get that

$$\sum_{i\leq j} b_{i,j} y_i y_j = \sum_{i\leq j} b_{i,j} Tr_1^n(\beta_i y) Tr_1^n(\beta_j y) = \sum_{k=0}^{n-1} Tr_1^n(\sum_{i\leq j} b_{i,j}\beta_i^{2^k} \beta_j y^{2^k+1}).$$

At first, let n=2m+1.

Notice that $Tr(y^2) = Tr(y)$, and $Tr(y^{2^{n-k}+1}) = Tr(y^{2^{n-k}+1+2^n-1}) = Tr(y^{2^{n-k}(2^k+1)})$

$= Tr(y^{2^k+1})$. So, when $1\leq k \leq m$, $Tr(y^{2^k+1}) = Tr(y^{2^{n-k}+1})$, and

$Tr_1^n(\sum\limits_{i,j} b_{i,j}\beta_i^{2^{n-k}} \beta_j y^{2^{n-k}+1}) = Tr_1^n((\sum\limits_{i,j} b_{i,j}\beta_i^{2^{n-k}} \beta_j y^{2^{n-k}+1})^{2^k}) = Tr_1^n(\sum\limits_{i,j} b_{i,j}\beta_i \beta_j^{2^k} y^{2^k+1})$.

So, when n=2m+1,

$$\sum_{i\leq j} b_{i,j} y_i y_j = Tr_1^n((\sum_{i\leq j} b_{i,j}\beta_i\beta_j)y^{2^0+1}) + \sum_{1\leq k\leq m} Tr_1^n((\sum_{i\leq j} b_{i,j}\beta_i^{2^k}\beta_j)y^{2^k+1}) + \sum_{1\leq k\leq m} Tr_1^n((\sum_{i\leq j} b_{i,j}\beta_i^{2^{n-k}}\beta_j)y^{2^{n-k}+1})$$

$$= Tr_1^n((\sum_{i\leq j} b_{i,j}\beta_i\beta_j)y^{2^0+1}) + \sum_{1\leq k\leq m} Tr_1^n((\sum_{i\leq j} b_{i,j}\beta_i^{2^k}\beta_j)y^{2^k+1}) + \sum_{1\leq k\leq m} Tr_1^n((\sum_{i\leq j} b_{i,j}\beta_i\beta_j^{2^k})y^{2^k+1})$$

$$= Tr_1^n(b_0 y) + \sum_{1\leq k\leq m} Tr_1^n(b_k y^{2^k+1}).$$

where $b_0 = \sum_{i<j} b_{i,j}(\beta_i\beta_j)^{2^{n-1}}$, $b_k = \sum_{i\leq j} b_{i,j}(\beta_i^{2^k}\beta_j + \beta_i\beta_j^{2^k})$ $(1\leq k\leq m)$.

Now we consider the case that n=2m. The proof is similar. Notice that there is only m elements in the coset of 2^m+1. We have the conclusion that:

$$\sum_{i<j} b_{i,j} y_i y_j = Tr_1^n(b_0 y) + \sum_{1\leq k\leq m-1} Tr(b_k y^{2^k+1}) + Tr_1^m(b_m y^{2^m+1}),$$

where $b_0 = \sum_{i<j} b_{i,j}(\beta_i\beta_j)^{2^{n-1}}$, $b_k = \sum_{i\leq j} b_{i,j}(\beta_i^{2^k}\beta_j + \beta_i\beta_j^{2^k})$ ($1\leq k\leq m-1$).

$$b_m = \sum_{i\leq j} b_{i,j}\beta_i^{2^m}\beta_j \quad , \quad \text{so} \quad (b_m)^{2^m} = (\sum_{i\leq j} b_{i,j}\beta_i^{2^m}\beta_j)^{2^m} = \sum_{i\leq j} b_{i,j}\beta_i^{2^{2m}}\beta_j^{2^m} =$$

$$\sum_{i\leq j} b_{i,j}\beta_i\beta_j^{2^m} \neq b_m, \ b_m \notin F_{2^m}.$$

So, whether n is odd or even, $b_i \in F_{2^n}, 0\leq i\leq m$.

Similiarly to the case in theorem2, it can be shown the polynomial is unique. This ends the proof.

3.1   The Number of Linearly Independent Bi-Affine Equations

Suppose the S-box is like: $y = S(x) = x^d$.

Assume the bi-affine equations are of the form:

$$\sum_{i,j} a_{i,j} x_i y_j + \sum_i b_i y_i + \sum_i c_i x_i + d = 0, \quad x_i, y_i, a_{i,j}, b_i, c_i, d \in F_2,$$

$$\sum_{i,j} a_{i,j} x_i y_j + \sum_i b_i y_i = \sum_i c_i x_i + d .$$

Let $g(x, y) = \sum_{i,j} a_{i,j} x_i y_j + \sum_i b_i y_i$.

By theorem1 and theorem2,

$$g(x, y) = \sum_{i,j} a_{i,j} x_i y_j + \sum_i b_i y_i = \sum_{0\leq k<n} Tr(y(c_k x^{2^k})) + Tr(ay) .$$

Replace y with x^d , and we can get:

$$g(x, y) = \sum_{0\leq k<n} Tr(x^d * (c_k x^{2^k})) + Tr(ax^d) = \sum_{0\leq k<n} Tr(c_k x^{2^k+d}) + Tr(ax^d) . \qquad (3)$$

The exponents of the monomials in one trace function, are in the same coset, and they are of the same degree. From the lemma above, the degrees of the component boolean functions of the trace function equal the weight of the exponents.

So the degree of $g(x, y)$ equals $\max(wt(d), wt(2^k + d))$. If $wt(d) = wt(2^k + d) = 1$, $g(x, y)$ equals some affine function in x. So we can get a bi-affine equation. Furthermore, the number of linear independent bi-affine equations equals that of linear independent boolean functions of $g(x, y)$.

The coefficient in one trace function can be any value in F_{2^n}, and we can get a bi-affine function from each value. So we can get 2^n equations in total. There are n linearly independent equations in the space. Besides that, if 2^{k_1+d} and 2^{k_2+d} belong to different coset, the bi-affine functions related to $Tr((c_{k_1} x^{2^{k_1+d}}))$ and $Tr(c_{k_2} x^{2^{k_2+d}})$ are in different spaces, and they are independent to each other.

The number of linearly independent equations should be the sum of the numbers of linearly independent equations in each space.

Note

Let index set S={ $2^k + d \mid wt(2^k + d) \geq 2, 0 \leq k < n$ }. If two elements in S belong to the same coset, we can get $2^{(2-1)n} = 2^n$ bi-affine functions. For example, if $2^{k_1} + a$ and $2^{k_2} + a$ $(k_1 \neq k_2)$ belong to one coset, i.e. $2^{k_1} + a = (2^{k_2} + a)2^l$ where $1 \leq l \leq n$. Then we can see the term of $Tr_1^n((b_{k_1} + b_{k_2}^{2^l})x^{2^{k_1}+a})$ that $b_{k_1} + b_{k_2}^{2^l}$ can be zero in 2^n ways. Analogously, if there are m elements in S belong to the same coset, we will get $2^{(m-1)n}$ equations, where (m-1)n ones are linearly independent.

3.2 The Number of Linear Independent Quadratic Equations

Suppose the equations are of the form:

$$\sum_{i,j} a_{i,j} x_i y_j + \sum_{i \leq j} b_{i,j} y_i y_j + \sum_{i<j} d_{i,j} x_i x_j + \sum_i e_i x_i + a = 0 \text{ (if } i = j, y_i y_j = y_i),$$

where $x_i, y_i, a_{i,j}, b_{i,j}, d_{i,j}, e_i, a \in F_2$.

$$\sum_{i,j} a_{i,j} x_i y_j + \sum_{i \leq j} b_{i,j} y_i y_j = \sum_{i<j} d_{i,j} x_i x_j + \sum_i e_i x_i + a .$$

Let $h(x, y) = \sum_{i,j} a_{i,j} x_i y_j + \sum_{i \leq j} b_{i,j} y_i y_j$.

From theorem2 and theorem3, when n=2m+1,

$$h(x, y) = \sum_{i,j} a_{i,j} x_i y_j + \sum_{i \leq j} b_{i,j} y_i y_j = \sum_{0 \leq k < n} Tr(y(c_k x^{2^k})) + Tr_1^n(b_0 y) + \sum_{1 \leq k \leq m} Tr_1^n(b_k y^{2^k+1})$$

Replace y with x^d,

$$h(x, y) = \sum_{0 \leq k < n} Tr(c_k x^{2^k+d}) + Tr_1^n(b_0 x^d) + \sum_{1 \leq k \leq m} Tr(b_k x^{(2^k+1)d}) . \qquad (4)$$

If n=2m,

$$h(x, y) = \sum_{i,j} a_{i,j} x_i y_j + \sum_{i \leq j} b_{i,j} y_i y_j$$

$$= \sum_{0 \leq k < n} Tr(y(c_k x^{2^k})) + (Tr_1^n(b_0 y) + \sum_{1 \leq k \leq m-1} Tr_1^n(b_k y^{2^k+1}) + Tr_1^m(b_m y^{2^m+1})).$$

Replace y with x^d,

$$h(x, y) = \sum_{0 \leq k < n} Tr(c_k x^{2^k+d}) + Tr_1^n(b_0 x^d) + \sum_{1 \leq k \leq m-1} Tr_1^n(b_k x^{(2^k+1)d}) + Tr_1^m(b_m x^{(2^m+1)d}). \quad (5)$$

The degree of $h(x, y)$ in (4) equals $\max(wt((2^k+1)d), wt(2^k+d))$. If $wt((2^k+1)d) = wt(2^k+d) = 2$, $g(x, y)$ equals some quadratic function in x, and we can get a quadratic equation. Furthermore, the number of linear independent quadratic equations equals that of linear independent boolean functions of $g(x, y)$. The situation about $h(x, y)$ in (5) is similar to that in (4).

Similarly we can get the results when n is even.

Note2

Let the set of exponents $S = \{ 2^k + d \mid wt(2^k + d) > 2, 0 \leq k < n \} \cup \{ (2^k+1)d \mid wt((2^k+1)d) > 2, 0 \leq k < n \}$. If two elements in S belong to the same coset, we can get $2^{(2-1)n} = 2^n$ bi-affine functions, which corresponding a vector space with dimension n. Similarly, if there are m elements in S belong to one coset, we will get $2^{(m-1)n}$ equations, where (m-1)n ones are linear independent.

Since note1 is a special case of note2, we should only consider the case in note2.

Algorithm(To count the number of linearly independent equations in note2)

```
Input n,d
Eqnum=0
For i=0 to n-1 do
  a[i]=pow(2,i)+d
for i=n to 2n-1 do
  a[i]=(pow(2,(i-n))+1)*d
temp=a[i]
for p=0 to n-1 do
   temp=(temp*2) % (int(pow(2,n)-1))
/*we can get all the members in one coset by cyclic shift
of the coset leader in binary*/
   for (int j=i+1;j<2*n;j++)
      if (temp==a[j]) eqnum+=n
  return eqnum;
```

Now we show how to find the number of linearly independent equations with low degree by two APN functions.

3.3 The Niho Exponent Case

Now we discuss about the Niho case. The number n is always odd this time.

Firstly, let us see how many linearly independent bi-affine equations the niho exponent satisfies.

When $m = 2t$ ($t > 0$), replacing d with $2^m + 2^{m/2} - 1$ in (3), and we will get that

$$g(x, y) = \sum_{0 \le k < n} Tr(c_k x^{2^k + 2^m + 2^{m/2} - 1}) + Tr(ax^{2^m + 2^{m/2} - 1}).$$

For $0 \le k < n$, $wt(2^k + 2^m + 2^{m/2} - 1) > 1$, $c_k = 0 (0 \le k < n)$.

At the same time, $wt(2^m + 2^{m/2} - 1) = 1 + m/2 > 1$, so $a = 0$.

Hence there does not exist non-trivial bi-affine equations.

If $m = 2t + 1$ ($m \ge 1$), with d replaced by $2^m + 2^{(3m+1)/2} - 1$,

$$g(x, y) = \sum_{0 \le k < n} Tr(c_k x^{2^k + 2^m + 2^{(3m+1)/2} - 1}) + Tr(ax^{2^m + 2^{(3m+1)/2} - 1}).$$

Since $wt(2^m + 2^{(3m+1)/2} - 1) = 1 + m > 1$, $a = 0$.

When k=m=1, $2^k + 2^m + 2^{(3m+1)/2} - 1 = 2^3 - 1$.

So when n=2m+1=3, $\forall x \in F_{2^3}$, $x \ne 0$, $x^{2^3 - 1} = 1$; if $x = 0$, $x^{2^3 - 1} = 0$.

When k=2, $2^k + 2^m + 2^{(3m+1)/2} - 1 = 9$, $x^9 \equiv x^2 \pmod{x^8 - x}$.

Thus c_1, c_2 can be any element in F_{2^3}. We have $2^3 * 2^3$ choices, and we can get 6 linearly independent bi-affine forms from them.

For other cases, since $wt(2^k + 2^m + 2^{(3m+1)/2} - 1) > 1$, $c_k = 0$.

There only exists trivial equations with degree not more than 1 this time.

Now we discuss about the number of quadratic equations that Niho exponent satisfies.

If $m = 2t$, with d replaced by $2^m + 2^{m/2} - 1$ in (5),

$$h(x, y) = \sum_{0 \le k < n} Tr(c_k x^{2^k + 2^m + 2^{m/2} - 1}) + \sum_{1 \le k \le m-1} Tr(b_k x^{(2^k + 1)(2^m + 2^{m/2} - 1)}) + Tr(b_0 x^{2^m + 2^{m/2} - 1})$$

$$+ Tr_1^m(b_m x^{(2^m + 1)(2^m + 2^{m/2} - 1)}).$$

As for $2^k + 2^m + 2^{m/2} - 1$, if k=0, wt=2; otherwise when $k = m = 2$, wt=2(n=2m+1=5); in the other cases, wt=2+m/2>2;

If n=5, we can get 10 linearly independent forms from them. If n is some other value, we will get n linear independent quadratic forms this time.

As for $(2^k + 1)(2^m + 2^{m/2} - 1)$, wt is always larger than 2.

For $Tr(b_0 x^{2^m + 2^{m/2} - 1})$, if m=2, $wt(2^m + 2^{m/2} - 1) = 2$, b_0 can be any value in F_{2^5},

and we can get 5 equations; otherwise $wt(2^m + 2^{m/2} - 1) > 2$.

So if m=2t, we have the following conclusions:

if n=5, there exists 15 linearly independent quadratic equations;

if n>5(n=2m+1,m=2t), there exists n linearly independent quadratic equations.

If $m = 2t + 1$, replacing d with $2^m + 2^{(3m+1)/2} - 1$ in (4), by discussion similar to above, we can get the following conclusions:

if n=3, there exists 9 linearly independent equations;

if n>3(n=2m+1,m=2t+1), there exists n linearly independent equations.

So in the Niho exponent case, we can get the number of equations with degree not more than 2(except those that are trivial) as below:

if n=3, there exists 12 linearly independent equations; 3 of them are linear, and the others are quadratic;

if n=5, there exists 15 linearly independent quadratic equations;

when n is some other value, we can get n linearly independent quadratic equations.

When n=7,m=3, we can get 7 additional linearly independent quadratic equations.

For niho exponent, since we can get additional equations when n=7, the conclusion in [8] is not perfect.

3.4 The Kasami Exponent Case

Now we discuss about Kasami exponent:

$$y = x^d : d = 2^{2k} - 2^k + 1, \gcd(k, n) = 1 (2 \le k \le \lfloor n/2 \rfloor)$$

Firstly we will discuss how many linearly independent bi-affine functions Kasami exponent satisfies.

Replace d with its expression in (3), and we can get

$$g(x, y) = \sum_{0 \le t < n} Tr(x^{2^{2k}-2^k+1} * (c_t x^{2^t})) + Tr(ax^{2^{2k}-2^k+1})$$

$$= \sum_{0 \le t < n} Tr(c_t x^{2^t + 2^{2k} - 2^k + 1}) + Tr(ax^{2^{2k}-2^k+1}).$$

Since $wt(2^t + 2^{2k} - 2^k + 1) > 1$, $wt(2^{2k} - 2^k + 1) > 1$, $c_t = 0(0 \le t < n)$, $a = 0$, We can only get a trivial equation.

Now let us discuss about the number of linearly independent quadratic equations derived from Kasami exponent.

If n=2m+1, replacing d with $2^{2k} - 2^k + 1$ in (4), and we can get

$$h(x, y) = \sum_{0 \le t < n} Tr(x^{2^{2k}-2^k+1} (c_t x^{2^t})) + Tr_1^n (b_0 x^{2^{2k}-2^k+1}) + \sum_{1 \le p \le m} Tr(b_p x^{(2^p+1)(2^{2k}-2^k+1)})$$

$$= \sum_{0 \le t < n} Tr(c_t x^{2^t + 2^{2k} - 2^k + 1}) + Tr_1^n (b_0 x^{2^{2k}-2^k+1}) + \sum_{1 \le p \le m} Tr(b_p x^{(2^p+1)(2^{2k}-2^k+1)}). \quad (6)$$

If n=2m, replacing d with $2^{2k} - 2^k + 1$ in (5),

$$h(x, y) = \sum_{0 \le t < n} Tr(c_k x^{2^t + 2^{2k} - 2^k + 1}) + Tr_1^n (b_0 x^{(2^{2k}-2^k+1)}) + \sum_{1 \le p \le m-1} Tr_1^n (b_p x^{(2^p+1)(2^{2k}-2^k+1)})$$

$$+ Tr_1^m (x^{(2^m+1)(2^{2k}-2^k+1)}). \quad (7)$$

For $\sum_{0 \le t < n} Tr(c_k x^{2^t + 2^{2k} - 2^k + 1})$, if t=k, $wt(2^t + 2^{2k} - 2^k + 1)$ =2; otherwise

$wt(2^t + 2^{2k} - 2^k + 1) > 2$.

For $\displaystyle\sum_{1\le p\le m-1} Tr_1^n (b_p x^{(2^p+1)(2^{2k}-2^k+1)})$, if $p=k$, $x^{(2^p+1)(2^{2k}-2^k+1)}=x^{2^{3k}+1}$.

When $3k<n(n=2m)$, $wt(2^{3k}+1)=2$;

when $3k\ge n$, $x^{2^{3k}+1}=x^{2^{3k}+1-(2^n-1)2^{3k-n}}=x^{2^{3k-n}+1}$, $wt(2^{3k-n}+1)\le 2$.

If $p\ne k$, $wt((2^p+1)(2^{2k}-2^k+1))>2$.

As for $Tr_1^m(x^{(2^m+1)(2^{2k}-2^k+1)})$, from $\gcd(k,n)=1$, n=2m, we can see $\gcd(k,m)=1$, $k\ne m$, so $wt((2^m+1)(2^{2k}-2^k+1))>2$.

At the same time $wt(2^{2k}-2^k+1)=k+1>2$.

So the formula in (6), (7) are quadratic under the following circumstances:

$$c_t=0(0\le t<n, t\ne k);$$
$$b_p=0(1\le p<n, p\ne k);$$
$$b_0=0;$$

where $c_k, b_k, b_0 \in F_{2^n}$.

For any k, b_k, c_k can be any value in F_{2^n}, so we can get 2n linearly independent equations.

We can get some additional linearly independent equations under following circumstances:

n	m	Additional number of equations
5	2	10
7	2	7
9	2	9
11	2	22

This also shows that the conslusions in [8] are not perfect.

3.5 To Find Out the Expressions of the Low Degree Equations

In the formulas, b_k, c_k can be expressed by $a_{i,j}, b_{i,j}, c_i$. Since the hamming degrees of the monomials in $h(x,y)$ do not exceed 2, we can find out the number of linearly independent equations. Furthermore, let b_k, c_k be the values of the basis in their spaces, we can extract $a_{i,j}, b_{i,j}, c_i$, and get $d_{i,j}, e_i, a$. That is to say, we can find out the expressions of the equations.

In the appendix, we show the method by two toy examples. In the result, we can see that the conclusions in [10] are not correct. We will find 9 bi-affine functions and 6 quadratic equations that the S-box of AES satisfies when n=3, which is consistent with the conclusions in [8].

3.6 To Compare the Safety of the Two APN Functions

Now we will discuss the security of the S-boxes constructed by Niho and Kasami exponents. $\Gamma = ((t-r)/n)^{\lceil (t-r)/n \rceil}$, (t-r)/n should not be less than 10 considering $\Gamma \geq 2^{32}$.

Practically n will not be too small. We can get n equations for Niho exponent, while 2n equations for Kasami exponent commonly.

The number of monomials $t = n^2 + n(n+1)/2$.

By (t-r)/n=n+(n-3)/2 $\geq$ 10, for Niho, $n \geq 7$. But when $n = 7$, wet can get 7 additional equations. So n should not be less than 9 (n is always odd for Niho exponent).

For Kasami exponent, $n \geq 8$.

Since Kasami exponent satisfy more equations than Niho exponent in the same field, the S-box constucted by Kasami exponent is more vulnerable than that by Niho exponent.

From the discussion above, we can see the method can be generalized to other S-boxes constructed by power functions.

4 Conclusion

We have proved the number of quadratic equations that can be derived from the APN power functions , such as Niho and Kasami exponent, and we estimate the security of the S-boxes constructed by the two exponents. From the discussion, it is shown that we can analyse the number of bi-affine or quadratic equations of S-boxes as long as they are based on the functions of $y = x^d$ in a similar way. We can estimate their security by this method.

References

1. Courtois, N.T., Meier, W.: Algebraic attacks on stream ciphers with linear feedback. In: Biham, E. (ed.) EUROCRYPT 2003. LNCS, vol. 2656, pp. 346–359. Springer, Heidelberg (2003)
2. Mihaljevie, M., Imai, H.: Cryptanalysis of Toyocrypt-HIS stream cipher. IEICE Transactions on Fundamentals E85-A, 66-73 (2002),
   http://www.csl.esat.sony.co.jp/atl/papers/IEICEjan02.pdf
3. Babbage, S.: Cryptanalysis of LILI-128. Technical report (January 2001),
   http://www.cosic.esat.kuleuven.ac.be/nessie/reports/
4. Meier, W., Pasalic, E., Carlet, C.: Algebraic attacks and decomposition of boolean functions. In: Cachin, C., Camenisch, J.L. (eds.) EUROCRYPT 2004. LNCS, vol. 3027, pp. 474–491. Springer, Heidelberg (2004)
5. Shannon, C.E.: Communication Theory of Secrecy System. Bell System Technical Journal 28, 656–715 (1949),
   http://netlab.cs.ucla.edu/wiki/files/shannon1949.pdf
6. Youssef, A.M., Gong, G.: Hyper-bent functions. In: Pfitzmann, B. (ed.) EUROCRYPT 2001. LNCS, vol. 2045, p. 406. Springer, Heidelberg (2001)

7. Cheon, J., Lee, D.: Resistance of S-boxes against algebraic attacks. In: Roy, B., Meier, W. (eds.) FSE 2004. LNCS, vol. 3017, pp. 83–94. Springer, Heidelberg (2004)
8. Cheon, J., Lee, D.H.: Quadratic equations from APN power functions. IEICE Transactions on Fundamentals of Electronics, Communications and Computer Sciences E89-A(1), 19–27 (2006)
9. Nawaz, Y., Gong, G., Gupta, K.C.: Upper bounds on algebraic immunity of Boolean power functions,
   http://www.cacr.math.uwaterloo.ca/techreports/2006/cacr2006-09.pdf
10. Courtois, N.T., Debraize, B., Garrido, E.: On exact algebraic [non-]immunity of S-boxes based on power functions, http://eprint.iacr.org/2005/203.ps
11. Courtois, N., Pieprzyk, J.: Cryptanalysis of block ciphers with overdefined systems of equations. In: Zheng, Y. (ed.) ASIACRYPT 2002. LNCS, vol. 2501, pp. 267–287. Springer, Heidelberg (2002)
12. Gong, G.: On existence and invariant of algebraic attack,
    http://www.cacr.math.uwaterloo.ca/techreports/2004/corr2004-17.pdf
13. Lidl, R., Niederreiter, H.: Introduction to finite fields and their applications. Cambridge University Press, Cambridge (ISBN 0-521-30706-6)
14. Dobbertin, H.: Almost perfect nonlinear power functions on GF(2^n): The Welch Case. IEEE Trans. Infrom. Theory 45(4), 1271–1275 (1999)

Appendix

To find out the equations with low degree that power functions satisfy

We will use Niho exponent for example. Let n=3, now $y = x^5$.

At first, we will find the bi-affine functions.

Let the bi-affine functions are like:

$$\sum_{i,j} a_{i,j} x_i y_j + \sum_i c_i x_i + d = 0 .$$

$$\sum_{i,j} a_{i,j} x_i y_j = \sum_{k=0}^{n-1} Tr_1^n (c_k x^{2^k} y), \ c_k = \sum_{i,j} a_{i,j} \beta_i^{2^k} \beta_j, \ c_k \in F_{2^3}, k = 0,1,2 .$$

Let $(\alpha_0, \alpha_1, \alpha_2)$ be a normal basis of F_{2^3}, and $(\alpha_0, \alpha_1, \alpha_2) = (1, a, a^2)$. Its dual basis is

$$(\beta_0, \beta_1, \beta_2) = (1, a^2, a) .$$

Replace β_i in c_k, and we can get that

$$\begin{cases} c_0 = (a_{00} + a_{12} + a_{21}) \cdot 1 + (a_{02} + a_{11} + a_{12} + a_{20} + a_{21}) \cdot a + (a_{01} + a_{10} + a_{11} + a_{22}) \cdot a^2 \\ c_1 = (a_{00} + a_{11} + a_{12} + a_{22}) \cdot 1 + (a_{02} + a_{10} + a_{12} + a_{21} + a_{22}) \cdot a + (a_{01} + a_{10} + a_{11} + a_{12} + a_{20} + a_{21}) \cdot a^2 \\ c_0 = (a_{00} + a_{11} + a_{21} + a_{22}) \cdot 1 + (a_{02} + a_{10} + a_{11} + a_{20} + a_{22}) \cdot a + (a_{01} + a_{12} + a_{20} + a_{21} + a_{22}) \cdot a^2 . \end{cases}$$

From the discussion above, when $c_0 = 0$, c_1, c_2 are any values in F_{2^3}, we can get bi-affine functions.

Let $c_0 = c_2 = 0$, c_1 be $1, a, a^2$ separately, and we can get 3 solutions of a_{ij}. By expanding $Tr(c_1 x^2 \cdot x^5)$, we can get 3 equations.

Analogously, let $c_0 = c_1 = 0$, c_2 be $1, a, a^2$ separately, and we can find 3 solutions of a_{ij}, and we can get another 3 equations.

Now we can get 6 bi-affine functions:

$$x_0 y_1 + x_1 y_0 + x_1 y_1 + x_2 y_0 + x_2 y_2 + x_2 = 0;$$

$$x_0 y_0 + x_1 y_1 + x_1 y_2 + x_2 y_2 + x_0 = 0;$$

$$x_0 y_2 + x_1 y_0 + x_1 y_1 + x_1 y_2 + x_2 y_1 + x_1 + x_2 = 0;$$

$$x_0 y_0 + x_1 y_1 + x_2 y_1 + x_2 y_2 + 1 = 0;$$

$$x_0 y_2 + x_1 y_0 + x_1 y_2 + x_2 y_0 + x_2 y_1 + x_2 y_2 = 0;$$

$$x_0 y_1 + x_1 y_1 + x_1 y_2 + x_2 y_0 + x_2 y_1 = 0.$$

Let the quadratic equations are like:

$$\sum_{i,j} a_{i,j} x_i y_j + \sum_{i<j} b_{i,j} y_i y_j + \sum_i c_i y_i + \sum_{i<j} d_{i,j} x_i x_j + \sum_i e_i x_i + a = 0$$

For $\sum_{i,j} a_{i,j} x_i y_j = \sum_{k=0}^{2} Tr_1^3(c_k x^{2^k} y)$, $c_0 = \sum_{i,j} a_{i,j} \beta_i \beta_j$, c_0 can be any value in F_{2^3}.

$$\sum_{i<j} b_{ij} y_i y_j = Tr_1^n(b_0 y) + Tr_1^3(b_1 y^{2^k+1}),$$

$$b_0 = \sum_{i<j} b_{i,j}(\beta_i \beta_j)^{2^2}, b_1 = \sum_{i<j} b_{i,j}(\beta_i^2 \beta_j + \beta_i \beta_j^2),$$

b_0, b_1 can be any value in F_{2^3}.

$$\sum_i c_i y_i = Tr_1^3(ay), a = \sum_i c_i \beta_i, a \text{ can be any value in } F_{2^3}.$$

Using the same method to find the bi-affine functions, we can find 9 quadratic equations:

$$x_0 y_2 + x_1 y_2 + x_2 y_1 + x_0 + x_1 + x_2 + x_1 x_2 = 0;$$

$$x_0 y_1 + x_1 y_0 + x_1 y_2 + x_2 y_1 + x_2 y_2 + x_0 x_1 + x_2 = 0;$$

$$x_0 y_2 + x_1 y_1 + x_2 y_0 + x_2 y_2 + x_0 x_2 + x_1 + x_2 = 0;$$

$$y_0 y_1 + y_2 + x_0 = 0; \quad y_0 y_1 + y_0 y_2 + y_1 + x_1 = 0; \quad y_1 y_2 + y_0 + y_1 + y_2 + x_0 = 0;$$

$$x_0 x_1 + x_0 x_2 + x_1 + y_2 = 0; \quad x_0 x_2 + x_1 + x_2 + y_1 = 0; \quad x_1 x_2 + x_0 + x_1 + x_2 + y_0 = 0.$$

It can be shown they are linearly independent.

The <plaintext, ciphertext> pairs of $y = x^5$ in F_{2^3} are as following:

(001,001), (010,111), (011,010), (100,011), (101,100), (110,101), (111,110).

We can verify that the equations above are correct.

In F_{2^3}, the S-box of AES satisfies 9 linearly independent bi-affine functions. By the same method as above, we can get the expressions:

$$x_0 y_2 + x_1 y_0 + x_1 y_2 + x_2 y_0 + x_2 y_1 + x_2 y_2 + x_2 = 0;$$

$$x_0 y_1 + x_1 y_1 + x_1 y_2 + x_2 y_0 + x_2 y_1 + x_1 = 0;$$

$$x_0 y_0 + x_1 y_1 + x_2 y_1 + x_2 y_2 + x_0 = 0;$$

$$x_0 y_0 + x_1 y_2 + x_2 y_1 + 1 = 0;$$

$$x_0 y_2 + x_1 y_1 + x_2 y_0 + x_2 y_2 = 0;$$

$$x_0 y_1 + x_1 y_0 + x_1 y_2 + x_2 y_1 + x_2 y_2 = 0;$$

$$x_0 y_0 + x_1 y_1 + x_1 y_2 + x_2 y_2 + y_0 = 0;$$

$$x_0 y_2 + x_1 y_0 + x_1 y_1 + x_1 y_2 + x_2 y_1 + y_1 = 0;$$

$$x_0 y_1 + x_1 y_0 + x_1 y_1 + x_2 y_0 + x_2 y_2 + y_1 + y_2 = 0.$$

In addition, the 6 quadratic equations are as follows:

$$x_0 y_2 + x_1 y_0 + x_1 y_1 + x_1 y_2 + x_2 y_1 + x_0 x_1 + x_2 = 0;$$

$$x_0 y_1 + x_1 y_0 + x_1 y_1 + x_2 y_0 + x_2 y_2 + x_0 x_1 + x_0 x_2 + x_1 = 0;$$

$$x_0 y_0 + x_1 y_1 + x_1 y_2 + x_2 y_2 + x_1 x_2 + x_0 + x_1 + x_2 = 0;$$

$$x_0 y_1 + x_1 y_1 + x_1 y_2 + x_2 y_0 + x_2 y_1 + y_0 y_1 + y_2 = 0;$$

$$x_0 y_1 + x_1 y_0 + x_1 y_1 + x_2 y_2 + x_0 y_2 + y_0 y_1 + y_0 y_2 + y_1 = 0;$$

$$x_0 y_0 + x_1 y_1 + x_2 y_1 + x_2 y_2 + y_1 y_2 + y_0 + y_1 + y_2 = 0.$$

It can be shown they are linearly independent.

The <plaintext,ciphertext> pairs are:

(001,001), (010,101), (011,110), (100,111), (101,010), (110,011), (111,100).

Easily to verify that the bi-affine and quadratic equations are correct.

Cryptanalysis of a Multivariate Public Key Encryption Scheme with Internal Perturbation Structure

Weiwei Cao and Lei Hu

State Key Laboratory of Information Security,
Graduate University of Chinese Academy of Sciences, Beijing 100049, China
{wwcao,hu}@is.ac.cn

Abstract. Recently, Wang et al proposed a new middle-field type scheme for multivariate public key encryption. There are three equations in the central map, so it is convenient to name it TH. They found that some linearization equations can be derived for TH and to overcome this defect, they combined the internal perturbation and plus methods to obtain an improved scheme which we call PTH+. They claimed that PTH+ can resist all known types of attacks, including differential attack, and to ensure it achieves a security level higher than 2^{80}, they suggested the parameter is taken as $(l, r, m) = (47, 6, 11)$. In this paper, we show that TH has a much weaker structure than what is analyzed by the inventors and it can be totally cracked by linearization attack. For PTH+, we propose a method to reduce the attack against PTH+ to an attack on TH+ (a plus variant of TH) using the property on its differentials, which was originally regarded as impossible by that authors. The total complexity of our attack is $2^{l+r+1}(2l)^w \approx 2^{72}$, which is independent on the number m of the additional random quadratic equations by the plus method and disproves the claim in their original paper that the larger is the m, the securer is PTH+.

Keywords: Multivariate public key encryption, internal perturbation, quadratic polynomial, differential attack.

1 Introduction

Public key cryptography (PKC) has opened a new era of cryptography since Diffie and Hellman delivered a new idea in their seminal paper in 1976 [DH76]. The classical trapdoors of PKC are based on the difficulty of integer factorization for RSA and discrete logarithm for ElGamal and ECC. However, with the arrival of quantum computer, integer factorization and discrete logarithm will be cracked by quantum computer attack [Sho97]. Therefore, one of the new public key cryptography, multivariate public key cryptosystem (MPKC) appeals more attention and has been a hot topic in the last years.

The public key of MPKC is a set of quadratic polynomials. Its security relies on the difficulty of solving systems of nonlinear polynomial equations, which has

H.Y. Youm and M. Yung (Eds.): WISA 2009, LNCS 5932, pp. 254–267, 2009.

been proved as an NP-hard problem. MPKC has become one of the promising alternatives for PKC implementation on small devices. As compared with RSA, the computation in MPKC can be implemented very fast since it is operated on a small finite field.

One of the first implementations of MPKC was suggested by Matsumoto and Imai [MI88]. It was broken by Patarin [Pat95] in 1995 by linearization attack. Linearization attack is to find equations satisfied by plaintext variables x_i and ciphertext variables z_j, of the form

$$\sum_{i,j} a_{ij} x_i z_j + \sum_i b_i x_i + \sum_j c_j z_j + d = 0.$$

They are called first order linearization equations, since they are linear in plaintext variables if ciphertext variables are fixed and of degree 1 in ciphertext variables. The first order linearization equation attack was extended to high order attack by Ding et al in [DH07], where they used linearization equations of the forms like

$$\sum_{i,j,k} a_{ijk} x_i z_j z_k + \sum_{i,j} b_{ij} x_i z_j + \sum_{j,k} c_{jk} z_j z_k + \sum_i d_i x_i + \sum_j e_j z_j + f = 0.$$

These equations are called second order linearization equations since they are of degree 2 in ciphertext variables and are also linear in plaintext variables if ciphertext variables are fixed.

Internal perturbation is a method proposed by Ding to resolve linearization equation attack for MPKC [Din04]. It is usually parameterized by a small integer r. As its application, the first MPKC scheme MI [MI88] was improved by internal perturbation to a variant PMI [Din04]. However, for some of its practical parameters, PMI can be efficiently broken by differential attack [FGS05]. Later, to counteract the differential attack, Ding and Gower repaired it by using a plus method, and proposed an enhanced version PMI+ [DG06]. This combination of internal perturbation and plus methods work well for MI, and PMI+ is still secure up to now.

Recently, a new MPKC scheme proposed by Wang et al in 2008 [WNZ08]. The scheme is designed based on a new central map, which belongs to the middle field type family [WYH06]. There are three equations in the central map, so it is convenient to name the basic version of this scheme as TH. TH was proved to be vulnerable to linearization attack like MI [WNZ08]. Following the same line for PMI+, the inventors of TH used both of the internal perturbation and plus methods to remedy the basic scheme TH. Here we call the plus variant of TH as TH+ and the internally perturbed and plus variant as PTH+. PTH+ is the final MPKC scheme constructed in [WNZ08].

In this paper, we show the basic scheme TH is far weaker than what the authors of [WNZ08] analyzed, and the weakness severely threats the safety of TH+ and PTH+. In fact, we will show that due to the weak structure of TH, TH+ can be totally cracked by linearization equation attack. Moreover, although the combination of perturbation and plus methods can resist the differential attack against PMI+, they can not work for PTH+. In this paper, we will make an observation of the rank distribution of the differentials of TH+ and then reduce the attack against PTH+ to an attack on TH+.

We organize the paper as follows. In Section 2, we describe the PTH+ scheme. In Section 3, we make a theoretical analysis that TH+ can be totally cracked by the linearization attack. In Sections 4 and 5, we explain our idea of reducing the attack against the PTH+ to an attack on the TH+ and discuss its complexity.

2   Description of PTH+

2.1   Basic Scheme TH+

The TH+ scheme follows the conventional design way of MPKC. Its encryption transformation is a composition of three maps L_1, ϕ, and L_2, where $L_1 : k^{2l} \rightarrow k^{2l}$ and $L_2 : k^{3l+m} \rightarrow k^{3l+m}$ are two invertible affine maps and kept as a private key, and the so-called central map $\phi : k^{2l} \rightarrow k^{3l+m}$ is constructed by composing of $3l + m$ quadratic polynomials in $2l$ variables, here $k = \mathbb{F}_2$ is the binary field. The composition map $E = L_2 \circ \phi \circ L_1$ is used as a public key, and it is an ordered set of $3l + m$ quadratic polynomials in $2l$ variables.

The central map ϕ is publicly known and is constructed as follows. Let K be an extension of k of degree l. Let $\pi : K \rightarrow k^l$ be the natural isomorphism, namely it is the coefficient vector of elements of K under a fixed basis. It is naturally extended to $\pi : K^2 \rightarrow k^{2l}$ and $\pi : K^3 \rightarrow k^{3l}$, also denoted by π. Let $\tilde{\phi}$ be the polynomial map from K^2 to K^3 defined by

$$\begin{cases} Y_1 = (X_1^2 X_2)^2 + \alpha_1 X_1^2 X_2 \\ Y_2 = (X_1 X_2)^2 + \alpha_2 X_1 X_2 \\ Y_3 = (X_1^2 X_2) + (X_1 X_2) \end{cases} \tag{2.1}$$

where α_1 and α_2 are randomly chosen elements in the field K and they are required to be nonzero and distinct for the purpose of successful decryption. Define a random quadratic polynomial map $p : k^{2l} \rightarrow k^m$. The central map of TH+ is then defined as

$$\phi = (\pi \circ \tilde{\phi} \circ \pi^{-1}, p).$$

It is a quadratic polynomial map from k^{2l} to k^{3l+m}, where p is the so-called plus polynomials added by the plus method [DG06]. To avoid direct algebraic attacks from XL [YC05] or Groebner Basis algorithm like F4 [Fag99] and F5 [Fag02], m should not be too big.

The key point of decryption is efficiently inverting $\tilde{\phi}$. Dividing the both sides of the first two equations of (2.1) by α_1^2 and α_2^2 respectively gives equations of the form $x^2 + x + \beta = 0$. It is well known this equation has solutions in K if and only if $\text{Tr}(\beta) = 0$ and in that case, there are two solutions and they differ 1, where Tr is the trace map from K to $\mathbb{F}_2$. Randomly choose an element δ in K such that $\text{Tr}(\delta) = 1$ (with a success probability of $1/2$). A solution is then explicitly given by [BSS99]

$$\tau(\beta) = \sum_{i=0}^{l-2} (\sum_{j=i+1}^{l-1} \delta^{2^j})\beta^{2^i}. \tag{2.2}$$

If β is computed from x_0 by $\beta = x_0^2 + x_0$, then this solution $\tau(\beta)$ is related to the original x_0 by

$$\tau(\beta) = x_0 + \mathrm{Tr}(\delta x_0). \tag{2.3}$$

(Inserting $\beta = x_0^2 + x_0$ into (2.2) and making a deduction gives the equality (2.3).)

Now, from the first two equations of (2.1) we can compute $X_1^2 X_2$ and $X_1 X_2$ from Y_1 and Y_2. More precisely, we compute $\alpha_1\tau(\alpha_1^{-2}Y_1)$ and $\alpha_2\tau(\alpha_2^{-2}Y_2)$ from Y_1 and Y_2. By the above formula,

$$\begin{cases} \alpha_1\tau(\alpha_1^{-2}Y_1) = X_1^2 X_2 + \alpha_1 \mathrm{Tr}(\delta\alpha_1^{-1}X_1^2 X_2) \\ \alpha_2\tau(\alpha_2^{-2}Y_2) = X_1 X_2 + \alpha_2 \mathrm{Tr}(\delta\alpha_2^{-1}X_1 X_2). \end{cases} \tag{2.4}$$

Utilizing the third equation of (2.1), we can test either of $\alpha_1\tau(\alpha_1^{-2}Y_1)$ and $\alpha_1\tau(\alpha_1^{-2}Y_1)+\alpha_1$ equals to $X_1^2 X_2$ and either of $\alpha_2\tau(\alpha_2^{-2}Y_2)$ and $\alpha_2\tau(\alpha_2^{-2}Y_2)+\alpha_2$ equals to $X_1 X_2$. Finally, it is easy to recover X_1 and X_2 from $X_1^2 X_2$ and $X_1 X_2$.

In the case of l being odd, exactly as in TH+ and PTH+, we can choose $\delta = 1$ since $\mathrm{Tr}(1) = l \cdot 1 = 1$, then we have

$$\tau(\beta) = \sum_{j=1}^{(l-1)/2} \beta^{2^{2j-1}}. \tag{2.5}$$

Obviously this solution $\tau(\beta)$ can be found by shift operation of vector representation of elements if a normal basis of K over k is used, which means the complexity of inverting the central map is very low.

2.2   Internally Perturbed Scheme PTH+

Taking notice of the fact that there are linearization euqaitons for TH+ due to the defect of ϕ (see Section 3 below), the authors in [WNZ08] proposed a perturbation type variant of TH+, which we call PTH+. In PTH+, perturbation is added into the central map of TH+. Let r be a small integer and define a quadratic polynomial map $\hat\phi = H \circ R$, where R: $k^{2l} \to k^r$ is an affine map, and H: $k^r \to k^{3l+m}$ is a random quadratic map. Since the last m coordinates of the quadratic map ϕ have been randomly chosen, the last m coordinates of the quadratic map H can be without loss of generality specified to 0. The central map of PTH+, denoted as $\hat\phi$, is then defined as $\hat\phi = \phi+H \circ R$. To achieve a reasonable decryption efficiency, the value of r must be small, for instance $r = 6$ like in [Din04]. Define the public key of PTH+ as $E' = L_2 \circ \hat\phi \circ L_1 = L_2 \circ (\phi+(H \circ R)) \circ L_1$. It is a quadratic polynomial map.

3   Linearization Equation Attack of TH+

The authors of [WNZ08] pointed out that there exist l first order linearization equations for TH+, which are also called tweakable polynomials in [DS08]. However, their theoretical analysis is not completely correct. Below we give a complete proof on the existence of linearization equations. Moreover, except these l linearly independent (first order) linearization equations, another l second order linearization equations still exist for TH+. That means we can find $2l$ linear equations in $2l$ plaintext variables once the $3l + m$ ciphertext variables are specified. So given a valid ciphertext, starting from these $2l$ equations, we can find the corresponding plaintext easily. Below, we explain how to derive $2l$ linearization equations.

Using the formula (2.4) we can get

$$\begin{cases} X_1^2 X_2 = \alpha_1 \tau(\alpha_1^{-2} Y_1) + \alpha_1 \mathrm{Tr}(\delta \alpha_1^{-1} X_1^2 X_2) \\ X_1 X_2 = \alpha_2 \tau(\alpha_2^{-2} Y_2) + \alpha_2 \mathrm{Tr}(\delta \alpha_2^{-1} X_1 X_2) \end{cases} \tag{3.1}$$

From (3.1), we get

$$\alpha_1 \tau(\alpha_1^{-2} Y_1) + \alpha_1 \mathrm{Tr}(\delta \alpha_1^{-1} X_1^2 X_2) = X_1 \cdot (\alpha_2 \tau(\alpha_2^{-2} Y_2) + \alpha_2 \mathrm{Tr}(\delta \alpha_2^{-1} X_1 X_2)).$$

The authors of [WNZ08] thought this is a linearization equation, however, it is not since it is of degree 3 in the variables corresponding to the X_i.

To get linearization equations, we insert (3.1) into the third equation of (2.1) and get

$$\alpha_1 \mathrm{Tr}(\delta \alpha_1^{-1} X_1^2 X_2) + \alpha_2 \mathrm{Tr}(\delta \alpha_2^{-1} X_1 X_2) = Y_3 + \alpha_1 \tau(\alpha_1^{-2} Y_1) + \alpha_2 \tau(\alpha_2^{-2} Y_2). \tag{3.2}$$

Taking the traces for the both sides of this equation gives

$$\mathrm{Tr}(\alpha_1) \cdot \mathrm{Tr}(\delta \alpha_1^{-1} X_1^2 X_2) + \mathrm{Tr}(\alpha_2) \cdot \mathrm{Tr}(\delta \alpha_2^{-1} X_1 X_2)$$
$$= \mathrm{Tr}(Y_3) + \mathrm{Tr}(\alpha_1 \tau(\alpha_1^{-2} Y_1)) + \mathrm{Tr}(\alpha_2 \tau(\alpha_2^{-2} Y_2)). \tag{3.3}$$

Assuming $(\mathrm{Tr}(\alpha_1), \mathrm{Tr}(\alpha_2)) \neq (0,0)$, from (3.2) and (3.3), we can linearly express $\mathrm{Tr}(\delta \alpha_1^{-1} X_1^2 X_2)$ and $\mathrm{Tr}(\delta \alpha_2^{-1} X_1 X_2)$ in terms of Y_3, $\alpha_1 \tau(\alpha_1^{-2} Y_1)$, $\alpha_2 \tau(\alpha_2^{-2} Y_2)$, $\mathrm{Tr}(Y_3)$, $\mathrm{Tr}(\alpha_1 \tau(\alpha_1^{-2} Y_1))$, and $\mathrm{Tr}(\alpha_2 \tau(\alpha_2^{-2} Y_2))$. Since τ is a linear operation over k, $\mathrm{Tr}(\delta \alpha_1^{-1} X_1^2 X_2)$ and $\mathrm{Tr}(\delta \alpha_2^{-1} X_1 X_2)$ are linear expressions of Y_1, Y_2 and Y_3. Let $Y = (Y_1, Y_2, Y_3)$. By (3.2) and (3.3) there exist linear expressions l_1 and l_2 on the corresponding components over k of Y such that $X_1^2 X_2 = l_1(Y)$ and $X_1 X_2 = l_2(Y)$. Then we have

$$l_1(Y) = X_1 l_2(Y). \tag{3.4}$$

Applying $\pi(Y) = L_2^{-1}(z_1, \cdots, z_{3l+m})$ and $\pi(X_1, X_2) = L_1(x_1, \cdots, x_{2l})$ to (3.4), and working on the base field k, we get l first order linearization equations over k of the form

$$\sum_{i,j} a_{ij} x_i z_j + \sum_i b_i x_i + \sum_j c_j z_j + d = 0, \tag{3.5}$$

where $x_1, \cdots, x_{2l}$ are the plaintext variables and $z_1, \cdots, z_{3l+m}$ are the ciphertext variables.

Having Equations (3.5) and (3.4), if we substitute the ciphertext variables with a fixed value in (3.5), then then the corresponding X_1 will be recovered from (3.4) and become a constant. Thus, the three equations in (2.1) will become linear equations in the variables of X_2. Again the variables of X_2 will be recovered. Consequently, TH+ can be totally cracked by the linearization equations of (3.5). The complexity of finding (3.5) is $\mathcal{O}(2l(3l+m)+5l+m+1)^w)$, $w \approx 2.732$, which is 2^{38} for the instance $(l, r, m) = (47, 6, 11)$ as proposed in [WNZ08].

As a matter of fact, we note that there still exist second order linearization equations for TH+. Multiplying $l_2(Y)^2$ to the both sides of the third equation of (3.1): $Y_3 = X_1^2 X_2 + X_1 X_2$, and associating with the relation in (3.4) give us

$$Y_3 l_2(Y)^2 = X_2 l_1(Y)^2 + X_2 l_1(Y) l_2(Y) \tag{3.6}$$

Applying L_1 and L_2^{-1} as for (3.4), (3.6) is transformed into l second order linearization equations over k of the form

$$\sum_{i,j,k} a_{ijk} x_i z_j z_k + \sum_{i,j} b_{ij} x_i z_j + \sum_{j,k} c_{jk} z_j z_k + \sum_i d_i x_i + \sum_j e_j z_j + f = 0. \tag{3.7}$$

Equations (3.4) and (3.6) are linearly independent over k since they are linear in X_1 and X_2 respectively once Y is fixed. So they are totaly $2l$ linearization equations in $2l$ plaintext variables. This means we can compute both X_1 and X_2 given by (3.5), (3.7), and a ciphertext. The complexity of finding these $2l$ linearization equations is $\mathcal{O}((2l(3l+m)(3l+m-1)/2+2l(3l+m)+(3l+m)(3l+m-1)/2+5l+m+1)^w)$.

4   Cryptanalysis of PTH+

In Section 3, we have shown that TH+ can be totally destructed by linearization equation attack. PTH+ is a perturbed variant of TH+. In this section we will show how to reduce the attack against PTH+ to an attack on TH+ scheme.

Let us recall the notations: $E := L_2 \circ \phi \circ L_1$ is a quadratic polynomial map corresponding to a public key of TH+; $E' = L_2 \circ \hat{\phi} \circ L_1$ is a quadratic polynomial map corresponding to a public key of PTH+, and $E' = E + (L_2 \circ H \circ R \circ L_1)$.

We define a linear space $\mathcal{K}$ as the kernel of $R \circ L_1$. Because L_1 is a bijection and typically rank$(R) = r$, we have generally that dim$(\mathcal{K}) = 2l - r$. For decryption efficiency, the value of r must be small. So $\mathcal{K}$ is a large linear space with dimension $2l - r$. According to the subspace $\mathcal{K}$, the whole linear space k^{2l} can be divided into 2^r cosets $C_0, \cdots, C_{2^r-1}$, and each coset C_i is a translate of $\mathcal{K}$ namely $C_i = w_i + \mathcal{K}$, where w_i is a representative of C_i, $0 \le i \le 2^r - 1$, and there are 2^r representatives which form an r-dimensional complementary space of $\mathcal{K}$. We name the complementary space as $\mathcal{W}$. Suppose $\mathcal{W}$ is spanned by r linearly independent vectors $w_1, \cdots, w_r \in k^{2l}$, we can linearly express $w_{r+1}, \cdots, w_{2^r-1}$ by $w_1, \cdots, w_r$. Let $w_0 = 0$.

Now we observe the perturbation part $L_2 \circ H \circ R \circ L_1$. It is easy to know that $L_2 \circ H \circ R \circ L_1$ maps the vectors in the same coset to a same vector, and maps 2^r different cosets to 2^r generally different vectors. This means, for the vectors in the same coset, we have $E' = E + c$, here c is a vector dependent on only the coset. In other words, PTH+ is degenerated into the TH+ scheme on each C_i, so we can still apply the linearization attack on each C_i by finding linearization equations on that C_i, $0 \leq i \leq 2^r - 1$. To recover the 2^r cosets, we need to recover $\mathcal{K}$.

The general strategy of attack is as follows:

1. Recover the $(2l - r)$-dimensinal subspace $\mathcal{K}$.
2. For $i = 0$ to $2^r - 1$, find a system of at least l linearly independent linearization equations which hold on C_i. This will be done by selecting $\mathcal{O}(2l(3l + m) + 5l + m + 1)$ plaintexts in C_i and their corresponding ciphertexts.
3. For a given ciphertext, we choose an i, $0 \leq i \leq 2^r - 1$, and the linearization equation system found for C_i in Step 2, and then replace its component values into the linearization equations and solve the resulted linear equation system on plaintext variables. If a solution is found, the algorithm terminates. If no solution is found, try another i until a solution is found.

The kernel space $\mathcal{K}$ is a $(2l - r)$-dimensional linear space. In the following Subsections 5.3-5.4, we can recover one of its sufficiently large subspaces, a $(2l - 2r)$-dimensional subspace $\mathcal{S}$, using the public key information of the cryptosystem. However, we can not find additional r linearly independent vectors which combine with $2l - 2r$ vectors of $\mathcal{S}$ to form a basis of $\mathcal{K}$. Fortunately, our experiment shows that the linearization equations holding over $\mathcal{S}$ are sufficient to mount a successful linearization attack on PTH+. But at this moment we can not explain this phenomenon theoretically. This results in the following attack strategy:

1. Recover a $(2l - 2r)$-dimensinal subspace $\mathcal{S}$ and an r-dimensional complementary space $\mathcal{W} = \{w_0, w_1, \cdots, w_{2^r - 1}\}$.
2. For $i = 0$ to $2^r - 1$, compose a subset $C_i' = w_i + \mathcal{S}$ and find a system of at least l linearly independent linearization equations which hold on C_i'. This will be done by selecting $\mathcal{O}(2l(3l + m) + 5l + m + 1)$ plaintexts in C_i' and their corresponding ciphertexts.
3. For a given ciphertext, we choose an i, $0 \leq i \leq 2^r - 1$, and the linearization equation system found for C_i' in Step 2, and then replace its component values into the linearization equations and solve the resulted linear equation system on plaintext variables. If a solution is found, the algorithm terminates. If no solution is found, try another i until a solution is found.

To test the validation of the above attack, we do an experiment with 16 different parameters where l is set to one of the four numbers 17, 19, 21, and 23, m is one of 5, 7, 9, and 11, and r is fixed to be 6, and we decrypt 100 valid ciphertexts for each of the 16 parameters using the linearizatione equations derived for the corresponding C_i', $0 \leq i \leq 2^r - 1$. Here we use relatively smaller parameters than $(l, m, r) = (47, 6, 11)$ since a usual personal computer can not implement a program of complexity exceeding 2^{50}. The experimental result shows that the attack for these parameters always succeeds.

Note that Steps 1 and 2 are a precomputation, namely they are only dependent on a public key but not dependent on a concrete ciphertext. The complexity of Step 2 is $\mathcal{O}((2l(3l+m)+5l+m+1)^w)$, as we have discussed in Section 3. Since r is a small integer, the complexity of Step 3 is quite small. In the next section, we only need to find a way to fulfill Step 1 and discuss its complexity. The main goal of Step 1 is to recover a sufficiently large subspace $\mathcal{S}$ and an r-dimensional complementary space $\mathcal{W}$.

5   Recovering $\mathcal{S}$ and $\mathcal{W}$

In this section, we recover $\mathcal{S}$ and $\mathcal{W}$ by the differentials of the central map of PTH+. Using some properties of the differentials, we can distinguish out a $(2l-2r)$-dimensional subspace $\mathcal{S}$ and an r-dimensional complementary space $\mathcal{W}$.

For any quadratic polynomial map $G: F_q^n \to F_q^m$, let

$$dG(a, x) = G(x + a) - G(x) - G(a) + G(0)$$

$dG(a, x)$ is bilinear and symmetric in a and x. If a is fixed, then $dG(a, x)$ is linear in x, and is also written as $dG_a(x)$. dG_a is called the differential of G with respect to a.

5.1   Differentials of TH+

For differentials of $\tilde{\phi}: K^2 \to K^3$, we consider the differentials of Y_1, Y_2 and Y_3. Let $X = (X_1, X_2)$, $A = (A_1, A_2) \in K^2$. By(2.1), the differentials are

$$\begin{aligned} dY_1(A, X) &= (X_1^2 A_2 + A_1^2 X_2)^2 + \alpha_1(X_1^2 A_2 + A_1^2 X_2) \\ dY_2(A, X) &= (X_1 A_2 + A_1 X_2)^2 + \alpha_2(X_1 A_2 + A_1 X_2) \\ dY_3(A, X) &= (X_1^2 A_2 + A_1^2 X_2) + (X_1 A_2 + A_1 X_2) \end{aligned} \qquad (5.1)$$

These three differentials compose as $d\tilde{\phi}_A(X)$.

Lemma 1. *Let $A = (A_1, A_2)$ be a nonzero vector in K^2, then the equation $d\tilde{\phi}_A(X) = 0$ has $|K| - 1$ nonzero solutions if $A_1 = 0$ or $A_2 = 0$ and has a unique nonzero solution as $X = A$ if $A_1 \neq 0$ and $A_2 \neq 0$.*

Proof. Let $t_1 = X_1^2 A_2 + A_1^2 X_2$, $t_2 = X_1 A_2 + A_1 X_2$, then $d\tilde{\phi}(X, A) = 0$ in (5.1) are exactly

$$\begin{cases} t_1^2 + \alpha_1 t_1 = 0 \\ t_2^2 + \alpha_2 t_2 = 0 \\ t_1 + t_2 = 0 \end{cases} \qquad (5.2)$$

This system of equations can be easily solved as $t_1 = t_2 = 0$. Then $X_1^2 A_2 + A_1^2 X_2 = 0$, and $X_1 A_2 + A_1 X_2 = 0$. The solutions on X_1 and X_2 are dependent on the parameters A_1 and A_2, as follows:

a) $A_1 = 0, A_2 \neq 0$: $X_1 = 0, X_2$ can be an arbitrary value in K.
b) $A_1 \neq 0, A_2 = 0$: $X_2 = 0, X_1$ can be an arbitrary value in K.
c) $A_1 \neq 0, A_2 \neq 0$: $X_1 = X_2 = 0$, or $X_1 = A_1$ and $X_2 = A_2$.

Cases a) and b) show that $d\tilde{\phi}_A(X) = 0$ has $|K|$ solutions, namely $|K|-1$ nonzero solutions, if and only if $A_1 = 0$ or $A_2 = 0$; Case c) shows that $d\tilde{\phi}_A(X) = 0$ has one zero and one nonzero solutions if and only if $A_1 \neq 0$ and $A_2 \neq 0$.   □

Corollary 1. *Let* $a = (a_1, a_2)$ *be a nonzero vector in* k^{2l}, *where* $a_1, a_2 \in k^l$. *If* $a_1 = 0$ *or* $a_2 = 0$ *then* $\mathrm{rank}(d(\pi \circ \tilde{\phi} \circ \pi^{-1})_a) = l$; *if* $a_1 \neq 0$ *and* $a_2 \neq 0$ *then* $\mathrm{rank}(d(\pi \circ \tilde{\phi} \circ \pi^{-1})_a) = 2l - 1$.

Proof. Let $0 \neq A = (A_1, A_2) \in K^2$ with $\pi(A_1) = a_1$ and $\pi(A_2) = a_2$. It is trivial that $d(\pi \circ \tilde{\phi} \circ \pi^{-1})_a(x) = 0$ has the same number of solutions with $d\tilde{\phi}_A(X) = 0$. By Lemma 1, this number is either $|K| = 2^l$ or 2 and the corollary follows.   □

Lemma 2. *Let* b *be a nonzero vector in* k^{2l} *and* $a = L_1(b) = (a_1, a_2)$ *with* $a_1, a_2 \in k^l$. *Let* E *be a public polynomial map of* TH+. *If* $a_1 = 0$ *or* $a_2 = 0$ *then* $\mathrm{rank}(dE_b) \leq l+m$ *and the equality holds with a probability almost to 1; if* $a_1 \neq 0$ *and* $a_2 \neq 0$ *then we have certainly that* $\mathrm{rank}(dE_b) = 2l - 1$.

Proof. Since L_1 and L_2 are bijections and $E = L_2 \circ \phi \circ L_1$, the ranks of differentials dE_b have the same distribution as that of $d\phi_a$. So we only need to analyze $d\phi_a$.

As previously defined in Section 2.1, $\phi = (\pi \circ \tilde{\phi} \circ \pi^{-1}, p)$. It is obvious that $x = a$ is always a nonzero solution for $d\phi_a(x) = 0$, so $\mathrm{rank}(d\phi_a) \leq 2l - 1$. Since $p : k^{2l} \to k^m$ is a randomly chosen quadratic polynomial map and $dp_a(x)$ can be considered as a set of random m linear equations in $2l$ variables, thus $\mathrm{rank}(dp_a) = m$ holds with a probability close to 1 (Fact 1 of Appendix).

Since $d\phi_a$ is a simple union of $d(\pi \circ \tilde{\phi} \circ \pi^{-1})_a$ and dp_a, it is trivial that $\mathrm{rank}(d(\pi \circ \tilde{\phi} \circ \pi^{-1})_a) \leq \mathrm{rank}(d\phi_a) \leq \mathrm{rank}(d(\pi \circ \tilde{\phi} \circ \pi^{-1})_a) + \mathrm{rank}(dp_a)$. If $a_1 = 0$ or $a_2 = 0$, then $\mathrm{rank}(d(\pi \circ \tilde{\phi} \circ \pi^{-1})_a) = l$ by Corollary 1, and using the above inequality we have $\mathrm{rank}(d\phi_a) \leq l + m$. Since dp_a is a random matrix, then by Fact 2 of Appendix, $\mathrm{rank}(d\phi_a) = l + m$ holds with a probability close to 1.

For another case that $a_1 \neq 0$ and $a_2 \neq 0$, we have $\mathrm{rank}(d(\pi \circ \tilde{\phi} \circ \pi^{-1})_a) = 2l-1$ again by Corollary 1 and hence $\mathrm{rank}(d\phi_a) \geq \mathrm{rank}(d(\pi \circ \tilde{\phi} \circ \pi^{-1})_a) = 2l-1$. Since we always have $\mathrm{rank}(d\phi_a) \leq 2l - 1$, and we conclude that $\mathrm{rank}(d\phi_a) = 2l - 1$.   □

5.2   Differentials of PTH+

Having Lemma 2, we find the following properties of the ranks of the differentials of PTH+ with respect a vector $b \in \mathcal{K}$ or not.

Lemma 3. *If* E' *is a public polynomial map of* PTH+, *and* $b \in \mathcal{K}$, *then we have* $\mathrm{rank}(dE'_b) = \mathrm{rank}(dE_b)$.

Proof. Since $E' = E + (L_2 \circ H \circ R \circ L_1)$, $dE'_b = dE_b + d(L_2 \circ H \circ R \circ L_1)_b$. Since L_1 and R are both linear, $R \circ L_1$ is also linear. Thus for $b \in \mathcal{K}$, $R \circ L_1(x + b) = R \circ L_1(x)$, and hence by definition, $d(L_2 \circ H \circ R \circ L_1)_b = 0$.   □

Lemma 4. *Let b be a nonzero vector in k^{2l} and $a = L_1(b) = (a_1, a_2)$. If E' is a public polynomial map of* PTH+ *and $b \notin \mathcal{K}$, then we have either* $\mathrm{rank}(dE'_b) \leq l + m + r - 1$ *if $a_1 = 0$ or $a_2 = 0$, or* $\mathrm{rank}(dE'_b) \geq 2l - r$ *if $a_1 \neq 0$ and $a_2 \neq 0$.*

Moreover, if $m + 2r < l + 1$ then $\mathrm{rank}(dE'_b) \leq l + m + r - 1$ *holds only if $a_1 = 0$ or $a_2 = 0$.*

Proof. We have $dE'_b = dE_b + d(L_2 \circ H \circ R \circ L_1)_b = dE_b + L_2 \circ dH_{R \circ L_1(b)}$. Since $x = b$ is always a nonzero solution for $dE'_b(x) = 0$, $\mathrm{rank}(dE'_b) \leq 2l - 1$. By definition, $H : k^r \to k^{3l+m}$ is a randomly chosen polynomial map and $x = R \circ L_1(b)$ is always a nonzero solution for $dH_{R \circ L_1(b)}(x) = 0$, by Fact 1 of Appendix, $\mathrm{rank}(dH_{R \circ L_1(b)}) = \mathrm{rank}(L_2 \circ dH_{R \circ L_1(b)}) = r - 1$ with a probability close to 1.

If $a_1 = 0$ or $a_2 = 0$, by Lemma 2 and simple linear algebra we have $\mathrm{rank}(dE'_b) \leq \mathrm{rank}(dE_b) + \mathrm{rank}(L_2 \circ dH_{R \circ L_1(b)}) \leq l + m + r - 1$.

For the other case that $a_1 \neq 0$ and $a_2 \neq 0$, by Lemma 2 we have $\mathrm{rank}(dE_b) = 2l - 1$. In this case, we always have $\mathrm{rank}(dE'_b) \geq 2l - r$ since otherwise from $dE_b = dE'_b + L_2 \circ dH_{R \circ L_1(b)}$ we will have $\mathrm{rank}(dE_b) \leq \mathrm{rank}(dE'_b) + \mathrm{rank}(L_2 \circ dH_{R \circ L_1(b)}) < (2l - r) + (r - 1) = 2l - 1$.

Finally, if $m + 2r < l + 1$ or equivalently if $l + m + r - 1 < 2l - r$, then we know that the rank dE'_b in the case of $a_1 = 0$ or $a_2 = 0$ is always less than that in the case of $a_1 \neq 0$ and $a_2 \neq 0$. This gives the last assertion of the lemma. $\square$

By Lemmas 3 and 2, we know that there exist some vectors in $\mathcal{K}$ making the differentials of E' diminish to dimension $l + m$, while by Lemma 4, there are some vectors not belonging to $\mathcal{K}$ making the differential of E' diminishing to dimension $l + m + r - 1$. This is a good property to distinguish out vectors in $\mathcal{K}$ and vectors not in $\mathcal{K}$. However, in order to find a big enough $\mathcal{S}$, we still need to analyze how many independent vectors can be obtained by the property in Lemmas 3 and 4. This will be analyzed in the following Section 5.3.

5.3   A $(2l - 2r)$-Dimensional Subspace $\mathcal{S}$ and Its r-Dimensional Complementary Space $\mathcal{W}$

Assume $m + 2r < l + 1$. Define two sets:

$$V_1 = \{b \in \mathcal{K} : \mathrm{rank}(dE'_b) \leq l + m\}$$

and

$$V_2 = \{b \notin \mathcal{K} : \mathrm{rank}(dE'_b) \leq l + m + r - 1\}.$$

As above defined, $\mathcal{K}$ is the kernel of $R \circ L_1$ and L_1 is invertible, so for a vector in k^{2l}, $b \in \mathcal{K}$ if and only if $a = L_1(b) \in \mathrm{kernel}(R)$. Let V_1, V_2 are mapped into U_1 and U_2 respectively by L_1. By Lemmas 3, 2, and 4, U_1 and U_2 are exactly as:

$$U_1 = \{(a_1, a_2) \in \mathrm{kernel}(R)|a_1 = 0 \text{ or } a_2 = 0\}$$

and

$$U_2 = \{(a_1, a_2) \notin \mathrm{kernel}(R)|a_1 = 0 \text{ or } a_2 = 0\}.$$

Obviously there are $2^{l+1} - 1$ vectors in $U_1 \cup U_2$, namely vectors $(a_1, a_2) \in k^{2l}$ with $a_1 = 0$ or $a_2 = 0$.

Since R is a random $(2l) \times r$ matrix of rank r and r is relatively very small to $2l$, then both of its most upper and most lower $r \times r$ submatrices are generally of full rank. So we can get $2l - 2r$ linearly independent vectors in kernel(R) as:

$$W_1 = \{\, w_i = (\overbrace{0,\cdots,0}^{l}, \overbrace{0,\cdots,0,1,0,\cdots,0}^{l-r}, \overbrace{*,\cdots,*}^{r}\,) \in k^{2l} : l+1 \le i \le 2l-r \,\}$$
$$\underset{i\text{th position}}{\uparrow}$$

and

$$W_2 = \{\, w_i = (\overbrace{*,\cdots,*}^{r}, \overbrace{0,\cdots,0,1,0,\cdots,0}^{l-r}, \overbrace{0,\cdots,0}^{l}\,) \in k^{2l} : r+1 \le i \le l \,\}.$$
$$\underset{i\text{th position}}{\uparrow}$$

Note that W_1 spans an $(l - r)$-dimensional linear subspace and it is the set of all vectors in kernel(R) with $a_1 = 0$. Similarly, W_2 spans another $(l - r)$-dimensional subspace and it is the set of all vectors in kernel(R) with $a_2 = 0$. So, U_1 is composed by the space spanned by W_1 and the space by W_2, and there are $2(2^{l-r} - 1) = 2^{l-r+1} - 2$ nonzero vectors in U_1. So $|V_1| = |U_1| = 2(2^{l-r} - 1) = 2^{l-r+1} - 2$. Since there are totally $2^{l+1} - 2$ nonzero vectors in $U_1 \cup U_2$, there are exactly $2^{l+1} - 2^{l-r+1}$ vectors in U_2 and $|V_2| = 2^{l+1} - 2^{l-r+1}$.

Define a $(2l - 2r)$-dimensional subspace $\mathcal{S}$ be the linear space spanned by the vectors in V_1 that are mapped into W_1 and W_2 by L_1. From V_2 we can find r linearly independent vectors to form a complementary space of $\mathcal{S}$, we let any such complementary space as $\mathcal{W}$. It is dependent on the search method.

5.4   Searching $\mathcal{S}$ and $\mathcal{W}$ and Its Complexity

Since V_1 is mapped by L_1 into W_1 and W_2, and V_2 is mapped by L_1 into U_2, then searching a $(2l - 2r)$-dimensional subspace $\mathcal{S}$ and an r-dimensional $\mathcal{W}$ equals to searching $2l - 2r$ independent vectors in W_1 and W_2, and r independent vectors in U_2. Let
$$P_1 = \Pr[a_1 = 0 \text{ or } a_2 = 0] = (2^{l+1} - 1)/2^{2l},$$
$$P_2 = \Pr[a \in \text{kernel}(R) | (a_1 = 0 \text{ or } a_2 = 0)] = (2^{l-r+1} - 1)/(2^{l+1} - 1),$$

and

$$P_3 = \Pr[a \notin \text{kernel}(R) | (a_1 = 0 \text{ or } a_2 = 0)] = (2^{l+1} - 2^{l-r+1})/(2^{l+1} - 1).$$

If we randomly choose a vector b in k^{2l}, and let $a = L_1(b)$, then
$$\Pr[b \in V_1] = \Pr[a \in U_1] = P_1 P_2 \approx 2^{-(l+r-1)},$$

and

$$\Pr[b \in V_2] = \Pr[a \in U_2] = P_1 P_3 \approx 2^{-(l-1)}.$$

Below we use rank$(dE_b') \le l + m$ and rank$(dE_b') \le l + m + r - 1$ as the decision condition of the search method since E' is publicly known public key polynomials.

1. Recover a $(2l - 2r)$-dimensional subspace $\mathcal{S}$. Randomly choose a vector b in k^{2l}, and verify if $\text{rank}(dE_b') \leq l + m$. If it is, keep it; if not, discard it and try a new b until $2l - 2r$ linear independent such vectors have been found. The complexity of this step is about $(2l - 2r) \cdot 2^{l+r-1}(2l)^w$.

2. Recover an r-dimensional complementary space $\mathcal{W}$. Randomly choose a vector b in k^{2l}, and verify if $\text{rank}(dE_b') \leq l + m + r - 1$. If it is, keep it; if not, discard it and try a new b until r linear independent such vectors have been found. The complexity of this step is about $r \cdot 2^{l-1}(2l)^w$.

Note that each time we choose a vector b, we need to compute $\text{rank}(dE_b')$, and its computation complexity is $(2l)^w$. It is obvious to see that the complexity of Step 1 is much larger than that of Step 2, so the complexity of the two steps totally is $(2l - 2r) \cdot 2^{l+r-1}(2l)^w \approx 2^{77}$ for the PTH+ parameter $(l, r, m) = (47, 6, 11)$.

The complexity shows that the instance is theoretically insecure. In fact, we can reduce the complexity by the following optimized searching strategy.

Optimized search. For a nonzero vector b in V_2, either the first or the second half of $L_1(b)$ is zero. For two nonzero vectors b_1 and b_2 evenly taken from V_2, the two halves of $L_1(b_1 + b_2) = L_1(b_1) + L_1(b_2)$ will fall in the following three cases: (1) the first one is zero and the second one is nonzero; (2) the first one is nonzero and the second one is zero; or (3) both halves are nonzero. The probabilities that they happen are $1/4$, $1/4$ and $1/2$. So there is a probability of $1/2$ such that one half of $L_1(b_1 + b_2)$ is zero namely $b_1 + b_2 \in V_1 \cup V_2$. Further, if b_1 and b_2 lie in the same coset of $\mathcal{K}$, then $b_1 + b_2$ are in V_1.

We can try to find about $2(2^r + l - r)$ vectors in V_2. Among them there will be about $2^r + l - r$ vectors such that the first half of $L_1(b)$ is zero and there are other about $2^r + l - r$ vectors such that the second half of $L_1(b)$ is zero. From the $2^r + l - r$ vectors with the first zero halves of their L_1-images, we can always choose $l - r$ pairs such that the two vectors are in the same coset of $\mathcal{K}$. Summing the two vectors in a pair results in $l - r$ vectors in V_1. Usually they are linearly independent. Similarly, we can get $l - r$ linearly independent vectors in V_1 and with the second zero halves of their L_1-images.

By the above analysis, we can find $\mathcal{S}$ and $\mathcal{W}$ as follows:

1. Randomly choose a vector b in k^{2l}, and verify if $\text{rank}(dE_b') \leq l + m + r - 1$. If it is, keep it; if not, discard it and try a new b until slightly more than $2^{r+1} + 2l - 2r$ linearly independent such vectors have been found. The complexity of this step is about $(2^{r+1} + 2l - 2r) \cdot 2^{l-1}(2l)^w$.

2. Choose r vectors from the above $2^{r+1} + 2l - 2r$ vectors to span $\mathcal{W}$.

3. Recover a $(2l - 2r)$-dimensional subspace $\mathcal{S}$. Choose a pair of two vectors b and b' from the vectors recovered in Step 1, and verify if $\text{rank}(dE_{b-b'}') \leq l + m$. If it is, keep $b - b'$; if not, choose another pair until $2l - 2r$ linearly independent such difference vectors have been found. The search of this step can be in the order of firstly fixing b and varying b' and then trying another fixed b and varying b'. The worst complexity of this step is about $\binom{2^{r+1}+2l-2r}{2} \times (2l)^w$.

Actually if we find a vector b in Step 1 that satisfies $\text{rank}(dE_b') \leq l + m$, we can keep it as a vector in $\mathcal{S}$. Since there is no necessary to choose such vectors in

Step 3 for computation, the number of pairs for subtraction will decrease and this lowers down the complexity of Step 3. Also, we can combine Step 1 and Step 3, that is when a b in V_2 have been found, do subtraction with other b's that have been recovered in V_2 and verify if $\text{rank}(dE'_{b-b'}) \leq l + m$. Once we have found $2l - 2r$ vectors for $\mathcal{S}$, then we stop. In this way we do not need to choose as much as $2^{r+1} + 2l - 2r$ vectors from V_2, and this will reduce the complexity of Step 1. For parameters $(l, r, m) = (47, 6, 11)$, the complexity of Step 1 is about 2^{72}, and the complexity of Step 3 is at most 2^{33}. So the total complexity is about 2^{72}.

The authors of [WNZ08] argued that if $l = 47, r = 6, m \geq 10$, PTH+ can resist the differential attack and in addition they argued that the larger is m, the more secure is the PTH+. However, the above complexity tells us that it is independent on m.

Finally we note that by Lemma 2, if $b \in V_1$, the decision condition $\text{rank}(dE'_b) \leq l + m$ will be typically satisfied with an equality sign when it holds.

6   Conclusion

In this paper, we utilize the property of the differential of PTH+ to reduce the attack against PTH+ to an attack on TH+. The total complexity of our attack on PTH+ is $(2l - 2r) \times 2^{l+r-1} \times (2l)^w \approx 2^{72}$ and it theoretically shows PTH+ is not secure. Our analysis also disproves the claim of the inventors of PTH+ that the number m of the plus polynomials in the design of PTH+ is an important measure for resisting differential attack.

Acknowledgement. The authors would like to thank anonymous referees for their helpful comments. The work of this paper was supported by the NSFC (60773134), the National 863 Program of China (2006AA01Z416) and the National Basic Research Program of China (2007CB311201).

References

[BSS99]   Blake, I., Seroussi, G., Smart, N.: Elliptic Curves in Cryptography. Cambridge Unversity Press, Cambridge (1999)

[CKP00]   Courtois, N., Klimov, A., Patarin, J., Shamir, A.: Efficient Algorithms for Solving Overdefined Systems of Multivariate Polynomial Equations. In: Preneel, B. (ed.) EUROCRYPT 2000. LNCS, vol. 1807, pp. 392–407. Springer, Heidelberg (2000)

[DH76]   Diffie, W., Hellman, M.: New Directions in Cryptography. IEEE Transactions on Information Theory 22, 644–654 (1976)

[DH07]   Ding, J., Hu, L., Nie, X., Li, J., Wagner, J.: High Order Linearization Equation (HOLE) Attack on Multivariate Public Key Cryptosystems. In: Okamoto, T., Wang, X. (eds.) PKC 2007. LNCS, vol. 4450, pp. 233–248. Springer, Heidelberg (2007)

[Din04]   Ding, J.: A New Variant of the Matsumoto-Imai Through Perturbation. In: Bao, F., Deng, R., Zhou, J. (eds.) PKC 2004. LNCS, vol. 2947, pp. 305–318. Springer, Heidelberg (2004)

[DG06]   Ding, J., Gower, J.: Inoculating Multivariate Schemes Against Differential Attacks. In: Yung, M., Dodis, Y., Kiayias, A., Malkin, T.G. (eds.) PKC 2006. LNCS, vol. 3958, pp. 290–301. Springer, Heidelberg (2006)

[DGS06]   Ding, J., Gower, J., Schmidt, D.: Multivariate Public-Key Cryptosystems. In: Advances in Information Security. Springer, Heidelberg (2006)

[DS05]   Ding, J., Schmidt, D.: Cryptanalysis of HEFV and the Internal Perturbation of HFE. In: Vaudenay, S. (ed.) PKC 2005. LNCS, vol. 3386, pp. 288–301. Springer, Heidelberg (2005)

[DS08]   Dinur, I., Shamir, A.: Cube Attacks on Tweakable Black Box Polynomials, http://eprint.iacr.org/2008/385

[Fag99]   Faugère, J.: A New Efficient Algorithm for Computing Gröebner Bases (F4). Journal of Applied and Pure Algebra 139, 61–88 (1999)

[Fag02]   Faugère, J.: A New Efficient Algorithm for Computing Gröebner Bases Without Reduction to Zero (F5). In: ISSAC, pp. 75–83. ACM Press, New York (2002)

[FGS05]   Pierre-Alain, F., Granboulan, L., Stern, J.: Differential Cryptanalysis for Multivariate Schemes. In: Cramer, R. (ed.) EUROCRYPT 2005. LNCS, vol. 3494, pp. 341–353. Springer, Heidelberg (2005)

[MI88]   Matsumoto, T., Imai, H.: Public Quadratic Polynomial-tuples for Efficient Signature Verification and Message Encryption. In: Günther, C.G. (ed.) EUROCRYPT 1988. LNCS, vol. 330, pp. 419–453. Springer, Heidelberg (1988)

[Pat95]   Patarin, J.: Cryptanalysis of the Matsumoto and Imai Public Key Scheme of Eurocrypt 1988. In: Coppersmith, D. (ed.) CRYPTO 1995. LNCS, vol. 963, pp. 248–261. Springer, Heidelberg (1995)

[Sho97]   Shor, P.: Polynomial-time Algorithms for Prime Factorization and Discrete Logarithms on a Quantum Computer. SIAM journal on computing 26, 1484–1509 (1997)

[WNZ08]   Wang, Z., Nie, X., Zheng, S., Yang, Y., Zhang, Z.: A New Construction of Multivariate Public Key Encryption Scheme through Internally Perturbed Plus. In: Gervasi, O., Murgante, B., Laganà, A., Taniar, D., Mun, Y., Gavrilova, M.L. (eds.) ICCSA 2008, Part I. LNCS, vol. 5072, pp. 1–13. Springer, Heidelberg (2008)

[WYH06]   Wang, L., Yang, B., Hu, Y., Lai, F.: A Medium-Field Multivariate Public Key Encryption Scheme. In: Pointcheval, D. (ed.) CT-RSA 2006. LNCS, vol. 3860, pp. 132–149. Springer, Heidelberg (2006)

[YC05]   Yang, B., Chen, J.: All in the XL Family: Theory and Practice. In: Park, C.-s., Chee, S. (eds.) ICISC 2004. LNCS, vol. 3506, pp. 67–88. Springer, Heidelberg (2005)

Appendix: Probabilities of the Ranks of Matrices

In this appendix we give some facts about the probabilities that matrices take some ranks. The matrices we discuss are defined over the binary field.

Fact 1: For a random binary $m \times u$ matrix M, $m \leq u$, the probability that $\text{rank}(M) = m$ is $\prod_{i=0}^{m-1}(1 - 2^{-(u-i)})$. For $m = 11, u = 2 \times 47$, the probability is larger than $1 - 2^{-80}$.

Fact 2: For a fixed binary $v \times u$ matrix M_1, $\text{rank}(M_1) = p$, and a random binary $m \times u$ matrix M_2, let $M = \binom{M_1}{M_2}$. If $p + m \leq u$, the probability that $\text{rank}(M) = p + m$ is $\prod_{i=0}^{m-1}(1 - 2^{-(u-p-i)})$. For $v = 3 \times 47 = 141, u = 2 \times 47, p = 47, m = 11$, the probability is larger than $1 - 2^{-33}$.

Towards Privacy Aware Pseudonymless Strategy for Avoiding Profile Generation in VANET *

Rasheed Hussain[1], Sangjin Kim[2], and Heekuck Oh[1]

[1] Hanyang University, Department of Computer Science and Engineering,
Republic of Korea
rasheed1984@gmail.com, hkoh@hanyang.ac.kr
[2] Korea University of Technology and Education,
School of Information and Media Engineering, Republic of Korea
sangjin@kut.ac.kr

Abstract. Inspiring from MANET (Mobile Ad hoc NETworks), VANET (Vehicular Ad hoc NETworks) employing vehicles as nodes, provide a wide range of applications in transportation system. The security of VANET has been a hot topic among the research community. VANETs must meet the basic security requirements such as authentication, integrity, confidentiality and privacy. In VANET, vehicles send beacon messages periodically every 100-300ms which carry speed and position information used for safe driving. The privacy of user is abused by profile generation where the adversary makes movement profiles against the vehicle using the identity information in the beacon. We outline the strategies using pseudonyms to provide privacy of user. After finding out deficiencies in pseudonym-based schemes, we propose a *pseudonymless* strategy to avoid *profilation*. In our scheme, we assume that each car is equipped with TRH (Tamper-Resistant Hardware) carrying out secure operations. Our proposed scheme assures the avoidance of profile generation without using mix zones and silent periods. We show that our proposed scheme is computationally efficient and less bandwidth consuming than other systems.

Keywords: VANET, Privacy, Profile Generation.

1 Introduction

VANETs are inherited case of MANETs where the movement of nodes (vehicles) is restricted by parameters like speed and direction. For example, the vehicles have more speed and they have directed movements in VANET [1]. The security of VANETs is a very important issue in implementation that is why most of research is carried out considering VANET security [2,3,4]. In many

* This research was supported by the Korea Research Foundation Grant funded by the Korean Government (KRF-2008-313-D01024).The research was also supported by Ministry of Knowledge Economy, Korea, under the HNRC (Home Network Research Center)- ITRC (Information Technology Research Center) support program supervised by the Institute of Information Technology Assessment.

H.Y. Youm and M. Yung (Eds.): WISA 2009, LNCS 5932, pp. 268–280, 2009.

VANET projects like NOW (Network on Wheel) [5], The EPFL Vehicular Networks Security [6], California Path [7], and SeVeCom [8], much efforts have been done regarding security. Amongst the basic security requirements authentication, integrity, confidentiality, and privacy, the user and location privacy in VANET environment is addressed by many researchers [9,10,11,12]. VANETs provide number of functionalities like safety, cooperative driving, and enabling driver to respond early to dangerous situations like accidents and find an alternate way in case of traffic jam [13]. In addition to all these VANET is also expected to respect the user itself and its location because this is confidential from the user point of view. The basic communication pattern of VANET is divided into two main parts, V2V (Vehicle to Vehicle) communication and V2I (Vehicle to Infrastructure) communication [14]. Normally in VANET, the nodes periodically send messages to each other and to the infrastructure as well for different purposes. For example, the vehicular nodes can share the speed and position information among each other and with the infrastructure that can be used for traffic safety. These periodically sent messages are called beacons [15]. Besides these beacons, there are also special type of messages like alarm signals by ambulance, stop signals by police van and ordinary safety messages by the infrastructure. Additionally, value-added services like internet, movies-on-demand, IPTV, and toll paying facilities may be provided by VANET [1]. The periodically sent beacons causes a serious problem called profile generation in which an adversary may keep track of the vehicle and make a movement profile based on the position of the user and use it for its own purpose which may abuse the user privacy [16,17,18,19]. Extensive research has been done in this area using different strategies. In this paper, we outline multiple pseudonyms strategy to avoid profilation in VANET. Then discussing the problems of pseudonym-based strategy for profilation, we propose a pseudonymless scheme to avoid profilation. We then tune our scheme to reduce the revocation cost. We also show that our scheme is privacy aware and computationally efficient than other systems.

We discuss some privacy preserving schemes by different authors in section 2 in detail. We outline the security framework for our scheme as a baseline in section 3. Section 4 gives brief introduction to pseudonyms and privacy and we consider multiple pseudonyms strategy for privacy and avoid profile generation. We suggest 'No Pseudonym' strategy in section 5 with grouping strategies to reduce the revocation cost. In section 6 we evaluate our scheme with respect to Scheuer et al. and Plobi et al. and finally in section 7, we give our concluding remarks.

2   Related Work

Currently, providing location and user privacy, and avoiding profile generation in VANET are a hot research topic [9,10-12,14,16-26]. Rass et al. categorized the periodically sent beacons into samples and trips to provide privacy [9]. They use separate pseudonyms, which are generated using ElGamal cryptosystems, for both samples and trips. Similarly, [17] outlines the pseudonymous strategy for privacy preservation, using public key cryptosystems. Generally, their schemes can be viewed as multiple pseudonyms schemes. [22,23] also proposed multiple pseudonym scheme based on public key cryptosystem. In their system, vehicles carry a large number of anonymous certificates. Ma et al. proposed two types of hierarchical pseudonyms and they also discussed refilling issues of multiple pseudonyms [24]. [11,14] suggest another pseudonym changing scheme to provide privacy and avoid profile generation in VANET. Their scheme uses mix zones [12], where a vehicle changes its pseudonym, to provide unlinkability between pseudonyms.

Chaurasia et al. proposed a similar technique to mix zone called anonymity zone [16,25]. They take a heuristic measures for anonymity. In their system, a vehicle needs to be silent for some time, based on the number of vehicles entering the anonymity zone, before changing its pseudonym. This silent period, however, may affect the VANET functionality, since VANET services depend on vehicles sending beacons every 100-300ms. Eichler also proposed a similar scheme using quiet periods [26]. This work however, focuses on pseudonym changing considering node re-interaction and quiet time.

Most of the authors suggested multiple pseudonyms and pseudonym changing based protocols to provide privacy but Schoch et al. discovered [18] that changing pseudonyms have bad effects on routing efficiency, and Liu et al. found that a malicious vehicle can easily generate a new ID before it is punished [19].

For the best of our knowledge, we believe that the most relevant work to our proposed scheme is done by Plobi et al. [27] and Scheuer et al. [21]. In both works, they assume the presence of TRH and use symmetric key cryptography for beaconing.

3 VANET Security Framework

In this section, we outline the security framework for VANET that is proposed by Plobi et al [27]. We consider this framework as a base for our proposed scheme. The participants in VANET security infrastructure are vehicular nodes, CA (Certification Authority) which issues certificates to vehicles (TRH), and GTTP (Geographically Distributed Trusted Third Party) which issues pseudonyms and keys to the vehicles and manages them. Additionally GTTP also has responsibility to revoke VRIs (Vehicle-Related Identity) if needed.

We assume the presence of a TRH inside the vehicle which provides security for the keys stored inside it [2]. There are mostly two terms used for TRH that are tamper-proof and tamper-evident. Tamper-proof is more strict term than tamper-evident and claims 100% security against tampering [3]. Due to the security sensitive nature of the keys, the vehicles should have tamper-proof devices. The root CA certificate, TRH's own certificate ($Cert_{TRH_i}$), vehicle's individual symmetric key (K_{V_i}), vehicle's VRI, common symmetric key (K_{all}), and group ID (G_{id}) are stored inside TRH and not possible to retrieve or change by any means except the TRH itself which is part of TRH operations. To reduce the risk of TRH compromise, it has its own battery that can be charged from the vehicle and clock that can be securely resynchronized when passing by roadside infrastructure. As it is clear from its name, TRH also has sensors that sense any tampering in the hardware. This sensing prevents compromise of any keys inside and TRH destroys all the keys when being tampered. Only authentic configuration is possible for the owner of the car at the time of initialization or in the case when vehicle is sold to other customer. We assume that all messages are assembled inside TRH and all the keys are kept secure in the TRH (at least until TRH is removed or replaced by another one). As a part of operations we also assume that TRH itself keeps on changing the symmetric keys when they are outdated. This updation is carried out by requesting the TTP (Trusted Third Party used in our proposed system) using the key and group ID updating protocol discussed in subsection 5.2. At the time of manufacturing, a smart card is associated to TRH that is also given to the customer by which the customer is able to configure the TRH. For complete initialization process, [27] may be referred.

4   Pseudonyms and Privacy

Pseudonyms are used to provide privacy in VANET. As described in section 2, for privacy and profiling, most of the work has been done based on pseudonyms and mix zones. Due to application requirements in VANET, vehicles send beacons every 100-300ms and normally beacons are digitally signed for security reasons. This strategy increases the cost both from signing and verification point of view. Most of the researchers have used asymmetric cryptography and digital signatures in their schemes. Scheuer et al. [21] suggested a scheme based on pseudonyms, in which they used symmetric cryptography for beaconing and for fast revocation. We extend the same concept of using symmetric cryptography for beaconing and consider the framework outlined in [27] by Plobi et al. Using a single pseudonym is effected by profile generation where the attacker, that can be either insider or outsider, can make movement profiles based on the position of the vehicle. For details about attacker's model and capabilities, [2] can be referred and references therein. To avoid such situations, mix zones and silent periods [12] were proposed by authors for providing privacy to VANET users. While being inside mix zone, the vehicles stop communicating with other vehicles and change their pseudonyms. The downside of mix zones and silent periods is that they have bad effects on the VANET functionalities like accident warning, traffic jam signals, and other security warnings. Therefore, we believe that the use of mix zones and silent periods is not a good solution for avoiding profile generation.

4.1   Multiple Pseudonyms

Using single pseudonym is a threat for privacy because of possibility of making movement profiles. This threat can be reduced if more than one temporary identifiers (pseudonyms) are used by a vehicle. We outlined in section 2 that most of the schemes which address privacy and profile generation issues, use multiple

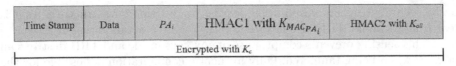

Fig. 1. Beacon message with multiple pseudonyms

pseudonyms strategies with changing nature. Almost every scheme uses asymmetric cryptography with digital signatures for beaconing. We consider multiple pseudonyms strategy for beaconing using symmetric cryptography. [21] uses this mechanism with single pseudonym.

Fig. 1 shows the beacon format with multiple pseudonyms P_{A_i} where $i \in \{1,2,3, \cdots, n\}$ and n is the total number of pseudonyms. Timestamp is used to assure the freshness of the beacon. Data contains the speed and position information. When vehicle wants to join the network, it initiates its request to GTTP for keys. In order to authenticate itself to GTTP, it has to provide its credentials that includes certificate $(Cert_{TRH_A})$ and VRI. After successful authentication, GTTP issues a set of pseudonyms (PA_i) while storing the relation of pseudonyms, VRI, and associated symmetric key $(K_{MAC_{PA_i}})$ which will be used for HMAC1 calculation. GTTP may give a single associated key for all pseudonyms or separate keys for each pseudonym. In addition to the individual key, GTTP also gives two more common keys to all VANET users $(K_{all}$ and $K_c)$ which are used for calculating HMAC2 and encryption of beacon, respectively. HMAC1 is used by GTTP for accountability of vehicle if needed. So every vehicle (more precisely its TRH) can decipher the message by K_c and can check the integrity of the message by calculating HMAC2. To ensure the integrity of HMAC1, which can be possibly changed by an adversary, we suggest that HMAC1 should be included in calculation of HMAC2. In this way the integrity of HMAC1 is preserved which is essential in revoking the anonymity of the message. Only the sending vehicle and GTTP has the individual symmetric key $(K_{MAC_{PA_i}})$, so no one else can calculate HMAC1. When a vehicle receives a beacon given in Fig. 1, the vehicle's TRH checks the integrity and authentication using HMAC2. HMAC1 is used by the GTTP only when it is necessary to revoke the user of the message. Pseudonyms (PA_i) keep changing after a specified amount of time. The method of beaconing using symmetric key cryptography is cost effective compared to schemes discussed in section 2 which use asymmetric and digital signatures. But there are still downsides of this approach. Using multiple pseudonyms has bad effect on the space requirements of both TRH and GTTP. Besides, an efficient refill strategy will be needed [24] and it will also cause inefficient bootstrapping.

5   Proposed Pseudonymless Beaconing

The main cause of profilation is the inclusion of such identities through which vehicles can be tracked. The most favorite candidates for profilation are the identities like pseudonyms. If all the vehicles anonymously send beacons to each

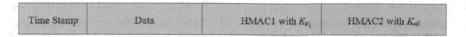

Fig. 2. Pseudonymless Beacon

other in the network without any identity and the functionality of the network is still maintained, then profile generation will not be possible. To the best of our knowledge, such identityless beaconing has not been proposed yet. We propose pseudonymless strategy for beaconing where the VRIs can still be revoked if required.

Fig. 2 shows the proposed beacon format without any identity information. The beacon contains timestamp for freshness, speed and position data, and the last two fields contain the calculated HMACs with K_{V_i} and K_{all}, respectively. Here, K_{V_i} is the individual secret key and K_{all} is the common key for all users. K_{all} is used by other vehicles to check the integrity of the entire message including HMAC1, while K_{V_i} is known only to the sending vehicle and TTP which issued the key. The complexity of this approach is $O(n)$ where n is the number of vehicles in VANET. Therefore, when there is a need for revocation of a VRI, TTP uses brute-force strategy. These situations will occur only in exceptional cases, for example, revocation may be used to determine the vehicle's responsibility in an accident. Also since the VRI of a vehicle can always be revoked, it is highly unlikely that it will be abused in normal cases. Therefore, we believe that this strategy can be practical. Using this approach, there is no need to use mix zones or silent periods which degrade the efficiency of the network.

5.1   TTP Instead of GTTP

Scheuer et al. [21] and Plobi et al. [27] employed a distributed strategy for pseudonym related operations in VANET. They proposed GTTP which covered small physical area employing symmetric cryptography operations. Vehicles receive new keys and pseudonym related information from a GTTP whenever they enter a new area covered by a different GTTP. GTTP stores the relation between keys, pseudonym, and VRI for revocation purpose. The effect of using GTTP is as follows.

First, it reduces information that must be searched when a revocation is needed. For example, if a message, which was sent in a specific area at a specific time, need to be revoked; the GTTP only has to search vehicles which were in that area at that particular time. Second, it limits the disclosed information when a security leak occurs. Since information received from one GTTP to another is independent, compromise of a key given by a GTTP in a certain area only affects message exchanged in that area. In other words, the consequences of security leak or compromise of a key are localized to the area which is under control of that GTTP.

In our proposed scheme, by grouping technique discussed in the next subsection, we do not require GTTP to reduce the revocation cost. Moreover, by updating keys, we also limit the amount of disclosed information when a key is compromised. Therefore, we use a single TTP, which manages key distribution, management, and revocation if required. For 'easy to access' and efficiency purpose, the TTP can be replicated to several physical locations and access through RSUs (Road-Side Unit) via high speed links. However, the number of replication will be very small compared to using GTTPs. How these TTPs be replicated and synchronized is out of scope of this paper.

5.2 Grouping

In pseudonymless beaconing we saw that the cost of revocation of a VRI is $O(n)$ where n is the number of vehicles entertained by TTP. If TTP arranges the vehicles to certain groups then the cost can be reduced to $O(g)$, where g is the size of the group. There should be certain limitations on group size. The group size should be large enough to provide sufficient privacy for the vehicle and small enough to provide efficient revocation by TTP. In other words, there is a tradeoff between privacy of vehicle and efficiency of TTP in regard to the size of the group. We consider two approaches: one using group secret key and other using individual secret key. We also discuss two grouping strategies. Note that we do not discuss the rules and circumstances under which TTP is supposed to revoke the VRI.

Group secret key. Each vehicle is allocated to some group by the TTP when the vehicle registers with the TTP. During registration, TTP issues group ID (G_{id}) and a group secret key (K_g) to the vehicle. Fig. 3 shows the format of the beacon. Here, G_{id} is the group ID of the vehicle, HMAC1 is calculated with K_g, and VRI is included in the calculation of HMAC1 so that TTP can recognize the vehicle in case of revocation. TTP can recognize the group by G_{id} and then the vehicle by calculating g number of HMACs. Thus the complexity of this approach is $O(g)$. The downside of this approach is that compromise of K_g affects the privacy of all group members.

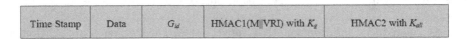

Time Stamp	Data	G_{id}	HMAC1(M‖VRI) with K_g	HMAC2 with K_{all}

Fig. 3. Beacon message using group secret key

Individual secret key. TTP organizes each vehicle into a group and gives G_{id} and K_{V_i} where K_{V_i} is individual secret key that is used to calculate HMAC1. In this method, the revocation cost is equal to $O(g)$. Fig. 4 shows the beacon format and in this method we do not need to include VRI in HMAC1 calculation. In this case, to revoke a VRI, the TTP computes HMAC1 with all the K_{V_i} of vehicles included in group G_{id}. Therefore, the complexity is still $O(g)$. The advantage of this method over the first one is that in case of key (K_{V_i}) compromise, only that individual vehicle's privacy is affected.

Time Stamp	Data	G_{id}	HMAC1(M) with K_{V_i}	HMAC2 with K_{all}

Fig. 4. Beacon message using individual secret key

Grouping Strategy. TTP can organize the groups in the following fashion.

Sequential Method. In sequential method, each entering vehicle is given the same group ID and TTP does this for certain amount of time. In other words, at a certain instant of time only one group will be populating. In the long term, this can be a threat for privacy because depending upon the traffic conditions if there are very few vehicles, let say only one vehicle entering a new group, then that vehicle can be easily tracked.

Random Method. TTP randomly assigns group ID to the vehicles as they enter the network. In this method, every group will be populating with equal probability. At any instant of time, all the groups will grow equally. This random fashion of grouping preserves privacy and provides anonymity for users.

Through our analysis, our scheme uses individual secret key with random grouping strategy.

Secret key and Group Updation. To increase the anonymity and privacy of VANET users, the individual secret key (K_{V_i}) and group ID (G_{id}) are periodically changed. This also reduces the amount of disclosed information when a key is compromised. TRH maintains a *counter* which defines the lifetime of K_{V_i} and group ID (G_{id}). When the counter becomes zero, TRH initiates request to TTP for new secret key and group ID. In this way, vehicles switch between groups. More precisely, at any instant of time, a vehicle will belong to one group and at other instant of time that vehicle may belong to another group with changed K_{V_i}. This will increase the anonymity of the users. We outline a set of requirements for key and group ID updation.

- Mutual authentication between TTP and TRH
- Confidentiality
- Integrity of Key (K'_{V_i})
- Availability of TTP
- Tamper resistance of TRH

In section 3, we assumed the presence of a secure TRH in every vehicle. As a result, we do not need to consider the security of stored keys in TRH at the time of initialization. However, the security of the keys at the time of updation is very important. Especially the integrity and confidentiality of the new secret key must be provided. Fig. 5 shows the key and group update protocol between a TRH and the TTP. Here, we do not consider a specific session key establishment protocol between a TRH and the TTP. We assume that a secure and authenticated session key $(K_{TRH-TTP})$ has already been established using a secure and

Vehicle(TRH)   **TTP**

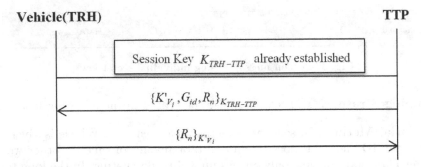

Fig. 5. Key and Group updating protocol

authenticated protocol. For example, secure Diffie-Hellman key exchange protocol may be used. After the $K_{TRH-TTP}$ is established between them, the new
individual secret key (K'_{V_i}), the new group ID (G_{id}), and a random number (R_n)
is sent to the TRH encrypted with $K_{TRH-TTP}$. We assume that this encrypted
message provides integrity of the inner content. This requirement can be satisfied
by various methods, for example, the hash value of the entire inner content may
be included inside the encrypted message. However, the acknowledgment to the
TTP is necessary. If TTP does not receive the acknowledgment to this message,
then TTP must re-execute the protocol again. TRH sends the acknowledgement
to the TTP by encrypting the random number (R_n) with the new updated key
(K'_{V_i}). TTP also keeps both the old and new keys in its database. TTP updates
it database only if it receives acknowledgment from TRH after sending new key
and group ID to the TRH. Note that if this updation is not securely and successfully carried out then the system will not work. In subsection 5.1, we already
discussed that TTP may be through RSU via high-speed links.

6   Evaluation

We Evaluate our proposed scheme according to following aspects.

– **Security**

We will discuss the security of both beacon messages and our key update protocol. Our beacon messages require integrity, privacy, and revocation. If the
privacy of the beacon message is preserved then information in a beacon is of no
value. In other words when the beacon is not carrying any identity information
by which it can be related to the user, the adversary cannot take any advantage
from the information contained in the beacon. That is why we do not consider
confidentiality of beacon messages. Moreover, due to much periodic nature of
the beacons, they do not need any strong type of authentication. From the integrity point of view, HMAC2 computed with K_{all} is included in our beacon
message. When a vehicle receives beacon, it verifies this MAC. Since the messages are assembled inside TRH, it is infeasible for anyone to abuse this MAC
except when TRH is tampered. However, this is not possible by our strong TRH

assumptions. So our beacon messages provide integrity. Our beacon messages also provide weak authentication. In other words, any outsider who does not possess valid TRH cannot generate valid beacon messages.

Now, we will consider the security of our key update protocol. Our key update protocol must provide integrity and confidentiality of the new updated key (K'_{V_i}). This requirement depends on the security of session key establishment protocol between the TTP and a TRH. It also depends on the integrity mechanism used when sending the new updated key (K'_{V_i}). As discussed in subsection 5.2, by our secure session key establishment protocol assumptions, this requirement is fulfilled.

Next, we will discuss the consequences of compromise of various keys. If K_{all} is compromised, it does not affect the privacy of the vehicles, while the compromise of K_{V_i} affects the security of that individual vehicle. More severe problem is that if an adversary obtained K_{all} somehow, it can insert bogus beacon messages into the network which may cause security problems. For example by inserting false position information in beacons, the traffic security messages may be abused. But due to our strong assumptions on TRH, we do not have to consider any kind of leakage of these keys from a TRH. The database of the central TTP is also critical to the security of our system. The security measures for that database are out of the scope of this paper. To provide the security against cryptoanalytic attacks, we change the K_{V_i} periodically.

- **Privacy**

Our proposed beacon messages carry HMAC1 which is calculated with K_{V_i} and can be used by TTP to revoke that corresponding message if needed. The complexity of this revocation is $O(g)$ because TTP has to calculate g number of HMAC1 against each K_{V_i} where g is the size of the group. It is not possible for any other party to revoke the message except when K_{V_i} of the concerned vehicle is obtained. By our TRH assumption, obtaining the key is not possible. If we remove the possibility of identifying the vehicles using HMAC1, there is no other information in the beacon message by which adversary can identify the vehicle. Therefore, our proposed beacon messages provide conditional anonymity of the vehicles.

- **Efficiency**

Now we show the bandwidth and computational efficiency provided by pseudonymless beaconing mechanism. Suppose SHA-256 with 192 bit key is used for HMAC calculation and AES with 192 bit key is used as symmetric cipher, it will provide adequate security. In [27] authors believe that 300 bytes are reasonable for beacon, warnings, and alarm signals. The security overhead in terms of size included in our scheme is 2 x 256 + 16bits (where G_{id} being 2 bytes) = 66 bytes and the total beacon including overhead will be 366 bytes and the security overhead involved is 18% of the message. In our proposed scheme TRH only calculates 2 HMACs for beaconing. Keys updation and management will be done depending upon the system parameter *counter*.

Table 1. Comparison with other systems

Scheme	Need for Mix Zones	Need for GTTP	Profile Generation	Revocation Cost	Beacon Size (Bytes)	Percentage Overhead	Computational Overhead
Scheuer et al.	✓	✓	x	O(n)	377	20%	$2H + 1E$
Plobi et al.	✓	✓	✓	O(1)	370	19%	$2H + 1E$
Our Scheme	x	x	x	O(g)	364	17.58%	$2H$

– Comparison with other Schemes

As discussed in section 2, most of the proposed schemes providing privacy, use mix zones and silent periods which have bad affects on smooth functionality of VANET. Our proposed pseudonymless beaconing mechanism do not use mix zone or silent period. Moreover, most of the schemes in section 2 are making use of asymmetric cryptography and digital signatures, which are computationaly heavier than symmetric cryptography in our scheme.

In table 1, we give a brief comparison of our scheme with Scheuer et al.'s scheme and Plobi et al.'s scheme. Our proposed pseudonymless beaconing mechanism is better than these two schemes in certain aspects like privacy which is not provided by Plobi et al.'s scheme although the revocation cost in Plobi et al.'s scheme is $O(1)$. In addition our proposed scheme is better than Scheuer et al.'s scheme with respect to bandwidth and computational cost. In table 1, H means HMAC calculation and E means symmetric encryption. As beacons are very important for safe driving features and other secure traffic management in VANET and also vehicles send beacons very often (100-300ms), our proposed scheme uses symmetric cryptography which is much faster than asymmetric cryptography and digital signatures. It reduces the overall security overhead thereby improving efficiency. Revocation cost in our scheme is $O(g)$, where g is the size of the group, comparing this cost with Scheuer et al.'s scheme, the cost is $O(n)$ where n is the number of vehicles under a single GTTP. We reduce the cost of revocation using single TTP in place, where in Scheuer et al.'s scheme many GTTPs are involved. In the case of Plobi et al.'s scheme, the cost is $O(1)$ but the scheme does not provide privacy against profile generation.

7   Conclusion

Profile generation is a serious threat to privacy in VANET. Most schemes employ digital signatures and asymmetric cryptography to protect the user privacy in VANET. We considered symmetric cryptography for periodically sent beacons which is computationally efficient and faster than the former approaches. Moreover, to overcome the deficiency of multiple pseudonym based strategy, we proposed pseudonymless strategy which is the best remedy for avoiding profile generation, because if the beacon does not carry any identity information then it is not possible to make movement profiles against a user. We also provide grouping techniques for TTP to reduce the revocation cost. The merits of our system includes better efficiency with respect to bandwidth and computational cost, less space requirements than Scheuer et al.'s scheme and not using any GTTP, mix zone, or silent period.

References

1. Li, C.-T., Hwang, M.-S., Chu, Y.-P.: A Secure and Efficient Communication Scheme with Authenticated Key Establishment and Privacy Preserving for Vehicular Ad Hoc Networks. Computer Communications 31, 2803–2814 (2008)
2. Raya, M., Hubaux, J.-P.: Securing Vehicular Ad Hoc Networks. J. Computer Security 15, 39–68 (2007)
3. Leinmuller, T., Schoch, E., Maihofer, C.: Security Requirements and Solution Concepts in Vehicular Ad Hoc Networks. In: Fourth Annual Conference on Wireless on Demand Network Systems and Services (WONS 2007), pp. 84–91 (2007)
4. Papadimitratos, P., Buttyan, L., Holczer, T., Schoch, E., Freudiger, J., Raya, M., Ma, Z., Kargl, F., Kung, A., Hubaux, J.-P.: Secure Vehicular Communication Systems: Design and Architecture. IEEE Communication Magazine 46, 110–118 (2008)
5. NOW: Network on Wheels Project, http://www.network-on-wheels.de/
6. EPFL Vehicular Networks Security Project, http://ivc.epfl.ch/
7. CALIFORNIA PATH (Partners for Advanced Transit and Highways), http://www.path.berkeley.edu/
8. SeVeCom (Secure Vehicle Communication) Project, http://www.sevecom.org/
9. Rass, S., Fuchs, S., Schaffer, M., Myamakya, K.: How to Protect Privacy in Floating Car Data Systems. In: Fifth ACM international workshop on Vehicular Inter NETworking, pp. 17–22 (2008)
10. Dotzer, F.: Privacy Issues in Vehicular Ad Hoc Networks. In: Danezis, G., Martin, D. (eds.) PET 2005. LNCS, vol. 3856, pp. 197–209. Springer, Heidelberg (2006)
11. Buttyan, L., Holczer, T., Vajda, I.: On the Effectiveness of Changing Pseudonyms to Provide Location Privacy in VANETs. In: Stajano, F., Meadows, C., Capkun, S., Moore, T. (eds.) ESAS 2007. LNCS, vol. 4572, pp. 129–141. Springer, Heidelberg (2007)
12. Beresford, A.R., Stajano, F.: Mix Zones: User Privacy in Location-aware Services. In: Second IEEE Annual Conference on Pervasive Computing and Communications Workshops, pp. 127–131 (2004)
13. Lo, N.-W., Tsai, H.-C.: Illusion Attack on VANET Applications - A Message Plausibility Problem. In: IEEE Globecom Workshops, pp. 1–8 (2007)
14. Burmester, M., Magkos, E., Chrissikopoulos, V.: Strengthening Privacy Protection in VANETs. In: IEEE International Conference on Wireless and Mobile Computing, pp. 508–513 (2008)
15. Schmidt, R.K., Leinmuller, T., Schoch, E., Held, A., Schafer, G.: Vehicle Behavior Analysis to Enhance Security in VANETs. In: Fourth Workshop on Vehicle to Vehicle Communications (V2VCOM 2008), (2008)
16. Chaurasia, B.K., Verma, S., Bhasker, S.M.: Maximizing Anonymity of a Vehicle. In: Fourth International Conference on Wireless Communication and Sensor Networks, pp. 95–98 (2008)

17. Chaurasia, B.K., Verma, S., Bhasker, S.M.: Message broadcast in VANETs using Group Signature. In: Fourth International Conference on Wireless Communication and Sensor Networks, pp. 131–136 (2008)
18. Schoch, E., et al.: Impact of Pseudonym Changes on Geographic Routing in VANETs. In: Buttyán, L., Gligor, V.D., Westhoff, D. (eds.) ESAS 2006. LNCS, vol. 4357, pp. 43–57. Springer, Heidelberg (2006)
19. Liu, B., Zhong, Y., Zhang, S.: Probabilistic Isolation of Malicious Vehicles in Pseudonym Changing VANETs. In: 7th IEEE International Conference on Computer and Information Technology, pp. 967–972 (2007)
20. Gerlach, M., Guttler, F.: Privacy in VANETs using Changing Pseudonyms - Ideal and Real. In: IEEE 65th Vehicular Technology Conference, pp. 2521–2525 (2007)
21. Scheuer, F., Plobi, K., Federrath, H.: Preventing Profile Generation in Vehicular Networks. In: IEEE International Conference on Wireless and Mobile Computing, pp. 520–525 (2008)
22. Fonseca, E., Festag, A., Baldessari, R., Aguiar, R.L.: Support of Anonymity in VANETs - Putting Pseudonymity into Practice. In: IEEE Wireless Communications and Networking Conference (WCNC 2007), pp. 3400–3405 (2007)
23. Calandriello, G., Papadimitratos, P., Hubaux, J.-P., Lioy, A.: Efficient and Robust Pseudonymous Authentication in VANET. In: Fourth ACM International Workshop on Vehicular ad Hoc Networks, pp. 19–28 (2007)
24. Ma, Z., Kargl, F., Weber, M.: Pseudonym-on-demand: A New Pseudonym Refill Strategy for Vehicular Communications. In: IEEE 68th Vehicular Technology Conference (VTC 2008), pp. 1–5 (2008)
25. Chaurasia, B.K., Verma, S.: Optimizing Pseudonym Updation for Anonymity in VANETS. In: IEEE Asia-Pacific Services Computing Conference, pp. 1633–1637 (2008)
26. Eichler, S.: Strategies for Pseudonym Changes in Vehicular Ad Hoc Networks depending on Node Mobility. In: IEEE Intelligent Vehicles Symposium, pp. 541–546 (2007)
27. Plobi, K., Federrath, H.: A Privacy Aware and Efficient Security Infrastructure for Vehicular Ad Hoc Networks. Computer Standards and Interfaces 30, 390–397 (2008)

A Selectable k-Times Relaxed Anonymous Authentication Scheme

Keita Emura, Atsuko Miyaji, and Kazumasa Omote

School of Information Science, Japan Advanced Institute of Science and Technology,
1-1, Asahidai, Nomi, Ishikawa, 923-1292, Japan
{k-emura,miyaji,omote}@jaist.ac.jp

Abstract. In a k-Times Anonymous Authentication (k-TAA) scheme,
Application Providers (APs) authenticates each group member in k times.
A user can preserve his/her own privacy and an AP can restrict the number
of used services to just k-times, since users are identified if they access an
AP more than k times. In all previous schemes, each AP assumes the same
k for all users. Added to this, total anonymity of unlinkability requires
large computation amount such as pairing computations compared with
non-anonymous schemes. In this paper, we propose a selectable k-TAA
scheme with relaxed anonymity. Relaxed anonymity is an intermediate
level of privacy required between total anonymity and linkability, where
authentication executions for the same AP are linkable, but an authentication execution with an AP v_0 and an authentication execution with an
AP v_1 ($v_0 \neq v_1$) are unlinkable. Our authentication algorithm is efficient
than existing ones thanks to this relaxed notion.

1 Introduction

In the recent information society, there are many types of applications. SaaS
(Software as a Service) [20], a kind of providing service, has become popular to
save cost of time such as an install and expense to use software. In the SaaS
environment, a program (which provides the service) runs on the server of the
Application Provider (AP), and the user is provided the service from the AP
using an on-line network. Recently, there are many SaaS-type services such as
the Oracle Database SaaS Platform [16], the IBM LotusLive [11], and so on. A
user is forced to manage many assumed names together with passwords since
each service requires separate account. Recently, OpenID has been introduced
(by Google, Microsoft, IBM, Verisign and so on) to reduce such a managing cost.
In OpenID, a user with only manages one account can be authenticated from
many APs. However, a new issue that access to services (these indicate a user's
tastes and habits) is exposed to different APs by the same OpenID. Therefore,
anonymous authentication schemes based on group signature schemes [2,5] are
required, where users are authenticated whether they are members of the group
or not. Group signatures are unlinkable, namely anybody cannot decide whether
an authenticated user is the same authenticated user or not. However, a simple
group signature scheme is not suitable in the SaaS environment, in which the

H.Y. Youm and M. Yung (Eds.): WISA 2009, LNCS 5932, pp. 281–295, 2009.

AP cannot demand a service fee by detecting each user. This is why k-Times Anonymous Authentication (k-TAA) schemes [1,13,14,17,18] are attractive for the SaaS environment, where each group member is anonymously authenticated by APs in k times. A user can preserve his/her own privacy and an AP can restrict the number of used services to just k-times, since users are identified if they access an AP more than k times.

Previous k-TAA Schemes: The first k-TAA scheme has been proposed in [17]. There are three entities in a k-TAA scheme, a Group Manager (GM), users and APs. The GM manages a group, and issues membership certificates for users. The first k-TAA scheme is constructed by a group signature scheme with the *tracing tag mechanism*: an AP opens k tag bases. A user makes a tag by using his/her secret key and one of the tag bases, which has not been used before. The user makes an authentication proof including the tag. The AP stores the authentication execution transcript. If the user attempts authentication more than k times, then he/her can be identified the same tag base. This tracing tag mechanism is used by other k-TAA schemes [1,13,14,18]. In all previous k-TAA schemes, each AP assumes the same k for all users. Therefore, an AP cannot decide the number of access of services for each user. Total anonymity of unlinkability requires large computation amount such as pairing computations compared with non-anonymous schemes. Furthermore, even if APs want to obtain the number of accesses by each user to improve the application providing strategy, e.g., long tail marketing [21], APs cannot obtain the number of accesses. For example, assume that any user accesses less than k-times and AP knows the total number of accesses is 10000, then AP cannot distinguish whether 10000 users access only once or 100 users access 100 times. On the other hands, by using a linkable authentication, access to services (these indicate a user's tastes and habits) is exposed among different APs. Therefore, an intermediate level of privacy required between total anonymity and linkability is necessary to preserve privacy and to satisfy information utilization, simultaneously.

Our Contribution: In this paper, we propose a selectable k-Times Relaxed Anonymous Authentication (k-TRAA) scheme that enables an allowable number to be assigned for each user. There are some suitable scenarios, where different allowable numbers are decided for each user. For example, a user who spends much money has to be given the right to more accesses to AP compared to other users. This is a natural requirement in the SaaS environment. We introduce a relaxed security notion called relaxed anonymity, which satisfies two kinds of anonymity: (1) two authentication executions with the same AP are linkable, but the AP cannot identify a user from the group of users, and (2) an authentication execution with an AP v_0 and an authentication execution with an AP v_1 ($v_0 \neq v_1$) are unlinkable. Relaxed anonymity is an intermediate level of privacy required between total anonymity and linkability. Under relaxed anonymity, a user's taste is not exposed among different APs. Otherwise, an AP can obtain the number of accesses of own application from each user. Our authentication algorithm is more efficient than previous k-TAA schemes such that no pairing computation is required and computation is independent to the allowable number k in authentication phase, where these large computations are concentrated in grant phase. Moreover, in our authentication algorithm, full off-line computation is available. We insist that these efficiencies are due to relaxed anonymity offering a tradeoff between privacy preservation and efficiency. To realize these efficiencies, our scheme is made from both a group signature scheme [6] and the newly introduced *sequence-of-zero-knowledge-proof mechanism*.

Organization: Some definitions are given in Section 2. Our scheme is presented in Section 3. The efficiency comparison is discussed in Section 4. Security analyses are presented in Section 5.

2 Definitions

In this section, we define complexity assumptions, the model of selectable k-TRAA and security requirements. Note that $x \in_R S$ means x is randomly chosen for a set S.

2.1 Bilinear Groups and Complexity Assumptions

Definition 1. (Bilinear Groups)

1. $\mathbb{G}_1$, $\mathbb{G}_2$ and $\mathbb{G}_3$ *are cyclic groups of prime order p.*
2. g_1 *and* g_2 *are generators of $\mathbb{G}_1$ and $\mathbb{G}_2$, respectively.*
3. ψ *is an efficiently computable isomorphism $\mathbb{G}_2 \to \mathbb{G}_1$ with $\psi(g_2) = g_1$.*
4. *e is an efficiently computable bilinear map $e : \mathbb{G}_1 \times \mathbb{G}_2 \to \mathbb{G}_3$ with the following properties.*
   - *Bilinearity : for all $u, u' \in \mathbb{G}_1$ and $v, v' \in \mathbb{G}_2$, $e(uu', v) = e(u, v)e(u', v)$ and $e(u, vv') = e(u, v)e(u, v')$.*
   - *Non-degeneracy : $e(g_1, g_2) \neq 1_{\mathbb{G}_3}$ ($1_{\mathbb{G}_3}$ is the $\mathbb{G}_3$'s unit).*

Our scheme is based on the q-strong Diffie-Hellman (q-SDH) [3], k-Power Computational Diffie-Hellman (k-PDDH) [10], and k-Power Decisional Diffie-Hellman (k-PDDH) [10] assumptions. For the security parameter λ, let $\epsilon = \epsilon(\lambda)$ be a negligible function, namely for every polynomial $poly(\cdot)$ and for sufficiently large λ, $\epsilon(\lambda) < 1/poly(\lambda)$.

Definition 2. (q-SDH assumption [3]) *The q-SDH problem in $(\mathbb{G}_1, \mathbb{G}_2)$ is a problem, for input of a $(q+2)$ tuple $(g, g', (g')^\xi, \cdots, (g')^{\xi^q}) \in \mathbb{G}_1 \times \mathbb{G}_2^{q+1}$, where $g = \psi(g')$, to compute a tuple $(x, g^{1/(\xi+x)})$. An algorithm $\mathcal{A}$ has an advantage ϵ in solving the q-SDH problem in $(\mathbb{G}_1, \mathbb{G}_2)$ if $\Pr[\mathcal{A}(g, g', (g')^\xi, \cdots, (g')^{\xi^q}) = (x, g^{1/(\xi+x)})] \geq \epsilon$. We say that the q-SDH assumption holds in $(\mathbb{G}_1, \mathbb{G}_2)$ if no PPT algorithm has an advantage of at least ϵ in solving the q-SDH problem in $(\mathbb{G}_1, \mathbb{G}_2)$.*

Definition 3. (k-PCDH [10]) *The k-PCDH problem in $\mathbb{G}_1$ is a problem, for input a tuple $(h', h, h^y, h^{y^2}, \ldots, h^{y^{k-1}}) \in \mathbb{G}_2 \times \mathbb{G}_1^k$, where $h = \psi(h')$, to compute h^{y^k}. An algorithm $\mathcal{A}$ has an advantage ϵ in solving the k-PCDH problem in $\mathbb{G}_1$*

if $\Pr[\mathcal{A}(h', h, h^y, \ldots, h^{y^{k-1}}) = h^{y^k}] \geq \epsilon$. *We say that the* k-*PCDH assumption holds in* $\mathbb{G}_1$ *if no PPT algorithm has an* ·*advantage of at least* ϵ *in solving the* k-*PCDH problem in* $\mathbb{G}_1$.

A decisional version of k-PCDH assumption (k-PDDH assumption) is simply defined such that $|\Pr[\mathcal{A}(h', h, h^y, \ldots, h^{y^k}) = 0] - \Pr[\mathcal{A}(h', h, h_1, \ldots, h_k) = 0]|$ is negligible, where $h_1, \ldots, h_k \in \mathbb{G}_1$. Note that a 3-PDDH problem instance $(h, h^y, h^{y^2}, h^{y^3})$ is a DDH (Decisional Diffie-Hellman[1]) tuple $(h, h^{y^2}, h^y, (h^{y^2})^y) = (h, h'', h^y, (h'')^y)$, where $h'' := h^{y^2}$. This means that if the DDH problem is easy to solve, then the k-PDDH problem ($k \geq 3$) is also easy to solve. Therefore, to hold the k-PDDH assumption in $\mathbb{G}_1$, we require that the k-PDDH assumption holds in $\mathbb{G}_1$. It is a stronger assumption than the eXternal Diffie-Hellman (XDH) assumption [4] which only requires that the DDH assumption holds in $\mathbb{G}_1$. We can use MNT curves [12], where there is no efficient isomorphism between $\mathbb{G}_1$ to $\mathbb{G}_2$.

2.2   Model of Selectable k-Times Relaxed Anonymous Authentication

Let λ be the security parameter, GM the group manager, AP the application provider, gpk the group public key, gsk the group secret key which is used for issuing a membership certificate, (mpk_i, msk_i) the member public/secret key of U_i ($i = 1, 2, \ldots, n$), LIST an identification list for tracing, Log_v the log list for keeping logs, and ID_v an identity of AP v. To simplify, we describe an allowable number k_i, although at times we describe $k_{v,i}$ for an AP v and a user U_i.

Definition 4. *System operations of a Selectable* k-*TRAA*

- GM-Setup(1^λ): *This algorithm takes as input* λ *and returns* gpk, gsk *and* LIST.
- AP-Setup(ID_v): *This algorithm takes as input* ID_v *and returns* Log_v.
- Join(Join-GM$\langle gpk, gsk, \mathsf{LIST}\rangle$, Join-U$\langle gpk\rangle$): *This algorithm takes as input* gpk, gsk, LIST *and* upk_i *from* GM, *and* gpk, upk_i *and* usk_i *from* U_i, *and returns* (mpk_i, msk_i), *and appends* mpk_i *and the user's identity* i *to* LIST *making the updated* LIST
- Grant(Grant-AP$\langle gpk, Log_v\rangle$, Grant-U$\langle gpk, msk_i, k_i\rangle$): *This algorithm takes as input* gpk *and* Log_v *from AP, and* gpk, msk_i *and* k_i *from* U_i. *The transcript of grants are recorded in* $reg_d \in Log_v$. *Note that* $d \in \mathbb{Z}_{>0}$ *is just an indexed number in the log, namely,* d *is independent of the user's identity* i.
- Auth(Verify$\langle gpk, Log_v\rangle$, Proof$\langle gpk, msk_i\rangle$): *Let* U_i *be assigned with the indexed number* d. *This algorithm takes as input* gpk *and* Log_v *from AP, and* gpk *and* msk_i *from* U_i, *and outputs* accept *when conditions (1) and (2) are satisfied: (1) the anonymous user is a member of the group, and (2) the anonymous user has already been authenticated less than* k *times, otherwise, outputs* reject. *When the output of this algorithm is* accept, *this algorithm records the transcript of authentications in* $reg_d \in Log_v$.

---

[1] Note that the DDH problem is a problem, for input a tuple $(g, g', g^u, (g')^v)$, where $g, g' \in \mathbb{G}_1$ and $u, v \in \mathbb{Z}_p^*$, to decide $u = v$ or not.

– Trace(gpk, Log_v): *This algorithm takes as input gpk and Log_v, and computes a user's public key mpk_i for a user who has already been authenticated more than k_i times. If the entry (i, mpk_i) is included in LIST, then it outputs i. Otherwise, the algorithm verifies all proofs included in Log_v. If all proofs are valid, then it outputs "GM". Otherwise, if a proof is invalid, then it outputs "AP".*

2.3 Security Definitions

In this subsection, we define relaxed anonymity, which is an intermediate level of privacy required between total anonymity and linkability. $\mathcal{A}$ is admitted to collude with users, GM and APs, respectively. Then $\mathcal{A}$ can play the role of these entities although accessible oracles are restricted. We define oracles as follows:

Oracles: We use oracles $\mathcal{O}_{\text{List}}$, $\mathcal{O}_{\text{Join-GM}}$, $\mathcal{O}_{\text{Join-U}}$, $\mathcal{O}_{\text{AP-Setup}}$, $\mathcal{O}_{\text{Proof}}$, $\mathcal{O}_{\text{Verify}}$, $\mathcal{O}_{\text{Grant-AP}}$, $\mathcal{O}_{\text{Grant-U}}$ and $\mathcal{O}_{\text{Query}}$. Each definition is as follows: the list oracle $\mathcal{O}_{\text{List}}$ [17,18] manages LIST. An adversary $\mathcal{A}$ is allowed to read LIST to call $\mathcal{O}_{\text{List}}$. If $\mathcal{A}$ colludes with some users, then $\mathcal{A}$ can write corresponding entries of LIST. In addition, if $\mathcal{A}$ colludes with the GM, then $\mathcal{A}$ can delete entries of LIST. $\mathcal{O}_{\text{Join-GM}}$ is the oracle which runs the Join-GM algorithm honestly. $\mathcal{O}_{\text{Join-U}}$ is the oracle which runs the Join-U algorithm on behalf of honest users. $\mathcal{O}_{\text{AP-Setup}}$ is the oracle which runs the AP-Setup algorithm honestly. $\mathcal{O}_{\text{Proof}}$ is the oracle which runs the Proof algorithm on behalf of honest user. $\mathcal{O}_{\text{Verify}}$ is the oracle which runs the Verify algorithm on behalf of honest APs. $\mathcal{O}_{\text{Grant-AP}}$ is the oracle which runs the Grant-AP algorithm on behalf of honest APs. $\mathcal{O}_{\text{Grant-U}}$ is the oracle which runs the Grant-U algorithm on behalf of honest users. $\mathcal{O}_{\text{Query}}$ is the challenge oracle which is defined in Definition 5. We describe situations when $\mathcal{A}$ is allowed to access the oracles as follows: $\mathcal{A}$ is always allowed to access $\mathcal{O}_{\text{List}}$, to access $\mathcal{O}_{\text{Join-GM}}$ if $\mathcal{A}$ does not collude with the GM, to access $\mathcal{O}_{\text{Join-U}}$, $\mathcal{O}_{\text{Grant-U}}$ and $\mathcal{O}_{\text{Proof}}$ if $\mathcal{A}$ does not collude with the user, and to access $\mathcal{O}_{\text{Grant-AP}}$ and $\mathcal{O}_{\text{Verify}}$ if $\mathcal{A}$ does not collude with the AP.

Next, we define the relaxed anonymity game. We refer to the L-anonymity game for linkable ring signature [19], namely, target users U_{i_0} and U_{i_1} are not input in $\mathcal{O}_{\text{Proof}}$ by an adversary.

Definition 5. Relaxed anonymity : *Relaxed anonymity requires that for all PPT $\mathcal{A}$, the advantage of $\mathcal{A}$ in the following game be negligible.*

An adversary $\mathcal{A}$ is allowed to collude with the GM, all APs, and all users except target users U_{i_0} and U_{i_1}, and to query oracles $\mathcal{O}_{\text{List}}$, $\mathcal{O}_{\text{Join-U}}$, $\mathcal{O}_{\text{Proof}}$, $\mathcal{O}_{\text{Grant-U}}$ and $\mathcal{O}_{\text{Query}}$. U_{i_0} and U_{i_1} are not input in $\mathcal{O}_{\text{Proof}}$ by $\mathcal{A}$, and can be input in $\mathcal{O}_{\text{Join-U}}$, and have not been granted by an AP v_0 and an AP v_1. Let k_{v_0,i_0} and k_{v_0,i_1} (resp. k_{v_1,i_0} and k_{v_1,i_1}) be allowable numbers of U_{i_0} and U_{i_1} for the AP v_0 (resp. the AP v_1). When $\mathcal{O}_{\text{Query}}$ is called by $\mathcal{A}$ for the first time, $\mathcal{O}_{\text{Query}}$ randomly chooses $b \in \{0, 1\}$, executes $\mathcal{O}_{\text{Grant-U}}(U_{i_0})$ with the AP v_0 and $\mathcal{O}_{\text{Grant-U}}(U_{i_b})$ with the AP v_1, and outputs the Log_{v_0} and Log_{v_1}. From the second call, $\mathcal{O}_{\text{Query}}$ executes both $\mathcal{O}_{\text{Proof}}(U_{i_0})$ with the AP v_0 and $\mathcal{O}_{\text{Proof}}(U_{i_b})$ with the AP v_1, and outputs the Log_{v_0}, Log_{v_1} and the transcript of the authentication protocol.

$\mathcal{O}_{\mathsf{Query}}$ can be used less than k times, where $k = Min\{k_{v_0,i_0}, k_{v_0,i_1}, k_{v_1,i_0}, k_{v_1,i_1}\}$. $\mathcal{A}$ outputs a bit b', and wins if $b' = b$. The advantage of $\mathcal{A}$ is defined as $Adv^{R\text{-}anon}(\mathcal{A}) = |\Pr(b = b') - \frac{1}{2}|$.

If $Adv^{R\text{-}anon}(\mathcal{A})$ is negligible, then $\mathcal{A}$ cannot distinguish authentications of U_{i_0} (namely $\mathcal{O}_{\mathsf{Proof}}(U_{i_0})$) with the AP v_0 from authentications of U_{i_b} (namely $\mathcal{O}_{\mathsf{Proof}}$ (U_{i_b})) with the different AP v_1. This means $\mathcal{A}$ cannot determine whether U_{i_b} is U_{i_0} or not. Note that the definition of relaxed anonymity does not guarantee that two authentications with the same AP are unlinkable. Therefore, definitions of relaxed anonymity captures the two properties (1) two authentications with the same AP are linkable, but AP v cannot determine a user from the group of users, and (2) an authentication with an AP v_0 and an authentication with an AP v_1 are unlinkable.

Next, we define the detectability game. This definition captures the fact that all users cannot execute Auth algorithm more than the allowable number of times, or they have to be detected by the Trace algorithm.

Definition 6. Detectability : *Detectability requires that for all PPT $\mathcal{A}$, the advantage of $\mathcal{A}$ in the following game be negligible.*

An adversary $\mathcal{A}$ is allowed to collude with all users. Let N be the number of users who colluded with $\mathcal{A}$, k_i be an allowable number of a user U_i ($i = 1, 2, \ldots, N$), and $K = \sum_{i=1}^{N} k_i$ be the total of allowable numbers. $\mathcal{A}$ wins if $\mathcal{A}$ can be accepted more than K times. The advantage of $\mathcal{A}$ is defined as the probability that $\mathcal{A}$ wins.

The difference between our detectability game and the previous detectability games (defined in [1,13,14,18]) is that the previous game uses $K = Nk$, since all users are forced to use the same allowable number k.

Next, we define the exculpability games for users, GM and AP. These definitions are the same as in [18].

Definition 7. Exculpability for users : *Exculpability for users requires that for all PPT $\mathcal{A}$, the advantage of $\mathcal{A}$ in the following game be negligible.*

An adversary $\mathcal{A}$ is allowed to collude with all entities except a target user U^*. $\mathcal{A}$ can run $\mathcal{O}_{\mathsf{Proof}}$ less than k^* times, where k^* is an allowable number of U^*. $\mathcal{A}$ wins if $\mathcal{A}$ can compute the authentication log, which is the input of the tracing algorithm that outputs the identification of U^*. The advantage of $\mathcal{A}$ is defined as the probability that $\mathcal{A}$ wins.

Definition 8. Exculpability for GM : *Exculpability for GM requires that for all PPT $\mathcal{A}$, the advantage of $\mathcal{A}$ in the following game be negligible.*

An adversary $\mathcal{A}$ is allowed to collude with all entities except the GM. $\mathcal{A}$ wins if $\mathcal{A}$ can compute the authentication log which is the input of the tracing algorithm that outputs GM. The advantage of $\mathcal{A}$ is defined as the probability that $\mathcal{A}$ wins.

Definition 9. Exculpability for AP : *Exculpability for AP requires that for all PPT $\mathcal{A}$, the advantage of $\mathcal{A}$ in the following game be negligible.*

An adversary $\mathcal{A}$ is allowed to collude with all entities except a target AP v^*. $\mathcal{A}$ wins if $\mathcal{A}$ can compute the authentication log which is the input of tracing algorithm that outputs AP v^*. The advantage of $\mathcal{A}$ is defined as the probability that $\mathcal{A}$ wins.

2.4 Proving Relations on Representations

By using the Fiat-Shamir heuristic [8], a digital signature scheme is constructed from zero-knowledge proofs of knowledge (ZK). These signatures are called SPKs. For a message M and secret values $(x_1, \ldots, x_n)$, we use the notation $SPK\{(x_1, \ldots, x_n) : R(x_1, \ldots, x_n)\}(M)$, which is a signature of M signed by a signer who has $(x_1, \ldots, x_n)$ satisfying the relation $R(x_1, \ldots, x_n)$. SPKs can be simulated without the knowledge of $(x_1, \ldots, x_n)$ in the random oracle model.

3 Proposed Scheme

In this section, we first introduce the underlying idea of our construction, and then propose a selectable k-TRAA scheme.

3.1 A Primitive Selectable k-TAA Scheme

Here we give a simple construction of a selectable k-TAA scheme based on the previous k-TAA scheme: The AP makes k_i tag bases, where k_i is an allowable number of a user U_i, and issues these bases to U_i regarding his/her secret keys for exclusive use. In this simple scheme, the number of the secret key of U_i depends on k_i. As another construction, the AP makes k_i tag bases by using a pseudo-random number for U_i, and opens these bases to U_i regarding his/her public keys for exclusive use. In this primitive scheme, the number of the public key depends on $N \sum_{i=1}^{N} k_i$, where N be the number of application group members. Our purpose is to achieve that the number of both the number of the secret key of each U_i and the number of the public key is independent of k_i and N.

3.2 Underlying Idea of Our Construction

In this subsection, we introduce a higher-level description of the key ideas called the sequence-of-zero-knowledge-proof mechanism. This idea is similar to a source authentication using hash chaining proposed in GR97 [9]. For the sake of clarity, we summarize the GR97 scheme as follows: Let H be a one-way hash function and Sign be a signing algorithm of a digital signature scheme: The sender splits a data to be signed to m blocks $Data_1, Data_2, \ldots, Data_m$, computes the hash chain $h_m := H(Data_m)$, $h_{m-1} := H(Data_{m-1} \| h_m)$, $\ldots$, $h_1 := H(Data_1 \| h_2)$, and makes a signature of h_1, $\sigma := \text{Sign}(h_1)$, and sends σ, h_1, $Data_1 \| h_2$, $\ldots$, $Data_{m-1} \| h_m$, $Data_m$. The receiver verifies σ, and checks $H(Data_1 \| h_2) \overset{?}{=} h_1$, $H(Data_2 \| h_3) \overset{?}{=} h_2 \ldots$, $H(Data_{m-1} \| h_m) \overset{?}{=} h_{m-1}$ and $H(Data_m) \overset{?}{=} h_m$. Only one signature verification is performed and only one hash is computed for every data block.

The concept of our scheme is the same as the GR97 scheme. Specifically, in the grant phase, a user proves that both (1) the user has a valid membership certificate by using a group signature (which requires pairing computations) and (2) a commitment is computed by using a part of the membership certificate (which requires the computations depend on the allowable number). In each authentication phase, the user only has to prove that a commitment is computed by using the same secret key which has already been used in the previous authentication phase.

3.3 Proposed Scheme

Let $NIZK$ be a Non-Interactive Zero-Knowledge proof and SPK be a Signatures based on Proofs of Knowledge. To detect a user U_i after k_i times authentication, a polynomial with degree $k_i + 2$ is applied. Let $f_{k_i}(X) = \prod_{j=1}^{k_i+2}(X+j) = \sum_{j=0}^{k_i+2} a_j X^j \in \mathbb{Z}_p[X]^2$. Let $M \in \{0,1\}^*$ be a message for deciding an allowable number k_i. It is assumed that the user cannot be identified by AP from M and k_i. The concurrently secure Join algorithm proposed in DP06 [6] is used in our Join algorithm, where all proofs are non-interactive using $NIZK$ and a signature scheme $DSig$. The verifying/signing key (upk_i, usk_i) of the signature scheme $DSig$ is used in Join.

- GM-Setup(1^λ)
  1. The GM selects cyclic groups of $\mathbb{G}_1$, $\mathbb{G}_2$ and $\mathbb{G}_3$ with λ-bits of prime order p, an isomorphism $\psi : \mathbb{G}_2 \to \mathbb{G}_1$, a bilinear map $e : \mathbb{G}_1 \times \mathbb{G}_2 \to \mathbb{G}_3$, and a hash function $\mathcal{H} : \{0,1\}^* \to \mathbb{Z}_p$.
  2. The GM chooses $\tilde{g} \in_R \mathbb{G}_1$ and a generator $g_2 \in \mathbb{G}_2$, and sets $g_1 = \psi(g_2)$.
  3. The GM chooses $\gamma \in_R \mathbb{Z}_p$, and computes $\omega = g_2^\gamma$.
  4. The GM outputs $gpk = (\mathbb{G}_1, \mathbb{G}_2, \mathbb{G}_3, e, \mathcal{H}, g_1, g_2, \tilde{g}, \omega)$ and $gsk = (\gamma)$.
- AP-Setup(ID_v)
  1. AP v outputs $Log_v = \emptyset$.
- Join(Join-GM$\langle gpk, gsk, \mathsf{LIST}, upk_i \rangle$, Join-U$\langle gpk, upk_i, usk_i \rangle$)
  1. U_i chooses $y_i \in_R \mathbb{Z}_p$, and computes $F_i = \tilde{g}^{y_i}$ and $\pi_1 = NIZK\{y_i : F_i = \tilde{g}^{y_i}\}$.
  2. U_i sends F_i and π_1 to the GM.
  3. The GM checks π_1. If π_1 is not valid, then aborts.
  4. The GM chooses $x_i \in_R \mathbb{Z}_p$, and computes $A_i = (g_1 F_i)^{1/(\gamma + x_i)}$, $B_i = e(g_1 F_i, g_2)/e(A_i, \omega)$, $E_i = e(A_i, g_2)$ and $\pi_2 = NIZK\{x_i : B_i = E_i^{x_i}\}$.
  5. The GM sends A_i, B_i, E_i and π_2 to U_i.
  6. U_i checks π_2. If π_2 is not valid, then aborts.
  7. U_i makes $S_{i,A_i} = DSig_{usk_i}(A_i)$, and sends S_{i,A_i} to the GM.
  8. The GM verifies S_{i,A_i} with respect to upk_i and A_i. If S_{i,A_i} is valid, then the GM sends x_i to U_i, and adds $(i, mpk_i = g_1^{x_i})$ to LIST.

---

[2] If k_i is known, then coefficients $\{a_j\}_{j=0}^{k_i+2}$ can be computed easily.

9. U_i checks $e(A_i, g_2)^{x_i} e(A_i, w) e(\tilde{g}, g_2)^{-y_i} \overset{?}{=} e(g_1, g_2)$ to verify the relation $A_i^{(x_i+\gamma)} = g_1 \tilde{g}^{y_i}$.

- Grant(Grant-AP$\langle gpk, Log_v \rangle$, Grant-U$\langle gpk, msk_i, k_i \rangle$)

1. U_i chooses $\alpha, \beta, \gamma \in_R \mathbb{Z}_p$, and computes $C_1 = A_i \tilde{g}^\alpha$, $C_2 = mpk_i(g_1^{f_{k_i}(y_i)})^\beta = g_1^{x_i+\beta \sum_{j=0}^{k_i+2} a_j y_i^j}$, $C_{3,1} = g_1^\beta$, $C_{3,2} = C_{3,1}^{y_i}$, $C_{4,1} = g_1^\gamma$, $C_{4,2} = C_{4,1}^{y_i}$, $C_{4,3} = C_{4,2}^{y_j}$, ..., $C_{4,k_i+2} = C_{4,k_i+1}^{y_j}$ and $h_j = g_1^{a_j \beta}$ $(j \in [0, k_i+2])$.

2. U_i sets $\tau = \alpha x_i$, $z_{i,j} = y_i^j$ $(j \in [0, k_i+2])$, and computes $\pi_c = SPK\{(\alpha, \beta, x_i, y_i, \tau, z_{i,1}, z_{i,2}, \ldots, z_{i,k_i+2})$: $\frac{e(C_1, \omega)}{e(g_1, g_2)} = \frac{e(\tilde{g}, g_2)^\tau \cdot e(\tilde{g}, g_2)^{y_i} \cdot e(\tilde{g}, \omega)^\alpha}{e(C_1, g_2)^{x_i}} \wedge C_2 = g_1^{x_i} \prod_{j=0}^{k_i+2} h_j^{z_{i,j}} \wedge C_{3,1} = g_1^\beta \wedge C_{3,2} = C_{3,1}^{y_i} \wedge C_{4,1} = g_1^\gamma \wedge C_{4,2} = C_{4,1}^{y_i} = C_{4,1}^{z_{i,1}} \wedge \ldots \wedge C_{4,k_i+2} = C_{4,k_i+1}^{y_i} = C_{4,1}^{z_{i,k_i+2}} \wedge h_0 = g_1^{a_0 \beta} \wedge \ldots \wedge h_{k_i+2} = g_1^{a_{k_i+2}\beta}\}(ID_v, k_i, M)$.

Concretely, U_i computes π_c as follows:

(a) U_i chooses $r_\alpha, r_\beta, r_\gamma, r_{x_i}, r_{y_i}, r_\tau, r_{z_{i,2}}, \ldots, r_{z_{i,k_i+2}} \in_R \mathbb{Z}_p$. Note that $r_{z_{i,0}}$ and $r_{z_{i,1}}$ are not necessary since $z_{i,0} = 1$ and $z_{i,1} = y_i$.

(b) U_i computes $R_1 = \frac{e(\tilde{g}, g_2)^{r_\tau} e(\tilde{g}, g_2)^{r_{y_i}} e(\tilde{g}, \omega)^{r_\alpha}}{e(C_1, g_2)^{r_{x_i}}}$, $R_2 = g_1^{r_{x_i}} h_0 h_1^{r_{y_i}} \prod_{j=2}^{k_i+2} h_j^{r_{z_{i,j}}}$, $R_{3,1} = g_1^{r_\beta}$, $R_{3,2} = C_{3,1}^{r_{y_i}}$, $R_{4,1} = g_1^{r_\gamma}$, $R_{4,2} = C_{4,1}^{r_{y_i}}$, $R_{4,3} = C_{4,2}^{r_{y_i}}$, $R'_{4,3} = C_{4,1}^{r_{z_{i,2}}}$, $R_{4,4} = C_{4,3}^{r_{y_i}}$, $R'_{4,4} = C_{4,1}^{r_{z_{i,3}}}$, ..., $R_{4,k_i+2} = C_{4,k_i+1}^{r_{y_i}}$, $R'_{4,k_i+2} = C_{4,1}^{r_{z_{i,k_i+2}}}$, $R_{5,0} = h_0^{r_\beta}$, $R_{5,1} = h_1^{r_\beta}$, ..., $R_{5,k_i+1} = h_{k_i+1}^{r_\beta}$ and $R_{5,k_i+2} = h_{k_i+2}^{r_\beta}$.

(c) U_i computes $c = \mathcal{H}(gpk, ID_v, k_i, M, C_1, C_2, C_{3,1}, C_{3,2}, C_{4,1}, \ldots, C_{4,k_i+2}, R_1, R_2, R_{3,1}, R_{3,2}, R_{4,1}, \ldots, R_{4,k_i+2}, R'_{4,3}, \ldots, R'_{4,k_i+2}, R_{5,1}, \ldots, R_{5,k_i+2}, h_0, \ldots, h_{k_i+2})$.

(d) U_i computes $s_\alpha = r_\alpha + c\alpha$, $s_\beta = r_\beta + c\beta$, $s_\gamma = r_\gamma + c\gamma$, $s_{x_i} = r_{x_i} + c x_i$, $s_{y_i} = r_{y_i} + c y_i$, $s_\tau = r_\tau + c\tau$, $s_{z_{i,2}} = r_{z_{i,2}} + c z_{i,2}$, ..., $s_{z_{i,k_i+2}} = r_{z_{i,k_i+2}} + c z_{i,k_i+2}$.

3. U_i sends $(k_i, M, C_1, C_2, C_{3,1}, C_{3,2}, C_{4,1}, \ldots, C_{4,k_i+2}, h_0, \ldots, h_{k_i+2}, \pi_c = (c, s_\alpha, s_\beta, s_\gamma, s_{x_i}, s_{y_i}, s_\tau, s_{z_{i,2}}, \ldots, s_{z_{i,k_i+2}}))$ to AP v.

4. If AP v cannot accept k_i and M, then it outputs reject. Otherwise, AP v checks π_c as follows:

(a) AP v computes $\tilde{R}_1 = \frac{e(\tilde{g}, g_2)^{s_\tau} \cdot e(\tilde{g}, g_2)^{s_{y_i}} \cdot e(\tilde{g}, \omega)^{s_\alpha}}{e(C_1, g_2)^{s_{x_i}}} \left(\frac{e(g_1, g_2)}{e(C_1, \omega)} \right)^c$, $\tilde{R}_2 = g_1^{s_{x_i}} h_0 h_1^{r_{y_i}} \prod_{j=2}^{k_i+2} h_j^{r_{z_{i,j}}} C_2^{-c}$, $\tilde{R}_{3,1} = g_1^{s_\beta} C_{3,1}^{-c}$, $\tilde{R}_{3,2} = C_{3,1}^{s_{y_i}} C_{3,2}^{-c}$, $\tilde{R}_{4,1} = g_1^{s_\gamma} C_{4,1}^{-c}$, $\tilde{R}_{4,2} = C_{4,1}^{s_{y_i}} C_{4,1}^{-c}$, $\tilde{R}_{4,3} = C_{4,2}^{s_{y_i}} C_{4,2}^{-c}$, $\tilde{R}'_{4,3} = C_{4,1}^{s_{z_{i,2}}} C_{4,1}^{-c}$, ..., $\tilde{R}_{4,k_i+2} = C_{4,k_i+1}^{s_{y_i}} C_{4,k_i+1}^{-c}$, $\tilde{R}'_{4,k_i+2} = C_{4,1}^{s_{z_{i,k_i+2}}} C_{4,1}^{-c}$, $\tilde{R}_{5,0} = h_0^{s_\beta} h_0^{-c}$, ..., $\tilde{R}_{5,k_i+2} = h_{k_i+2}^{s_\beta} h_{k_i+2}^{-c}$.

(b) AP v checks $c \overset{?}{=} \mathcal{H}(gpk, ID_v, k_i, M, C_1, C_2, C_{3,1}, C_{3,2}, C_{4,1}, \ldots, C_{4,k_i+2}, \tilde{R}_1, \tilde{R}_2, \tilde{R}_{3,1}, \tilde{R}_{3,2}, \tilde{R}_{4,1}, \ldots, \tilde{R}_{4,k_i+2}, \tilde{R}'_{4,3}, \ldots, \tilde{R}'_{4,3}, \tilde{R}_{5,0}, \ldots, \tilde{R}_{5,k_i+2}, h_0, \ldots, h_{k_i+2})$.

5. If π_c is a valid proof, then AP v adds $reg_{d+1} = \{k_i, (C_1, C_2), (C_{3,1}, C_{3,2}),$ $(C_{4,1}, \ldots, C_{4,k_i+2}, h_0, \ldots, h_{k_i+2}), \pi_c\}$ to Log_v. Otherwise, it outputs reject.

U_i makes an NIZK proof that the ciphertext C_2 is a ciphertext against the valid identification mpk_i by using a part of membership certificate y_i. For $\ell = 0, 1, \ldots, k_i + 1$, the discrete logarithm of $C_{4,\ell+1}$ based on $C_{4,\ell}$ is y_i and the discrete logarithm of $C_{4,\ell+1}$ based on $C_{4,1}$ is $z_{i,\ell}$ ($C_{4,\ell+1} = C_{4,\ell}^{y_i} = C_{4,1}^{z_{i,\ell}}$). $z_{i,1} = y_i$ holds since $C_{4,2} = C_{4,1}^{z_{i,1}} = C_{4,1}^{y_i}$, and $z_{i,2} = y_i^2$ holds since $C_{4,3} = C_{4,1}^{z_{i,1}} = C_{4,2}^{y_i} = (C_{4,1}^{y_i})^{y_i} = C_{4,1}^{y_i^2}$. To repeat the above procedure, $z_{i,\ell} = y_i^{\ell}$ ($\ell = 1, \ldots, k_i + 2$) holds. Therefore, $C_2 = g_1^{x_i} \prod_{j=0}^{k_i+2} h_j^{z_{i,j}} = g_1^{x_i + \beta \sum_{j=0}^{k_i+2} a_j y_i^j} = mpk_i(g_1^{f_{k_i}(y_i)})^{\beta}$ hold.

- Auth(Verify$\langle gpk, ask_v, Log_v \rangle$, Proof$\langle gpk, msk_i \rangle$)
  On the ℓ-th authentication ($1 \leq \ell \leq k_i + 1$), U_i and AP execute Auth as follows: Note that $C_{3,\ell} = (g_1^{y_i^{\ell-1}})^{\beta}$ and $C_{3,\ell+1} = (g_1^{y_i^{\ell}})^{\beta}$ have already been computed in the ($\ell - 1$)-th authentication. Let Grant algorithm be the 0-th authentication.
  1. U_i computes $C_{3,\ell+2} = C_{3,\ell+1}^{y_i}(= (g_1^{y_i^{\ell}})^{\beta})$.
  2. U_i computes $\pi_\ell = SPK\{(y_i) : C_{3,\ell+1} = C_{3,\ell}^{y_i} \wedge C_{3,\ell+2} = C_{3,\ell+1}^{y_i}\}(ID_v)$. Concretely, U_i computes π_ℓ as follows:
     (a) U_i chooses $r_{y_i} \in_R \mathbb{Z}_p$.
     (b) U_i computes $R_1 = C_{3,\ell}^{r_{y_i}}$, $R_2 = C_{3,\ell+1}^{r_{y_i}}$, $c = \mathcal{H}(gpk, \ell, ID_v, R_1, R_2)$ and $s_{y_i} = r_{y_i} + c y_i$.
  3. U_i sends $\pi_\ell = (\ell, c, s_{y_i}, C_{3,\ell}, C_{3,\ell+1}, C_{3,\ell+2})$ to AP v.
  4. AP v searches $(C_{3,\ell}, C_{3,\ell+1})$ from Log_v. There are three cases as follows:

     Case-1 : If there exist $C_{3,\ell}$ and $C_{3,\ell+1}$ on reg_d and $\ell \leq k_i$, then AP v computes $\tilde{R}_1 = C_{3,\ell}^{s_{y_i}} C_{3,\ell+1}^{-c}$ and $\tilde{R}_2 = C_{3,\ell+1}^{s_{y_i}} C_{3,\ell+2}^{-c}$, and checks $c \stackrel{?}{=} \mathcal{H}(gpk, \ell, ID_v, \tilde{R}_1, \tilde{R}_2)$. If the checking condition c holds, then AP v adds $C_{3,\ell+2}$ and π_ℓ to $reg_d \in Log_v$, and outputs accept.

     Case-2 : If there exist $C_{3,\ell}$ and $C_{3,\ell+1}$ on reg_d and $\ell = k_i + 1$, then AP v outputs reject, and executes Trace.

     Case-3 : Otherwise, neither $C_{3,\ell}$ and $C_{3,\ell+1}$ exist nor the proof is valid, then AP v outputs reject.

In the ℓ-th authentication, U_i only computes $C_{3,\ell+2} = C_{3,\ell+1}^{y_i}$, where $C_{3,\ell+1} = (g_1^{\beta})^{y_i^{\ell}}$, and an NIZK proof π_ℓ that the discrete logarithm of $C_{3,\ell+2}$ based on $C_{3,\ell+1}$ is the same as the discrete logarithm of $C_{3,\ell+1}$ based on $C_{3,\ell}$. We explain the relation between grant messages and the 1st authentication in Fig. 1. In Grant, the knowledge of the discrete logarithm of $C_{3,2}$ based on $C_{3,1}$ has already been proven. Therefore, if $\pi 1$ is valid, then the discrete logarithm of $C_{3,3}$ based on $C_{3,2}$ is y_i, which is a part of a valid membership

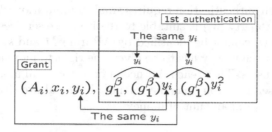

Fig. 1. The relation between Grant and the 1st authentication

certificate. This is called the sequence-of-zero-knowledge-proof mechanism whose concept is the same as the GR97 scheme.

- Trace(gpk, Log_v)

  Let $reg_d = \{k_i, (C_1, C_2), (C_{3,1}, \ldots, C_{3,k_i+3}), (C_{4,1}, \ldots, C_{4,k_i+2}, h_0, \ldots, h_{k_i+2}),$
  $\pi_c, (\pi_1, \ldots, \pi_{k_i+1})\}$ be the log in the $(k_i + 1)$-th authentication.

  1. AP v computes $C_2 / \prod_{j=1}^{k_i+3} C_{3,j}^{a_{j-1}} = g_1^{x_i}$.
  2. If there exists $(i, g_1^{x_i}) \in$ LIST, then output i. Otherwise, if $g_1^{x_i}$ is not included in LIST, then AP v verifies π_c and all π_j $(j = 1, 2, \ldots, k_i+1)$. If there is an invalid proof, then AP v outputs "AP". Otherwise, all proofs are valid, then AP v outputs "GM".

3.4   Application of Our Scheme

Our scheme is characterized by (1) the allowable number selectability with constant size secret keys, (2) relaxed anonymity, and (3) an efficient authentication algorithm Auth. As a natural requirement in the SaaS environment, a user who spends much money has to be given the right to more accesses to AP compared to other users. It can be achieved by the property (1). To improve the application providing strategy, e.g., long tail marketing [21], APs want to obtain the number of accesses by each user. However, from the viewpoint of users, accesses to services do not want to be linked to preserve one's own tastes and habits. The property (2) can achieve these different requirements, simultaneously. There is a situation in which each authentication requires small computations although the first certificate issuing requires large computations. We assume that a power-restricted device, e.g., a smart card, a cell phone, and so on, is used for authentication. First, a membership certificate is embedded in this power-restricted device. When users want to be granted a service, they go to a service counter, insert the power-restricted device into a high-spec machine, and execute the Grant algorithm by using the computational power of the high-spec machine. This is a natural situation, e.g., prepaid rail pass cards with built-in IC chips, cell phones equipped with an electronic money function, and so on. For example, SUICA (Super Urban Intelligent CArd) [7] is known as prepaid rail pass cards in Japan. These cards can be refilled using more money at card vending machines in train stations, and can be read by a card reader to enter the station. In this kind of services, each authentication requires small computations. Therefore, our scheme (with the property (3)) is suitable to realize these services. In Grant, we assume that a user U_i cannot be identified by AP from M and k_i. For example, a user who spends much money has to be given the right to more accesses to AP compared to other users. Then, each AP needs both the total amount of money and whether the user is a group member or not. In the above situation, M can be considered as the total amount of money.

4   Comparison

In this section, we compare our scheme with previous k-TAA schemes. In our scheme, the size of Log_v is $O(K)$, where $K = \sum_{i=1}^{N} k_i$ and N is the number of application group members. The log size of the previous k-TAA schemes [1,13,14,17,18] is $O(Nk)$, namely, the size of Log_v is similar to that of the previous schemes. The TFS04 [17] and the NS05 [14] scheme do not enable the constant proving cost. Although the TS06 scheme and the ASM06 scheme enable the constant proving cost, each AP has to publish k signatures. The Nguyen06 [13] scheme enable the constant proving cost without $O(k)$ public keys. Note that the number of public keys of the Nguyen06 scheme is $O(|AG|)$ where $|AG|$ is the number of users in application group managed by an AP, since this scheme enable the dynamic property using an accumulator [4]. If the dynamic property is deleted from the Nguyen06 scheme (namely update algorithm is deleted), then this scheme enable the constant proving cost with constant size public key. However, these schemes assume the same k for all users. In the selectable allowable number setting, the number of the secret key of U_i depends on k_i under the simple construction (See Section 3.1). Our selectable scheme enables the constant proving cost with constant size public key and secret key without increasing the number of secret and public key. In addition, our scheme enables a higher efficiency compared with previous k-TAA schemes. Concretely, in our Auth algorithm, a user requires 3 exponentiations and 1 hash function, and an AP requires 4 exponentiations and 2 scalar multiplications, whereas, the Nguyen06 scheme (without update algorithm), a user requires 22 exponentiations, 27 scalar multiplications and 1 hash function, and an AP requires 21 exponentiations, 20 scalar multiplications, 6 pairings and 1 hash function. (See Section 4.3 in [13] for details). In the previous schemes constructed by the tracing tag mechanism, the AP sends a challenge number for each authentication execution to extract a user's public key stored in LIST. Therefore, the user will be able to compute a part of the proof after she receives the challenge number. Our scheme enables full off-line computation that ALL proofs are pre-computable before executing Auth. In [1,13], the length of the proof of knowledge sent by the user is 500 Bytes, and it is 700 Bytes in [18]. On the other hands, in our scheme, the length of the proof of knowledge sent by the user is only 100 Bytes. We insist that these efficiencies are due to relaxed anonymity offering a tradeoff between privacy preservation and efficiency. This result due to the fact that the efficiency of group signature schemes can be drastically improved when unlinkability is not taken into account [15].

5 Security

In this section, we show the sketch of security proofs.

Theorem 1. *The proposed scheme satisfies relaxed anonymity under the $(k+2)$-PDDH assumption and the q-SDH assumption in the random oracle model.*

Proof. In our scheme, C_1 is a group signature [6] using a SDH-tuple member certificate. If $\mathcal{A}$ can guess $b \in \{0,1\}$, then $\mathcal{A}$ can also break the anonymity of the group signature scheme [6]. Next, we construct an algorithm $\mathcal{B}$ to break the $(k+2)$-PDDH problem by using an adversary $\mathcal{A}$ which breaks relaxed anonymity. Let $(h'', h', h_1', \ldots, h_{k+2}')$ be a $(k+2)$-PDDH instance, where $k = Min\{k_{v_0,i_0}, k_{v_0,i_1}, k_{v_1,i_0}, k_{v_1,i_1}\}$ which is defined in the relaxed anonymity game. $\mathcal{B}$ sets $g_2 := h''$ and $g_1 = h'$, and chooses other secret keys, the same as in the real scheme. For U_{i_0}, $\mathcal{B}$ sets $y_{i_0} := \log_{h'} h_1'$. For U_{i_1}, $\mathcal{B}$ chooses $y_{i_1} \in_R \mathbb{Z}_p^*$. For $\mathcal{O}_{\mathsf{Join-U}}(U_{i_0})$ and $\mathcal{O}_{\mathsf{Grant-U}}(U_{i_0})$ with the AP v_0 (which is executed in $\mathcal{O}_{query}$), $\mathcal{B}$ computes the simulated NIZK proof which includes the backpatch of the hash function without knowing y_{i_0}. Note that $C_{3,1} = g_1^\beta = (h')^\beta$ and $C_{3,2} := (h_1')^\beta$ where $\beta \in_R \mathbb{Z}_p^*$. If $b = 0$, then the above simulation is executed twice for $\mathcal{O}_{\mathsf{Grant-U}}(U_{i_0})$ with the AP v_1. If $b = 1$, then $\mathcal{B}$ simulates $\mathcal{O}_{\mathsf{Join-U}}(U_{i_1})$ and $\mathcal{O}_{\mathsf{Grant-U}}(U_{i_1})$ with the AP v_1 the same as the real scheme. For $\mathcal{O}_{\mathsf{Proof}}(U_{i_0})$ in the ℓ-th authentication, $\mathcal{B}$ computes the simulated NIZK proof which includes the backpatch of the hash function without knowing y_{i_0}, and sets $C_{3,\ell+1} := (h_\ell')^\beta$ and $C_{3,\ell+2} := (h_{\ell+1}')^\beta$. For $1 \le \ell \le k$, if $h_\ell' = (h')^{(\log_{h'} h_1')^\ell} = (h')^{y_{i_0}^\ell}$ and $h_{\ell+1}' = (h')^{y_{i_0}^{\ell+1}}$, then this simulation is the same as the real scheme. Therefore $\mathcal{A}$ has an advantage. Otherwise, if h_ℓ' and $h_{\ell+1}'$ are random values, then $\mathcal{A}$ has no advantage. Therefore, if $\mathcal{A}$ outputs b' and $b' = b$, then $\mathcal{B}$ outputs 1. Otherwise, $\mathcal{B}$ outputs 0. □

Theorem 2. *The proposed scheme satisfies detectability under the q-SDH assumption and the $(k+1)$-PCDH assumption, where $k \in \mathbb{Z}_{>0}$ is the largest allowable number of users.*

Proof. There are three cases, as follows: (1) U_i who is assigned in $reg_d \in Log_v$ executes the authentication protocol more than k_i times. This case does not happen because the number of accesses is restricted by only k_i times in Grant. (2) U_i who is assigned in $reg_d \in Log_v$ executes the authentication protocol more than k_i times by using a transcript of the authentication execution of U_j $(i \ne j)$. In this case, U_i has to compute $C_{3,\ell+2} = C_{3,\ell+1}^{y_j}$ from $(C_{3,1}, C_{3,2}, \ldots, C_{3,\ell+1}) = (g_1^\beta, (g_1^\beta)^{y_j}, \ldots, (g_1^\beta)^{y_j^\ell})$ where a random number $\beta \in_R \mathbb{Z}_p^*$ chosen by U_j and $y_j \in \mathbb{Z}_p^*$ which is a membership certificate of U_j. Then the $(\ell+1)$-PCDH assumption does not hold. (3) U_i with a valid membership certificate (A_i, x_i, y_i) executes the Grant protocol by using (A, x, y) which is a forged membership certificate. If U_i can forge a membership certificate, then the SDH assumption does not hold. □

Theorem 3. *The proposed scheme satisfies exculpability under the q-SDH assumption and the $(k+1)$-PCDH assumption, where $k \in \mathbb{Z}_{>0}$ is the largest allowable number of users.*

Proof. **Exculpability for users** : If $\mathcal{A}$ can output the authentication log for the identification of non-corrupted user U^*, then at least one commitment $C_{3,\ell}$ (which has not appeared in the outputs of $\mathcal{O}_{\mathsf{Proof}}(U^*)$) is included in the log. Then the $(k+1)$-PCDH assumption does not hold since $\ell \le k + 2$.

Exculpability for AP : This is clearly satisfied.

Exculpability for GM : If a public key g_1^x, where the result of the decryption of $C_2 = g_1^{x+\beta \sum_{j=0}^{k+2} a_j y^j}$, is not included in LIST, then there exists a forged membership certificate $A = (g_1 \tilde{g}^y)^{\frac{1}{x+\gamma}}$, where γ is the group secret key, since $C_1 = (g_1 \tilde{g}^y)^{\frac{1}{x+\gamma}} \tilde{g}^\alpha$ and $C_2 = g_1^{x+\beta \sum_{j=0}^{k+2} a_j y^j}$ have to be computed for the same values x and y on Grant. Then the SDH assumption does not hold since a new valid membership certificate A can be computed without being issued from the GM. □

6   Conclusion

In this paper, we propose for the first time a selectable k-TRAA that enables an allowable number to be assigned for each user and the number of the secret key of each U_i does not depend on k_i. We introduce relaxed anonymity, and our scheme enables a higher efficiency compared with previous k-TAA schemes. This result due to the fact that the efficiency of group signature schemes can be drastically improved when unlinkability is not taken into account [15]. Under relaxed anonymity, user information, such as access to services, is not exposed among different APs. Otherwise, an AP can obtain the number of accesses of own application from each user. This information is important to improve application-providing strategy. This means that relaxed anonymity is useful security requirement to preserve privacy and to satisfy information utilization, simultaneously.

References

1. Au, M.H., Susilo, W., Mu, Y.: Constant-size dynamic -TAA. In: De Prisco, R., Yung, M. (eds.) SCN 2006. LNCS, vol. 4116, pp. 111–125. Springer, Heidelberg (2006)
2. Bellare, M., Micciancio, D., Warinschi, B.: Foundations of group signatures: Formal definitions, simplified requirements, and a construction based on general assumptions. In: Biham, E. (ed.) EUROCRYPT 2003. LNCS, vol. 2656, pp. 614–629. Springer, Heidelberg (2003)
3. Boneh, D., Boyen, X.: Short signatures without random oracles and the sdh assumption in bilinear groups. J. Cryptology 21(2), 149–177 (2008)
4. Boneh, D., Boyen, X., Shacham, H.: Short group signatures. In: Franklin, M. (ed.) CRYPTO 2004. LNCS, vol. 3152, pp. 41–55. Springer, Heidelberg (2004)
5. Chaum, D., van Heyst, E.: Group signatures. In: Davies, D.W. (ed.) EUROCRYPT 1991. LNCS, vol. 547, pp. 257–265. Springer, Heidelberg (1991)

6. Delerablée, C., Pointcheval, D.: Dynamic fully anonymous short group signatures. In: Nguyên, P.Q. (ed.) VIETCRYPT 2006. LNCS, vol. 4341, pp. 193–210. Springer, Heidelberg (2006)
7. East Japan Railway Company Homepage. About Suica, http://www.jreast.co.jp/e/suica-nex/suica.html
8. Fiat, A., Shamir, A.: How to prove yourself: Practical solutions to identification and signature problems. In: Odlyzko, A.M. (ed.) CRYPTO 1986. LNCS, vol. 263, pp. 186–194. Springer, Heidelberg (1987)
9. Gennaro, R., Rohatgi, P.: How to sign digital streams. In: Kaliski Jr., B.S. (ed.) CRYPTO 1997. LNCS, vol. 1294, pp. 180–197. Springer, Heidelberg (1997)
10. Golle, P., Jarecki, S., Mironov, I.: Cryptographic primitives enforcing communication and storage complexity. In: Blaze, M. (ed.) FC 2002. LNCS, vol. 2357, pp. 120–135. Springer, Heidelberg (2003)
11. IBM. Lotuslive, https://www.lotuslive.com/
12. Miyaji, A., Nakabayashi, M., Takano, S.: New explicit conditions of elliptic curve traces for FR-reduction. IEICE Transactions 84-A(5), 1234–1243 (2001)
13. Nguyen, L.: Efficient dynamic -times anonymous authentication. In: Nguyên, P.Q. (ed.) VIETCRYPT 2006. LNCS, vol. 4341, pp. 81–98. Springer, Heidelberg (2006)
14. Nguyen, L., Safavi-Naini, R.: Dynamic k-times anonymous authentication. In: Ioannidis, J., Keromytis, A.D., Yung, M. (eds.) ACNS 2005. LNCS, vol. 3531, pp. 318–333. Springer, Heidelberg (2005)
15. Ohtake, G., Fujii, A., Hanaoka, G., Ogawa, K.: On the theoretical gap between group signatures with and without unlinkability. In: Preneel, B. (ed.) AFRICACRYPT 2009. LNCS, vol. 5580, pp. 149–166. Springer, Heidelberg (2009)
16. Oracle. SaaS Platform, http://www.oracle.com/technologies/saas/index.html
17. Teranishi, I., Furukawa, J., Sako, K.: k-times anonymous authentication (Extended abstract). In: Lee, P.J. (ed.) ASIACRYPT 2004. LNCS, vol. 3329, pp. 308–322. Springer, Heidelberg (2004)
18. Teranishi, I., Sako, K.: k-times anonymous authentication with a constant proving cost. In: Yung, M., Dodis, Y., Kiayias, A., Malkin, T.G. (eds.) PKC 2006. LNCS, vol. 3958, pp. 525–542. Springer, Heidelberg (2006)
19. Tsang, P.P., Wei, V.K.: Short linkable ring signatures for E-voting, E-cash and attestation. In: Deng, R.H., Bao, F., Pang, H., Zhou, J. (eds.) ISPEC 2005. LNCS, vol. 3439, pp. 48–60. Springer, Heidelberg (2005)
20. Wikipedia. Software as a service, http://en.wikipedia.org/wiki/Software_as_a_service
21. Wikipedia. The Long Tail, http://en.wikipedia.org/wiki/The_Long_Tail

PUF-Based Authentication Protocols – Revisited*

Heike Busch, Stefan Katzenbeisser, and Paul Baecher

Darmstadt University of Technology, Germany
h.busch@me.com, katzenbeisser@seceng.informatik.tu-darmstadt.de,
pbaecher@gmail.com
www.seceng.informatik.tu-darmstadt.de

Abstract. Physical Unclonable Functions (*PUF*) are physical objects
that are unique and unclonable. *PUF*s were used in the past to construct authentication protocols secure against physical attackers. However, in this paper we show that known constructions are not fully secure
if attackers have raw access to the *PUF* for a short period of time. We
therefore propose a new, stronger, and more realistic attacker model.
Subsequently, we suggest two constructions of authentication protocols,
which are secure against physical attackers in the new model and which
only need symmetric primitives.

1 Introduction

Classical authentication protocols, where one communication partner proves its
identity to another participant, are commonly based on cryptographic primitives. Their security usually relies on a computationally hard problem. Most
constructions are based on the possession of a secret key, which is assumed not
to fall in the hands of an adversary. However, this assumption may be violated if
an adversary has physical access to the device that performs authentication for
a short time. In this period, the adversary may read the whole memory of the
device including all secret information, unless hardware security measures are
taken. With this information, the adversary can finally impersonate a person.
Such an attack is usually outside of the considered attacker model in classical
cryptography. However, in many practical authentication scenarios, this attack
is realistic. Consider for example a situation, where a waiter in a restaurant carries a credit card away from the table for billing. During this short time the card
is not under full control of the owner and an adversary may read the data of
the card memory in order to extract the secret information. Moreover, the card
reader is potentially under full control of the adversary. Thus, data stored in the
memory of the reader is potentially at risk as well. Classical cryptography does
not provide a secure way of authentication in this attacker model.

Physical Unclonable Functions (*PUF*) were proposed as a building block for
authentication schemes that resist physical attacks. *PUF*s, as introduced by

---

* This work was supported by CASED (www.cased.de).

H.Y. Youm and M. Yung (Eds.): WISA 2009, LNCS 5932, pp. 296–308, 2009.

Pappu [1, 2], are physical objects which are unique and unclonable [3, 4, 1, 2, 5, 6]. Technically speaking, a *PUF* responds to a stimulus with a physical output (which can be measured and encoded as a bit string) and has the following properties: First, it is impossible to clone a *PUF* even with highly complex equipment. Second, it is infeasible to predict the output for a chosen stimulus without physically evaluating the *PUF*, and third, the output looks random. These properties of a *PUF*, i.e., unclonability, unpredictability and the pseudo-random output, inspired researchers to build authentication protocols relying on *PUF challenge-response-pairs* (*CRP*) [6, 7, 8]. In these protocols, the server issues a challenge in form of a stimulus, which the client has to answer by measuring its *PUF*. If this *PUF* response matches a pre-recorded response held at the server, the client is authenticated. If the number of *CRP*s of a *PUF* is large and an adversary cannot measure all *PUF* responses, he will most likely not be able to answer the challenge of the server even if he had physical access to the *PUF* for a short time.

In this work we show that current *PUF*-based authentication protocols are not fully secure in the above-mentioned attacker model where the attacker has physical control of the *PUF* and the corresponding reader during a short time. In particular, we revisit the authentication protocol of Tuyls et al., called *TAP* in the sequel [7]. The authors proposed a *PUF*-based challenge-response authentication protocol in a bank setting: A *PUF* is embedded in a personalized smart card in a non-separable way. Since the *PUF* is unclonable and its response is unpredictable, the owner can use it to authenticate itself against a server. More precisely, the protocol consists of two phases, an enrollment phase and a verification phase. During the enrollment phase, the *PUF* is embedded inseparably in the smart card. Then, the server challenges the *PUF* with several stimuli and stores the resulting challenge-response pairs in its database. Afterwards the smart card handed out to the owner to the person. During verification, the holder of the card inserts it into the card reader. The server chooses a random *PUF* challenge and sends it to the reader. The reader measures the *PUF* and returns the response back to the server. If the measured *PUF* response matches the recorded response in the server's database, the server is convinced that the reader has physical access to the *PUF*. Thus, the holder of the smart card is authenticated. Furthermore, both parties can derive a session key from the *PUF* output, which can subsequently be used to establish an encrypted communication channel between the server and the card reader.

While this protocol assures that an adversary cannot impersonate the client once he has physical access to the card, the adversary gets enough information to impersonate the server: Consider an adversary, who has access to the reader as well as to the *PUF* for a short time period. In this time it can read the whole memory of the *PUF* and the card reader including all secret information. Moreover, the adversary can challenge the *PUF* with any stimuli at will in order to collect a number of challenge-response-pairs. This information is enough to impersonate the server: The adversary engages in the *TAP* protocol and chooses a valid challenge of its collected *CRP*s and sends it to the card reader. The reader measures the output of the *PUF* for the given challenge and forwards the corresponding response to the adversary. The reader will be unaware that it has authenticated to (and established a key with) the adversary instead of the server, as it cannot distinguish a challenge chosen by the server from one issued by the attacker.

Contribution. We lay open a weakness in current *PUF*-based authentication protocols. In particular, we show that they are not fully secure when an adversary has physical access to the *PUF* as well as the reader during a short time. We propose a new, stronger and more realistic attack scenario and design two authentication protocols which are secure in the new model. Both protocols use only symmetric primitives and thus lend themselves well for implementation on power-constrained devices. The main idea of both approaches is to enable the reader to distinguish challenges issued by the server and the adversary in a secure way. We suggest two solutions, which rely on Bloom filters [9, 10] and hash trees [11]. Moreover, we will compare both approaches.

Organization. We first give in Section 2 an overview of *PUF*s, Bloom filters, and hash trees. In Section 3 we recall the *PUF*-based authentication protocol of Tuyls et al. and describe in Section 4 the weakness of the protocol as well as the resulting attack scenario. Afterwards, in Section 5 and Section 6, we introduce our solutions based on Bloom filters and hash trees, which prevent the impersonation attack.

2   Preliminaries

2.1   Physical Unclonable Functions

Informally, a Physical Unclonable Function (*PUF*) is a physical object which reacts with a response r to a stimulus c. Such a pair (c, r) of a stimulus and response is called a *challenge-response-pair* (*CRP*). Furthermore, a *PUF* satisfies the following properties: (1) It is impossible to build another *PUF* that has the same response behavior. (2) It is hard to predict the output of a *PUF* for a given input without performing and measurement. (3) The output looks random. We can distinguish between *strong* and *weak* *PUF*s. Strong *PUF*s have a large number of challenge-response-pairs such that an efficient adversary, measuring only a few *CRP*s, cannot predict the response for a random challenge with high probability. If the number of different *CRP*s is rather small, we speak of a weak *PUF*. In the following we consider only strong *PUF*s [12, 1, 5] as building blocks for our protocols.

2.2   Fuzzy Extractors and Helper Data

The responses of *PUF*s are noisy by nature. Therefore, the output of a *PUF* cannot directly be used for the authentication process. When a *PUF* is measured with a challenge c_i, it produces a corresponding output r_i, which is usually noisy and not uniformly distributed. However, for cryptographic applications, a completely noisy-free output with a perfect uniform distribution is required. Possible solutions to handle the noisy outputs are *Fuzzy Extractors* or *Helper Data Algorithms* [13]. A Fuzzy Extractor consists of a pair of algorithms (G, W). During an enrollment phase the algorithm G takes as input a *PUF* response r and generates as output a secret s together with helper data w. The algorithm W takes as input a noisy response r' and helper data w. It reconstructs the secret s unless the noise level in r' is too high. For further details we refer the reader to [13, 8, 14, 7].

2.3 Bloom Filters

A Bloom filter $\mathcal{B}$ is a probabilistic data structure that encodes a set of elements $\mathcal{X}$ into a ℓ bit array B in order to allow fast set membership tests [9, 10, 15]. The idea is to encode the elements by using several hash functions. More precisely, the Bloom filter $\mathcal{B}$ consists of ℓ bits $B[0], \ldots, B[\ell-1]$, where all entries are initially set to 0, and a set $\mathcal{H}$ of k independent collision-secure hash functions $h_i : \{0,1\}^* \rightarrow \{0, \ldots, \ell-1\}$. In order to encode a set $\mathcal{X} = \{x_1, \ldots, x_m\}$ into the Bloom filter, the elements of $\mathcal{X}$ are added sequentially into the array B according to the following rule: For each element $x \in \mathcal{X}$, we set $B[h_i(x)]$ to one for all $1 \le i \le k$.

The algorithm CHECK$(x', \mathcal{B}, \mathcal{H})$ verifies if a given element x' belongs to the Bloom filter $\mathcal{B}$ and works as follows: CHECK$(\cdot)$ evaluates all k hash functions on x' and outputs 1 iff $B[h_i(x')] = 1$ for all $1 \le i \le k$. If one of the bits is set to 0, the algorithm outputs 0 since x' is clearly not in $\mathcal{B}$. Note that the length ℓ of the array B has to be chosen carefully (as well as the number of hash functions k). In case that many bits in B are set to 1, the probability that an arbitrary element $\hat{x}$ is recognized as a member of $\mathcal{X}$ increases. We call the event that an element $\hat{x} \notin \mathcal{X}$ is falsely recognized as member of $\mathcal{X}$ by the Bloom filter, e.g. CHECK$(\hat{x}, \mathcal{B}, \mathcal{H}) = 1$, as a *false positive*. The parameters of the Bloom filter have to be chosen in a way that makes this error very unlikely.

2.4 Hash Trees

A hash tree (or *Merkle tree*) $\mathcal{T}$ is a complete binary tree used to prove set membership [11, 16]. The hash tree consists of a root node v_r, several internal nodes v_i, and leaf nodes v_l. Each leaf node represents a data value d_i. Furthermore, each internal node stores a hash value of the concatenation of the values stored in its children. The hash values are computed with a collision-resistant hash function $h : \{0,1\}^* \rightarrow \{0,1\}^n$. Thus, if an internal node v_{i+1} has a left child $v_{i,0}$ storing the value $x_{i,0}$ and a right child $v_{i,1}$ storing the value $x_{i,1}$, then v_{i+1} stores the value $x_{i+1} \leftarrow h(x_{i,0}||x_{i,1})$. It is well known that it is, given a tree $\mathcal{T}$, infeasible to find another path that yields to the same root node v_r.

The algorithm CHECK$(x_r, d_i, \langle \tau_1, \ldots, \tau_t \rangle)$ verifies if a given element d_i belongs to the hash tree $\mathcal{T}$. Instead of exposing all leaf nodes, the algorithm is passed an *authentication path*, which consists of the hash values $\langle \tau_1, \ldots, \tau_t \rangle$ of the siblings of all nodes along the path from the leaf node v_l storing d_i to the root node v_r. Let x_i be the hash value of node v_i along the authentication path. We set $x_1 \leftarrow v_l$ and compute the remaining hash values $x_2, \ldots, x_t$ as follows: Let $i_1, \ldots, i_t$ be the binary representation of tree index i. If $i_j = 0$, we set $x_{j+1} \leftarrow h(x_j || \tau_j)$, and otherwise, if $i_j = 1$, we set $x_{j+1} \leftarrow h(\tau_j || x_j)$. The algorithm outputs 1 iff $x_{t+1} = x_r$, i.e., if the root hash matches, and 0 otherwise.

3 PUF-Based Authentication and Key Establishment

Tuyls et al. [7] proposed a token-based protocol to authenticate a credit card against a central bank authority. In the following we refer to the bank as server and the ATM as reader. Furthermore, for the sake of simplicity, we describe the details of the protocol in the two party scenario where a single smart card is authenticated to a central server.

In the enrollment phase, the server issues a smart card including a PUF together with a current identifier ID_{PUF}. It generates a set of random challenges $\mathcal{C} = \{c'_1, \ldots, c'_k\}$ for the PUF and measures for each c'_i the corresponding responses r'_i. Furthermore, for each challenge c'_i a random secret s'_i is chosen from a set of random secrets $\mathcal{S}'$ and a helper data $w'_i \in \mathcal{W}$ is computed by solving $s'_i = W(r'_i, w'_i)$. The rewritable non-volatile memory on the card stores the identifier ID_{PUF}, a usage counter n, indicating how many times the authentication protocol ran, and the current hash value $m = h(rd)$, where rd is a random string generated by the server and h is a one-way hash function. The central server holds in its database the card identifier ID_{PUF}, both values n' and $m' = rd$, as well as a list of challenges $\mathcal{C}$ with the particular corresponding secrets $\mathcal{S}'$ and helper data $\mathcal{W}$. Once a card is ready for use, it is initialized with $n' = n = 0$, $m' = rd$ and $m = h(rd)$.

The PUF-based authentication and session key establishment protocol, depicted in Figure 1, consists of the following steps: The user inserts its card into a card reader. The reader sends an initialization message consisting of a nonce α, the usage counter n, and the identifier ID_{PUF} to the server. The server checks if $n \geq n'$. If this condition does not hold, then the server generates an error message and aborts. Otherwise, the server computes $M = h^{n-n'}(m')$, where h^n denotes the n-th composition of h. Furthermore, the server computes a temporary key $K'_1 = h(M || ID_{PUF})$. It then generates a nonce β, selects randomly a challenge $c'_i \in \mathcal{C}$ and computes the value $T = \alpha || c'_i || w'_i || \beta$ where w'_i is the helper data corresponding to c'_i. The server authenticates the quadruple $(\alpha, c'_i, w'_i, \beta)$ by computing a MAC with key K'_1. It encrypts $T || MAC_{K'_1}(T)$ using K'_1 and sends the result to the reader. The reader derives the temporary key $K_1 = h(m || ID_{PUF})$, decrypts $Enc_{K_1}[T || MAC_{K_1}(T)]$ using K_1, and validates the MAC by using K_1. If the MAC is invalid or if the decrypted nonce α is wrong, then the reader aborts with an error message. Otherwise, the reader challenges the PUF with c'_i, which produces a corresponding response r_i. The reader executes the helper data algorithm G with input r_i and helper data w_i in order to obtain the secret s_i. Now, the reader computes a session key $K = h(K_1 || s_i)$ based on the temporary key K_1 and the secret s_i. Finally, the reader generates a MAC based on

Server	Reader+PUF
$ID_{PUF}, n', m' = rd, \{\mathcal{C}, \mathcal{W}, \mathcal{S}'\}$	$ID_{PUF}, n, m = h^n(rd)$

Server		Reader+PUF
	$\xleftarrow{\quad \alpha, n, ID_{PUF} \quad}$	Generate nonce α
Check $n \geq n'$		
Compute $M' = h^{n-n'}(m')$		$K_1 = h(m \| ID_{PUF})$
$K_1' = h(M' \| ID_{PUF})$		
$c_i' \in_R \mathcal{C}$		
Generate nonce β		
$T = \alpha \| c_i' \| w_i' \| \beta$	$\xrightarrow{\quad Enc_{K_1'}[T \| MAC_{K_1'}(T)] \quad}$	Check MAC
		$r_i \leftarrow PUF(c_i)$
		$s_i = G(r_i, w_i)$
$K' = h(K_1' \| s_i')$	$\xleftarrow{\quad MAC_K(\beta) \quad}$	$K = h(K_1 \| s_i)$
Check MAC		
$n' \to n+1, \; m' \to h(m')$	$\xleftarrow{\quad \text{Use } K \quad}$	$n \to n+1, \; m \to h(m)$
Remove c_i' from database		

Fig. 1. *PUF*-based authentication and key establishment protocol of [7]

the nonce β using K and sends the MAC to the server. The server computes its session key $K' = h(K_1' \| s_i')$ based on the temporary key K_1' as well as the secret s_i' corresponding to c_i, which was generated in the enrollment phase and is stored in the database. Furthermore, the server verifies that the MAC is a valid tag on the nonce β (with respect to the secret derived from the response and the helper data). If the MAC is invalid, then the server aborts with an error message. Otherwise the server is convinced that the reader has physical access to the *PUF* and the holder of the smart card is authenticated. Subsequently, both parties use the symmetric key $K' = K$ as a session key in order to set up a secure channel.

4   How to Impersonate the Server

In the above protocol, an adversary who has access to the smart card and the *PUF* for a short period of time can impersonate the server, unless the communication between the bank and the reader is authenticated. We assume that the initial enrollment phase of the protocol is secure, ranging from the *PUF* fabrication up to the point where the user physically receives the smart card including the *PUF*. We now turn to the description of the adversary.

Let $\mathcal{A}$ be an adversary who has once access to the smart card including the *PUF* for a certain time. The algorithm $\mathcal{A}$ selects a small number of challenges $\mathcal{C}^*$ and measures the corresponding responses $\mathcal{R}^*$. Moreover, it reads the identifier ID_{PUF}, the usage counter n and the current hash value m stored on the card memory. With this information, the adversary can impersonate the server in subsequent runs of the authentication protocol as follows: The adversary computes the value $M^* = h^{n-n^*}(m)$, which is possible, because $\mathcal{A}$ obtained the usage counter n and the hash value $m = h(M)$ from the card memory. Now, $\mathcal{A}$ calculates $K_1^* = h(M^*||ID_{PUF})$ and generates a random nonce β^*. The adversary chooses a challenge $c_i^* \in \mathcal{C}^*$ and runs the algorithm W with input r_i in order to get helper data $w_i^* \in \mathcal{W}^*$ as well as a secret $s_i^* \in \mathcal{S}^*$. Furthermore, $\mathcal{A}$ produces a MAC on the quadruple $(\alpha, c_i^*, w_i^*, \beta^*)$ using the key K_1^*, encrypts the MAC with K_1^* and sends the resulting value $Enc_{K_1^*}[(\alpha||c_i^*||w_i^*||\beta^*)||MAC_{K_1^*}(\alpha||c_i^*||w_i^*||\beta^*)]$ to the reader. The reader subsequently computes $K_1 = h(m||ID_{PUF})$, decrypts $Enc_{K_1^*}[(\alpha||c_i^*||w_i^*||\beta^*)||MAC_{K_1^*}(\alpha||c_i^*||w_i^*||\beta^*)]$ and checks whether the MAC is valid. Since the MAC and the decrypted nonce α are valid, the protocol does not abort with an error message. Consequently, the reader challenges the PUF with c_i^*, which produces a corresponding response r_i^*. Running the algorithm G with input r_i^* as well as w_i^*, the reader extracts the secret s_i^* from the output of the PUF. At last, the reader computes a MAC on the nonce β^* using the session key K. Afterwards, it sends the MAC to the adversary $\mathcal{A}$. Finally, $\mathcal{A}$ computes its session key $K^* = h(K_1^*||s_i^*)$ by hashing the temporary key K_1^* and the secret s_i^*. Thus, the attack succeeded and the adversary is able to impersonate the server successfully. Moreover, the key K^* (respectively K) is a symmetric key established between the reader and the adversary.

5 PUF-Based AKE-Protocol Based on Bloom Filters

The main reason why the attack works is that the adversary gets physical access to the smart card including the PUF at least once for a short period. During this time, the adversary measures the PUF and uses the obtained challenge-response-pairs in subsequent authentication runs. Due to the symmetric nature of the protocol, the reader cannot decide whether a given challenge during a run of the protocol was initially measured by the server or subsequently by the adversary. This weakness can be solved – without requiring an authenticated link between the server and reader – by storing a subset of "valid" challenges $\mathcal{V}$ initially measured by the server in the read-only memory on the smart card. However, storing all challenges is too expensive. Moreover, an adversary that reads the memory of the smart card would learn the subset of legal challenges. Thus, the set $\mathcal{V}$ has to be stored on the card in a compact form, which does not allow a computational bounded adversary to gain information on legal challenges. In this case impersonation can be prevented, as the adversary does not know the set of valid challenges and will most likely present an invalid challenge, which will not be accepted by the reader (this requires a strong PUF since the number of challenges must be large). We propose two solutions for compactly storing legal PUF challenges, one relies on Bloom filters [9] and the other one on hash trees [11].

5.1 AKE-Protocol Based on Bloom Filter

The modified protocol, based on Bloom filters, is depicted in Figure 2 and contains the following modifications: During the secure enrollment phase, the server computes a subset of valid challenges $\mathcal{V} \subseteq \mathcal{C}$ by choosing a certain number of challenges $c_i \in \mathcal{C}$ at random. Afterwards, the server stores the challenges c_i of $\mathcal{V}$ with the corresponding responses in its database. Furthermore, the server computes a Bloom filter $\mathcal{B}$ of size ℓ and stores all x challenges using k hash functions in $\mathcal{B}$. The rewritable non-volatile memory on the smart card stores the identifier ID_{PUF}, the usage counter n, the current hash value m, and the Bloom filter $\mathcal{B}$. If the reader receives a challenge c_i, the reader verifies that it receives a challenge that was initially chosen by the server by checking whether $c_i \in \mathcal{B}$, e.g., whether all array locations $B[h_j(c_i)]$ for $1 \le j \le k$ are set to 1. If any check fails, then clearly c_i is not a member of $\mathcal{V}$ and the reader aborts. Otherwise, the reader follows the protocol steps as described in Section 3. This way, the reader can be sure that the server initially selected the challenge, unless an adversary succeeded in guessing a valid challenge.

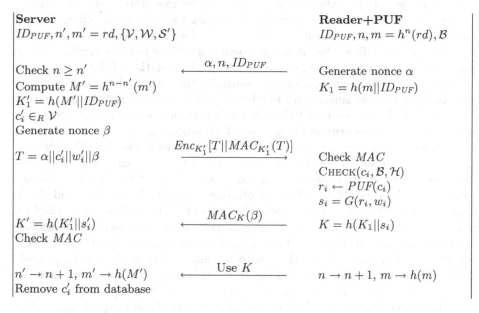

Fig. 2. *PUF*-based authentication and key establishment based on Bloom filters

5.2   Security Trade Off between Space and False Positives

In this section we take a closer look at the parameters of the Bloom filter. Since Bloom filters always have a false positive probability, which in our protocol results in the fact that the card reader accepts invalid challenges, the goal is to find a trade off between the space required to store the valid challenges on the card, the number of hash functions, and the false positive probability of the Bloom filter. Recall that a false positive occurs if an element is accepted to be in the Bloom filter, although it is not in the set [9, 10, 15]. The probability of a false positive f can be computed as

$$f = \left(1 - \left(1 - \frac{1}{\ell}\right)^{kx}\right)^k \approx (1 - e^{-kx/\ell})^k, \tag{1}$$

where x is the number of challenges in $\mathcal{V}$, ℓ is the number of bits in the array of the Bloom filter and k is the number of hash functions. We have to choose the parameters of a Bloom filter appropriately in order to find a trade off between the computation time (corresponds to the number k of hash functions), the size (corresponds to the number ℓ of bits in the Bloom filter array), and the probability of error (corresponds to the false positive probability f), which we will discuss in Section 6.3.

5.3   Analysis of the *AKE*-Protocol Based on Bloom Filters

We analyze in this section the extended *PUF*-based protocol depicted in Figure 2. The main idea of the extension is to let both parties store a subset of valid challenges $\mathcal{V}$ of the challenge space $\mathcal{C}$. The reader then can check efficiently if a received challenge c is a member of $\mathcal{V}$. If $c \notin \mathcal{V}$, the reader aborts. Otherwise, if $c \in \mathcal{V}$, the reader follows the further protocol steps. Since only the server and the smart card know the subset $\mathcal{V}$, we prevent the impersonation of the server by an adversary. As a consequence, we have a mutual authentication between the server and the holder of the smart card. Furthermore, the protocol is resistant against replay attacks because each *PUF* challenge is used only once. The protocol also retains backwards-security of the original *TAP* protocol.

Now, let us consider the subset $\mathcal{V}$ of valid challenges in more detail. Since we use a strong *PUF* we cannot draw conclusions about the elements of $\mathcal{V}$. Moreover, the subset $\mathcal{V}$ of valid challenges is encoded as a Bloom filter in the read-only memory on the card. If an adversary obtains the ℓ-bit Bloom filter and the k hash functions, it cannot deduce the x elements of $\mathcal{V}$. This follows from the fact that the hash functions are chosen independently as well as that hash functions are collision-resistant. However, there is still the probability that the adversary guesses a valid challenge, i.e., the adversary manages to find a challenge such that the testing algorithm of the Bloom filter outputs 1. The probability that the adversary $\mathcal{A}$ guesses such an element (event win) can be upper bounded by the probability that it guesses a challenge being in the set $\mathcal{V}$ and the probability that an invalid challenge is accepted. This probability can be computed as follows:

$$\mathsf{Prob}[\mathcal{A}\,\mathsf{win}] \leq \frac{|\mathcal{V}|}{|\mathcal{C}|} + \left(1 - \left(1 - \frac{1}{\ell}\right)^{kx}\right)^k. \tag{2}$$

Finally, the protocol is very efficient because it only requires symmetric cryptographic primitives.

6   The *PUF*-Based *AKE*-Protocol Based on Hash Trees

In this section we propose an alternative solution based on hash tress. The benefit of this approach is that we only need to store a constant amount of data in the read-only memory of the smart card regardless of the number of elements of $\mathcal{V}$. Although this approach reduces storage, it induces an additional communication overhead.

6.1 *AKE*-Protocol Based on Hash Trees

The extended *PUF*-based authentication and key establishment protocol based on hash trees is depicted in Figure 3. During the secure enrollment phase, the server computes a subset of valid challenges $\mathcal{V} \subseteq \mathcal{C}$ by choosing uniformly a certain number of challenges of $\mathcal{C}$. Furthermore, let h be an one-way hash function. The server computes a hash tree $\mathcal{T}$, based on the elements of $\mathcal{V}$, as follows: Let us assume that the number of challenges of the subset is a power of how: $|\mathcal{V}| = 2^\nu$. To authenticate the challenges $c_0, ..., c_{|\mathcal{V}|}$, the server places each challenge at the leaf nodes of a binary tree of depth $\mathcal{V}$. Moreover, the root node and each internal node of the binary tree are computed as hashes of its two child nodes (see Section 2.4). The root hash value x_τ is stored in the memory of the smart card, whereas the server stores the hash tree $\mathcal{T}$. To authenticate a challenge c_i, the server discloses i, the corresponding path τ, between c_i and the root node and all necessary sibling nodes, and sends the additional information to the reader. The reader now runs the algorithm $\mathrm{CHECK}(x_\tau, c_i, \tau)$ in order to verify the validity of the received path with its stored root value x_r. If the function returns 0, then the verification is not successful and the reader aborts with an error message. Otherwise, the reader knows that the challenge c_i is a member of the set $\mathcal{V}$ of

Server		Reader+PUF
$ID_{PUF}, n', m' = rd, \{\mathcal{C}, \mathcal{W}, \mathcal{S}'\}, T$		$ID_{PUF}, n, m = h^n(rd), x_\tau$
	$\xleftarrow{\quad \alpha, n, ID_{PUF} \quad}$	
Check $n \geq n'$		Generate nonce α
Compute $M' = h^{n-n'}(m')$		$K_1 = h(m \| ID_{PUF})$
$K_1' = h(M' \| ID_{PUF})$		
$c_i' \in_R \mathcal{C}$		
Generate nonce β and		
Compute $\tau = \langle \tau_1, ..., \tau_d \rangle$		
$T = \alpha \| c_i' \| w_i' \| \beta \| i \| \tau$	$\xrightarrow{\quad Enc_{K_1'}[T \| MAC_{K_1'}(T)] \quad}$	Check MAC
		$\mathrm{CHECK}(x_\tau, c_i, \tau)$
		$r_i \leftarrow PUF(c_i)$
		$s_i = G(r_i, w_i)$
$K' = h(K_1' \| s_i')$	$\xleftarrow{\quad MAC_K(\beta) \quad}$	$K = h(K_1 \| s_i)$
Check MAC		
$n' \to n + 1, m' \to h(M')$	$\xleftarrow{\quad \text{Use } K \quad}$	$n \to n + 1, m \to h(m)$
Remove c_i' from database		

Fig. 3. *PUF*-based authentication and key establishment protocol based on hash trees

valid challenges. Subsequently, the reader follows the protocol steps as described in Section 3.

6.2 Communication Overhead of Hash Trees

The benefit of hash trees is that we only have to store the root value of the hash tree in the read-only memory of the smart card. Unfortunately, this solution involves additional communication overhead. The server has to send additional data besides the quadruple $(\alpha, c_i, w_i, \beta)$ in order to allow the reader to verify validity of c_i. Let us assume that the hash function h maps its input c_i to an λ-bit output v_i. Now the server has to transmit for each level of the tree the particular sibling node. Thus, the server has to send $\nu \cdot \lambda = \log_2 |\mathcal{V}| \cdot \lambda$ additional bits to the reader.

6.3 Analysis of the *PUF*-Based *AKE*-Protocol Based on Hash Trees

The security properties follow analogously to Section 5.3, except for the assumptions about the Bloom filter. Here, the subset of valid challenges $\mathcal{V}$ is encoded as a hash tree. The security of hash trees relies on the one-way property and on the collision-resistance of the hash function h. Moreover, it is well known, that this data structure authenticates the elements in the hash tree, i.e., it is computationally infeasible to find a valid path given only the root node v_r. Finally, since only hash values are transmitted, and because the adversary is unable to invert the one-way hash function h, no information about the other challenges in the set (authenticated through the same tree) is leaked.

Comparison of Bloom filters and hash trees. Both of our above-mentioned approaches can be implemented on a smart card. Table 1 summarizes the storage and communication overhead of Bloom filters and hash trees. Recall that $\mathcal{V}$ is the set of challenges, ℓ the length of the Bloom filter, and λ the output length of the hash function.

Our solution based on Bloom filter has the advantage that we do not need any additional communication overhead. However, in order to reduce the false-positives, the length ℓ has to be chosen appropriately. On the other hand, if we want to optimize the storage capacity on the smart card, then the solution based on hash trees is the better choice because we only have to store the ℓ-bit hash value of the root node. Although this solution is very storage efficient, we get an addition communication overhead of $\lambda \cdot \log_2 |\mathcal{V}|$ bits.

Table 1. Comparison of the storage and communication overhead of Bloom filters and hash trees

	Storage on the Credit Card	Communication Overhead
Bloom filter	ℓ	0
Hash tree	λ	$\lambda \cdot \log_2 |\mathcal{V}|$

7 Conclusion

In this paper, we have described a weakness in current *PUF*-based authentication protocols and have proposed a new, stronger and more realistic attacker model. We have provided two constructions of mutual authentication protocols, which are secure against physical attackers. Both approaches use space-efficient data structures, which are used to encode valid *PUF* challenges. One is based on Bloom filters and the other one on hash trees. Finally, we compared the storage and the communication overhead of both approaches.

Acknowledgments. We thank Dominique Schröder and the anonymous reviewers for their valuable comments.

References

[1] Pappu, R.: Physical One-Way Functions. PhD thesis, Massachusetts Institute of Technology (2001)

[2] Pappu, R., Recht, B., Taylor, J., Gershenfeld, N.: Physical one-way functions. Science 297(5589), 2026–2030 (2002)

[3] Gassend, B.: Physical random functions. Master's thesis, Massachusetts Institute of Technology (2003)

[4] Hammouri, G., Sunar, B.: PUF-HB: A tamper-resilient HB based authentication protocol. In: Bellovin, S.M., Gennaro, R., Keromytis, A.D., Yung, M. (eds.) ACNS 2008. LNCS, vol. 5037, pp. 346–365. Springer, Heidelberg (2008)

[5] Tuyls, P., et al.: Information-theoretic security analysis of physical uncloneable functions. In: Patrick, A.S., Yung, M. (eds.) FC 2005. LNCS, vol. 3570, pp. 141–155. Springer, Heidelberg (2005)

[6] Tuyls, P., et al.: Anti-Counterfeiting. In: Security, Privacy and Trust in Modern Data Management, pp. 293–312. Springer, Heidelberg (2007)

[7] Škorić, B., Tuyls, P.: Strong Authentication with Physical Unclonable Functions. In: Security, Privacy and Trust in Modern Data Management, pp. 133–148. Springer, Heidelberg (2007)

[8] Guajardo, J., Kumar, S.S., Schrijen, G.J., Tuyls, P.: FPGA intrinsic PUFs and their use for IP protection. In: Paillier, P., Verbauwhede, I. (eds.) CHES 2007. LNCS, vol. 4727, pp. 63–80. Springer, Heidelberg (2007)

[9] Bloom, B.H.: Space/time trade-offs in hash coding with allowable errors. Communications of the ACM 13(7), 422–426 (1970)

[10] Broder, A., Mitzenmacher, M.: Network applications of bloom filters: A survey. Internet Mathematics 1(4), 485–509 (2004)

[11] Merkle, R.C.: Protocols for public key cryptosystems. IEEE Symposium on Security and Privacy 122 (1980)

[12] Guajardo, J., Škorić, B., Tuyls, P., Kumar, S.S., Bel, T., Blom, A.H., Schrijen, G.J.: Anti-counterfeiting, key distribution, and key storage in an ambient world via physical unclonable functions. Information Systems Frontiers 11(1), 19–41 (2009)

[13] Dodis, Y., Reyzin, L., Smith, A.: Fuzzy extractors: How to generate strong keys from biometrics and other noisy data. In: Cachin, C., Camenisch, J.L. (eds.) EUROCRYPT 2004. LNCS, vol. 3027, pp. 523–540. Springer, Heidelberg (2004)
[14] Škorić, B., Tuyls, P.: Robust key extraction from physical unclonable functions. In: Ioannidis, J., Keromytis, A.D., Yung, M. (eds.) ACNS 2005. LNCS, vol. 3531, pp. 407–422. Springer, Heidelberg (2005)
[15] Mitzenmacher, M., Upfal, E.: Probability and Computing: Randomized Algorithms and Probabilistic Analysis. Cambridge University Press, New York (2005)
[16] Micali, S., Reyzin, L.: Min-round resettable zero knowledge in the public key model. In: Preneel, B. (ed.) EUROCRYPT 2000. LNCS, vol. 1807, pp. 373–393. Springer, Heidelberg (2000)

Mediated Ciphertext-Policy Attribute-Based Encryption and Its Application

Luan Ibraimi[1,2], Milan Petkovic[2], Svetla Nikova[1], Pieter Hartel[1], and Willem Jonker[1,2]

[1] Faculty of EEMCS, University of Twente, The Netherlands
[2] Philips Research, The Netherlands

Abstract. In Ciphertext-Policy Attribute-Based Encryption (CP-ABE), a user secret key is associated with a set of attributes, and the ciphertext is associated with an access policy over attributes. The user can decrypt the ciphertext if and only if the attribute set of his secret key satisfies the access policy specified in the ciphertext. Several CP-ABE schemes have been proposed, however, some practical problems, such as attribute revocation, still needs to be addressed. In this paper, we propose a mediated Ciphertext-Policy Attribute-Based Encryption (mCP-ABE) which extends CP-ABE with instantaneous attribute revocation. Furthermore, we demonstrate how to apply the proposed mCP-ABE scheme to securely manage Personal Health Records (PHRs).

1 Introduction

Modern distributed information systems require flexible access control models which go beyond discretionary, mandatory and role-based access control. Recently proposed models, such as attribute-based access control, define access control policies based on different attributes of the requester, environment, or the data object. On the other hand, the current trend of service-based information systems and storage outsourcing require increased protection of data including access control methods that are cryptographically enforced. The concept of Attribute-Based Encryption(ABE) fulfills the aforementioned requirements. It provides an elegant way of encrypting data such that the encryptor defines the attribute set that the decryptor needs to posses in order to decrypt the ciphertext. Since Sahai and Waters [1] proposed the basic ABE scheme, several more advanced schemes have been developed, such as most notably Ciphertext-Policy ABE schemes (CP-ABE) [2,3,4]. In these schemes, a ciphertext is associated with an access policy and the user secret key is associated with a set of attributes. A secret key holder can decrypt the ciphertext if the attributes associated with his secret key satisfy the access policy associated with the ciphertext. For example, consider a situation when two organizations, a Hospital and a University, conduct research in the field of neurological disorders. The Hospital wants to allow access to their research results to all staff from the University who have the role Professor and belong to the Department of Neurology (DN). To enforce the policy, the Hospital encrypts the data according to the access policy

H.Y. Youm and M. Yung (Eds.): WISA 2009, LNCS 5932, pp. 309–323, 2009.

$\tau_{Results}$=(University Professor $\wedge$ Member of DN). Only users who have a secret key associated with a set of attributes ω=(University Professor, Member of DN) can satisfy the access policy $\tau_{Results}$ and be able to decrypt the ciphertext.

The state-of-the-art CP-ABE schemes provide limited support for revocation of attributes, a feature, which is becoming increasingly important in modern access control systems. In general, attribute revocation may happen due to the following reasons: 1) an attribute is not valid because it has expired, for instance, the attribute "project manager-January 2009" is valid until January 2009, or 2) a user is misusing her secret key associated with a set of attributes, for instance, Alice might give a copy of her secret key to Bob who is not a legitimate user. In particular, attribute revocation is an important requirement in the domain of access control to personal health data, which is our application field for attribute-based encryption.

Contribution. In this paper, we propose a new scheme for attribute revocation in CP-ABE called mediated Ciphertext-Policy Attribute-Based Encryption (mCP-ABE). Previous CP-ABE systems proposed to use a system where attributes are valid within a specific time frame [5]. However, the drawback of this approach is that there is no way to revoke an attribute before the expiration date. In our scheme the secret key is divided into two shares, one share for the mediator and the other for the user. To decrypt the data, the user must contact the mediator to receive a decryption token. The mediator keeps an attribute revocation list (ARL) and refuses to issue the decryption token for revoked attributes. Without the token, the user cannot decrypt the ciphertext, therefore the attribute is implicitly revoked. In our scheme we assume that each user has a unique identifier I_u (in CP-ABE the user is identified only with a set of attributes) and may have many attributes. The identifier is used by the mediator to check if there are revoked attributes related to I_u. Different users having different identifiers, may have the same attribute set. For example, Alice with an identifier I_{Alice}, and Bob with an identifier I_{Bob}, may have the same attribute set $\omega = (att_1, att_2)$. The technique of splitting the attribute components of the secret key into two shares, and the technique of using an identifier I_u for each user, helps us to achieve the following attribute revocations: i) revoking an attribute from a single user without affecting other users, and ii) revoking an attribute from the system where all users are affected.

We also define a security model for the proposed scheme which formalizes the security attacks and provide a security proof under the generic group model. Finally, we demonstrate the applicability of the proposed scheme to securely manage Personal Health Records (PHRs).

1.1 Related Work

Attribute-Based Encryption. Sahai and Waters in their seminal paper [1] introduce the concept of ABE. There are two types of ABE schemes: Key-Policy ABE schemes (KP-ABE) [6] and Ciphertext-Policy ABE schemes (CP-ABE)[2,3,4]. In KP-ABE, a ciphertext is associated with a set of attributes and a user secret key is associated with an access policy. A secret key holder can decrypt the ciphertext if the attributes associated with the ciphertext satisfy the access policy associated with the secret key. In CP-ABE the idea is reversed. A ciphertext is associated with an access policy and the user secret key is associated with a set of attributes. A secret key holder can decrypt the ciphertext if the attributes associated with the secret key satisfy the access policy associated with the ciphertext.

Mediated Cryptography. Boneh et al.[7,8] introduce a method for fast revocation of public key certificates and security capabilities in a RSA cryptosystem called mediated RSA (mRSA). The method uses an online semi-trusted mediator (SEM) which has a share of each users secret key, while the user has the remaining share of the secret key. To decrypt or sign a message, a user must first contact SEM and receive a message-specific token. Without the token, the user cannot decrypt or sign a message. Instantaneous user revocation is obtained by instructing SEM to stop issuing tokens for future decrypt/sign requests. Thus, in mediated cryptography the Trusted Authority (TA) responsible to generate a user key pair, does not deliver the full decryption key to users, but it delivers only a share of it. This method achieves faster revocation of user's security capabilities compared to previous certification techniques such as Certificate Revocation List (CRL) and Online Certificate Status Protocol (OCSP). Libert and Quisquater [9] show that the architecture for revoking security capabilities can be applied to several existing public key encryption schemes including the Boneh-Franklin scheme, and several signature schemes including the GDH scheme. Nali et al. [10] present a mediated hierarchical identity-based encryption and signature scheme. The hierarchical nature of the schemes and the instant revocation capability offered by the SEM architecture allows to enforce access control cryptographically in hierarchically structured communities of users whose access privileges change dynamically. Nali et al. [11] also show how to extend the Libert and Quisquater mediated identity-based cryptographic scheme to allow the enforcement of role-based access control (RBAC).

Revocation. Credential revocation is a critical issue for access control systems. For ABE systems, Pirretti et al. [5] propose to use user attributes for a limited time period. After a specific time period the attribute would become invalid. However, in such systems an attribute cannot be revoked before the expiration date. This approach also requires the list of keys that correspond to attributes to be updated regularly, which would also require the users secret keys to be updated regularly. Boldyreva et al. [12] proposes a revocable IBE scheme. The proposed idea for the revocation is based on binary tree data structure, proposed previously in the PKI setting [13,14]. Boldyreva approach is an improvement to Boneh and Franklin [15] idea, however, when the number of revoked users increases, then the advantage of the proposed scheme is lost over that proposed by Boneh and Franklin, especially when the number of revoked users r becomes close to to the total number of users n in the system. Even if r is less than n, still the key update complexity is bounded by $O(r \log(\frac{n}{r}))$ while in a realistic solution, the key update complexity should depend on the number of revoked users. Ostrovsky et al.[16] proposes a Key-Policy Attribute-Based Encryption scheme, where the user secret key may be associated with a non-monotonic access policy. The non-monotonic access policy can be represented by a boolean formula such as AND, OR, NOT, and Out Of (threshold) operations. The main drawback of the scheme is that the size of attributes in the ciphertext is fixed, which restricts the expressivity of the scheme. We note that the concept of revoking an attribute is similar to the concept of revoking an identity. Hence, one can revoke an identity of the user instead of revoking an attribute in an access structure. In this paper we propose a mediated CP-ABE scheme which is not limited to the fixed size of attributes which can be revoked.

The rest of this paper is organized as follows. Section 2 provides background information. In Section 3 we give a formal definition of the mCP-ABE scheme. Section 4 describes the construction of mCP-ABE scheme. In Section 5 we apply the mCP-ABE scheme and describe a general architecture for secure management of Personal Health Records (PHRs). The last section concludes the paper.

2   Background

In this section, we briefly review the basics of bilinear pairing and the security proof in the generic group model, and give a formal definition of CP-ABE.

2.1   Bilinear Pairing

Let $\mathbb{G}_0$ and $\mathbb{G}_1$ be two multiplicative groups of prime order p, and let g be a generator of $\mathbb{G}_0$. A pairing (or bilinear map) $\hat{e} : \mathbb{G}_0 \times \mathbb{G}_0 \rightarrow \mathbb{G}_1$ satisfies the following properties [15]:

1. Bilinear: for all $u, v \in \mathbb{G}_0$ and $a, b \in \mathbb{Z}_p^*$, we have $\hat{e}(u^a, v^b) = \hat{e}(u, v)^{ab}$.
2. Non-degenerate: $\hat{e}(g, g) \neq 1$.

$\mathbb{G}_0$ is said to be a bilinear group if the group operation in $\mathbb{G}_0$ and the bilinear map $\hat{e} : \mathbb{G}_0 \times \mathbb{G}_0 \rightarrow \mathbb{G}_1$ can be computed efficiently. Note that the map is symmetric since $\hat{e}(g^a, g^b) = \hat{e}(g, g)^{ab} = \hat{e}(g^b, g^a)$.

2.2   Security in the Generic Group Model

We prove the security of the scheme based on the generic group model, introduced by Shoup [17]. A proof in the generic group model is based on the fact that the discrete logarithm and the Diffie-Hellman problem are hard to solve as long as the order of the group is a large prime number. The same applies to a group with bilinear pairing where finding the discrete logarithm is a hard problem. In the generic group model group elements are encoded as unique random strings, in such a way that the adversary cannot test any property other than equality.

We prove the security of the mCP-ABE scheme based on the argument that no adversary that acts generally on the groups can break the security of our scheme. This means that if there is an efficient adversary who can discover vulnerabilities in our scheme, then these vulnerabilities can be used to exploit mathematical properties of groups used in the scheme. In the generic model, the adversary has access to the oracles that compute group operations in $\mathbb{G}_0$, $\mathbb{G}_1$, and to the oracle that performs non-degenerate paring $\hat{e}$, while the adversary can test the equality by itself.

2.3   Formal Definition of CP-ABE

The building block of our construction is a CP-ABE scheme. In CP-ABE a message is encrypted under an access policy τ over the set of possible attributes, and a user secret key sk_ω is associated with an attribute set ω. A secret key sk_ω can decrypt the message encrypted under the access policy τ, if and only if the user attribute set ω satisfies the access policy τ. CP-ABE scheme consists of two entities: TA and users. The four algorithms: Setup, Keygen, Encrypt and Decrypt are defined as follows [2]:

- Setup(k): run by the TA, this algorithm takes as input a security parameter k and outputs the public key pk and a master key mk.
- Keygen(ω, mk): run by the TA, this algorithm takes as input the master key mk and a set of attributes ω. The algorithm outputs a secret key sk_ω associated with ω.
- Encrypt(m, τ, pk): run by the encryptor, this algorithm takes as input a public key pk, a message m, and an access policy represented by an access tree τ. The algorithm returns the ciphertext c_τ such that only users who have the secret key shares associated with attributes that satisfy the access tree τ will be able to decrypt the message.
- Decrypt(c_τ, sk_ω): run by the decryptor, this algorithm takes as input a ciphertext c_τ, a secret key sk_ω associated with ω, and it outputs a message m, or an error symbol $\perp$ when the attribute set ω does not satisfy the access tree τ.

3   Mediated Ciphertext-Policy Attribute-Based Encryption (mCP-ABE)

In this section, first, we give a formal definition of our proposed scheme, and later we give the security model in which our scheme is proven to be secure.

3.1   Formal Definition of mCP-ABE

The mCP-ABE scheme consists of three entities: TA, a mediator and users. The TA uses the master key to generate a user secret key, which is then divided into two shares such that the first share of the user secret key is sent to the mediator and the second share of the user secret key is sent to a user. The mediator has to stay online all the time, while the TA can be put off-line once it has generated secret keys for all users. The mCP-ABE scheme consists of five algorithms: Setup, Keygen, Encrypt, m-Decrypt, and Decrypt (Setup and Encrypt algorithms are same as in CP-ABE scheme):

- Keygen(mk, ω, I_u): run by the TA, this algorithm takes as input the master key mk, the user attribute set ω, and the user identifier I_u. The algorithm outputs two secret key shares associated with ω and I_u : $sk_{\omega I_u,1}$ and $sk_{\omega I_u,2}$. The first share of the secret key $sk_{\omega I_u,1}$ is delivered to the mediator, and the second share of the secret key $sk_{\omega I_u,2}$ is delivered to the user. The secret key shares are delivered through a secure channel to the mediator and to the user.
- m-Decrypt($c_\tau, I_i, sk_{\omega I_i,1}$) : run by the mediator, this algorithm takes as input a ciphertext c_τ, the identifier I_i and the secret key $sk_{\omega I_i,1}$, and outputs a message $\hat{c}_\tau$, or an error symbol $\perp$ when the non-revoked attributes from the set ω do not satisfy the access tree τ.
- Decrypt($\hat{c}_\tau, sk_{I_i\omega,2}$): run by the message receiver, this algorithm takes as input a ciphertext $\hat{c}_\tau$, and a secret key $sk_{I_i\omega,2}$, and outputs a message m, or an error symbol $\perp$ when the non-revoked attributes from the set ω does not satisfy the access tree τ.

In practice, there might be multiple entities acting as mediators, and a global entity acting as TA. For example, a healthcare organization may choose Proxy$_1$ as its mediator and a government organization may choose Proxy$_2$ as its mediator, where each mediator has the first share of the secret key for registered users in the hospital organization, respectively in the government organization. Vanrenen et al. [18] propose the use of peer-to-peer networking (P2P) which would allow users to require a decryption token from every mediator, such as the mediator either tries to compute a decryption token by itself, or forwards the request to its neighbors.

3.2   Security Model

In our scheme the TA is a fully trusted entity which stores securely the master key. We skip discussions about the key escrow problem, since different existing threshold schemes [19,20] can be applied to solve this problem. A mediator is a semi-trusted entity, namely, it should issue decryption tokens to users, but it is not trusted in the sense that it should not obtain information about the plaintaixt. We define semantic security of mCP-ABE scheme following the security model of Libert and Quisquater [9]. For an encryption scheme to be semantically secure the adversary must not learn anything about the plaintext when the ciphertext and the public key used to create the ciphertext are given. In the security game, the challenger simulates the game and answers adversary $\mathcal{A}$ queries as follows:

1. Setup. The challenger runs the Setup algorithm to generate (pk, mk) and gives the public key pk to the adversary $\mathcal{A}$.
2. Phase1. $\mathcal{A}$ performs a polynomially bounded number of queries:
   - Keygen$^1(\omega, I_u)$. $\mathcal{A}$ asks for a secret key for the attribute set ω and identifier I_u, and receives the mediator share of the secret key $sk_{\omega I_u,1}$.
   - Keygen$^2(\omega, I_u)$. $\mathcal{A}$ asks for a secret key for the attribute set ω and identifier I_u, and receives the user share of the secret key $sk_{\omega I_u,2}$.

3. Challenge. $\mathcal{A}$ sends to the challenger two messages m_0, m_1, and the challenge access policy τ^*, such that none of the full secret keys $sk_{\omega I_u}$ (both $sk_{\omega I_u,1}$ and $sk_{\omega I_u,2}$) generated from the interaction with Keygen^1 and Keygen^2 oracles satisfies τ^*. The challenger picks a random bit $b \in (0,1)$ and returns $c_{\tau^*} =$ Encrypt(m_b, τ^*, pk).
4. Phase2. $\mathcal{A}$ can continue querying with the restriction that none of the full secret keys $sk_{\omega I_u}$ generated from the interaction with Keygen^1 and Keygen^2 oracles satisfies τ^*.
5. Guess. $\mathcal{A}$ outputs a guess $b' \in (0,1)$.

Definition 1. *The mCP-ABE scheme is said to be semantically secure if any polynomial-time adversary has only a negligible advantage in the security game, where the advantage is defined to be $|\Pr[b' = b] - \frac{1}{2}|$.*

Note that the security game formally captures the following security requirements:

- Resistance against secret key collusion, where different users cannot combine their attribute sets to extend their decryption power. For example, suppose there is a message encrypted under the access tree $\tau = (a_1 \wedge a_2 \wedge a_3)$. Suppose Alice has a secret key $sk_{\omega_A I_A}$ associated with an attribute set $\omega_A = (a_1, a_2)$, and Bob has a secret key $sk_{\omega_B I_B}$ associated with an attribute set $\omega_B = (a_3, a_4)$. Neither Alice's secret key, nor Bob's secret key satisfies the access tree τ. But, if Alice and Bob combine their attribute sets $\omega_A \cup \omega_B = (a_1, a_2, a_3, a_4)$, then the combined attribute sets satisfies the access tree τ. Therefore in the security game we allow the adversary to make secret key queries associated with different attribute sets, say ω_1 and ω_2, such that neither ω_1, nor ω_2 alone can satisfy the challenge access policy τ^*, but $\omega_1 \cup \omega_2$ can satisfy τ^*.
- Resistance against malicious cooperation between the mediator and some users to decrypt the ciphertext associated with an access policy, when the users secret key does not satisfy the access policy. For example, even if a user with attribute set $\omega = (a_1, a_2)$ collude with the mediator, the user should not be capable to decrypt a ciphertext encrypted under a challenge access policy $\tau^* = (a_1 \wedge a_2 \wedge a_3)$, since ω does not satisfy τ^*. Therefore in the security game the adversary is allowed to ask the mediator share (first share of the secret key $sk_{\omega I_u,1}$) and the user share of a secret key (second share of the secret key $sk_{\omega I_u,2}$) for any set of attributes which does not satisfy the challenge access policy τ^*.

4   mCP-ABE Scheme

In this section, we give a description of the access policy associated with the ciphertext, and then we give the construction of the scheme. Finally, we analyze the security of the scheme and describe how to revoke user attributes.

4.1   Access Policy

In mCP-ABE scheme, an access policy is represented by an access tree τ, in which inner nodes are either $\wedge$ (and) or $\vee$ (or) boolean operators, and leaf nodes are attributes. The access tree τ specifies which combination of attributes the decryptor needs to posses in order to decrypt the ciphertext. Figure 1 presents an example of an access tree τ representing an access policy: $(a_1 \wedge a_4) \vee (a_3 \vee a_5)$. To decrypt an encrypted message under the access tree τ, the decryptor must possess a secret key which is associated with the attribute set which satisfies τ. Attributes are interpreted as logic variables, and possessing a secret key associated with an attribute makes the corresponding logical variable *true*. There are several different sets of attributes that can satisfy the access tree τ presented in Figure 1, such as the attribute set (a_1, a_4), the attribute (a_3), or the attribute (a_5). In our scheme, we assume that attributes are ordered in the access tree e.g index(a_1)=1, index(a_4)=2, index(a_3)=3 and index(a_5)=4.

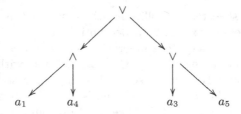

Fig. 1. Access tree $\tau = (a_1 \wedge a_4) \vee (a_3 \vee a_5)$

4.2   Main Construction

1. Setup(k) : On input of the security parameter k, the algorithm generates a group $\mathbb{G}_0$ of prime order p with a generator g and a bilinear map $\hat{e}$: $\mathbb{G}_0 \times \mathbb{G}_0 \rightarrow \mathbb{G}_1$. The algorithm generates the system attribute set $\Omega = (a_1, a_2, \ldots a_n)$, for some integer n, and for each $a_j \in \Omega$ chooses a random elements $t_j \in \mathbb{Z}_p$. Let $y = \hat{e}(g, g)^\alpha$, where α is chosen at random from $\mathbb{Z}_p$, and $\{T_j = g^{t_j}\}_{j=1}^n$. The public key is published as:

$$pk = (g, y, \{T_j\}_{j=1}^n)$$

The master secret key consists of the following components:

$$mk = (\alpha, \{t_j\}_{j=1}^n)$$

2. Keygen(mk, ω, I_u) : To generate a secret key for the user with an attribute set ω and an identifier I_u, the Keygen algorithm performs as follows:
   (a) Compute the base component of the secret key: $d_0 = g^{\alpha - u_{id}}$ where $u_{id} \in_R \mathbb{Z}_p$ (for each user with an identifier I_u a unique random value u_{id} is generated).

(b) Compute the attribute component of the secret key. For each attribute $a_j \in \omega$, choose $u_j \in_R \mathbb{Z}_p$ and compute $d_{j,1} = g^{\frac{u_j}{t_j}}$ and $d_{j,2} = g^{\frac{u_{id} - u_j}{t_j}}$.

The secret key of the form: $sk_{\omega I_u,1} = \{d_{j,1}\}_{a_j \in \omega}$ is delivered to the mediator, and the secret key of the form: $sk_{\omega I_u,2} = (d_0, \{d_{j,2}\}_{a_j \in \omega})$ is delivered to the user.

3. $\mathsf{Encrypt}(m, \tau, pk)$: To encrypt a message $m \in \mathbb{G}_1$ the algorithm proceeds as follows:

   – Select a random element $s \in \mathbb{Z}_p$ and compute:

$$c_0 = g^s$$
$$c_1 = m \cdot y^s = m \cdot \hat{e}(g,g)^{\alpha s}$$

   – Set the value of the root node of τ to be s, mark all nodes as un-assigned, and mark the root node assigned. Recursively, for each assigned non-leaf node, suppose its value is s, do the following.
      • If the symbol is $\wedge$ and its child nodes are marked un-assigned, let n be the number of child nodes, set the value of each child node, except the last one, to be $s_i \in_R \mathbb{Z}_p$, and the value of the last node to be $s_n = s - \sum_{i=1}^{n-1} s_i \mod p$ (i represents the index of an attribute in the access tree). Mark this node assigned.
      • If the symbol is $\vee$, set the values of its child nodes to be s. Mark this node assigned.
   – For each leaf attribute $a_{j,i} \in \tau$, compute $c_{j,i} = T_j^{s_i}$.

   Return the ciphertext $c_\tau = (\tau, c_0, c_1, \{c_{j,i}\}_{a_{j,i} \in \tau})$.

4. $\mathsf{m\text{-}Decrypt}(c_\tau, sk_{\omega I_i,1}, I_i)$: when receiving the ciphertext c_τ, the recipient I_i firstly chooses the smallest set $\omega' \subseteq \omega$ that satisfies τ and forwards to the mediator (c_τ, ω', I_i). The mediator checks the Attribute Revocation List (ARL) if any $a_j \in \omega'$ is revoked either from system attribute set Ω or from the user attribute set ω.

   (a) If an attribute is revoked, the mediator returns an error symbol $\bot$ and does not perform further computations.
   (b) If no attribute is revoked, the mediator computes $\hat{c}_\tau$ as follows:

$$\hat{c}_\tau = \prod_{a_j \in \omega'} \hat{e}(T_j^{s_i}, g^{\frac{u_j}{t_j}})$$
$$= \hat{e}(g,g)^{\sum_{a_j \in \omega'} u_j s_i}$$

   Sends $\hat{c}_\tau$ to the recipient.

5. $\mathsf{Decrypt}(\hat{c}_\tau, sk_{\omega I_i,2})$: To decrypt the ciphertext the recipient proceeds as follows:

(a) compute:

$$c''_\tau = \prod_{a_j \in \omega'} \hat{e}(T_j^{s_i}, g^{\frac{u_{id}-u_j}{t_j}})$$

$$= \prod_{a_j \in \omega'} \hat{e}(g^{t_j s_i}, g^{\frac{u_{id}-u_j}{t_j}})$$

$$= \hat{e}(g,g)^{\sum_{a_j \in \omega'}(u_{id}-u_j)s_i}$$

(b) compute:

$$\hat{e}(c_0, d_0) \cdot \hat{c}_\tau \cdot c''_\tau = \hat{e}(g^s, g^{\alpha-u_{id}}) \cdot \hat{e}(g,g)^{\sum_{a_j \in \omega'} u_j s_i} \cdot \hat{e}(g,g)^{\sum_{a_j \in \omega'}(u_{id}-u_j)s_i}$$

$$= \hat{e}(g^s, g^{\alpha-u_{id}}) \cdot \hat{e}(g,g)^{u_{id}s}$$

$$= \hat{e}(g^s, g^\alpha)$$

(c) return m, where

$$m = \frac{c_1}{\hat{e}(g^s, g^\alpha)}$$

$$= \frac{m \cdot \hat{e}(g,g)^{\alpha s}}{\hat{e}(g^s, g^\alpha)}$$

Efficiency. Our scheme is similar to the work of Cheung and Newport [3] on ciphertext-policy attribute-based encryption, however we make major changes in the Key Generation phase, Encryption phase and Decryption phase in order to improve the expressivity of the scheme (the scheme in [3] supports only access policies with logical conjunction), and we improve the efficiency of the scheme (in [3] the size of the ciphertext and secret key increases linearly with the total number of attributes in the system). In our proposed scheme, the size of the shares of the secret key $sk_{\omega I_u,1}$ and $sk_{\omega I_u,2}$ depend on the number of attributes the user has and consists of $|\omega| + 1$ group elements in $\mathbb{G}_0$ ($|\omega|$ is the cardinality of a set ω). The size of the ciphertext c_τ depends on the size of the access policy τ and has $|\tau| + 1$ group elements in $\mathbb{G}_0$, and one group element in $\mathbb{G}_1$. In the m-Decrypt phase, the mediator has to compute ω' pairing operations, where $\omega' \subseteq \omega$ is the attribute set which satisfies the access policy τ. In the decryption phase, to reveal the message, the user has to compute $\omega' + 1$ pairing operations.

For the sake of simplicity, we mentioned only access policies which consist of $\wedge$ (and) and $\vee$ (or) nodes. Note that our scheme, in addition to $\wedge$ (and) and $\vee$ (or) nodes, can support threshold nodes or Out Of nodes. For example, the encryptor may specify the access policy 2 Out Of (a_1, a_2, a_3), which implies that the user must have at least two attributes in order to satisfy the access policy and be able to decrypt. If the access policy contains threshold nodes, then the attribute shares s_i can be generated using threshold secret sharing techniques e.g. using Shamir's secret sharing technique. This would require, to use Lagrange basis polynomials in the decryption phase in order to reconstruct the value s.

4.3 Security Analysis

We give a brief discussion about the security of the proposed scheme. A full formal security proof using the generic group model is given in the full version of this paper [21]. To decrypt a ciphertext without satisfying the access policy, the adversary has to construct $\hat{e}(g^s, g^\alpha)$, and then divide c_1 with $\hat{e}(g^s, g^\alpha)$ to obtain m. To obtain $\hat{e}(g^s, g^\alpha)$, the adversary must first obtain $\hat{e}(g, g)^{su_{id}}$, which can be calculated by pairing the components of the secret key $g^{\frac{u_{id} - u_j}{t_j}}$ with the components of the ciphertext $g^{t_j s_i}$, and then multiply the result with the decryption token $\hat{e}(g, g)^{\sum_{a_j \in \omega'} u_j s_i}$ received from the mediator. However, $\hat{e}(g, g)^{su_{id}}$ can be computed only if the adversary has enough attributes which satisfy the access policy, otherwise this would not be possible. Also note that, if a user acting as an adversary is revoked, then the user will not get the decryption token $\hat{e}(g, g)^{\sum_{a_j \in \omega'} u_j s_i}$ from the mediator, and as a result of this the user cannot reconstruct $\hat{e}(g, g)^{su_{id}}$ even if the user has a secret key with attributes which satisfy the access policy. If we assume that the adversary is able to compromise the mediator, then the adversary will be able to learn the mediator share of user secret key $sk_{\omega I_u, 1}$, and be able to compute the decryption token $\hat{e}(g, g)^{\sum_{a_j \in \omega'} u_j s_i}$. However, the decryption token will not help the adversary to decrypt the ciphertext since the adversary does not know the second share of the user secret key $sk_{\omega I_u, 2}$.

The very important security of mCP-ABE scheme is a collusion resistance of user secret keys - it should not be possible for different users to combine their secret keys in order to extend their decryption power. Therefore, to prevent collusion, the Keygen algorithm of our scheme generates a random value u_{id} for each user, which is embedded in each component of the user secret key. Users cannot combine components of the secret key since different users have different random value in their secret keys.

4.4 Attribute Revocation

As already mentioned in section 1, there can be many reasons why an attribute can be revoked. We assume that the mediator maintains an Attribute Revocation List (ARL) which simply has information about attributes revoked from the system attribute set Ω, and attributes revoked from user attribute set ω. The basic idea is that, when an attribute a_j is revoked from the system attribute set Ω, the TA removes a_j from the system attribute set, and notifies the mediator to stop performing decryption tasks for all users whose attribute secret key involves a_j. When an attribute is revoked from a specific user I_u, TA notifies the mediator to stop helping the user I_u to perform decryption tasks for the attribute a_j. Therefore, the attribute revocation is achieved immediately after the revocation decision is made. We assume that there is a policy of revocation authorization maintained by the mediator that describes who is responsible to revoke system or user attributes. At least, the TA should be able to revoke the system and user attributes, and the owner of the attribute should be able to revoke its attribute because the owner may be the first to notice the compromise of her secret key.

5   Applying mCP-ABE in Practice

In this section we describe an application of mCP-ABE. We propose to use mCP-ABE to securely manage Personal Health Records (PHRs). This application demonstrate the practicality and usefulness of our scheme.

5.1   Using mCP-ABE to Securely Manage PHRs

Issues around the confidentiality of health records are considered as one of the primary reason for the lack of the deployment of open interoperable health record systems. Health data is sensitive: inappropriate disclosure of a record can change a patient's life, and there may be no way to repair such harm financially or technically. Although, access to health data in the professional medical domain is tightly controlled by existing legislations, such as the U.S. Health Insurance Portability and Accountability Act (HIPAA) [22], private web PHR systems stay outside the scope of this legislation. Therefore, a number of patients might hesitate to upload their sensitive health records to web PHR systems such as the Microsoft Health Vault, Google Health or WebMD. The scheme presented in this paper allows patients to store their sensitive health records on web PHR systems in an encrypted form, while still giving them control to share their data with healthcare providers and/or with their family members. The reason is because the data is encrypted according to an access policy, and the policy moves with the encrypted data. Thus, even if the server which stores health records gets compromised, the confidentiality of the data is preserved since the data is encrypted, and the attacker cannot decrypt the encrypted data without having a secret key. Figure 2 illustrates a general architecture of a PHR system that uses mCP-ABE. The architecture consists of a publishing server, a data repository that includes a security mediator (Proxy), TA and several data users. The publishing server can be implemented on a home PC of the data source (a patient) or as a trusted service. Its role is to protect and publish health records. The data repository stores encrypted health records, while Proxy is used in the data consumption phase for revocation. The TA is used to set up the keys. Note that the TA and the publishing server do not have to be always online (the TA is needed only in the set-up phase while the publishing server can upload the protected data in an ad-hoc way). There are three basic processes in the management of PHRs:

1. Setup: The steps of this phase are depicted with number 1 in Figure 2. In this phase, the TA distributes the keys to the patients, users and Proxy.
2. Data protection (upload of data to the PHR): When a patient wants to upload protected data to the repository, she contacts the TA to check which attributes are allowed to be used as a policy. Then she creates her access control policy and encrypts the data with the keys corresponding to that policy. Then the data is uploaded to the repository. If she wants to change the policy she can re-encrypt the data and update the repository. All this can be done by a publishing server on behalf of the patient who specifies the access control policy.

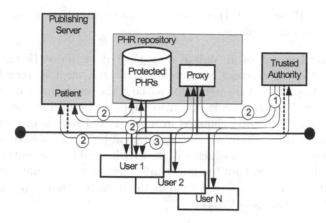

Fig. 2. Secure Management of PHRs

3. Data consumption (doctor's request-response) and revocation: When a user wants to use patient data he contacts the PHR repository and downloads encrypted data. The user makes a request to Proxy for a decryption token. The request contains the encrypted data and a set of user attributes which satisfy the access policy associated with the encrypted data. Proxy checks if any attribute from the user request is not revoked, and, if so, Proxy generates the decryption token and send it to the user. After receiving the decryption token the user decrypts the patient data using the keys corresponding to the appropriate attributes which satisfy the access tree. The steps of this phase are depicted with number 3 in Figure 2.

An addition advantage of an online semi-trusted mediator (Proxy) is that the mCP-ABE scheme can be used to enforce context attributes such as: system date and time or the location from where the request comes from. This is useful for healthcare applications which require context-aware access control where access to patients data depends not only on user roles, but also on the context information. Suppose there is an access policy $\tau = (Location = Hospital \wedge (A \wedge B))$ which says that a doctor (we assume that a doctor is identified with attributes A and B) can view patients health records only if doctor's request comes from inside the hospital. Outside the hospital, no user should be able to decrypt the ciphertext encrypted under the access tree τ, even if the user may own a secret key associated with the attributes (A, B). Using mCP-ABE scheme, the enforcement of τ, which contain context attributes, is done as follows:

- a patient encrypts her health record according to the access policy $(A \wedge B)$, and then uploads his data to a PHR repository. Hence, part of the access policy is enforced in the Encryption phase by the patient.
- a doctor download encrypted data and request from Proxy a decryption token.
- Proxy checks the context attribute inside τ and issues decryption token only if the request comes from inside the hospital (e.g. only if the request comes

from a specific IP address), therefore the context attribute is enforced by Proxy in the m-Decrypt phase.

Note that the involvement of an online semi-trusted mediator (Proxy) plays a crucial role in the enforcement of context attributes, as it is very hard or rather impossible to enforce these attributes without the involvement of an online component. The mCP-ABE scheme can also support the off-line use of data. Then the architecture is slightly changed in a way that Proxy is distributed to the users or their domains within which the data will be used. As a consequence there will be a number of proxies which will be coordinated by the central Proxy. The above defined process will not fundamentally change, except that the central Proxy will update the local ones and that in the data consumption phase, the user will contact only the local Proxy.

6 Conclusion

We propose a mediated Ciphertext-Policy Attribute-Based Encryption scheme that supports revocation of user attributes. If an attribute is revoked, the user cannot use it in the decryption phase. The scheme allows the encryptor to encrypt a message according to an access policy over a set of attributes, and only users who satisfy the access policy and whose attributes are not revoked can decrypt the ciphertext. Furthermore, we demonstrate how to use the proposed scheme to solve very important problems in managing Personal Health Records (PHRs). A possible extension to this work would be to provide a scheme which would have a security proof under standard complexity assumptions.

References

1. Sahai, A., Waters, B.: Fuzzy identity-based encryption. In: Cramer, R. (ed.) EUROCRYPT 2005. LNCS, vol. 3494, pp. 457–473. Springer, Heidelberg (2005)
2. Bethencourt, J., Sahai, A., Waters, B.: Ciphertext-policy attribute-based encryption. In: Proceedings of the 2007 IEEE Symposium on Security and Privacy, pp. 321–334. IEEE Computer Society Press, Washington (2007)
3. Cheung, L., Newport, C.: Provably secure ciphertext policy ABE, pp. 456–465. ACM, New York (2007)
4. Ibraimi, L., Tang, Q., Hartel, P., Jonker, W.: Efficient and provable secure ciphertext-policy attribute-based encryption schemes. In: Bao, F., Li, H., Wang, G. (eds.) ISPEC 2009. LNCS, vol. 5451, pp. 1–12. Springer, Heidelberg (2009)
5. Pirretti, M., Traynor, P., McDaniel, P., Waters, B.: Secure attribute-based systems. In: Proceedings of the 13th ACM Conference on Computer and Communications Security, pp. 99–112 (2006)
6. Goyal, V., Pandey, O., Sahai, A., Waters, B.: Attribute-based encryption for fine-grained access control of encrypted data. In: Proceedings of the 13th ACM Conference on Computer and Communications Security, pp. 89–98. ACM Press, New York (2006)

7. Boneh, D., Ding, X., Tsudik, G., Wong, C.M.: A method for fast revocation of public key certificates and security capabilities. In: SSYM 2001: Proceedings of the 10th conference on USENIX Security Symposium, Berkeley, CA, USA, p. 22. USENIX Association (2001)
8. Boneh, D., Ding, X., Tsudik, G.: Fine-grained control of security capabilities. ACM Transactions on Internet Technology 4, 60–82 (2004)
9. Libert, B., Quisquater, J.J.: Efficient revocation and threshold pairing based cryptosystems. In: Proceedings of the twenty-second annual symposium on Principles of distributed computing, pp. 163–171. ACM, New York (2003)
10. Nali, D., Miri, A., Adams, C.: Efficient revocation of dynamic security privileges in hierarchically structured communities. In: Proceedings of the 2nd Annual Conference on Privacy, Security and Trust (PST 2004), Fredericton, New Brunswick, Canada, pp. 219–223 (2004)
11. Nali, D., Adams, C., Miri, A.: Using mediated identity-based cryptography to support role-based access control. In: Zhang, K., Zheng, Y. (eds.) ISC 2004. LNCS, vol. 3225, pp. 245–256. Springer, Heidelberg (2004)
12. Boldyreva, A., Goyal, V., Kumar, V.: Identity-based encryption with efficient revocation. In: Proceedings of the 15th ACM conference on Computer and communications security, pp. 417–426. ACM, New York (2008)
13. Aiello, W., Lodha, S., Ostrovsky, R.: Fast digital identity revocation (extended abstract). In: Krawczyk, H. (ed.) CRYPTO 1998. LNCS, vol. 1462, pp. 137–152. Springer, Heidelberg (1998)
14. Naor, M., Nissim, K.: Certificate revocation and certificate update. In: USENIX Security Symposium, pp. 561–570. IEEE, Los Alamitos (1998)
15. Boneh, D., Franklin, M.: Identity-based encryption from the weil pairing. In: Kilian, J. (ed.) CRYPTO 2001. LNCS, vol. 2139, p. 213. Springer, Heidelberg (2001)
16. Ostrovksy, A., Sahai, A., Waters, B.: Attribute based encryption with non-monotonic access structures. In: ACM conference on Computer and Communications Security, pp. 195–203. ACM, New York (2007)
17. Shoup, V.: Lower bounds for discrete logarithms and related problems. In: Fumy, W. (ed.) EUROCRYPT 1997. LNCS, vol. 1233, pp. 256–266. Springer, Heidelberg (1997)
18. Vanrenen, G., Smith, S.: Distributing security-mediated PKI. In: Katsikas, S.K., Gritzalis, S., López, J. (eds.) EuroPKI 2004. LNCS, vol. 3093, pp. 218–231. Springer, Heidelberg (2004)
19. Shamir, A.: How to share a secret. Communications of the ACM 22(11), 612–613 (1979)
20. Desmedt, Y., Frankel, Y.: Threshold cryptosystems. In: Brassard, G. (ed.) CRYPTO 1989. LNCS, vol. 435, pp. 307–315. Springer, Heidelberg (1990)
21. Ibraimi, L., Petkovic, M., Nikova, S.I., Hartel, P.H., Jonker, W.: Mediated ciphertext-policy attribute-based encryption and its application (extended version). Technical Report TR-CTIT-09-12, Enschede (April 2009)
22. The US Department of Health and Human Services. Summary of the HIPAA privacy rule (2003)

Securing Remote Access Inside Wireless Mesh Networks

Mark Manulis

Cryptographic Protocols Group
Department of Computer Science
TU Darmstadt & CASED, Germany
mark@manulis.eu

Abstract. Wireless mesh networks (WMNs) that are being increasingly deployed in communities and public places provide a relatively stable routing infrastructure and can be used for diverse carrier-managed services. As a particular example we consider the scenario where a mobile device initially registered for the use with one wireless network (its home network) moves to the area covered by another network inside the same mesh. The goal is to establish a secure access to the home network using the infrastructure of the mesh.

Classical mechanisms such as VPNs can protect end-to-end communication between the mobile device and its home network while remaining transparent to the routing infrastructure. In WMNs this transparency can be misused for packet injection leading to the unnecessary consumption of the communication bandwidth. This may have negative impact on the cooperation of mesh routers which is essential for the connection establishment.

In this paper we describe how to establish remote connections inside WMNs while guaranteeing secure end-to-end communication between the mobile device and its home network *and* secure transmission of the corresponding packets along the underlying multi-hop path. Our solution is a provably secure, yet lightweight and round-optimal remote network access protocol in which intermediate mesh routers are considered to be part of the security architecture. We also sketch some ideas on the practical realization of the protocol using known standards and mention extensions with regard to forward secrecy, anonymity and accounting.

1 Introduction and Motivation

The increasing deployment of wireless networks in urban areas set up in private households and in public places offers high potential for ubiquitous communication to mobile clients. A promising approach for combining the power of different wireless networks, possibly under distinct administrative control, while expanding their overall coverage area is the composition into a *wireless mesh network (WMN)* through the deployment of dynamic routing protocols such as AODV [24], DSR [18], LQSR [13]. In this way a WMN consists of individual nodes (routers) which communicate with each other over wireless links in a multi-hop fashion. Many current WMNs are based on WLAN standards (IEEE 802.11a/b/g) so that mesh routing is performed at the network layer. A more efficient WMN infrastructure with mesh routing at the link layer is currently being specified within IEEE 802.11s. In contrast to a more general form of ad-hoc networks

H.Y. Youm and M. Yung (Eds.): WISA 2009, LNCS 5932, pp. 324–338, 2009.

WMNs enjoy *static* topology with mostly stable routing infrastructure. This infrastructure allows for various carrier-managed services, which may also apply amongst wireless networks that compose the *same* mesh. One of them is the remote network access.

A typical urban WMN connects not only residential networks but also networks of institutions of public interest such as civil administration, police station, doctor's office/surgery, etc. Such networks (which we refer to as *home networks*) can be accessed by employees through the mobile devices initialized for this purpose while being in the proximity of the corresponding access points. On the other hand, it is desirable to allow remote access to these networks from locations covered by other wireless networks, which are still part of the same mesh. For example, doctors visiting patients at home or employees of a local health authority while inspecting restaurants in the neighborhood may wish to remotely access databases and application servers in their institutions in order to perform tasks that they would usually do within their native environment. Observe that these scenarios do not assume high mobility of the clients and the established WMN routing path will likely remain unchanged for the whole duration of the session.

An important security goal of a remote network access protocol is to protect the end-to-end communication between the device and its home network, in particular to protect the application content from being eavesdropped or modified during its transmission. Another significant problem in WMNs is that intermediate mesh routers are under different administrative control and that their cooperation is crucial for the establishment and maintenance of the remote session. A promising approach to increase this cooperation and so improve the robustness of the remote connection is to deploy additional incentive mechanisms [9], e.g. let the home network reimburse the cooperating mesh routers for resources that they allocate to establish and maintain the remote session. It is clear that traditional end-to-end security mechanisms (such as IPsec VPNs) that are transparent to the underlying routing infrastructure do not adequately reflect the additional security needs of cooperating mesh routers along the remote path.

1.1 Refining End-to-End Protection with the Concept of Path Security

With traditional authentication and key establishment protocols for end-to-end security the mobile device and its home network can compute the secret *end-to-end key* for the session after authenticating each other. This key is sufficient to protect the application data exchanged between the device and its remotely accessed home network.

Nevertheless, the communication nature of WMNs allows the adversary to inject packets into the wireless channel and although the deployed end-to-end protection will prevent the end points from accepting such rogue packets the cooperating mesh routers on the path will not be able to distinguish them from the "good" packets originated by the end points. As a consequence, mesh routers will forward rogue packets, thus wasting their own (costly) communication resources. This illustrates the need of stronger protection mechanisms that would ensure authentication at the packet level as well.

We observe that the above problem comes from the missing binding between the end-to-end protected (application) channel and the underlying multi-hop WMN channel. In this work we investigate how to efficiently bind these independent channels for the duration of the remote session.[1] Our solution is to refine the traditional concept of end-to-end security by achieving similar goals for the underlying path (*path security*). More precisely, we provide all entities involved into the remote session, i.e. the mobile device, its home network, and intermediate mesh routers, with the ability to compute an additional *path key* as a by-product of an authentication protocol for the establishment of the end-to-end protected remote connection. In order to ensure security of this path key we also require authentication of intermediate mesh routers and the end points during the execution of this protocol. In this way the application content exchanged between the mobile device and its home network can be still protected using the end-to-end key whereas the path key will protect the multi-hop transmission of corresponding packets along the established path, thus thwarting possible resource consumption attacks on the mesh routers. The existence of the path key shared amongst the end points and the mesh routers may also offer further application-specific benefits once the remote connection is established. For example, it can be used to protect control messages used in quality-of-service (QoS) mechanisms in some real-time applications such as VoIP.

1.2 Related Work

The security architecture for WLANs specified in IEEE 802.11i provides an authenticated network access of a mobile device to a wireless network with additional computation of a session key. The authentication can be based on pre-shared keys (WPA, WPA2) or use the IEEE 802.1X specification based on the EAP framework [2], the latter is often used in combination with a RADIUS authentication server. However, IEEE 802.11i has been designed for single-hop connections and is, therefore, not directly applicable for WMNs. A direct extension of IEEE 802.11i to a multi-hop scenario has been described in [22]. Further, with PANA [14] there exists an authentication carrying protocol that can encapsulate EAP messages and transmit them in a multi-hop fashion across WMNs as described in [10, 19]. In this way any EAP method, e.g. using username/passwords [11], shared keys [7], or public-key certificates [25,17], can be applied. Alternatively, end-to-end security between the mobile client and its home network can be established using IPsec VPN tunnels.

The previously mentioned solutions, especially using current EAP methods, have been designed to achieve the classical end-to-end security goals. Therefore, they do not include any mechanisms that would be sufficient for the requirements of path security. In particular, they offer no protection against the impersonation of mesh routers and resource consumption attacks. On the other hand, with LHAP [29] there exists a lightweight authentication protocol (based on one-way hash functions) that can be used to obtain packet authentication in multi-hop communication scenarios. But LHAP was not designed to establish an end-to-end secure connection which is inherent to our setting. Moreover, LHAP does not offer mutual authentication between the mesh routers and the accessed home network.

---

[1] We remark that for networks with frequent route updates and for applications with highly mobile clients it would be rather undesirable to bind the application session to the underlying routing path. Yet, *inside* WMNs where the routes are rather stable and for remote sessions in which the clients remain static (as in our application examples) this binding is a promising practical method to provide stronger security guarantees for the remote connection.

Finally, we mention that the notion of path authentication is frequently used in the literature on routing protocols, especially in the context of secure route updates and announcements [27,28,26]. However, this process is oblivious to the actual application and is therefore not applicable in our case.

1.3 Organization

In Section 2 we describe our remote network access protocol, called SERENA, that simultaneously achieves the requirements of end-to-end and path security through the corresponding binding of the application channel to the underlying mesh path. In addition we provide some remarks on its efficiency and describe ideas on how to realize the protocol in practice using known authentication standards. In Section 3 we specify a single formal model for a secure remote network access protocol inside the wireless mesh capturing the mutual authentication and key establishment goals within the concepts of end-to-end and path security. The concept of path security is thereby modeled through the consideration of mesh routers as an inherent part of the security architecture. We utilize the classical modeling techniques from [5] which we extend to the multi-hop setting of WMNs. Following the model, in Section 4 we formally argue on the security of our protocol. Finally, in Section 5 we briefly discuss some further extensions regarding forward secrecy of the computed keys, anonymity of the mobile client with respect to the mesh network infrastructure, and accounting between the networks.

2 SERENA: A Single Protocol for End-to-End and Path Security

Here we introduce SERENA — our protocol for the secure remote access of the mobile device $\mathcal{M}$ to its wireless home network $\mathcal{H}$ using intermediate mesh routers $\{\mathcal{R}_i\}_i$, all belonging to the same mesh.

2.1 Notations and Building Blocks

SERENA uses several (well-known) cryptographic primitives:

- A *pseudo-random function* PRF : $\{0,1\}^\kappa \times \{0,1\}^* \to \{0,1\}^*$ which is used for the purpose of key derivation and can be realized using block-ciphers or keyed one-way hash functions. By $\mathrm{Adv}_{\mathrm{PRF}}^{\mathrm{prf}}(\kappa)$ we denote the maximum advantage over all PPT adversaries (running within time κ) in distinguishing the outputs of PRF from those of a random function better than by a random guess. We use PRF to derive various keys and sometimes we assume that the output is split in two parts, e.g. PRF with *expansion* such as the one defined within the TLS standard and analyzed in [15] can be used for this purpose.
- An *asymmetric encryption scheme* satisfying the property of indistinguishability under (adaptive) chosen-ciphertext attacks (IND-CCA2) whose encryption and decryption operations are denoted Enc and Dec, respectively. By $\mathrm{Adv}_{(\mathrm{Enc},\mathrm{Dec})}^{\mathrm{ind\text{-}cca2}}(\kappa)$ we denote the maximum advantage over all PPT adversaries (running within time κ) in breaking the IND-CCA2 property of (Enc, Dec) better than by a random guess; The property of IND-CCA2 security is for example preserved in several encryption schemes including RSA-OAEP [16] and DHIES [1].

- A *digital signature scheme* with existential unforgeability under chosen message attacks (EUF-CMA) whose signing and verification operations are denoted $\mathtt{Sig}$ and $\mathtt{Ver}$, respectively. By $\mathrm{Succ}_{(\mathtt{Sig},\mathtt{Ver})}^{\mathrm{euf\text{-}cma}}(\kappa)$ we denote the maximum success probability over all PPT adversaries (running within time κ) given access to the signing oracle in finding a forgery; examples of such schemes include DSS [23] and PSS [6].
- A *sequential aggregate signature scheme* with existential unforgeability under chosen message attacks (EUF-CMA) whose aggregate signing and verification operations are denoted $\mathtt{ASig}$ and $\mathtt{AVer}$, respectively. By $\mathrm{Succ}_{(\mathtt{ASig},\mathtt{AVer})}^{\mathrm{euf\text{-}cma}}(\kappa)$ we denote the maximum success probability over all PPT adversaries (running within time κ) given access to the aggregate signing oracle of an uncorrupted signer in finding an aggregate signature forgery; for formal security definitions and construction examples we refer to [21, 20].
- A *message authentication code* function $\mathtt{MAC}$ with *weak unforgeability against chosen message attacks* (WUF-CMA) [4], e.g., the popular function HMAC [3] can be used for this purpose. By $\mathrm{Succ}_{\mathtt{MAC}}^{\mathrm{wuf\text{-}cma}}(\kappa)$ we denote the maximum success probability over all PPT adversaries (running within time κ) given access to the $\mathtt{MAC}$ oracle in finding a $\mathtt{MAC}$ forgery.

2.2 Initialization of SERENA

We assume that prior to the execution of Π the involved parties are in possession of the following long-lived keys: LL_i of each mesh router $\mathcal{R}_i$ consists of a private/public signature/verification key pair (sk_i, vk_i) which is suitable for the deployed aggregate signature scheme ($\mathtt{ASig}$, $\mathtt{AVer}$) and a decryption/encryption key pair (dk_i, ek_i); $LL_{\mathcal{H}}$ consists of a private/public signature/verification key pair $(sk_{\mathcal{H}}, vk_{\mathcal{H}})$ which is suitable for the digital signature scheme ($\mathtt{Sig}$, $\mathtt{Ver}$) and a pair $(\mathcal{M}, (k_{\mathcal{M}}, \alpha_{\mathcal{M}}))$ where $(k_{\mathcal{M}}, \alpha_{\mathcal{M}})$ is a high-entropy secret key consisting of a PRF key $k_{\mathcal{M}}$ and a MAC key $\alpha_{\mathcal{M}}$ shared with the hosted $\mathcal{M}$; $LL_{\mathcal{M}}$ consists of $(k_{\mathcal{M}}, \alpha_{\mathcal{M}})$. In practice it is sufficient for $\mathcal{H}$ and $\mathcal{M}$ to share $k_{\mathcal{M}}$ and derive the corresponding MAC key $\alpha_{\mathcal{M}}$ as $\mathrm{PRF}_{k_{\mathcal{M}}}(l)$ for some publicly fixed label l. Further, we assume that the public keys of all routers of the same mesh (including the home network $\mathcal{H}$) are known. Note that due to the static infrastructure of wireless mesh networks this assumption is feasible (this is clearly in contrast to (dynamic) ad-hoc networks where such assumptions are undesirable).

2.3 Lightweight SERENA — Specification

The detailed specification of lightweight SERENA is illustrated in Figure 1. For the clarity of presentation we assume that the path between $\mathcal{M}$ and $\mathcal{H}$ through the mesh network is given by a sequence of intermediate mesh routers $\mathcal{R}_1, \ldots, \mathcal{R}_n$ whereby each $\mathcal{R}_i$ knows the next hop router $\mathcal{R}_{i+1}$ as a result of underlying routing mechanism. In the following we highlight how SERENA merges end-to-end protection with path security.

SERENA uses distinct *labels* l_i, $i = 1, \ldots, 4$, which are fixed in advance, as input to PRF for the derivation of different keys. Also, during the execution each party $P \in \{\mathcal{M}, \{\mathcal{R}_i\}_i, \mathcal{H}\}$ computes own session id sid_P as a concatenated bit string $\mathcal{M}|r_{\mathcal{M}}| \{\mathcal{R}_i|r_i\}_i|\mathcal{H}|r_{\mathcal{H}}$ where r_P denotes a *random nonce* chosen by P. For this parties use auxiliary variables $usid$ and $dsid$ which contain respective substrings of concatenated identities and nonces, depending on the position of P in the path.

Both $\mathcal{H}$ and $\mathcal{M}$ derive their end-to-end key K_e using the shared key $k_\mathcal{M}$ and their session ids. The equality of their session ids is ensured upon the verification of the corresponding MAC values $\mu_\mathcal{H}$ and $\mu_\mathcal{M}$, whereby bits 0 and 1 are used to guarantee that $\mu_\mathcal{H} \neq \mu_\mathcal{M}$.

The difficult part of the protocol from the security point of view is the computation of the path key K_p and the mutual authentication between $\mathcal{H}$ and each mesh router $\mathcal{R}_i$. The main idea is to let the home network choose the secret key material k'_p which will be securely transported through the mesh network. Since the encryption of k'_p by $\mathcal{H}$ for each router $\mathcal{R}_i$ would lead to the linear increase of the ciphertext we apply a better approach — *hop-by-hop re-encryption*, i.e. k'_p originally encrypted by $\mathcal{H}$ for $\mathcal{R}_n$ is re-encrypted by $\mathcal{R}_n$ for $\mathcal{R}_{n-1}$ and so on. The challenge of such re-encryption is to allow each $\mathcal{R}_i$ to independently verify that decrypted k'_p is essentially the same as the one originally sent by $\mathcal{H}$, i.e. without trusting the next-hop router. The trick is to use the PRF output θ_p, which is never sent but is signed within $\sigma_\mathcal{H}$, which in turn is forwarded along the mesh path until it reaches $\mathcal{R}_1$. The verification of this $\sigma_\mathcal{H}$ implicitly requires each $\mathcal{R}_i$ to recompute θ_p and also to hold the same session id. Since θ_p and k_p (used to derive the path key K_p) are both derived from k'_p, the validity of the signature $\sigma_\mathcal{H}$ verified by $\mathcal{R}_i$ immediately implies the equality of k'_p received by $\mathcal{R}_i$ from $\mathcal{R}_{i+1}$ and the original k'_p sent by $\mathcal{H}$. Since the validity of $\sigma_\mathcal{H}$ also implies that $\mathcal{R}_i$ and $\mathcal{H}$ share the same session id, it is easy to see that at the end of the protocol each mesh router computes the same path key. At the same time the validity of $\sigma_\mathcal{H}$ is used for the authentication of $\mathcal{H}$ towards each $\mathcal{R}_i$. The mobile device $\mathcal{M}$ computes the same path key since it can derive k'_p directly using $k_\mathcal{M}$. The authentication of each $\mathcal{R}_i$ towards $\mathcal{H}$ is performed via the aggregate signature $\bar{\sigma}_{1..n}$. Each $\mathcal{R}_i$ contributes to the computation of this signature by executing the ASig algorithm using its own signing key sk_i, the aggregate signature $\bar{\sigma}_{1..i-1}$ previously received from $\mathcal{R}_{i-1}$, and the message $sid_i|\theta_p|\mu_\mathcal{M}$. The successful verification of $\bar{\sigma}_{1..n}$ by $\mathcal{H}$ authenticates the established path and ensures $\mathcal{H}$ that the mesh routers $\{\mathcal{R}_i\}_i$ hold the same session id and the path key as $\mathcal{H}$.

Note that during the protocol execution parties perform various checks. We assume that if at some point the corresponding check fails then the party immediately aborts. If parties do not abort then the execution of SERENA was successful, meaning that from now on $\mathcal{M}$ can be granted access to $\mathcal{H}$. In practice, $\mathcal{M}$ will be assigned an IP address from the domain of $\mathcal{H}$ and treated by $\mathcal{H}$ as one of its native devices.

2.4   Efficiency of SERENA

We stress that SERENA is lightweight in the sense that the mobile device $\mathcal{M}$ does *not* need to perform any costly public-key operations. Hence, SERENA can also be used with performance-constraint devices such as PDAs and smart phones provided they have a wireless IP interface. The public-key cryptography is used in SERENA for the hop-by-hop re-encryption of k'_p and the mutual authentication between $\{\mathcal{R}_i\}_i$ and $\mathcal{H}$. On the other hand, hop-by-hop re-encryption and the use of aggregate signatures allows to significantly reduce the (wireless) communication overhead which is the main bottleneck

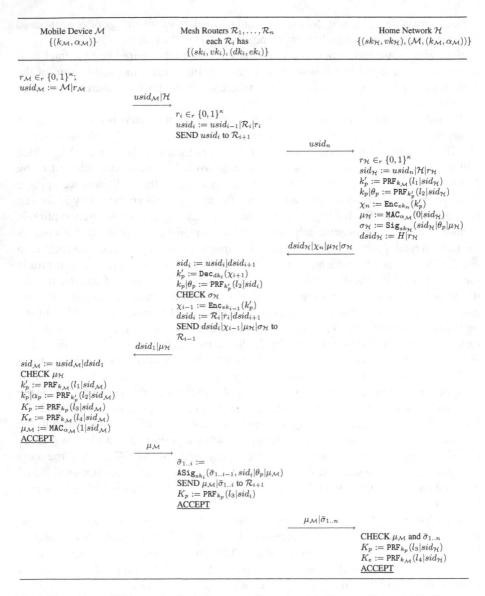

Fig. 1. Execution of lightweight SERENA protocol between $\mathcal{M}$, $\mathcal{R}_1, \ldots, \mathcal{R}_n$, and $\mathcal{H}$. Computations of intermediate mesh routers are described per each $\mathcal{R}_i$ and are triggered upon the delivery of the expected message from $\mathcal{R}_{i-1}$ or $\mathcal{R}_{i+1}$; indices $i = 0$ and $i = n + 1$ refer to $\mathcal{M}$ and $\mathcal{H}$, respectively.

in WMNs (in contrast to the usually rich computation resources available to the mesh routers). Note also that SERENA requires three communication rounds — by one round we mean a complete message flow between $\mathcal{M}$ and $\mathcal{H}$ routed through $\{\mathcal{R}_i\}_i$, which is optimal to achieve mutual authentication (with key confirmation) based on the deployed challenge-response technique.

Remark 1. The modularity of SERENA allows to completely remove public-key operations (and the corresponding long-lived keys) resulting in a more efficient protocol that would nevertheless still ensure the traditional end-to-end security between M and H. However, this modification will no longer allow for a secure computation of the path key and the mutual authentication between the home network and the mesh routers.

2.5 Implementation of SERENA Using Authentication Standards

In order to realize SERENA with existing authentication standards we have to consider the following connection links: the wireless link between M and R_1 and the wireless multi-hop path from R_1 to H over the last-hop router R_n.

We assume that R_1 combines the routing functionality with that of an access point for M. Prior to the execution of SERENA the mobile device M would be typically connected to R_1 at the data link layer, which is preferable as no IP assignment for M is necessary in this case. Then, on the link between M and R_1 our protocol can be implemented as a new EAP method within IEEE 802.1X framework. Along the multi-hop path from R_1 to H (which combines the functionality of a mesh router and the authentication/application server) we can further implement SERENA using encapsulated EAP messages within an appropriate carrying protocol, for example with PANA [14].

Once SERENA is successfully executed H can allocate an IP address for M (either as a parameter within SERENA or via DHCP). This would complete the establishment of a secure remote network access between M and H. In the course of subsequent communication between M and H intermediate mesh routers $R_1, \ldots, R_n$ can continue acting as layer 2 bridges until the session is finished.

The end-to-end protection of the session content between M and H can be achieved traditionally using the Authentication Header (AH) or Encapsulation Security Payload (ESP) mechanisms of IPsec in the *tunnel mode*, whereby the session key for this should be derived from K_e. The additional packet protection that should prevent resource consumption attacks via packet injection can be also performed using AH and ESP, however, this time in the *transport mode* and using the session key derived from K_p.

3   Security Model

Our model extends the two-party model from [5] towards the multi-hop communication nature of WMNs and the additional path security goals.

3.1   Participants and Communication

Mobile Device and Home Network. The goal of the protocol, denoted abstractly by Π, is to establish a secure remote network access between the mobile device M and its wireless home network H. The identity M is assumed to be unique within H. Both M and H hold their corresponding long-lived keys LL_M and LL_H. By $\kappa \in \mathbb{N}$ we denote the *security parameter* such that all secrets used in Π are polynomially bounded in κ.

Intermediate Mesh Routers. We assume that $\mathcal{H}$ is part of the WMN and that the connection between $\mathcal{M}$ and $\mathcal{H}$ will be established along a multi-hop path consisting of some intermediate WMN routers $\mathcal{R}_1, \ldots, \mathcal{R}_n$. For the ease of presentation, we consider $\mathcal{R}_1$ as the first-hop router for $\mathcal{M}$ and $\mathcal{R}_n$ as the last-hop router before $\mathcal{H}$. Typically, such route will be defined in the connection establishment phase by the underlying routing protocol. Each intermediate mesh router $\mathcal{R}_i$ has a long-lived key LL_i.

Instances, Sessions, Partnering, Session Keys. In order to model participation of $\mathcal{M}$, $\mathcal{H}$, and $\{\mathcal{R}_i\}_i$ in distinct sessions of Π we consider an unlimited number of instances: By $[P, s]$ we denote the *s-th instance of* $P \in \{\mathcal{M}, \mathcal{H}, \{\mathcal{R}_i\}_i\}$ where $s \in \mathbb{N}$. Each instance $[P, s]$ is initialized with the party's long-lived key and invoked for one session which will be identified through a unique (public) *session id* sid_P^s. Instances of $\mathcal{M}$, $\mathcal{H}$ and $\{\mathcal{R}_i\}_i$ with identical session ids are *partnered*, i.e. participate in the same session.

Prior to the termination of Π an instance $[P, s]$ *accepts* if the protocol execution was successful (from the perspective of $[P, s]$) or *aborts*. A mesh router $\mathcal{R}_i$ accepts after it successfully authenticates $\mathcal{H}$ and computes the *session path key* $K_p \in \{0, 1\}^\kappa$; the home network $\mathcal{H}$ accepts after it successfully authenticates the mobile device and the intermediate mesh routers on the path, and computes the *session end-to-end key* $K_e \in \{0, 1\}^\kappa$ (in addition to K_p); and the mobile device $\mathcal{M}$ accepts after it successfully authenticates $\mathcal{H}$ and computes K_e and K_p.

3.2 Adversary and Security Definitions

Adversarial Model. Adversary $\mathcal{A}$ is a PPT machine with complete control over the communication. It can mount attacks via the following set of queries:

- Invoke(P, m): This is the protocol invocation query for some entity $P \in \{\mathcal{M}, \mathcal{H}, \{\mathcal{R}_i\}_i\}$. In response, a new instance $[P, s]$ is created and its first protocol message is given to $\mathcal{A}$. The optional input m indicates the message expected by the instance to start the execution; for the initiator of the protocol m is supposed to be empty.
- Send(P, s, m): This query models communication control by $\mathcal{A}$ and contains a message m which should be delivered to the s-th instance of $P \in \{\mathcal{M}, \mathcal{H}, \{\mathcal{R}_i\}_i\}$. In response, $\mathcal{A}$ receives the outgoing message of $[P, s]$, or an empty string if $[P, s]$ terminates having processed m.
- Corrupt(P): This query models corruptions of $P \in \{\mathcal{M}, \mathcal{H}, \{\mathcal{R}_i\}_i\}$. In response, $\mathcal{A}$ receives LL_P.
- Reveal_Ke(P, s): This query models independence of end-to-end keys computed by the instances of $P \in \{\mathcal{M}, \mathcal{H}\}$ in different sessions. In response, $\mathcal{A}$ is given K_e held by the instance; the query is answered only if $[P, s]$ has accepted.
- Reveal_Kp(P, s): This query models independence of path keys computed by the instances of $P \in \{\mathcal{M}, \mathcal{H}, \{\mathcal{R}_i\}_i\}$ in different sessions. In response, $\mathcal{A}$ is given K_p held by the instance; the query is answered only if $[P, s]$ has accepted.
- Test_Ke(P, s): This query will be used to define the AKE-security of the end-to-end key K_e and can be asked only for $P \in \{\mathcal{M}, \mathcal{H}\}$. It is answered only if $[P, s]$ has accepted and the answer is based on some secret bit $b \in \{0, 1\}$ chosen in advance: If $b = 1$ then $\mathcal{A}$ is given K_e, otherwise a randomly chosen value from $\{0, 1\}^\kappa$.

- Test_Kp(P, s): This query will be used to define the AKE-security of the path key K_p and can be asked for $P \in \{\mathcal{M}, \mathcal{H}, \{\mathcal{R}_i\}_i\}$. It is answered only if $[P, s]$ has accepted and the answer is based on some secret bit $b \in \{0, 1\}$ chosen in advance: If $b = 1$ then $\mathcal{A}$ is given K_p, otherwise a randomly chosen value from $\{0, 1\}^\kappa$.

Definition 1 (Correctness). *A protocol Π is* correct *if the invoked instances of $\mathcal{M}$, $\mathcal{H}$, and $\{\mathcal{R}_i\}_i$ terminate having accepted and **all** of the following holds: $\mathcal{M}$ and $\mathcal{H}$ hold the same end-to-end key K_e; $\mathcal{M}$, $\mathcal{H}$, and $\{\mathcal{R}_i\}_i$ hold the same path key K_p.*

End-to-End Security. Our first definition is about mutual authentication between $\mathcal{M}$ and $\mathcal{H}$. It also ensures that partnered instances of $\mathcal{M}$ and $\mathcal{H}$ accept holding the same end-to-end key K_e and the same path key K_p.

Definition 2 (MA between $\mathcal{M}$ and $\mathcal{H}$). *Given a correct protocol Π by $\text{Game}_\Pi^{\text{ma-m-h}}(\mathcal{A}, \kappa)$ we denote the interaction between the instances of $\mathcal{M}$, $\mathcal{H}$ and $\{\mathcal{R}_i\}_i$ with a PPT adversary $\mathcal{A}$ that is allowed to query* Invoke, Send, Corrupt, Reveal_Ke, *and* Reveal_Kp. *$\mathcal{A}$ wins if at some point during the interaction:*

*(1) an uncorrupted instance of $\mathcal{M}$ accepts but there is **no** uncorrupted partnered instance of $\mathcal{H}$, **or***

*(2) an uncorrupted instance of $\mathcal{H}$ accepts but there is **no** uncorrupted partnered instance of $\mathcal{M}$, **or***

*(3) uncorrupted partnered instances of $\mathcal{M}$ and $\mathcal{H}$ accept without holding the same session end-to-end key K_e, **or***

(4) $\{\mathcal{R}_i\}_i$ are uncorrupted and uncorrupted partnered instances of $\mathcal{M}$ and $\mathcal{H}$ accept without holding the same session path key K_p.

The maximum probability of this event over all adversaries (running in time κ) is denoted

$$\text{Succ}_\Pi^{\text{ma-m-h}}(\mathcal{A}, \kappa) = \max_{\mathcal{A}} |\Pr[\mathcal{A} \text{ wins in } \text{Game}_\Pi^{\text{ma-m-h}}(\mathcal{A}, \kappa)]|.$$

Π provides mutual authentication between $\mathcal{M}$ and $\mathcal{H}$ *if this probability is negligible in κ.*

The secrecy of the established end-to-end key K_e is modeled through the classical notion of authenticated key exchange (AKE) security adopted to our setting. Recall that the basic idea behind AKE-security is to require the indistinguishability of the key computed in some *test* session from some randomly chosen value by any *outsider* adversary. Observe that in case of K_e we must allow the adversary to corrupt the intermediate mesh routers $\{\mathcal{R}_i\}_i$ also during the test session.

In order to define the AKE-security of K_e (denoted as *e-AKE-security*) we first specify the auxiliary notion of *e-freshness* for the instances of $\mathcal{M}$ and $\mathcal{H}$. It provides conditions under which $\mathcal{A}$ can be treated as an outsider with respect to the test session for which it has to distinguish K_e.

Definition 3 (e-Freshness). *In the execution of Π an instance $[P, s]$ with $P \in \{\mathcal{M}, \mathcal{H}\}$ is* e-fresh *if **none** of the following holds:*

- $\mathcal{A}$ *asks* Corrupt(P);
- IF $P = \mathcal{M}$: $\mathcal{A}$ *asks* Reveal_Ke($\mathcal{M}, s$) *after* $[\mathcal{M}, s]$ *has accepted or* Reveal_Ke($\mathcal{H}$, t) *after* $[\mathcal{H}, t]$ *has accepted and* $[\mathcal{M}, s]$ *and* $[\mathcal{H}, t]$ *are partnered;*
- IF $P = \mathcal{H}$: $\mathcal{A}$ *asks* Reveal_Ke($\mathcal{H}, s$) *after* $[\mathcal{H}, s]$ *has accepted or* Reveal_Ke($\mathcal{M}$, t) *after* $[\mathcal{M}, t]$ *has accepted and* $[\mathcal{H}, s]$ *and* $[\mathcal{M}, t]$ *are partnered.*

Definition 4 (e-AKE-Security). *Given a correct protocol Π, a uniformly chosen bit b, and a PPT adversary $\mathcal{A}$ with access to the queries* Invoke, Send, Corrupt, Reveal_Ke, Reveal_Kp, *and* Test_Ke, *by* $\mathsf{Game}_{\Pi}^{\text{e-ake},b}(\mathcal{A}, \kappa)$ *we denote the following interaction between the instances of $\mathcal{M}$, $\mathcal{H}$ and $\{\mathcal{R}_i\}_i$ with $\mathcal{A}$:*

- $\mathcal{A}$ *interacts with instances via queries;*
- *at some point $\mathcal{A}$ asks a* Test_Ke *query to an instance $[P, s]$ which has accepted and is* **e-fresh** *(and remains such by the end of the interaction);*
- $\mathcal{A}$ *continues interacting with instances and when $\mathcal{A}$ terminates, it outputs a bit, which is then set as the output of the interaction.*

$\mathcal{A}$ *wins if the output of* $\mathsf{Game}_{\Pi}^{\text{e-ake},b}(\mathcal{A}, \kappa)$ *is identical to b. The maximum probability of the adversarial advantage over the random guess of b, over all adversaries (running in time κ) is denoted*

$$\mathsf{Adv}_{\Pi}^{\text{e-ake}}(\mathcal{A}, \kappa) = \underset{\mathcal{A}}{\max} |2 \Pr[\mathsf{Game}_{\Pi}^{\text{e-ake},b}(\mathcal{A}, \kappa) = b] - 1|.$$

If this advantage is negligible in κ then Π provides e-AKE-security.

Path Security. Here we formalize the additional path security requirements from Section 1.1. First, we define mutual authentication between the home network $\mathcal{H}$ and each mesh router $\mathcal{R}_i$ on the path, which also captures the equality of session path keys computed by $\mathcal{H}$ and each $\mathcal{R}_i$. Note also that winning conditions in this case require the instance of $\mathcal{R}_i$ to be uncorrupted, however, the adversary can still corrupt any other $\mathcal{R}_j$, $j \neq i$. This models attacks of malicious routers aiming to disrupt the mutual authentication process and the computation of identical path keys between the (honest) intermediate mesh routers and the home network.

Definition 5 (MA between $\mathcal{H}$ and $\mathcal{R}_i$). *Given a correct protocol Π by* $\mathsf{Game}_{\Pi}^{\text{ma-h-r}}(\mathcal{A}, \kappa)$ *we denote the interaction between the instances of $\mathcal{M}$, $\mathcal{H}$ and $\{\mathcal{R}_i\}_i$ with a PPT adversary $\mathcal{A}$ that is allowed to query* Invoke, Send, Corrupt, Reveal_Ke, *and* Reveal_Kp. *$\mathcal{A}$ wins if at some point during the interaction:*

*(1) an uncorrupted instance of $\mathcal{R}_i$ accepts but there is **no** uncorrupted partnered instance of $\mathcal{H}$, **or***

*(2) an uncorrupted instance of $\mathcal{H}$ accepts but there is **no** uncorrupted partnered instance of $\mathcal{R}_i$, **or***

(3) $\mathcal{M}$ is uncorrupted and uncorrupted partnered instances of $\mathcal{H}$ and $\mathcal{R}_i$ accept without holding the same session path key K_p.

The maximum probability of this event over all adversaries (running in time κ) is denoted

$$\mathsf{Succ}_{\Pi}^{\text{ma-h-r}}(\mathcal{A}, \kappa) = \underset{\mathcal{A}}{\max} |\Pr[\mathcal{A} \text{ wins in } \mathsf{Game}_{\Pi}^{\text{ma-h-r}}(\mathcal{A}, \kappa)]|.$$

Π provides mutual authentication between $\mathcal{H}$ and $\mathcal{R}_i$ *if this probability is negligible in κ.*

Recall that the instances of participants are seen as partnered if they hold the same session ids. Hence, if Π provides both the mutual authentication between $\mathcal{M}$ and $\mathcal{H}$ *and* the mutual authentication between $\mathcal{H}$ and each $\mathcal{R}_i$ then the execution of Π in which $\mathcal{H}$ accepts implies the partnering for the corresponding instances of $\mathcal{M}$ and each $\mathcal{R}_i$, which in turn implies that the instances of $\mathcal{M}$, $\{\mathcal{R}_i\}_i$, and $\mathcal{H}$ computed the same K_p.

Further, we model the AKE-security of the path key K_p (denoted *p-AKE-security*) modeled using the auxiliary notion of *p-freshness* that specifies the conditions under which $\mathcal{A}$ can be treated as an outsider with respect to the test session for which it has to distinguish K_p.

Definition 6 (p-Freshness). *In the execution of Π an instance $[P, s]$ with $P \in \{\mathcal{M}, \mathcal{H}, \{\mathcal{R}_i\}_i\}$ is p-fresh if none of the following holds:*

- $\mathcal{A}$ *asks* Corrupt(P);
- IF $P = \mathcal{M}$: $\mathcal{A}$ *asks* Reveal_Kp($\mathcal{M}, s$) *after* $[\mathcal{M}, s]$ *has accepted or* Reveal_Kp(P', t) *for* $P' \in \{\mathcal{H}, \{\mathcal{R}_i\}_i\}$ *after* $[P', t]$ *has accepted and* $[\mathcal{M}, s]$ *and* $[P', t]$ *are partnered;*
- IF $P = \mathcal{H}$: $\mathcal{A}$ *asks* Reveal_Kp($\mathcal{H}, s$) *after* $[\mathcal{H}, s]$ *has accepted or* Reveal_Kp(P', t) *for* $P' \in \{\mathcal{M}, \{\mathcal{R}_i\}_i\}$ *after* $[P', t]$ *has accepted and* $[\mathcal{H}, s]$ *and* $[P', t]$ *are partnered;*
- IF $P = \mathcal{R}_i$: $\mathcal{A}$ *asks* Reveal_Kp($\mathcal{R}_i, s$) *after* $[\mathcal{R}_i, s]$ *has accepted or* Reveal_Kp(P', t) *for* $P' \in \{\mathcal{M}, \mathcal{H}, \{\mathcal{R}_j\}_{j \neq i}\}$ *after* $[P', t]$ *has accepted and* $[\mathcal{R}_i, s]$ *and* $[P', t]$ *are partnered.*

Definition 7 (p-AKE-Security). *Given a correct protocol Π, a uniformly chosen bit b, and a PPT adversary $\mathcal{A}$ with access to the queries* Invoke, Send, Corrupt, Reveal_Ke, Reveal_Kp, *and* Test_Kp, *by* $\mathsf{Game}_\Pi^{\mathsf{p\text{-}ake},b}(\mathcal{A}, \kappa)$ *we denote the following interaction between the instances of $\mathcal{M}$, $\mathcal{H}$ and $\{\mathcal{R}_i\}_i$ with $\mathcal{A}$:*

- $\mathcal{A}$ *interacts with instances via queries;*
- *at some point $\mathcal{A}$ asks a* Test_Kp *query to an instance $[P, s]$ which has accepted and is p-fresh (and remains such by the end of the interaction);*
- $\mathcal{A}$ *continues interacting with instances and when $\mathcal{A}$ terminates, it outputs a bit, which is then set as the output of the interaction.*

$\mathcal{A}$ *wins if the output of* $\mathsf{Game}_\Pi^{\mathsf{p\text{-}ake},b}(\mathcal{A}, \kappa)$ *is identical to b. The maximum probability of the adversarial advantage over the random guess of b, over all adversaries (running in time κ) is denoted*

$$\mathsf{Adv}_\Pi^{\mathsf{p\text{-}ake}}(\mathcal{A}, \kappa) = \max_{\mathcal{A}} |2 \Pr[\mathsf{Game}_\Pi^{\mathsf{p\text{-}ake},b}(\mathcal{A}, \kappa) = b] - 1|.$$

If this advantage is negligible in κ then Π provides p-AKE-security.

4 Security Analysis of SERENA

The following theorems (see full version for proofs) show that SERENA provides both end-to-end and path security. By q we denote the total number of the invoked sessions.

Theorem 1 (MA between $\mathcal{M}$ and $\mathcal{H}$). *Given a WUF-CMA secure* MAC *the protocol* SERENA *specified in Figure 1 provides mutual authentication between the participating mobile device and its home network in the sense of Definition 2, and*

$$\text{Succ}_{\text{SERENA}}^{\text{ma-m-h}}(\kappa) \leq \frac{2q^2}{2^\kappa} + 2\text{Succ}_{\text{MAC}}^{\text{wuf-cma}}(\kappa).$$

Theorem 2 (e-AKE). *Given a WUF-CMA secure* MAC *and a pseudo-random* PRF *the protocol* SERENA *specified in Figure 1 provides e-AKE-security in the sense of Definition 4, and*

$$\text{Adv}_{\text{SERENA}}^{\text{e-ake}}(\kappa) \leq \frac{4q^2}{2^\kappa} + 4\text{Succ}_{\text{MAC}}^{\text{wuf-cma}}(\kappa) + 4q\text{Adv}_{\text{PRF}}^{\text{prf}}(\kappa).$$

Theorem 3 (MA between $\mathcal{H}$ and $\mathcal{R}_i$). *Given EUF-CMA secure* (Sig, Ver) *and* (ASig, AVer) *the protocol* SERENA *specified in Figure 1 provides mutual authentication between each participating mesh router $\mathcal{R}_i$ and the home network $\mathcal{H}$ in the sense of Definition 5, and*

$$\text{Succ}_{\text{SERENA}}^{\text{ma-h-r}}(\kappa) \leq \frac{(n+1)q^2}{2^\kappa} + \text{Succ}_{(\text{Sig},\text{Ver})}^{\text{euf-cma}}(\kappa) + n\text{Succ}_{(\text{ASig},\text{AVer})}^{\text{euf-cma}}(\kappa).$$

Theorem 4 (p-AKE). *Given EUF-CMA secure* (Sig, Ver) *and* (ASig, AVer)*, a IND-CCA2 secure* (Enc, Dec)*, and a pseudo-random* PRF *the protocol* SERENA *specified in Figure 1 provides p-AKE-security in the sense of Definition 7, and*

$$\text{Adv}_{\text{SERENA}}^{\text{e-ake}}(\kappa) \leq \frac{2(n+2)q^2}{2^\kappa} + 4\text{Succ}_{\text{MAC}}^{\text{wuf-cma}}(\kappa) + 2\text{Succ}_{(\text{Sig},\text{Ver})}^{\text{euf-cma}}(\kappa) +$$

$$+ 2n\text{Succ}_{(\text{ASig},\text{AVer})}^{\text{euf-cma}}(\kappa) + 2q\text{Adv}_{(\text{Enc},\text{Dec})}^{\text{ind-cca2}}(\kappa) + 8q\text{Adv}_{\text{PRF}}^{\text{prf}}(\kappa).$$

5   Forward Secrecy, Anonymity, and Accounting

SERENA can be extended in a modular way to address forward secrecy, anonymity of mobile devices, and accounting between the cooperating networks of the same mesh.

Forward Secrecy of End-to-End and Path Keys. The common way to achieve *forward secrecy*, i.e. to ensure that AKE-security holds even if $\mathcal{A}$ gains access to the long-lived keys of participants after their instances have accepted in the test session, is to derive the key from some ephemeral secret information which is valid only for one particular session.

   Forward secrecy of K_e can be obtained using the classical Diffie-Hellman technique by deriving K_e from an ephemeral secret $g^{x_\mathcal{M} x_\mathcal{H}}$ (where $\mathbb{G}$ is a cyclic group of prime order q and $x_\mathcal{M}, x_\mathcal{H} \in \mathbb{Z}_q$), e.g. as an output of $\text{PRF}_{f(g^{x_\mathcal{M} x_\mathcal{H}})}(l_3|sid)$ with some randomness extractor f [12]. For this, $\mathcal{M}$ and $\mathcal{H}$ must exchange $g^{x_\mathcal{M}}$ and $g^{x_\mathcal{H}}$ as part of their first protocol messages. In order to ensure authentication, $\mathcal{H}$ and $\mathcal{M}$ must include $g^{x_\mathcal{M}}$ and $g^{x_\mathcal{H}}$ into the computation of $\mu_\mathcal{H}$ and $\mu_\mathcal{M}$, respectively. The AKE-security of such K_e would then rely on the Decisional Diffie-Hellman Problem.

Forward secrecy of K_p can be obtained similarly using the Generalized Diffie-Hellman technique [8] by deriving K_p from $g^{x_\mathcal{M}(\prod_i x_i)x_\mathcal{H}}$ (where $x_\mathcal{M}, x_i, x_\mathcal{H} \in \mathbb{Z}_q$ and each x_i is chosen by $\mathcal{R}_i$). Note that in this case there is no need to apply hop-by-hop encryption of k_p' so that all computations involving k_p' become obsolete and the key pairs (dk_i, ek_i) can be dismissed from the routers' long-lived keys. The AKE-security of K_p would then rely on the Group Decisional Diffie-Hellman Problem. Note that achieving forward secrecy increases the protocol costs and if forward secrecy is required for K_e and K_p simultaneously then exponents $x_\mathcal{M}$ and $x_\mathcal{H}$ should be independent for each of the above techniques.

Anonymity of $\mathcal{M}$. If necessary SERENA can be extended to achieve anonymity of $\mathcal{M}$ and unlinkability of its sessions simply by encrypting the identity $\mathcal{M}$ in the first protocol message using the IND-CCA2 secure encryption scheme $(\mathcal{E}, \mathcal{D})$. Here we assume that $\mathcal{H}$ holds own decryption/encryption key pair $(dk_\mathcal{H}, ek_\mathcal{H})$ as part of its long-lived key, which is already the case as $\mathcal{H}$ is one of the networks of the mesh. This implies that the session id would be computed using the encryption of $\mathcal{M}$. We remark that this solution increases the costs by one public key operation for both $\mathcal{M}$ and $\mathcal{H}$.

Accounting between $\mathcal{H}$ and $\mathcal{R}_i$. In SERENA each mesh router obtains $\sigma_\mathcal{H}$ computed by $\mathcal{H}$ amongst other parameters on the session id which also includes the identity of each $\mathcal{R}_i$. This signature can be extended with the time-stamp T and used for the purpose of accounting between $\mathcal{H}$ and $\mathcal{R}_i$, e.g. as an incentive mechanism for the cooperation of mesh routers. Note that since the packets are routed in real-time, both $\mathcal{H}$ and each $\mathcal{R}_i$ can independently keep track on the size of the transmitted packets or on the duration of the session. Further, any signature $\sigma_\mathcal{H}'$ computed by $\mathcal{H}$ for some time interval $[T, T']$ could serve as a cryptographically protected acknowledgement of $\mathcal{H}$ for the reimbursement of $\mathcal{R}_i$.

References

1. Abdalla, M., Bellare, M., Rogaway, P.: The Oracle Diffie-Hellman Assumptions and an Analysis of DHIES. In: Naccache, D. (ed.) CT-RSA 2001. LNCS, vol. 2020, pp. 143–158. Springer, Heidelberg (2001)
2. Aboba, B., Blunk, L., Vollbrecht, J., Carlson, J., Levkowetz, H.: Extensible Authentication Protocol (EAP). RFC 3748, IETF (2004)
3. Bellare, M., Canetti, R., Krawczyk, H.: Keying Hash Functions for Message Authentication. In: Koblitz, N. (ed.) CRYPTO 1996. LNCS, vol. 1109, pp. 1–15. Springer, Heidelberg (1996)
4. Bellare, M., Namprempre, C.: Authenticated Encryption: Relations among Notions and Analysis of the Generic Composition Paradigm. In: Okamoto, T. (ed.) ASIACRYPT 2000. LNCS, vol. 1976, pp. 531–545. Springer, Heidelberg (2000)
5. Bellare, M., Rogaway, P.: Entity Authentication and Key Distribution. In: Stinson, D.R. (ed.) CRYPTO 1993. LNCS, vol. 773, pp. 232–249. Springer, Heidelberg (1994)
6. Bellare, M., Rogaway, P.: The Exact Security of Digital Signatures - How to Sign with RSA and Rabin. In: Maurer, U.M. (ed.) EUROCRYPT 1996. LNCS, vol. 1070, pp. 399–416. Springer, Heidelberg (1996)
7. Bersani, F., Tschofenig, H.: The EAP-PSK Protocol: A Pre-Shared Key EAP Method. RFC 4764, IETF (2007)

8. Bresson, E., Chevassut, O., Pointcheval, D., Quisquater, J.-J.: Provably Authenticated Group Diffie-Hellman Key Exchange. In: CCS 2001, pp. 255–264. ACM, New York (2001)
9. Buttyán, L., Hubaux, J.-P.: Security and Cooperation in Wireless Networks. Cambridge Univ. Press, Cambridge (2008)
10. Cheikhrouhou, O., Laurent-Maknavicius, M., Chaouchi, H.: Security Architecture in a Multi-Hop Mesh Network. In: SAR 2006 (2006)
11. Clancy, T., Arbaugh, W.: EAP Password Authenticated Exchange. RFC 4746, IETF (2006)
12. Dodis, Y., Gennaro, R., Håstad, J., Krawczyk, H., Rabin, T.: Randomness Extraction and Key Derivation Using the CBC, Cascade and HMAC Modes. In: Franklin, M. (ed.) CRYPTO 2004. LNCS, vol. 3152, pp. 494–510. Springer, Heidelberg (2004)
13. Draves, R., Padhye, J., Zill, B.: Comparison of Routing Metrics for Static Multi-Hop Wireless Networks. In: SIGCOMM 2004, pp. 133–144. ACM, New York (2004)
14. Forsberg, D., Ohba, Y., Patil, B., Tschofenig, H., Yegin, A.: Protocol for Carrying Authentication for Network Access (PANA). RFC 5191, IETF (2008)
15. Fouque, P.-A., Pointcheval, D., Zimmer, S.: HMAC is a Randomness Extractor and Applications to TLS. In: Safavi-Naini, R., Seberry, J. (eds.) ACISP 2003. LNCS, vol. 2727, pp. 180–191. Springer, Heidelberg (2003)
16. Fujisaki, E., Okamoto, T., Pointcheval, D., Stern, J.: RSA-OAEP Is Secure under the RSA Assumption. Journal of Cryptology 17(2), 81–104 (2004)
17. Funk, P., Blake-Wilson, S.: Extensible Authentication Protocol Tunneled Transport Layer Security Authenticated Protocol Version 0 (EAP-TTLSv0). RFC 5281, IETF (2008)
18. Johnson, D., Hu, Y., Maltz, D.: The Dynamic Source Routing Protocol (DSR) for Mobile Ad Hoc Networks for IPv4. RFC 4728, IETF (2007)
19. Khan, K., Akbar, M.: Authentication in Multi-Hop Wireless Mesh Networks. PWASET 16, 178–183 (2006)
20. Lu, S., Ostrovsky, R., Sahai, A., Shacham, H., Waters, B.: Sequential Aggregate Signatures and Multisignatures without Random Oracles. In: Vaudenay, S. (ed.) EUROCRYPT 2006. LNCS, vol. 4004, pp. 465–485. Springer, Heidelberg (2006)
21. Lysyanskaya, A., Micali, S., Reyzin, L., Shacham, H.: Sequential Aggregate Signatures from Trapdoor Permutations. In: Cachin, C., Camenisch, J.L. (eds.) EUROCRYPT 2004. LNCS, vol. 3027, pp. 74–90. Springer, Heidelberg (2004)
22. Moustafa, H., Bourdon, G., Gourhant, Y.: Authentication, Authorization and Accounting (AAA) in Hybrid Ad hoc Hotspot's Environments. In: WMASH 2006, pp. 37–46. ACM, New York (2006)
23. NIST. Digital Signature Standard (DSS). FIPS PUB 186-2 (2000)
24. Perkins, C., Belding-Royer, E., Das, S.: Ad hoc On-Demand Distance Vector (AODV) Routing. RFC 3561, IETF (2003)
25. Simon, D., Aboba, B., Hurst, R.: The EAP-TLS Authentication Protocol. RFC 5216, IETF (2008)
26. Xu, S., Mu, Y., Susilo, W.: Online/Offline Signatures and Multisignatures for AODV and DSR Routing Security. In: Batten, L.M., Safavi-Naini, R. (eds.) ACISP 2006. LNCS, vol. 4058, pp. 99–110. Springer, Heidelberg (2006)
27. Zhao, M., Smith, S.W., Nicol, D.M.: Aggregated Path Authentication for Efficient BGP Security. In: ACM CCS 2005, pp. 128–138. ACM, New York (2005)
28. Zhu, H., Bao, F., Li, T., Wu, Y.: Sequential Aggregate Signatures for Wireless Routing Protocols. In: IEEE WCNC 2005, pp. 2436–2439. IEEE, Los Alamitos (2005)
29. Zhu, S., Xu, S., Setia, S., Jajodia, S.: LHAP: A Lightweight Network Access Control Protocol for Ad Hoc Networks. Ad Hoc Networks 4(5), 567–585 (2006)

Detecting Ringing-Based DoS Attacks on VoIP Proxy Servers

Dae Hyun Yum[1], Sun Young Kim[1], HoKun Moon[2],
Mi-Yeon Kim[2], Jae-Hoon Roh[2], and Pil Joong Lee[1]

[1] Information Security Lab, POSTECH, Republic of Korea
{dhyum,kimsy,pjl}@postech.ac.kr
[2] Central R&D Lab, KT, Republic of Korea
{hkmoon,miyeon,jhroh}@kt.co.kr

Abstract. NGN (Next Generation Network) is often called all-IP network as the general idea behind NGN is that one network transports all information and services. When NGN is deployed in a large scale, VoIP will eventually replace PSTN, the traditional model of voice telephony. While VoIP promises both low cost and a variety of advanced services, it may entail security vulnerabilities. Unlike PSTN, intelligence is placed at the edge and the security measures are not incorporated into the network. VoIP-specific attacks have already been introduced, of which the ringing-based DoS attack belongs. In this paper, we propose a detection system of the ringing-based DoS attacks. We model the normal traffic of legitimate users with the gamma distribution and then quantify the discrepancy between the normal traffic and the attack traffic with Pearson's chi-square statistic. Simulation results show that the proposed detection system can reliably detect the ringing-based DoS attacks.

Keywords: VoIP, SIP, DoS attack, detection system.

1 Introduction

VoIP (Voice over Internet Protocol) is a way to make and receive phone calls using voice communications over IP networks such as the Internet or other packet-switched networks rather than the traditional PSTN (Public Switched Telephone Network). VoIP applications have grown rapidly and continue to enjoy fast growth due to the cost advantages and wide range of advanced services. Many simple and efficient products (e.g., Skype, Net2Phone, and Google talk) have already appeared in the market and are in heated race to expand market presence. One of the key innovations driving the current evolution of VoIP systems is SIP (Session Initiation Protocol) [25,32]. SIP is an application-layer signalling protocol, addressing two fundamental problems in establishing real-time communication sessions; it helps two parties wanting to communicate find each other on the Internet (i.e., rendezvous) and allows those parties to negotiate how they are going to communicate (i.e., session negotiation). SIP is an open standard developed by the IETF (Internet Engineering Task Force) and

H.Y. Youm and M. Yung (Eds.): WISA 2009, LNCS 5932, pp. 339–353, 2009.

is being used to construct residential telephony services, peer-to-peer systems, PBX (Private Branch eXchange) replacement systems, and large-scale carrier next-generation networks, such as the IMS (IP Multimedia Subsystem) of the 3GPP (3rd Generation Partnership Project).

In PSTN, security, reliability, and availability rely on a closed networking environment dedicated to a single service (namely, voice). On the other hand, VoIP is based on an open environment such as the Internet, which simplifies mounting an attack (e.g., on a VoIP server) [11]. This is due to the fact that VoIP services are based on open standardized protocols for signalling (e.g., SIP) and transport protocol technologies (e.g., RTP: Real-time Transport Protocol [27]), using servers reachable through the Internet and often provided over general purpose computing hardware. As a result, these technologies are not designed mainly with security features in mind. Therefore, a malicious user can exploit any possible misconfiguration in the aforementioned signaling or media transport protocols, attempting to disturb or disrupt VoIP services. Additionally, such services inherit numerous vulnerabilities from the utilization of the underlying transport protocols like TCP, IP, and UDP. For example, instead of generating thousands of costly voice calls as required in PSTN, the attacker can easily and in a similar manner generate and send thousands of VoIP signalling messages to attack VoIP servers. Various aspects of VoIP security can be found in [3,11,13,30,37].

DoS (Denial-of-Service) attack or DDoS (Distributed Denial-of-Service) attack aims to make a computer resource unavailable to its intended users temporarily or indefinitely by saturating the target machine commonly with external communications requests. A quantitative estimate of worldwide DoS attack frequency found 12,000 attacks over a three-week period in 2001 [22]. The 2004 CSI/FBI Computer Crime and Security Survey listed DoS attacks among the most financially expensive security incidents. To counter DoS attacks, many defense mechanisms have been proposed (e.g., [16,18,19]), which usually degrade the end-to-end TCP performance. All the overheads of DoS defense systems become superfluous in the absence of DoS attacks. An alternative choice is to deploy DoS detection systems (e.g., [9,23,34]) and then bring defense systems into play when necessary. The main goal of detection systems is to detect and distinguish malicious traffic from legitimate traffic [4]. When large amounts of traffic from legitimate clients suddenly arrive at a server, it is called a flash event. Clearly, legitimate user activity like a flash event can be easily confused with a DoS attack, and vice versa. All detection methods define an attack as an abnormal and noticeable deviation of some statistic of the monitored network traffic workload.

To detect DoS attacks on VoIP servers, one should consider VoIP-specific DoS attacks as well as traditional TCP/IP based DoS attacks. First, to mount a DoS attack on VoIP servers, attackers can flood the target server with different patterns of malformed packets and hope that at least one of them would trigger a bug in the target. This type of attack is called a fuzzing attack and can be detected by checking the following five categories of rules: incorrect grammar, oversized field values, invalid message or field name, redundant or repetitive header field, and invalid semantic [6]. Flooding attacks with invalid messages can also be detected with state transition models in the SIP transaction layer [5]. Combining rule matching and state transition models was proposed to detect more SIP attacks [29]. Second, attackers can overwhelm the target server by sending huge amounts of valid SIP messages. As the flooding packets are valid messages, this type of attack cannot be detected by using above-mentioned rules. To solve the problem, Hellinger distance based detection system was proposed [28]; after quantifying the normal correlation among protocol attributes such as SIP request and response messages, attacking traffic is differentiated from normal traffic based on Hellinger distance. The potential risk of DoS attacks on a PSAP (Public Safety Answering Point) was studied in [10]; to distinguish between attacks and flash events, incoming emergency calls are examined separately by their originating network domain that is PSTN, mobile networks, or VoIP. The load characteristics for the calls of each originating domain are compared to each other and used to detect suspicious call patterns. Third, instead of flooding attacks, semantic attacks can be launched, which exploits a specific feature of some protocol or application installed at the victim in order to consume excess amounts of its resources [21]. A ringing-based attack exploits the semantics of the SIP protocol without requiring a high traffic rate from the attackers [7]; malicious or compromised user agents stay in the ringing state for an unusually long amount of time by not picking up incoming calls. While RET (Random Early Termination) algorithm can mitigate the damage of the ringing-based attack by dropping suspected transactions, it cannot distinguish attack traffic from normal traffic. Currently, no detection system is known for ringing-based DoS attacks.

Our contribution. To distinguish ringing-based attack traffic from normal user traffic, we study the statistical characteristics of normal ringing patterns. A popular choice to model the characteristics of communication network patterns is the CUSUM (CUmulative SUM) scheme that belongs to non-parametric tests or distribution-free tests assuming no knowledge whatsoever about the distributions of the underlying populations, except perhaps that they are continuous [2,20,31,34]. However, there are a number of disadvantages associated with non-parametric tests. Primarily, they do not utilize all the information provided by the sample data, and thus a non-parametric test will be less efficient than the corresponding parametric procedure when both methods are applicable. This means that, to achieve the same power, a non-parametric test requires a larger sample size than does the corresponding parametric test [33]. Moreover, non-parametric tests may mistake high-rate normal traffic (e.g., flash events) for attack traffic because they do not take into consideration the normal but sudden variation of network traffic [23]. Possibilities of adopting parametric tests for detection systems of TCP SYN flooding attacks were explored with several probability distributions in [23]. We propose a parametric test for detecting ringing-based DoS attacks by modeling the distribution of normal ringing patterns with a gamma distribution. It is assumed that parametric tests can be applicable to detect ringing-based attacks, because normal ringing patterns are mainly dependent on human behaviors (i.e., behaviors of users picking up incoming calls) and not on network conditions; ringing begins once invite signal arrives at the receiver and the network condition is irrelevant afterwards. Recent researches on human behavior show that human mobility patterns are surprisingly predictable [12,14] and telephone call holding times can be described with a Pareto distribution [8]. Detecting SPIT (SPam over Internet Telephony) calls and fraud by checking human communication patterns was studied in [1,24]. Our experimental results indicate that the proposed detection system can distinguish attack traffic and normal traffic efficiently. We also discuss how to obtain a trade-off between the detection latency and the false alarm rate of the proposed system.

2   Ringing-Based DoS Attacks

In this section, we review VoIP call scenario and ringing-based DoS attacks mainly by following [7] and [26].

2.1   Session Initiation Protocol

The session initiation protocol SIP [25] is an application-layer signaling protocol that can establish, modify, and terminate multimedia sessions (e.g., VoIP calls) among Internet endpoints. A SIP network consists of UA (User Agent), proxy server, redirect server, registrar server and location server. A UA is a software embedded within an end point device (e.g., VoIP phone) that is capable of generating requests or sending a response to a request. There are two types of UAs, both residing within the same devices but providing different functions: UAC (User Agent Client) and UAS (User Agent Server). The UAC generates a request (the beginnings of a session establishment) when a subscriber dials his or her phone, or uses some other communication applications. The UAS responds to a request from a UAC as the recipient of the request. A proxy server receives SIP requests and responses and forwards them through the network on behalf of UAs. The proxy is also responsible for determining how to route requests and responses. We can categorize proxies into two types: stateless proxies and stateful proxies. A stateless proxy does not have any concept of a dialog or any knowledge of the state for any dialog or session. The stateless proxy is nothing more than a simple router responsible for the forwarding of a request or response. It does not maintain any knowledge of previous requests or responses, and it does not maintain any routing data other than a routing table of its own used to determine how to route messages based on the URI (Uniform Resource Identifier) provided in the SIP message. A stateful proxy maintains the state of each and every dialog that it is part of. As a request is sent to the proxy, it saves the call identifier for the session, as well as other pertinent identification data. This means that in order for the stateful proxy to know when a session has changed status or has ended, it must receive all responses. This is why networks that use stateful proxies must route all messages for a session through the same route.

This ensures that all SIP responses use the same path as the request, so that the stateful proxy will receive all of the responses and any additional requests for a session. One reason for using a stateful proxy rather than a stateless proxy is to enable the deployment of richer functionality within the proxy. A redirect server is used to provide alternate addresses for a request. There are many reasons why this would be desirable. For example, the network operator may wish to send alternate addresses for routing of a request when proxies become busy as a means of load sharing. A registrar server is used to authenticate and record the current location of a subscriber device. When a device is turned on or changes location (resulting in a new IP address being assigned), the device will send a REGISTER message to the registrar to provide its new address. A location server provides subscriber addresses to other proxies based on the registration process. This function is not necessarily a stand-alone function but rather integrated within a SIP server providing registrar and proxy services.

SIP uses a request/response mechanism for establishing a session. This means that any entity wishing to establish a session with another entity must first initiate a request that contains the parameters for establishing the connection, such as resources required. The receiving entity must then respond to the request, either accepting or rejecting the request. The UAC within the calling device is responsible for initiating a request to the UAS at the destination device. If the UAS is going to accept the request, it will send a response back to the UAC, which will then return an acknowledgment to the UAS. The final act of sending the acknowledgment is what establishes the session. Figure 1 depicts the basic

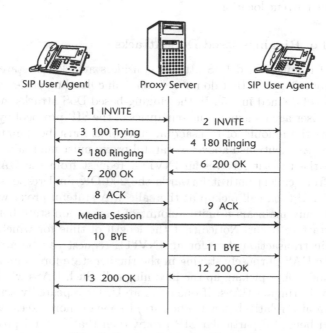

Fig. 1. Session Initiation Protocol

call scenario that we are interested in. The UAC (i.e., caller on the left side) requests the initiation of a communication session with the UAS (i.e., callee on the right side) by sending an INVITE request via the proxy. Upon receiving the INVITE at the proxy, the proxy will forward the INVITE to the UAS and, in the meantime, the proxy will send a 100 Trying provisional response to the UAC to prevent unnecessary retransmissions of the INVITE request. Upon receiving the INVITE at the UAS, the UAS will immediately send a 180 Ringing provisional response to the UAC via the proxy. Similar to a telephone, the UAS is now in the ringing state. Once the UAS picks up the call, it sends a 200 OK final response to the UAC via the proxy. The UAC will acknowledge reception of the 200 OK final response with an ACK request. An interactive communication session has now been established between the UAC and UAS. Now the two entities begins to send RTP [27] voice streams directly to each other based on the negotiated media session parameters that were contained in the message bodies of the INVITE and 200 OK. An established session can be terminated when either the UAC or UAS sends a BYE request to the other party. The other party will then acknowledge reception of the BYE request by responding with a 200 OK final response. In Figure 1, the proxy can be thought of as a server that handles all incoming calls for the UAS. For example, a company using VoIP might issue logical SIP URIs to its employees (e.g., sip:alice@company.com). However, since employees might be at different locations throughout the day, it will be necessary to map the logical SIP URIs to actual contact IP addresses (e.g., sip:alice@128.174.254.29) when an incoming call is received. In conjunction with a registrar and location service, the company can use SIP proxies to route incoming INVITE requests to the UAS at its current location.

2.2   Model of Ringing-Based DoS Attacks

The model of ringing-based DoS attacks considers attacks on stateful proxies originating from attackers that do not rely on traffic flooding or deviations from the SIP protocol defined in [25]. In the ringing-based DoS attacks, malicious or compromised user agents exploit the semantics of the SIP protocol by intentionally increasing the amount of transaction state that must be maintained by a stateful proxy. As mentioned above, a stateful proxy must maintain transaction context from the time it receives an INVITE request from the UAC (i.e., incoming call from caller) until it forwards the 200 OK final response from the UAS to the UAC (i.e., call pickup at the callee). A stateful proxy with limited resources can only maintain a finite amount of transaction state before service disruptions start to occur. Notice that the length of time for which the proxy must maintain transaction state for an INVITE request can be artificially increased by the UAS purposely staying in the ringing state for an unusually long amount of time before picking up the incoming call. Such UASs will be referred to as excessively ringing UASs. If one or more UACs repeatedly send INVITE requests (even at a relatively low traffic rate) to one or more excessively ringing UASs located behind a particular SIP proxy, then that stateful proxy can potentially become the target of what is called a ringing-based DoS attack, which is named after the "180 Ringing" provisional response that indicates the UAS is in the ringing state. In a ringing-based DoS attack, excessively ringing UASs consume resources at the stateful proxy for a longer period of time, which makes these resources unavailable for other incoming calls. As shown in [7], even a relatively low rate of incoming calls to excessively ringing UASs can significantly disrupt overall service. Many possibilities exist for a user agent to become a willing or unwilling participant in a ringing-based DoS attack. First, attacking UACs might collude with disgruntled insiders within an enterprize that willingly act as excessively ringing UASs. Second, attacking UACs might automatically scan a list of SIP URIs to detect unresponsive callees. These unresponsive callees might unwillingly become excessively ringing UASs during an attack since they will not pick up the incoming calls from the attacking UACs. Lastly, attacking UACs might utilize a botnet located behind a target SIP proxy to act as excessively ringing UASs. In the Web application domain, such botnets have previously been rented to disrupt online businesses [17]. Evidence of using VoIP networks to build botnets has already been reported [15].

3 Detecting Ringing-Based DoS Attacks

3.1 Model of the Normal Ringing Patterns

The most common statistical choice to model a physical phenomenon is the Gaussian[1] distribution that is a continuous probability distribution that describes data that clusters around a mean or average [33,35]. The graph of the associated probability density function is bell-shaped and symmetric, with a peak at the mean, and is known as the Gaussian function or bell curve. The mathematical equation for the density function of the Gaussian random variable X depends upon the two parameters μ and σ, its mean and standard deviation, and is given by $f(x; \mu, \sigma) = \frac{1}{\sqrt{2\pi}\sigma} e^{-(1/2)[(x-\mu)/\sigma]^2}$ where $\pi = 3.14159\cdots$ and $e = 2.71828\cdots$. The Gaussian distribution can be used to describe, at least approximately, any variable that tends to cluster around the mean. Due to the central limit theorem, any random variable that is the sum of a large number of independent factors usually follow a Gaussian distribution. Even though the Gaussian distribution has a wide range of applicability, it dose not seem to be a good candidate to describe the normal ringing patterns because the real-life data of ringing time, which will be given later, follow a skewed or asymmetric distribution.

In statistics, there are many asymmetric distributions that are frequently used: beta distribution, gamma distribution, log-normal distribution, Maxwell-Boltzmann distribution, Pareto distribution, Rice distribution, Snedecor's F-distribution, Student's t-distribution, Weibull distribution, etc. After trying several asymmetric distributions, we found that gamma distribution was fit for our

---

[1] Gaussian distribution is also called the normal distribution. In this paper, we stick to the former term and use "normal" to mean the absence of attack traffic as in "normal traffic."

purpose fairly well. The gamma distribution is a two-parameter family of continuous probability distributions and derives its name from the well-known gamma function $\Gamma(\alpha) = \int_0^\infty x^{\alpha-1}e^{-x}dx$, studied in many areas of mathematics [33]. The gamma function satisfies $\Gamma(\alpha) = (\alpha - 1)\Gamma(\alpha - 1)$ for $\alpha > 1$ and $\Gamma(n) = (n-1)!$ for a positive integer n. The continuous random variable X has a gamma distribution, with parameters α and β if its density function is given by

$$f(x; \alpha, \beta) = \frac{1}{\beta^\alpha \Gamma(\alpha)} x^{\alpha-1} e^{-x/\beta} \quad \text{for } x > 0$$

where $\alpha > 0$ and $\beta > 0$. A special gamma function for which $\alpha = 1$ is called the exponential distribution that describes the times between events in a Poisson process, i.e. a process in which events occur continuously and independently at a constant average rate. Another important special case of the gamma distribution is obtained by letting $\alpha = \nu/2$ and $\beta = 2$, where ν is a positive integer. The result is called the chi-square distribution, where ν is called the degree of freedom. To find the mean of the gamma distribution, we write

$$\mu = E(X) = \int_0^\infty x \cdot f(x; \alpha, \beta) \, dx = \int_0^\infty x \cdot \frac{1}{\beta^\alpha \Gamma(\alpha)} x^{\alpha-1} e^{-x/\beta} \, dx.$$

Now, let $y = x/\beta$, to give

$$\mu = \frac{\beta}{\Gamma(\alpha)} \int_0^\infty y^\alpha e^{-y} \, dy = \frac{\beta}{\Gamma(\alpha)} \cdot \Gamma(\alpha + 1) = \frac{\beta}{\Gamma(\alpha)} \cdot \alpha\Gamma(\alpha) = \alpha\beta.$$

The variance of the gamma distribution can be found by proceeding as above to obtain

$$\sigma^2 = E(X^2) - \mu^2 = \frac{\beta^2}{\Gamma(\alpha)} \cdot \Gamma(\alpha + 2) - \mu^2 = (\alpha + 1)\alpha\beta^2 - \alpha^2\beta^2 = \alpha\beta^2.$$

The standard deviation of the gamma distribution easily follows.

$$\sigma = \sqrt{\sigma^2} = \sqrt{\alpha\beta^2}.$$

To carry out a survey on the distribution of normal ringing time, we actually made 181 phone calls in the daytime, measuring the ringing time by the second.[2] The result is shown in Figure 2, which is roughly an asymmetric bell-shaped distribution with mean 16.54 and standard deviation 11.14. Figure 2 depicts the result only up to 60 seconds of ringing time; normal calls with ringing time more than 60 seconds are very scarce. The maximum ringing time, after which trying to call is aborted automatically, is 60 seconds for mobile phones and 120 seconds for landline phones.

To describe the normal ringing patterns of Figure 2 with a gamma distribution, we have to choose curves (i.e., the parameters α and β) that fit the data with

---

[2] We could not find the distribution of ringing times at commercial VoIP servers. The ringing time was not recorded in the log file of commercial servers, because only data related with charging purpose were recorded.

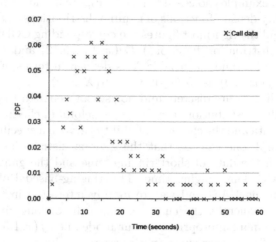

Fig. 2. Ringing time distribution

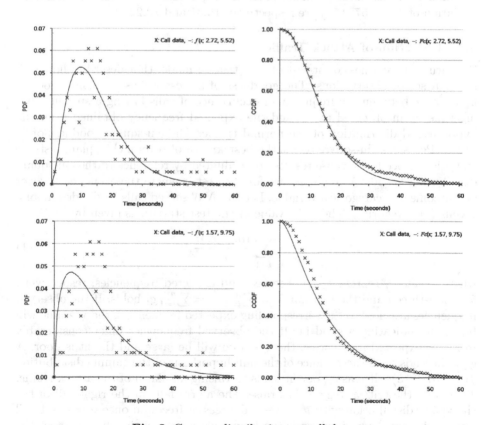

Fig. 3. Gamma distribution vs. call data

small errors. Two example choices are given in Figure 3; the leftmost figures contain the probability density functions of gamma distributions $f(x; 2.72, 5.52)$ and $f(x; 1.57, 9.75)$ and the rightmost figures the corresponding CCDFs (complementary cumulative distribution functions) $F_C(x; 2.72, 5.52)$ and $F_C(x; 1.57, 9.75)$, where the CCDF of a random variable X is defined in terms of the probability density function $f(x; \alpha, \beta)$ as $F_C(x; \alpha, \beta) = \Pr(X > x) = 1 - \int_{-\infty}^{x} f(t; \alpha, \beta) \, dt$. As the survey call data are discrete and not smooth enough to perfectly match a known probability distribution, the CCDF is helpful in checking the similarity between the call data and the chosen probability curve in a specific interval. From CCDFs of Figure 3, one can see that the gamma distribution $f(x; 2.72, 5.52)$ matches well with the data of short ringing time and the gamma distribution $f(x; 1.57, 9.75)$ with long ringing time. The ringing-based DoS attack is due to excessively ringing UASs and we are mainly interested in the calls of long ringing time. Consequently, the curves fitting calls of long ringing time with smallest errors are more appropriate for our model, i.e., $f(x; 1.57, 9.75)$ is better than $f(x; 2.72, 5.52)$ or other curves with errors equally scattered in the entire interval. After extensive search for good curves, we decided $f(x; 1.57, 9.75)$ as the model for the normal ringing patterns. Note that the mean and standard deviation of $f(x; 1.57, 9.75)$ are respectively 15.31 and 12.22.

3.2 Detection of Attack Traffic

We turn to a goodness-of-fit test of a statistical model that describes how well it fits a set of observations. The goodness-of-fit test is based on how good a fit we have between the frequency of occurrence of calls having a specific ringing time in an observed traffic and the expected frequency obtained from the hypothesized distribution of the normal traffic. The chi-square goodness-of-fit test or Pearson's chi-square test is the best-known of several chi-square tests — statistical procedures whose results are evaluated by reference to the chi-square distribution [33,36]. A common case for the test is where the events each cover an outcome of a categorical variable. Let the N observed calls be divided among n cells (or categories). Then, the value of the test statistic is given by

$$\chi^2 = \sum_{i=1}^{n} \frac{(o_i - e_i)^2}{e_i}, \tag{1}$$

where o_i and e_i represent the observed and expected frequencies, respectively, for the i-th cell and the relation $N = \sum_{i=1}^{n} o_i = \sum_{i=1}^{n} e_i$ holds. If the observed frequencies are close to the corresponding expected frequencies, the χ^2-value will be small, indicating a good fit. If the observed frequencies differ considerably from the expected frequencies, the χ^2-value will be large and the fit is poor. A good fit leads to the acceptance of the null hypothesis of the gamma distribution $f(x; 1.57, 9.75)$ whereas a poor fit leads to raising an alarm of the attack traffic. Therefore, the critical region that raises the alarm falls in the right tail of the chi-square distribution with $\nu = n - 1$ degrees of freedom; once $\nu = n - 1$ cell frequencies are determined, so is the frequency for the n-th cell. To detect the ringing-based DoS attacks, we divide calls into three cells (i.e., $n = 3$).

- $cell_1$ = {calls of ringing time t seconds, where $0 < t \leq 15$}
- $cell_2$ = {calls of ringing time t seconds, where $15 < t \leq 31$}
- $cell_3$ = {calls of ringing time t seconds, where $31 < t$}

The criterion of the cell selection is as follows. Around 15–16 and 30–31 seconds, the CCDF of the survey call data is almost equal to the CCDF of the hypothesized gamma distribution $f(x; 1.57, 9.75)$. If the boundaries of cells are chosen around 15–16 and 31–32, then $f(x; 1.57, 9.75)$ can simulate the call data with small errors. The first cell $cell_1$ includes calls of ringing time less than or equal to 15 seconds. We do not divide $cell_1$ into subcells, because attack traffic is related with excessively ringing calls and thus variations in calls of short ringing time are not important. Actually, dividing $cell_1$ and $cell_2$ into subcells may widen the gap between the gamma distribution and the call data. Besides 15–16 and 30–31 seconds, the CCDF of the call data roughly coincides with that of $f(x; 1.57, 9.75)$ around 35, 50, and 55 seconds. Therefore, if we divide $cell_3$, which includes most of excessively ringing calls, into subcells with boundaries around those values, attack traffic can be detected more sensitively without increasing the modeling gap. However, the chi-square test breaks down if expected frequencies are too low. It will normally be acceptable so long as no more than 10% of the cells have expected frequencies below 5. As we intend to use the chi-square test for samples of various sizes, we refrain from dividing $cell_3$ into more subcells.

In order to evaluate the performance of the proposed detection method, we experimented with $N = 1000$ daytime call traffic in Table 1. The total number of calls was fixed as $N = 1000$ and the ratio of AC to NC+AC was varied from 0% to 20%, where AC and NC denote the attack call and the normal call, respectively. For each ratio, the experiment was repeated ten times and the resulting mean and minimum/maximum of the chi-square test values are given in Table 1. According to [7], ringing times of attack calls were scattered uniformly between 30 to 120 seconds. Expected frequencies of Equation 1 are computed by

$$e_1 = N \int_0^{15} f(x; 1.57, 9.75)\, dx,$$

$$e_2 = N \int_{15}^{31} f(x; 1.57, 9.75)\, dx,$$

$$e_3 = N \int_{31}^{\infty} f(x; 1.57, 9.75)\, dx.$$

Let δ be the threshold of the chi-square test value to raise the alarm; if the chi-square value is greater than δ, the test regards it as an abnormal situation. The probabilities of false alarm Pr_{FA} for various values of δ are given in Table 2 that can be obtained by the following command of Mathematica with $t = \delta$.

$$1.0 - \texttt{CDF[ChiSquareDistribution[2], t]}$$

The threshold δ can be adjusted by the policy of system administrators. If the policy is to detect most of the attack traffic (e.g., 5% ratio of attack traffic), the

Table 1. Chi-square values of experiment with $N = 1000$ call traffic

AC (attack call)	NC (normal call)	$\frac{AC}{NC+AC} \times 100$ (%)	χ^2 (chi-square value)		
			m (min)	μ (mean)	M (max)
0	1000	0	0.27	3.08	5.72
50	950	5	9.89	16.52	23.41
100	900	10	58.06	74.98	87.10
200	800	20	272.84	324.33	350.77

Table 2. Probability of False Alarm

δ	5	6	7	8	9	10	15	20	30	40
Pr_{FA}	0.082	0.050	0.030	0.018	0.011	0.0067	0.00055	0.000045	3.1×10^{-7}	2.1×10^{-9}

Table 3. Chi-square values of experiment with $N = 500$ call traffic

AC (attack call)	NC (normal call)	$\frac{AC}{NC+AC} \times 100$ (%)	χ^2 (chi-square value)		
			m (min)	μ (mean)	M (max)
0	500	0	0.05	2.11	5.64
25	475	5	6.85	12.09	22.69
50	450	10	31.10	42.96	62.07
100	400	20	142.31	169.30	209.45

threshold is set as $\delta = 8$ that gives the false alarm rate $Pr_{FA}(8) = 0.018$. If the system administrator is interested in detecting attacks of more than 10% ratio of attack traffic, the threshold can be set as $\delta = 40$ that gives extremely low false alarm rate $Pr_{FA}(40) = 2.1 \times 10^{-9}$.

Now, we discuss the detection latency. After 1000 calls of the test traffic arrive at the server for t seconds, we only have to determine to which cell (i.e., $cell_1$, $cell_2$, and $cell_3$) each call belongs and thus the detection latency is about $t + 31$ seconds. The detection latency of flooding-based DoS attack is usually more than 1 minute. For commercial VoIP servers, t is expected to be just a few seconds and thus the detection latency of the proposed method is relatively short. We can also reduce the detection latency by testing fewer calls. Experiment result with fewer test calls are given in Table 3 and Table 4. As the number of test calls decreases, the detection latency decreases but the threshold should also be lowered; therefore, there is a trade-off between the detection latency (i.e., the number of test calls) and the false alarm rate (i.e., the threshold).

We finally remark that any detection system itself should not be vulnerable to DoS attacks. The proposed detection method monitors a fixed number of test calls (say 1000 calls) and then decides whether there is a ringing-based DoS attack or not. If the call traffic (from either legitimate users or attackers or from both) is very high, it is recommended that a part of total traffic be examined. In this case, one may refer to the theory of statistical inference that deals with

Table 4. Chi-square values of experiment with $N = 300$ call traffic

AC (attack call)	NC (normal call)	$\frac{AC}{NC+AC} \times 100$ (%)	χ^2 (chi-square value)		
			m (min)	μ (mean)	M (max)
0	300	0	0.48	1.92	4.63
15	285	5	5.36	10.28	14.56
30	270	10	20.50	30.56	40.22
60	240	20	84.26	106.15	134.89

the methods estimating a population parameter based on information obtained from a random sample selected from the population [33].

4   Conclusion

We proposed a parametric test for detecting ringing-based DoS attacks with the gamma distribution model and the chi-square goodness-of-fit statistic. Simulation results showed that it could distinguish between attack traffic and normal traffic. While the total calls of legitimate users may change abruptly (e.g., flash events), the distribution of ringing time does not change rapidly. To obtain a more precise model of the distribution of the normal ringing pattern, the analysis of a large-scale empirical data is required and the parameters of the gamma distribution (i.e., α and β) should reflect the behaviors of users according to the date and time. For example, it is expected that people will pick up incoming calls in the daytime more quickly than at the midnight. We invite readers to the refinement of the proposed model.

Acknowledgement

We would like to thank Jae Woo Seo for his help for data analysis. This research was supported by the KT and the BK21.

References

1. Abidogun, O.A.: Data mining, fraud detection and mobile telecommunications: Call pattern analysis with unsupervised neural networks. Master's thesis, University of the Wester Cape (August 2005)
2. Basseville, M., Nikiforov, I.V.: Detection of Abrupt Changes: Theory and Application. Prentice-Hall, Englewood Cliffs (1993)
3. Benini, M., Sicari, S.: Assessing the risk of intercepting VoIP calls. Computer Networks 52(12), 2432–2446 (2008)
4. Carl, G., Kesidis, G., Brooks, R.R., Rai, S.: Denial-of-service attack-detection techniques. IEEE Internet Computing 10(1), 82–89 (2006)
5. Chen, E.Y.: Detecting DoS attacks on SIP systems. In: 1st IEEE Workshop on VoIP Management and Security, pp. 53–58. IEEE, Los Alamitos (2006)

6. Chen, E.Y., Itoh, M.: Scalable detection of SIP fuzzing attacks. In: SECURWARE, pp. 114–119. IEEE, Los Alamitos (2008)
7. Conner, W., Nahrstedt, K.: Protecting SIP proxy servers from ringing-based denial-of-service attacks. In: ISM, pp. 340–347. IEEE Computer Society, Los Alamitos (2008)
8. Dang, T.D., Sonkoly, B., Molnar, S.: Fractal analysis and modeling of VoIP traffic. In: Telecommunications Network Strategy and Planning Symposium, Networks 2004, pp. 123–130 (2004)
9. Feinstein, L., Schnackenberg, D., Balupari, R., Kindred, D.: Statistical approaches to DDoS attack detection and response. In: DISCEX, vol. (1), pp. 303–314. IEEE Computer Society, Los Alamitos (2003)
10. Fuchs, C., Aschenbruck, N., Leder, F., Martini, P.: Detecting VoIP based DoS attacks at the public safety answering point. In: ASIACCS, pp. 148–155. ACM, New York (2008)
11. Geneiatakis, D., Dagiuklas, T., Kambourakis, G., Lambrinoudakis, C., Gritzalis, S., Ehlert, S., Sisalem, D.: Survey of security vulnerabilities in session initiation protocol. IEEE Communications Surveys and Tutorials 8(1-4), 68–81 (2006)
12. Gonzalez, M.C., Hidalgo, C.A., Barabási, A.-L.: Understanding individual human mobility patterns. Nature 453, 779–782 (2008)
13. Gupta, P., Shmatikov, V.: Security analysis of Voice-over-IP protocols. In: CSF, pp. 49–63. IEEE Computer Society, Los Alamitos (2007)
14. Heger, M.: Human travel patterns surprisingly predictable. IEEE Spectrum Magazine (June 2008),
http://www.spectrum.ieee.org/telecom/wireless/
human-travel-patterns-surprisingly-predictable
15. Hines, M.: Attackers get chatty on VoIP. PCWorld (May 2007),
http://www.pcworld.com/businesscenter/article/132389/
16. Juels, A., Brainard, J.G.: Client puzzles: A cryptographic countermeasure against connection depletion attacks. In: NDSS.The Internet Society (1999)
17. Kandula, S., Katabi, D., Jacob, M., Berger, A.: Botz-4-sale: Surviving organized DDoS attacks that mimic flash crowds. In: NSDI. USENIX (2005)
18. Lee, H., Park, K.: On the effectiveness of probabilistic packet marking for IP trace-back under denial of service attack. In: INFOCOM, pp. 338–347. IEEE, Los Alamitos (2001)
19. Lemon, J.: Resisting SYN flood DoS attacks with a SYN cache. In: BSDCon, pp. 89–97. USENIX (2002)
20. Leu, F.-Y., Yang, W.-J.: Intrusion detection with CUSUM for TCP-based DDoS. In: Enokido, T., Yan, L., Xiao, B., Kim, D.Y., Dai, Y.-S., Yang, L.T. (eds.) EUC-WS 2005. LNCS, vol. 3823, pp. 1255–1264. Springer, Heidelberg (2005)
21. Mirkovic, J., Reiher, P.L.: A taxonomy of DDoS attack and DDoS defense mechanisms. Computer Communication Review 34(2), 39–53 (2004)
22. Moore, D., Shannon, C., Brown, D.J., Voelker, G.M., Savage, S.: Inferring internet denial-of-service activity. In: USENIX Security Symposium. USENIX (2001)
23. Ohsita, Y., Ata, S., Murata, M.: Detecting distributed denial-of-service attacks by analyzing TCP SYN packets statistically. IEICE Transactions 89-B(10), 2868–2877 (2006)
24. Quittek, J., Niccolini, S., Tartarelli, S., Stiemerling, M., Brunner, M., Ewald, T.: Detecting SPIT calls by checking human communication patterns. In: ICC, pp. 1979–1984. IEEE, Los Alamitos (2007)

25. Rosenberg, J., Schulzrinne, H., Camarillo, G., Johnston, A., Peterson, J., Sparks, R., Handley, M., Schooler, E.: SIP: Session Initiation Protocol. RFC 3261, IETF (June 2002)
26. Russell, T.: Session Initiation Protocol (SIP): Controlling Convergent Networks. McGraw-Hill Osborne Media, New York (2008)
27. Schulzrinne, H., Casner, S.L., Frederick, R., Jacobson, V.: RTP: A transport protocol for real-time applications. RFC 3550 (July 2003)
28. Sengar, H., Wang, H., Wijesekera, D., Jajodia, S.: Detecting VoIP floods using the Hellinger distance. IEEE Trans. Parallel Distrib. Syst. 19(6), 794–805 (2008)
29. Seo, D., Lee, H., Nuwere, E.: Detecting more SIP attacks on VoIP services by combining rule matching and state transition models. In: SEC. IFIP, vol. 278, pp. 397–411. Springer, Heidelberg (2008)
30. Sicker, D.C., Lookabaugh, T.D.: VoIP security: Not an afterthought. ACM Queue 2(6), 56–64 (2004)
31. Siris, V.A., Papagalou, F.: Application of anomaly detection algorithms for detecting SYN flooding attacks. Computer Communications 29(9), 1433–1442 (2006)
32. Sparks, R.: SIP: basics and beyond. ACM Queue 5(2), 22–33 (2007)
33. Walpole, R.E., Myers, R.H., Myers, S.L., Ye, K.: Probability and Statistics for Engineers and Scientists. Prentice-Hall, Englewood Cliffs (2006)
34. Wang, H., Zhang, D., Shin, K.G.: Detecting SYN flooding attacks. In: INFOCOM, IEEE, Los Alamitos (2002)
35. Wikipedia. Normal distribution,
    http://en.wikipedia.org/wiki/Normal_distribution
36. Wikipedia. Pearson's chi-square test,
    http://en.wikipedia.org/wiki/Pearson%27s_chi-square_test
37. Zhang, R., Wang, X., Yang, X., Farley, R., Jiang, X.: An empirical investigation into the security of phone features in SIP-based VoIP systems. In: Bao, F., Li, H., Wang, G. (eds.) ISPEC 2009. LNCS, vol. 5451, pp. 59–70. Springer, Heidelberg (2009)

USN Middleware Security Model

Mijoo Kim[1,*], Mi Yeon Yoon[1], Hyun Cheol Jeong[1], and Heung Youl Youm[2]

[1] KISA(Korea Internet & Security Agency),
IT Venture Tower, Jungdaero 135(Garak-dong 78),
Songpa-gu, Seoul, Republic of Korea
[2] Soonchunhyang University,
Eupnae-ri, Shinchang-myun, Asan-si,
Chungcheongnam-do, Republic of Korea
{mijoo.kim,myyoon,hcjung}@kisa.or.kr, hyyoum@sch.ac.kr

Abstract. In recent years, USN technology has become one of the major issues in business and academia. This paper analyzes USN services from a security point of view and proposes a security model of USN middleware that plays a bridge role in linking heterogeneous sensor network to USN applications. In USN environment, sensor networks transmit data sensed from all distributed sensor nodes to USN middleware. And then USN middleware processes the received sensing data, makes information for supporting specific application service, and transmits the information to appropriate USN application. Like this, USN middleware plays a significant role in USN service. If the USN middleware is compromised, USN service might be not functioning normally. Hence, security in USN middleware is very important. In this paper, we propose security functions in USN middleware and specify a USN middleware security model that the security functions are reflected in.

Keywords: USN middleware security, USN security, Ubiquitous Sensor Network security.

1 Introduction

USN is a shortened word of "Ubiquitous Sensor Network" and USN service is a service that uses wireless sensor network technology. It is provided based on the data sensed by many sensors on the sensor network. USN service is applicable to various service areas, such as construction, distribution, harbors, etc. So, sensor network technology contributes to make people's life convenient. Also, it saves power consumption and has low cost. For this reason, these days sensor network technology is catching on as a core technology to implement green IT environment.

USN service architecture is composed of USN application, USN middleware and sensor network. USN application provides interface for administrators and users. In case of USN middleware, there are two cases. If the dedicated sensor network is used

---

* This research performed by authors except for Heung Youl Youm was supported by the TTA(Telecommunications Technology Association) (2009-P1-32-08J46).

H.Y. Youm and M. Yung (Eds.): WISA 2009, LNCS 5932, pp. 354–365, 2009.

for service, USN application may use the dedicated sensor network in its own way. However, in case USN application uses open sensor network, USN middleware is needed for linking a sensor network to appropriate USN application. USN middleware supports heterogeneous sensor networks and various USN applications. So USN middleware makes data sensed by sensor connect to appropriate USN application. Like this, USN middleware plays a pivotal role in USN service environment. And this means that all data is delivered through USN middleware. This can be a weak point. Because attacker can control sensor network and achieve all information for attacking USN middleware.

Hence, security in USN middleware must be dealt with. This paper proposes security functions which should be provided in USN middleware and shows its relationships with existing USN middleware model.

This paper is organized as follows. In section 2, we describe security issues related to USN middleware. In section 3, as related works, standardization activities in USN middleware and security are described. In section 4, we propose security functions and security model for USN middleware. And finally, we conclude it in section 5.

2 Security Issues in USN Middleware

This section describes security threats related to USN middleware. USN middleware is located between USN application and Sensor Network in the USN service model. It processes data sensed by sensors and sends the processed data to appropriate USN application. It means all data used for providing USN service must pass through USN middleware. Like this, USN middleware plays an important part in USN service. Therefore, if security threats related to USN middleware happen, it causes a service disrupt, misuse, system failure, and so on.

Security threats related to USN middleware are as followings:

- Security threats targeting service entity:
  This kind of security threat attacks service entities which are application, sensors and middleware itself.

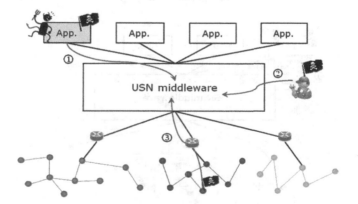

Fig. 1. Unauthorized access to USN middleware

Attacker can deliver an attack against on USN middleware itself by accessing illegally. Also there is a close correlation among application, middleware, and sensor network. So, attacking sensors and applications can have an effect on USN middleware security and USN service providing. Because attacker can access to USN middleware with compromised legal sensors and applications identities. Figure 1 shows security threat targeting USN middleware.

● Security threat targeting data: [7]
Data-related security threats means in case USN middleware is received unreliable data from application or sensor network or the data stored in USN middleware is leaked and modified illegally. Since USN services are provided based on the data, the data that is collected, processed and forwarded by USN middleware is very important. There are two kinds of security threat related to data. One thing is that sensitive data is leaked in the middle of data transmission by sensor network and application or sensitive date stored in USN middleware is leaked. Figure 2 shows data leakage. Another is that malicious or abnormal traffic is flowed into USN middleware. Figure 3 shows malicious or abnormal traffic flowing into USN middleware.

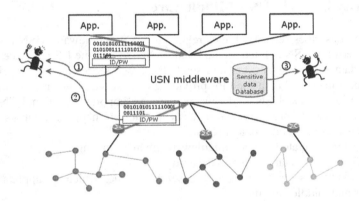

Fig. 2. Data leakage

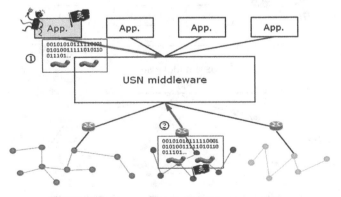

Fig. 3. Malicious or abnormal traffic transmission

- Security threats targeting communication channel:
  This security threat is that communication between two entities is attacked. There are two kinds of communication in USN service model. One is communication between application and USN middleware and another is communication between USN middleware and sensor network. Attacker can eavesdrop on communication between two entities and abuse the information collected by eavesdropping. Figure 4 show eavesdropping on communications

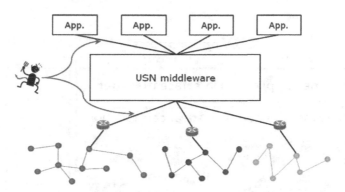

Fig. 4. Eavesdropping on communications

3 Related Works

This section gives existing works relating to USN middleware.

3.1 Existing Works

Relating to USN middleware, several middleware has been introduced. In this section, some of them are described. MiLAN[8] middleware aims to ensure application QoS requirement. This can be implemented by controlling sensor network directly. DSWare[9] middleware supports SQL-like query language and group-base management for sensor nodes. Impala[10] middleware supports a dynamic update for sensor node functions. TinyDB[11] middleware is operated over Tiny OS and supports SQL-like query languages. It regards sensor network as a virtual distributed database. Cougar[12] middleware stores all sensed data in server, and then user can access to DB and achieve sensed data. Most of existing middleware is to enhance performance and functionality for transaction between applications and sensor networks. But security in middleware has been not considered deeply.

3.2 USN Middleware Standardization [3]

Basic model for USN middleware is defined by Q.25/16. The title of the work item is "Service description and requirements for USN middleware" and also called as F.usn-mw. And Q.25/16 is a study group which develops standards on USN applications and services. F.usn-mw describes USN services and defines requirements for USN middleware to support various types of USN services. Also, it defines a functional model of USN middleware.

F.usn-mw said that each USN application may access the sensor network in a dedicated manner, in case open sensor network environments. But there is a common application platform for USN applications to use sensor network, a sensor network can be shared among different applications.

Figure 5 shows the functional model of USN middleware defined by Q.25/16.

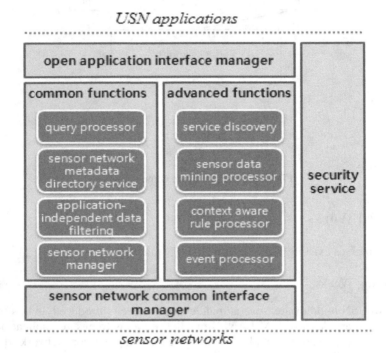

Fig. 5. Functional model of USN middleware

USN middleware is composed of open application interface manager, common functions, advanced functions, sensor network common interface manager, and security service.

Detailed functions for each component are as followings:

- Open application interface manager: This component provides application interface for applications to access USN middleware via web service.

- Sensor network common interface manager: This component provides common sensor network interface to access USN middleware.

- Common functions: Common functions are a group of functions that is consisted of four sub-functions which are query processor, sensor network metadata directory service, application-independent data filtering, and sensor network manager.

- Query processor: This offers query scheduling function for multiple USN applications and sensor networks, and query routing function to sensor node. And it provides filtering function for application-dependent RFID tag data and sensed data. It also aggregate and integrate sensed data, based on an application policy.

- Sensor network metadata directory service: It has a responsibility of registration of USN metadata and discovery of required sensor networks.

- Application-independent data filtering: It checks validation for sensed data regarding associated measurement units, data types and value ranges. It also offers a RFID tag data filtering function.

- Sensor network manager: This manages sensor network, sensor gateways, and RFID readers. And it performs a remote upgrade for sensor node software and manages topology.

● Advanced functions: Advanced functions are a group of functions that provides more intelligent service.

- Service discovery: This has a responsibility of registration and discovery for USN middleware service and USN service.

- Sensor data mining processor: This function is to detect outlier, analyze patterns, and predict potential events.

- Context aware rule processor: It applies application-dependent context aware rule on collected sensed data.

- Event processor: It generates events using output of context aware rule processing and processes events.

● Security service: Security service is to avoid an unauthorized access and protect information.

3.3 USN Security Standardization [4][5][6]

ITU-T Q.6/17, which has a responsibility of standard development in ubiquitous telecommunication service security, has been taking the lead in USN security standard development. For USN security, there are three kinds of work items which are security framework for ubiquitous sensor network (X.usnsec-1)[4], USN middleware security guideline (X.usnsec-2)[5], and Secure routing mechanisms for wireless sensor network (X.usnsec-3)[6]. This paper specifies X.usnsec-1 and X.usnsec-2 briefly.

First of all, X.usnsec-1 describes overall issues in USN security. It defines security model and threat model for USN environment. According to X.usnsec-1, security threats in USN are classified under two kinds which are threats in Sensor network and threats in IP network. For threats in IP network, the threat model developed in ITU-T X.805 can be applied [2]. Threats in Sensor network can be divided into two groups [1][2][4].

One is general threats. Detailed threats are as followings:
- Destruction of information and/or other resources
- Corruption or modification of information
- Theft, removal or loss of information and/or other resources
- Disclosure of information
- Interruption of services
- Sensor node compromise
- Eavesdropping
- Privacy of sensed data
- DOS attacks
- Malicious commodity networks

Another is routing-specific threats [13][14][15][16]. Detailed threats are as followings:
- Spoofed, altered, replayed routing information
- Selective forwarding
- Sinkhole attack
- Sybil attacks
- Wormhole attacks
- HELLO flood attacks
- Acknowledgement spoofing

X.usnsec-1 also defines security requirement to deal with abovementioned security threats. Detailed security requirements are as followings:
- Data Confidentiality
- Data Authentication
- Data Integrity
- Access control
- Non-repudiation
- Communication security
- Availability
- Privacy

And X.usnsec-1 provides security technologies for USN as countermeasures, which are key management, authenticated broadcast, secure data aggregation, data freshness, tamper resistant module, USN middleware security and IP network security.

Next, X.usnsec-2 deals with security issues in USN middleware and provides guidelines for USN middleware security. It also defines security threats on USN middleware and requirements for USN middleware security. Security threats and requirements are specified throughout this paper. So, detailed explanation is omitted.

4   USN Middleware Security Model

This section describes security functions and security models for providing secure USN service.

4.1   Security Functions for USN Middleware

Figure 6 shows security functions that USN middleware should be provided for trustworthy USN service. Security functions can be divided into five sub-functions which

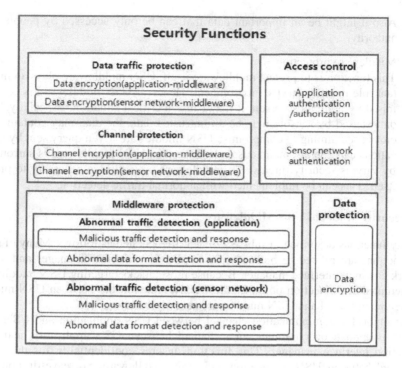

Fig. 6. Security functions for USN middleware security

are data traffic protection, channel protection, access control, data protection, and middleware protection.

- Data traffic protection:
  This function is to protect sensitive data exchanged between application and middleware and sensitive data exchanged between sensor network and middleware. The sensitive data can be an authentication data, such as a password, secret information, etc.

- Channel protection:
  This function is to protect communication channel between application and middleware and communication channel between sensor network and middleware.

- Access control:
  This function is to prevent unauthorized access from application and sensor network. It can be implemented with authentication for application and sensor network. Especially, authorization is required for application. Because application have different privileges for accessing to specific resources (e.g., sensed data, etc).

- Data protection:
  This function is to ensure confidentiality for data stored in USN middleware. The data can be an authentication data of applications and sensor networks.

Also it might be an important data that can be only accessed by people with authority.

- Middleware protection:
  This function is to protect middleware itself. USN middleware plays an important role in USN service environment. Hence, if USN middleware is compromised by malicious person, it may cause a critical situation. Especially, since data sensed by sensors may be an untrusted data, the data may contain malicious code aiming to compromise USN middleware. Even query sent by applications can contain malicious code. Like this, USN middleware is surrounded by various security threats. So, middleware protection is necessary to protect itself. This can be implemented with abnormal traffic detection.

4.2 Security Model of USN Middleware

Security functions described in sub-section 4.1 are operated as follows. Many of security functions are related to both of open application interface manager and sensor network common interface manager. Because most attacks targeting USN middleware are attempted at connection point. Figure 7 shows relationship between USN middleware reference model and USN middleware security function.

Encryption for communication channel between USN application and USN middleware is applied on communication channel between USN application and open application interface manager. This function ensures confidentiality for the traffic exchanged between USN application and USN middleware. So, eavesdropping on communication channel can be prevented.

Data encryption for communication between USN application and USN middleware is to protect sensitive data of USN application or USN middleware. So, this function is implemented at the open application interface manager. And USN application also has this function. By doing this, malicious person cannot find original data even they achieve or hijack traffic exchanged over communication channel. This function guarantees confidentiality for sensitive data of USN middleware or USN application.

Middleware protection is applied to open application interface manager and sensor network common interface manager. Procedure of middleware protection applied to open application interface manager is same with middleware protection function applied to sensor network common interface manager. But they check traffic or data sent by application or sensor network, based on different data format criteria. Because data format in application is different from data format in sensor network.

Application authentication/authorization is applied to open application interface manager. If mutual authentication is supported, USN application also has an authentication function. This function is to avoid that untrusted USN applications access to USN middleware illegally. In addition, even though illegal access is happened, it is not able to change anything on middleware configuration and to access sensitive data stored at middleware database. For avoiding this situation, authorization function should be provided.

Data encryption is applied to database in USN middleware. The database stores various data containing sensitive data. Sensitive data means information used for

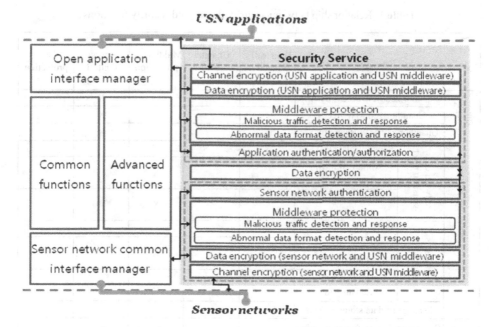

Fig. 7. Security model of USN middleware and its relationship

entity authentication and authorization. So this function is performed with application authentication/authorization function and sensor network authentication function.

Sensor network authentication is applied to sensor network common interface manager. If mutual authentication is supported, sensor network also has an authentication function. This function is to avoid that untrusted sensor networks access to USN middleware illegally.

Data encryption for communication between sensor network and USN middleware is to protect sensitive data of sensor network or USN middleware. So, this function is implemented at the sensor network common interface manager. And sensor network also has this function. By doing this, malicious person cannot find original data even they achieve or hijack traffic exchanged over communication channel. This function guarantees confidentiality for sensitive data of USN middleware or sensor network.

Encryption for communication channel between sensor network and USN middleware is applied on communication channel between sensor network and sensor network common interface manager. This function ensures confidentiality for the traffic exchanged between sensor network and USN middleware. So, eavesdropping on communication channel can be prevented.

4.3 Relationship between Security Threats and Security Functions

Table 1 shows the relationship between security threats described in clause 2 and security functions.

Table 1. Relationship between security threats and security functions

			Data traffic protection	Channel protection	Middleware protection	Access control	Data protection
Service entity		Unauthorized MW access by App.				X	
		Unauthorized MW Access				X	
		Unauthorized MW access by SN				X	
Data	Data transmission by App.	Sensitive data leakage	X				
		Abnormal traffic transmission			X		
		Malicious traffic transmission			X		
	Data transmission by SN	Sensitive data leakage	X				
		Abnormal traffic transmission			X		
		Malicious traffic transmission			X		
		Leakage of data stored in MW					X
Communication channel		Eavesdropping on communication App.-MW		X			
		Eavesdropping on communication MW-SN		X			

Unauthorized USN middleware accesses by USN application and sensor network could be prevented with access control mechanisms, such as authentication and authorization. Using this, USN middleware can protect itself from illegal accesses.

Relating to data, there are three types of security threats. Two of them are in case data is transmitted by application and sensor network. In those cases, they have three kinds of data-related security threats which are sensitive data leakage, abnormal traffic transmission and malicious traffic transmission. Sensitive data leakage could be prevented with data traffic protection mechanisms, like an encryption for sensitive data. And abnormal traffic transmission and malicious traffic transmission could be dealt with middleware protection means, such as abnormal data format detection and malicious traffic detection, etc. Security threat relating to data stored in USN middleware could be prevented with data protection mechanisms, like a data encryption function. Finally, eavesdropping on communication between USN application and USN middleware and communication between sensor network and USN middleware could be prevented with channel protection mechanisms, such as encryption for communication channel.

5 Conclusion

USN middleware plays a significant role in USN service. It supports heterogeneous sensor networks and various USN applications. And all data used in USN service pass through USN middleware. So, if USN middleware is compromised, USN service might be not functioning normally. Hence, USN middleware security is a necessary part and it should be handled very carefully. This paper proposes USN middleware security model. And it was proven that our model is secure in terms of unauthorized access, data leakage, malicious/abnormal traffic transmission, eavesdropping, etc. We expect that the security model of USN middleware is used as a basis when USN middleware is designed and deployed. But it is just a theoretical model, because we could not apply this to actual USN middleware. It remains as a further work together with performance evaluation.

References

1. ITU-T X.800, Security architecture for Open Systems Interconnection for CCITT applications
2. ITU-T X.805, Security architecture for systems providing end-to-end communications
3. ITU-T F.usn-mw, Service description and requirements for USN middleware
4. ITU-T X.usnsec-1, security framework for ubiquitous sensor network
5. ITU-T X.usnsec-2, USN middleware security guideline
6. ITU-T X.usnsec-3, Secure routing mechanisms for wireless sensor network
7. Sabbah, E., Majeed, A., Kang, K.-D., Liu, K., Abu-Ghazaleh, N.: An Application-Driven Perspective on Wireless Sensor Network Security, Q2SWinet 2006 (2006)
8. Heinzelman, W., Murphy, A., Carvalho, H., Perillo, M.: Middleware to Support Sensor Network Applications. IEEE Network Magazine Special Issue (January 2004)
9. Li, S., Son, S.H., Stankovic, J.A.: Event detection services using data service middleware in distributed sensor networks. In: Zhao, F., Guibas, L.J. (eds.) IPSN 2003. LNCS, vol. 2634, pp. 502–517. Springer, Heidelberg (2003)
10. Liu, T., Martonosi, M.: Impala: A Middleware System for Managing Autonomic, Parallel Sensor Systems In: Proc. ACM SIGPLAN Symp. Principles and Practice of Parallel Programming, pp. 107–118 (2003)
11. Madden, S.R., Franklin, M.J., Hellerstein, J.M.: TinyDB: An Acquisitional Query Processing System for Sensor Networks. ACM TODS 30(1), 122–173 (2005)
12. Yao, Y., Gehrke, J.E.: The Cougar Approach to In-Network Query Processing in Sensor Networks. SIGMOD RECORD 31(3) (2002)
13. Karlof, C., Wagner, D.: Secure Routing in Wireless Sensor Networks: Attacks and Countermeasures. Ad Hoc Networks 1(2-3), 293–315 (2003)
14. Loo, C., Ng, M., Leckie, C., Palaniswami, M.: Intrusion Detection for Routing Attacks in Sensor Networks. International Journal of Distributed Sensor Networks 2(4), 313–332 (2006)
15. Karlof, C., Wagner, D.: Secure Routing in Wireless Sensor Networks: Attacks and Countermeasures. Elsevier's AdHoc Networks Journal (2002)
16. Undercoffer, J., Avancha, S., Joshi, A., Pinkston, J.: Security for Sensor Networks. CADIP Research Symposium (2002)

Practical Broadcast Authentication Using Short-Lived Signatures in WSNs

Chae Hoon Lim

Dept. Computer Sciences and Engineering, Sejong University, Seoul, Korea
chlim@sejong.ac.kr

Abstract. Efficient broadcast authentication in wireless sensor networks has been a long-lasting hard problem, mainly due to the resource constraint on sensor nodes. Though extensive research has been done in past years, there seems to exist no satisfactory solution to date. In this paper we propose a practical approach to the problem using short-lived digital signatures, in which a base station makes use of a short RSA modulus of limited lifetime, say, RSA-512 with 20-min lifetime, for authenticated broadcast with Rabin signatures giving message recovery. For this, we present an efficient and robust protocol using a one-way key chain to periodically distribute short RSA moduli to all sensor nodes in an authentic and loss-tolerant way. We also provide conservative lifetime estimation for short RSA moduli based on the state-of-the art factoring experiments and apply a number of possible optimizations in algorithms and parameters. The proposed scheme overcomes most drawbacks of existing schemes such as μTESLA and one-time signatures and turns out to be very efficient and practical. It can also be extended to provide secure failover of base stations and authentication delegation to mobile users.

1 Introduction

Broadcast is a main communication primitive in wireless sensor networks (WSNs). A base station queries sensor nodes to collect environmental data by broadcasting commands and data to the whole or part of the network. Since WSNs often operate unattended, possibly in hostile environments, it is essential to ensure the authenticity of such commands and data for reliable operations. However, providing broadcast authentication in large-scale distributed senor networks is a nontrivial task, mainly due to the resource constraint on sensor nodes. Thus a lot of research efforts have been devoted to developing and improving symmetric key-based techniques such as μTESLA and its variants [24,25,18,20] and one-time signatures [30,21,4,15]. However, each of these approaches has been known to have some serious drawbacks.

μTESLA makes use of timed one-way key chain and delayed disclosure of authentication keys, thus requiring network-wide time synchronization and buffering of received packets until the authentication key is disclosed. In particular, the buffering requirement at the receiver side introduces a serious vulnerability to DoS (denial-of-service) attacks exhausting the limited storage of sensor nodes.

H.Y. Youm and M. Yung (Eds.): WISA 2009, LNCS 5932, pp. 366–383, 2009.

One-time signatures have been known for more than two decades and brought back to research interest in recent years as an alternative to digital signatures for broadcast/multicast authentication. Though much progress has been made in recent years, they are still not very practical for sensor network application due to the large signature/public key size and the high cost for public key update.

On the other hand, it has been shown in recent years [10,26,36,34,19,33] that public key cryptography, such as ECC (elliptic curve cryptography) and RSA signatures, could be performed quite efficiently even on resource-constrained sensor nodes. Thus it may be plausible to consider the use of standard digital signatures for broadcast authentication. However, an RSA signature is too large to fit in a single packet, requiring fragmentation into multiple packets, and an ECC signature incurs too much processing delay in verification on sensor nodes. Such excessive signature size or processing delay makes them extremely vulnerable to DoS attacks. Note that fragmented signatures cannot be verified until all the fragments are received and thus transmission of unauthenticated signature fragments is completely vulnerable to DoS attacks. Thus, standard digital signatures are not very practical for broadcast authentication either.

In this paper, we propose a very practical solution for broadcast authentication in sensor networks using short-lived RSA/Rabin signatures. Our primary observation is that it suffices to require that signatures should remain unforgeable only for a short period of time since we are only interested in authentication rather than non-repudiation. This observation leads us to consider the use of short RSA moduli of limited lifetime to maximize computation and communication efficiency on resource-constrained sensor nodes. For example, we may use a fresh instance of RSA-512 in every hour instead of using RSA-1024 over the entire network lifetime, provided that factoring RSA-512 in a few hours is infeasible. The key size and lifetime should be chosen carefully but can be adjusted immediately in the next time interval to cope with any new threat discovered.

A main problem in using short-lived signatures for broadcast authentication is to periodically distribute varying public keys to all sensor nodes in an authentic and loss-tolerant way. As a natural solution, we may use a multi-level public key system (much like multi-level μTESLA [18]) by constructing a multi-level public key hierarchy with varying key sizes and lifetimes in each level. The base station then uses long-term higher-level public key signatures to authentically distribute short-term lower-level public keys and the lowest level public keys can be used to authenticate actual broadcast messages. For example, we may use two-level public keys with (RSA-1024, RSA-512) or three-level public keys with (ECC-160, RSA-768, RSA-512). The use of strong signatures in higher levels may not incur too much overhead in computation and communication, since they are used infrequently only for authentic distribution of lower-level public keys. However, this approach requires additional measures to protect high-level signatures against DoS attacks (e.g., see [22,35,11]), either due to time-consuming operations in ECC or long signatures requiring fragmentation in RSA.

If protection mechanisms against DoS attacks are required anyway, we need not to resort to expensive long-lived signatures to distribute short public keys.

For better energy efficiency, we thus propose a hybrid technique in this paper. More specifically, an efficient and robust mechanism is proposed for authentic and loss-tolerant distribution of short RSA moduli using timed one-way key chains (a kind of DoS-resistant, long-interval μTESLA). Then actual broadcast authentication can be done using Rabin signatures with message recovery. To further improve the computation and communication efficiency, we make use of various optimization techniques in algorithms/parameters based on detailed security and efficiency considerations. We also examine up-to-date factoring experiments and provide conservative estimates for lifetimes of short RSA moduli.

The proposed scheme turns out to be very efficient in both computation and communication. For example, with RSA-512, the base station only needs to broadcast 64 bytes of signature for typical broadcast messages and a sensor node only needs to perform a single squaring modulo RSA-512 for verification. Contrary to existing schemes such as μTESLA and one-time signatures, it does not require network-wide time synchronization, allows immediate authentication of broadcast messages and provides lightweight key management with almost negligible overhead. It is also easy to extend the proposed scheme to provide secure failover of base stations and authentication delegation to mobile users.

This paper is organized as follows. In Section 2 we describe our design choices for the underlying algorithms and parameters used in the proposed scheme. In Section 3 we provide our estimation on the lifetime of short RSA moduli based on the latest factoring experiments. The detailed description and analysis of the proposed scheme is presented in Section 4 and some possible extensions of the basic scheme is briefly mentioned in Section 5. We finally conclude in Section 6.

2   Design Choices

2.1   One-Way Function, MAC and Hash

Hash, MAC(message authentication code) and one-way functions are basic cryptographic primitives required in almost all security protocols in sensor networks. To save program memory and implementation effort, we propose to build all these primitives from a single block cipher AES with 128-bit key. AES has been shown to be very efficient even on 8-bit microprocessors in both speed and code size. For example, an optimized assembly implementation of AES on ATmega128, which is available under the GPL license, takes less than 0.5msec to encrypt one 128-bit block (including the key setup time) only using 1570 bytes of memory on 8 MHz MicaZ mote [27] (see also [28]).

Let $E_K(x)$ denote AES encryption of message x under key K. To build a hash function H from AES, we can use the Matyas-Meyer-Oseas construction, which has been shown to be secure under the black-box analysis [3]. Let $m = m_1\|m_2\|\cdots\|m_t$ be the message to be hashed, where the last block m_t should be length padded with the bit-length of m (before padding) modulo 2^{32}. Then, set $h_0 = IV$ (a fixed initial vector), iterate the computation $h_i = E_{h_{i-1}}(m_i)\oplus m_i$ for $i = 1, 2, \cdots, t$ and output $H(m) = h_t$ as the hash value for m. The resulting hash value is of length 128 bits, secure enough for our short-lived signatures. With an implementation of AES in [27], it would take about 2msec on MicaZ mote to hash a 64-byte message. For comparison, we note that typical (unoptimized) C implementation of SHA1 on the same platform takes about 7.5msec to hash one 64-byte block and has much larger code size [23,15].

The above hash function can also be used as a one-way function F for key chain construction: $F(K) = T_k(H(K))$ for k-bit key chain values, where $T_k(x)$ is a simple left k-bit truncation of x. Note that one-way function with $k \leq 96$ can be evaluated only using a single block encryption. For our purpose, it suffices to take $k = 80$. For message authentication, we can use CBC-MAC or its enhancement CMAC standardized by NIST [37].

2.2 Rabin Signatures Giving Message Recovery

Though ECC signatures are very compact in size, their heavy verification overhead on sensor nodes makes it still vulnerable to DoS attacks even with short ECC parameters (e.g., ECC-112). The main drawback of large signature size in RSA, however, can be mostly eliminated by using a short RSA modulus (e.g., RSA-512) and a signature scheme giving message recovery. We thus decide to use the probabilistic Rabin signature scheme giving message recovery proposed by Bellare and Rogaway [2] (the encoding format described below follows IEEE P1363a [40] with slight modification for our short-lived modulus environment).

Let n be an RSA modulus of length l_n bits. Let H be a secure hash function of output length l_h and G be a function stretching l_h-bit input into $(l_n - l_h - 9)$-bit output. The function G on input x can be implemented by concatenating $E_x(c_i)$ for $i = 1, 2 \cdots$, where c_i is a counter represented as a 128-bit string. Suppose that a message m consists of a recoverable part m_1 and a non-recoverable part m_2. Let l_1 be the bit-length of m_1 in 32 bits and let ϕ be all-zero string. The signature for m is a random square root of $y = 0\|m^\star\|w\|bc_{16}$, where $m^\star = G(w) \oplus m'$, $m' = \phi\|1\|m_1\|r$, $w = H(l_1\|m_1\|H(m_2)\|r)$ and r is a random salt of length l_r (0/1 is represented as a single bit). The salt length l_r can be taken as $l_r = \log_2 q_{sig}$ bits, where q_{sig} represents the maximum number of signatures that can be generated for a given modulus [6]. Note that the maximum length of a recoverable message part is $l_n - l_h - l_r - 10$ bits:

The 802.15.4 radio standard specifies the maximum MAC-layer payload length as 102 bytes in the 2003 version [38] and 118 bytes in the 2006 revision [39]. Thus a signature with modulus of length up to 768 bits can still fit in a single packet.[1] We thus decide to use short RSA moduli between 512 and 768 bits of lifetimes at most 1 day (typically less than an hour; see Section 3). For such short-lived signatures, it would be safe to take $l_h = 128$ and $l_r = 16$. Then the recoverable message part can be as long as $l_n - 154$ bits. That is, the signature overhead

---

[1] Note however that the actual modulus length that can be used may vary, depending on the additional protocols adopted. In addition to TinyOS header (1 or 2 bytes) and application header (3 bytes in our proposed scheme), additional overheads may be introduced, e.g., due to link layer security enabled (9 bytes for the mandatory-to-implement AES-CCM-64; see [29]) and/or network layer protocols such as 6LoWPAN (7-12 bytes for typical UDP/IPv6 datagrams; see [12]).

can be reduced to 20 bytes for long messages (of length 44 bytes or more). For typical short messages, we need to send a fixed l_n-bit signature.

2.3 Energy-Efficient RSA Key Generation

For further energy efficiency on sensor nodes, we can use a special modulus n such that some pre-specified bits appear in the high-order bits of n (e.g., see [16] for comprehensive survey and analysis). The pre-specified bits P may contain ID information and security parameters, etc. We may also include a special bit-string (i.e., all ones) in the leading part of P for fast modular reduction.

To embed a pre-specified bits P of length $t < l_n/2 - \delta$ into a modulus n, we first generate a random prime p of length $l_n/2$, choose a random k of length δ bits and test $\hat{q} = P2^{l_n-t}/p + k$ for primality, where '/' denotes integer division. If $\hat{q}$ is not prime, increase k by 2 and test the new $\hat{q}$ again. This process is repeated until a prime q is found. If all δ random bits are consumed, then we can restart the whole process with a new random prime p. It is easy to see that $n = pq = P2^{l_n-t} + kp - r$ for some $r < p$ and thus the pre-specified string P of length t appears in the high-order bits of n if $t < l_n/2 - \delta$. This method can pre-specify up to $l_n/2$ high-order bits of n (even up to $2l_n/3$ bits can be pre-specified with a more sophisticated method [14]). For easy generation of primes, however, we may take δ at least larger than $\log_2(l_n/2)$ according to the prime number theorem (e.g., $\delta = 8 \sim 16$). There is no known security problem for this kind of modulus as well surveyed in [16].

In general, modular reduction is more complicated and time-consuming than multiplication. The most popular Montgomery reduction is not very efficient for just a few modular multiplications due to the heavy overhead for pre- and post-transformations required (note that we only need a single modular squaring for verification of Rabin signatures). However, the modular reduction process can be substantially simplified with a special form of RSA modulus.

Let n be represented in base b notation (e.g., $b = 2^8$ for ATmega128 and $b = 2^{16}$ for MSP430) as $n = n_{k-1}b^{k-1} + \cdots + n_1 b + n_0$, $0 \le n_i \le b - 1$. To simplify the modular reduction process, we would like to use a special modulus of the form $n = b^k - n'$ with $n' < b^{k-u}$ (by pre-specifying the leading u digits of n as all 1 string). Suppose that we want to reduce a $2k$-digit integer x modulo n. Let $u = w + \delta$. Prepend zeros to x so that x is $k + k'$ digits long with k' a multiple of w. Since we have $b^k = n' \bmod n$ and n' is $k - w - \delta$ digits long, we can reduce x by w digits at a time using w digit by $k - w - \delta$ digit multiplication as follows (here $x[i, j] = x_i b^{i-j} + \cdots + x_j$ and $x+ = y$ means $x \leftarrow x + y$).

```
for i=k+k'-1 to i= k+w-1 step i=i-w
    x[i-w, i-w-k+1] += x[i, i-w+1] × n'[k-u-1, 0];
    if(final carry)
        x[i-w, i-w-k+1] += n'[k-u-1, 0];
```

If $x[k-1, 0] > n$ at the end of the algorithm, we then need to subtract n to obtain the final result (i.e., compute $x[k - 1, 0] + n'$ and neglect the final carry bit). Since the probability of the 'if' condition being true is expected to be less than

$b^{-\delta}$ on average, we can avoid the conditional addition in most cases by taking $\delta \geq 2$. This algorithm is very efficient, just requiring $k(k-u)$ multiplications for modular reduction. Further performance improvement comes from its ability to process w digits at a time as it significantly decreases the number of memory accesses. We may take $(u = 8, w = 4 \sim 6)$ for ATmega128 and $(u = 4, w = 2)$ for MSP430 by considering the number of general-purpose registers available (see [10,36,34,32]). This kind of a special RSA modulus is also communication- and storage-efficient since we only need to communicate and store n' instead of n.

We may have some concern about the security of the above special modulus against the NFS (number field sieve) factoring algorithm [17]. For a given n, if one can find an integral polynomial f of degree d (usually between 3 and 10) such that $f(m) = 0 \mod n$ for some integer m and all coefficients of f is substantially smaller than the d-th root of n, then the much faster special number field sieve (SNFS) algorithm can be applied to factor n. This would be the case for $n = 2^k - r$ with very small r. However, as noted in [16], if r behaves as a random number of at least about k/d bits, then the probability is negligible that such a polynomial f with unusually small coefficients exists. In our special modulus, at most the higher half of n can be pre-specified and only a small portion of the pre-specified bits consists of all 1's (e.g., 64 bits for $u = 8$). Thus it is highly unlikely that the NFS can factor our special moduli faster than regular RSA moduli.

2.4   Mechanism to Mitigate DoS Attacks

To mitigate DoS attacks for unauthenticated messages, we make use of weak authentication based on timed one-way key chains (see [22] for details). Let $\{I_i\}$ be a sequence of time intervals with duration d and let $\{K_i\}$ be a one-way key chain (maintained by a base station) such that $K_{i-1} = F(K_i)$ for $i = J$ to 1. Each key K_i is assigned to time interval I_i and the key K_0 is predistributed to all sensor nodes in the network. Prior to sending a message m_i in I_i, the base station exhaustively searches for a message-specific puzzle solution S_i such that $Int(H(K_i, m_i, S_i)) = 0 \mod 2^l$, where $Int(s)$ means the bitstring s is interpreted as an integer and l is a fixed parameter determining the strength of weak authentication. Thus the equation simply means the left l bits of the hash output must be all zeros. The base station then broadcasts (K_i, m_i, S_i) in I_i and a sensor node accepts m_i only if K_i is the right key assigned to I_i and $Int(H(K_i, m_i, S_i)) = 0 \mod 2^l$ (see Section 4.2 for more details).

The *weak* authentication property of the above scheme mainly comes from the asymmetry in computing times available to the sender and the attacker. To find a valid S_i, the attacker needs the same amount of computation as the sender (i.e., 2^l hash operations on average) but can start the computation only after the key K_i is disclosed. Since K_i remains valid only for duration d, the attacker has at most time d for attack, while the sender may have plenty of time for precomputation before transmission. Thus it is expected that the attacker can produce only a limited number of forgeries within the available time. This mechanism is particularly effective for applications where only a few predetermined messages need to be transmitted over a long time interval. This is the case with our application to public key update (see Section 4.2). Note that in case of a large d, the message may be released at the point $d - \delta$ of the time interval, so that the attacker has only δ time units for forgery.

The parameter l should be determined carefully by considering the capability of the base station and the available time for precomputation. In particular, the base station should be able to find a solution S_i surely before transmission. Let $P_c(l)$ be the probability of failing to find a puzzle solution after 2^{l+c} trials for a fixed c. Then we have $P_c(l) = (1 - 2^{-l})^{2^{l+c}}$, assuming that the hash function H behaves pseudorandomly. Since the function $f(x) = (1 - x^{-1})^{\alpha x}$ is strictly increasing for $x \geq 1$ and has the limit $e^{-\alpha}$ as x approaches infinity, we have $P_c(l) < e^{-2^c}$ for any $l \geq 0$. This shows that the failure probability becomes negligible for $c \geq 5$, as $P_5(l) < 2^{-46.2}$ and $P_6(l) < 2^{-92.3}$. Therefore, it would be sufficient to take a value l such that about 2^{l+6} hash operations can be executed on the base station within the allowed time.

3   Lifetime Estimation of Short RSA Moduli

In this section, we examine up-to-date factoring experiments and research results and suggest some conservative lifetime estimates for short RSA moduli.

The fastest known algorithm for factoring general RSA moduli is the general number field sieve (GNFS) [17]. The GNFS algorithm consists of four major steps: polynomial selection, sieving, matrix reduction and square root. The most time-consuming steps are sieving and matrix reduction, whose complexity determines the asymptotic running time of GNFS

$$L(n) = e^{c(\ln n)^{1/3}(\ln \ln n)^{2/3}}, \tag{1}$$

where $c = (64/9)^{1/3} \simeq 1.923$ (by neglecting the $o(1)$ term). Some observations on the algorithmic/experimental behaviors of GNFS may help:

- The first two steps, polynomial selection and sieving, allows virtually unlimited parallelization and thus can be easily distributed to an arbitrary number of independent processors to reduce the running time.
- The matrix step however allows only a limited degree of parallelism and needs to run on tightly coupled processors (dedicated clusters/supercomputers or PC clusters connected by Gigabit Ethernet). Thus this step can easily become a bottleneck of the whole NFS process in large factorization efforts. As the number of processors increases, per-processor utilization drops fast, due to the heavy inter-processor communications overhead (e.g., see [1] for some matrix step implementation experiments).
- After sieving, some nontrivial substeps for matrix building and filtering are needed to build a matrix from raw sieved data and filter out less useful relations to reduce the matrix size for the matrix step (by a factor of up to several tens). These substeps usually require a centralized computation with massive memory and thus can hardly be parallelized.

- The square root step is the least-time consuming but requires some complex computations typically on a single computer.

A number of NFS-based factoring records have been reported so far (e.g., see Zimmermann's integer factoring records page at http://www.loria.fr/ zimmerma/ records/factor.html). The latest factored RSA challenge number is RSA-640, factored on November 2, 2005. It is reported that the factoring effort took about five months of calendar time on a cluster of 80 2.2GHz-Opteron CPUs (3 months for sieving and 1.5 months for matrix solving, but the polynomial selection time not included).

More recently, Chen et al. [5] report a faster factoring result for RSA-512 using two supercomputers, HP cluster and IBM p595 SMP(symmetric multiprocessing).[2] We will use their result as a benchmark for our lifetime estimation as it employs more advanced computing facilities and provides detailed timings for each step. Table 1 summarizes detailed timing data for each step.

Table 1. Statistics for factoring RSA-512 in [5]

Polynomial selection	24 hr	50 cores of HP cluster
Sieving	68.9 hr	
Matrix building	7.3 hr	one CPU of HP cluster
Matrix filtering	2.2 hr	24 cores of IBM p595
Matrix solving	37.5 hr	
Square root	2.98 hr	one CPU of HP cluster
Total	142.9 hr (about 6 days)	

Since polynomial selection and sieving can be easily parallelizable, these two steps could be done in less than 20 hours on the full capacity (426 cores) of the HP cluster. Then the complete factorization could be completed in less than 3 days, as the authors estimated in [5]. Since a computer cluster of a few hundreds of nodes is readily available in large organizations, it is reasonable to assume that RSA-512 can be factored in a few days (say, 1.5 days). We then apply a 100-fold safety margin. This gives us 20 minutes as the lifetime of RSA-512. The lifetimes for larger moduli can be obtained using the traditional extrapolation

$$\text{Lifetime}(n) = \frac{L(n)}{L(2^{512})} \times \text{Lifetime(RSA-512)}. \tag{2}$$

This yields lifetimes of 2 hours for RSA-576 and 1 day for RSA-640. Roughly, this estimation is also consistent with the RSA-640 factoring experiment; we

---

[2] The HP cluster consists of 106 nodes connected by InfiniBand DDR switch, where each node is a HP Proliant DL Server with two Woodcrest dual core 3.0GHz CPUs. The IBM p595 SMP has 64 Power5+ 1.9GHz CPUs and 256GB RAM.

can obtain 1 day lifetime for RSA-640 by assuming that RSA-640 can be factored in 100 days more conservatively and applying the same 100-fold safety margin.

We believe that the above lifetime estimation is quite conservative (in particular, for RSA-512). The sieving time may be shrunk to a relatively small portion of the overall running time in a large factorization project, but we can hardly achieve linear speedup for the matrix step. The running times for the remaining steps may not be linearly scaled down either, since they largely depend on a single processor speed. However, with dedicated hardware, we may expect more drastic runtime reduction in the major bottlenecks, in particular as the modulus length increases. We thus consider applicability of dedicated hardware to the major NFS steps and derive even more conservative estimates.

Substantial research work for dedicated hardware designs to speed up NFS factoring has been conducted so far, mainly in hardware designs for fast sieving such as TWIRL [31] (see also [8] for a brief survey) and linear algebra circuits to do the most time-consuming part (i.e., a large number of matrix-by-vector multiplications) of the matrix step (e.g., see [9]). No such hardware design has been verified or built yet, but it might be realizable especially for short RSA moduli in the future. To get more conservative lifetime estimation, assume that hardware siever and matrix solver for RSA-640 exist which can complete the two major jobs in minutes (we do not consider smaller moduli in this case as they may be no more secure). Even in this case, it is extremely unlike that RSA-640 could be factored within 20 minutes, since there still remain many other hard-to-parallelizable steps, such as the matrix building/filtering substeps, the remaining substep of the matrix step (i.e., the Berlekamp-Massey step) and the square root step. We thus believe that 20 minutes of lifetime for RSA-640 will be safe even in a fairly long-term perspective. From this, we obtain 2 hours and 1 day for RSA-704 and RSA-768, respectively, as before.

The lifetimes for short RSA moduli estimated so far are summarized in Table 2. For example, for a fixed lifetime of 20 minutes, we can start with RSA-512 and migrate to RSA-640 through incremental changes as factoring experiences indicate any potential danger of the current modulus length. Of course, the modulus size may be further extended up to RSA-768, if necessary. Though obtained in a very conservative way, these lifetimes need to be re-evaluated periodically as technology advances. It is also very important to keep up to date with advances in algorithms and dedicated hardware. For comparison, we note that RFC 4359 recommends 1 week lifetime for RSA-768 in the context of IP security protocols.

Table 2. Lifetime estimation for short RSA moduli

Lifetime	20 min	2 hr	1 day	
Modulus	512	576	640	short-term use
length (bits)	640	704	768	long-term migration

4   The Proposed Scheme

To use short-lived signatures for broadcast authentication, we need a robust and loss-tolerant mechanism to periodically update a short RSA modulus. A novel method based on a one-way key chain is presented for this purpose.

4.1   Key Generation and Initialization

Key Chain Generation. Let the lifetime of the sensor network be divided into time intervals $\{J_1, J_2, \cdots\}$ with a fixed duration d, called TESLA intervals (*T-interval*, for short). Build a sequence of longer time intervals $\{I_1, I_2, \cdots\}$ of duration $D = \lambda d$, called RSA intervals (*R-interval*, for short), by naming the i-th λ consecutive T-intervals as I_i, i.e., $I_i = \{J_{(i-1)\lambda+1}, \cdots, J_{i\lambda}\}$. Let T_i be the start time of J_i, so I_i begins at time $T_{(i-1)\lambda+1}$. The parameters λ and d can be determined based on the lifetime of RSA moduli used.

We divide each R-interval into two parts determined by the parameter λ_1: the *commit interval* consisting of the first λ_1 T-intervals and the *reveal interval* consisting of the remaining $\lambda - \lambda_1$ T-intervals. That is, I_i-commit interval $= \{J_{(i-1)\lambda+1}, \cdots, J_{(i-1)\lambda+\lambda_1}\}$ and I_i-reveal interval $= \{J_{(i-1)\lambda+\lambda_1+1}, \cdots, J_{i\lambda}\}$. We will assign to each R-interval I_i a fresh RSA modulus n_i of lifetime at least $(2\lambda - \lambda_1)d$ and to each T-interval J_j a key chain value K_j of lifetime d. The next R-interval modulus n_{i+1} will be committed in the I_i-commit interval and the actual value n_{i+1} will be revealed in the I_i-reveal interval. As example parameters, we may have $d = 1$ minute, $\lambda = 15$ and $\lambda_1 = 10$ for RSA-512 of lifetime 20 minutes.

Now the network control center (NCC) randomly picks K_I and generates a one-way key chain $\{K_1, K_2, \cdots, K_I\}$ such that $K_{i-1} = F(K_i)$ for $i = I, \cdots, 2, 1$ and assigns the key K_i to the T-interval J_i. The final value K_0 serves as a commitment to the key chain and should be pre-distributed to each sensor node before deployment. The number I, together with the T-interval duration d, determines the network lifetime.

RSA Modulus Generation. The base station uses a fresh RSA modulus $n_i = p_i q_i$ of length l_{n_i} for each R-interval I_i. The modulus n_i thus needs to be generated in I_{i-2} and distributed in I_{i-1}. The modulus length l_{n_i} may vary in each R-interval (but very rarely in practice).

We use RSA moduli with pre-specified bits for efficiency, as suggested in Section 2.3. Let P_i be a set of parameters to be embedded into the high-order bits of n_i. We use the following general format for P_i of length 21 bytes (actual values for the present case appear in parenthesis):

$$P_i = SID \parallel IID \parallel sTime \parallel pLen \parallel mLen \parallel zLen \parallel K, \qquad (3)$$

where SID is a 2-byte sender ID ($=$ base station ID), IID is a 2-byte Issuer ID ($=$ NCC ID), $sTime$ is a 4-byte start time of the validity period of n_i ($= (i-1)\lambda+1$), $pLen$ is a 1-byte validity period of n_i ($= \lambda$), $mLen$ is a 1-byte modulus length ($= l_{n_i}/8$), $zLen$ is a 1-byte parameter for weak authentication, and K is a 10-byte key chain value associated with some T-interval in I_{i-1} (see Section 4.2). If any field is not defined (as for n_1) or unnecessary, it should be filled with random bits.

For further efficiency, we also want to force the leading 8 bytes of n_i to have all 1's. Then n_i should have the special form $n_i = 2^{l_{n_i}} - n_i'$ with $n_i' = P_i \parallel r_i$ for some r_i of length $\ell_{n_i} - 232$. That is, n_i should have the pre-specified bits $11 \cdots 1 (64 \text{ ones}) \parallel \overline{P_i}$ in its high-order 232 bits, where $\overline{x}$ denotes one's complement of x. The base station then distributes n_i' of length $l_{n_i} - 64$ and sensor nodes only need to store n_i' and can use various parameters in n_i' during the protocol.

Initialization. Before deployment, the NCC determines a set of fixed global parameters $\Phi = \{BID, d, \lambda, \lambda_1, K_0, T_1\}$ and the first modulus n_1 as described above. Here BID is the base station ID. The NCC then initializes the base station with Φ, K_I (and some intermediate values for better efficiency) and RSA key (n_1', p_1, q_1). Each sensor node is initialized with Φ and n_1'.

4.2 Public Key Update

A TESLA-like protocol is proposed to distribute a fresh RSA modulus in each R-interval. This protocol also makes use of weak authenticators to mitigate DoS attacks very effectively (see Section 2.4).

The protocol consists of three phases: *KDM-precom, KDM-commit* and *KDM-reveal* (here KDM stands for key distribution message). Let m be a KDM message and K be a key chain value associated with m. Let $\sigma = MAC_K(m)$ be a MAC for m under key K and ω be a weak authenticator for σ. Roughly speaking, the base station precomputes ω during the KDM-precom phase, send a KDM-commit message $\{\omega, \sigma\}$ in the KDM-commit phase and reveals $\{m, K\}$ in the KDM-reveal phase.

All broadcast messages have a common format $\{Type, SID, m\}$. The 1-byte *Type* code identifies the type of message m and lets the receiver know how to process m. *SID* is the sender ID to support multiple senders and used to retrieve the parameters associated with the sender. Suppose that the base station wants to distribute n_i for I_i. The detailed public key update process is as follows.

KDM-precom. In this phase (during I_{i-2}), the base station generates a new modulus n_i for I_i and prepares a KDM-commit message. For this, it first determines a set of parameters for P_i (see eq. (3)). In particular, it randomly chooses J_a in the I_{i-1}-commit interval and J_b in the I_{i-1}-reveal interval, and computes the associated key chain values K_a and K_b. Note that $a \in [(i-2)\lambda+1, (i-2)\lambda+\lambda_1]$ and $b \in [(i-2)\lambda + \lambda_1 + 1, (i-1)\lambda]$. The key K_b is placed in the last part of P_i. Other parameters in P_i are changed as needed.

Once n_i is generated, the base station computes $\sigma_i = H(n_i')$ (truncated to 8 bytes) and exhaustively searches for a random 8-byte number S_a such that

$$Int(H(K_a, \sigma_i, S_a)) = 0 \mod 2^{Z_{i-1}}, \tag{4}$$

where Z_{i-1} is the parameter $zLen$ contained in n_{i-1}'. The pair (K_a, S_a) serves as a weak authenticator for σ_i, which will be revealed in the KDM-commit phase.

KDM-commit. At a random point within the T-interval J_a, the base station broadcasts the prepared KDM-commit message

$$\{Type,\ SID,\ a,\ K_a,\ \sigma_i,\ S_a\}. \tag{5}$$

To eliminate potential errors during verification due to clock discrepancies between the base station and sensor nodes (their maximum value denoted by Δ), the base station may place enough guard band around the boundary of T-intervals and broadcast the message within the safe region. This is possible since the T-interval duration d is long enough (e.g., 1 minute) and clock synchronization within seconds can be trivially done.

Suppose that a sensor node u receiving the message (5) at time T maintains the index-key pair (u, K_u). Then, to verify the message, u performs the following:

(1) verify the correctness of the received T-interval index a: $\lceil \frac{T+\Delta-T_1}{d} \rceil = a$?
(2) compute the current R-interval index $\alpha = \lceil \frac{a}{\lambda} \rceil$ and verify that the T-interval index a belongs to the I_α-commit interval: $a \leq (\alpha - 1)\lambda + \lambda_1$?
(3) verify the authenticity of the revealed key K_a: $F(K_a)^{a-u} = K_u$?
(4) verify the weak authenticator (K_a, S_a): $int(H(K_a, \sigma_i, S_a)) = 0 \bmod 2^{Z_u}$?

Node u rejects the message if any of the checks fails. If all the checks succeed, u replaces (u, K_u) with (a, K_a) and stores (SID, α, σ_i) in the KDM-commit message buffer (in fact $\alpha = i - 1$). At this time, u also erases outdated commit messages, if any, with index smaller than $\alpha - 1$ (not α; see KDM-reveal phase below). Note that the first two checks are the most crucial for security: KDM-commit messages should be received only in the commit interval. The last two steps are to filter out potential forgeries received. There may be a limited number of σ's that can pass these checks, depending on the strength of weak authenticators and the attacker's capability. These forgeries, however, will be definitely rejected in the KDM-reveal phase.

KDM-reveal. At any time during the I_{i-1}-reveal interval, the base station broadcasts the modulus n_i':

$$\{Type,\ SID,\ b,\ n_i'\}. \tag{6}$$

We can suppress SID since it appears as the first two bytes of n_i'. On receiving the message (6), a sensor node u having (u, K_u) performs the following:

(1) parse the first 21 bytes of n_i' as $P_i = SID \parallel IID \parallel S \parallel P \parallel L \parallel Z \parallel K_b$.
(2) reject if the bit-length of the received n_i' is not equal to $8L - 64$.
(3) compute the assumed R-interval index β from the received b as $\beta = \lceil \frac{b}{\lambda} \rceil$ and check if there exists an entry containing (SID, β) in the KDM-commit message buffer. If it exists, retrieve the corresponding σ_i. reject otherwise.
(4) check that $H(n_i') = \sigma_i$ and $F(K_b)^{b-u} = K_u$.

If all the checks succeed, u updates (u, K_u) with (b, K_b) and stores (SID, n_i') in the public key buffer.

In fact, the KDM-reveal message can be transmitted/received at any time after the I_{i-1}-commit interval. Its verification can be done successfully anytime, as far as a sensor node has received the associated MAC σ_i in the I_{i-1}-commit interval. Such a node, if it failed to receive n_i' in the I_{i-1}-reveal interval, may even make a local query (or a query to the base station) to get the modulus. This is the reason why we did not explicitly check the validity of the index b based on the local clock. However, as noted in the KDM-commit phase, any outdated commit message (SID, j, σ_{j+1}) with $j < i-1$ can be safely removed from the buffer in the interval I_i.

Authenticated Broadcast. Actual messages broadcast by the base station in R-interval I_i has the format

$$\{Type,\ SID,\ rSign_i(m)\}. \tag{7}$$

A sensor node identifies it by the $Type$ code, retrieves n_i' using SID from the public key buffer, and recovers m from the message recovery signature $rSign_i(m)$.

4.3   Security and Efficiency

The proposed key management scheme can be thought of as a strengthened variant of TESLA with long interval length. The R-interval actually corresponds to the time interval in TESLA and smaller T-intervals are introduced additionally for DoS protection purpose. Its stripped-down version can be stated as: broadcast a MAC σ_i first (under some degree of DoS protection) and then reveal the message and key n_i' later. The large enough R-interval length enables us to carry out both commit and reveal in the same R-interval, which reduces key update latency. Unlike to TESLA, only a short MAC value is committed first in our scheme and the longer message is revealed later together with the associated key. This has advantage in that the more critical commit message can be made short and of fixed length.

Assuming that it is hard to factor the modulus n_i within the time $(2\lambda - \lambda_1)d$ and the functions F and H are secure, it is then clear that the key update protocol is secure as far as the authenticator σ_i is received before the associated key (included in n_i') is revealed. This security condition is guaranteed by the first two checks in the KDM-commit phase. The security checks are very robust since we do not need network-wide clock synchronization due to the large interval length. Note that it is also very important to check the length of the received n_i' in the KDM-reveal phase (see step (2)), since otherwise an attacker may generate a valid MAC after appending arbitrary bits to n_i'. This may be possible since we generate a MAC by simply hashing a keyed message.

The mechanism for mitigating DoS attacks is also quite effective due to the large asymmetry (at least a factor of λ) in computing times available to the sender and the attacker. Thus, even an attacker as powerful as the base station would be able to produce only a limited number of forgeries. This would be sufficient to deter potential DoS attempts for the commit message. Furthermore, to achieve stronger DoS resistance or to reduce the base station's workload, we may use the more resource-rich backend server in NCC to precompute and send a weak authenticator to the base station in each R-interval. The strength of weak authentication can be re-adjusted by increasing the parameter $zLen$, as technology advances.

The proposed scheme is proactive in nature. Each modulus is used only for a short period of time, which can limit damage upon being broken. Furthermore, the modulus length can be increased in any time interval to cope with a new threat discovered. The overall network security can be further enhanced even in the case of base station failure/compromise if the NCC keeps all key chain values and periodically sends part of them to the base station (e.g., once in an R-interval). If any abnormal sign in the base station is detected, the NCC may stop sending necessary keys. We will detail this issue in the full paper, including secure fail-over of base stations.

Loss tolerance in public key update is also very important due to the long R-interval. If a sensor node fails to receive an authentic RSA modulus in any R-interval, then the node will not be able to authenticate broadcast messages during the next R-interval. However, the failed node can get back to the authentic state anytime later if it receives a modulus correctly. For better reliability, the base station may broadcast KDM messages as many times as necessary, depending on the network condition, to reduce the packet loss rate to a sufficient level. This does not cause much overhead since only two KDM messages need to be transmitted over a long period of time (at least 15 minutes). Since the KDM-commit message is more critical, it would be better to take $\lambda_1 > \lambda/2$.

Efficiency. The proposed scheme is very efficient in both computation and communication. Suppose that we have 80-bit key chain values, 64-bit MACs (σ_i's) and key chain parameters ($d = 1\text{min}$, $\lambda = 15$, $\lambda_1 = 10$) for RSA-512 of 20-minute lifetime. The key management overhead amortized over time is almost negligible in the proposed scheme. The communicated bits include 33 bytes for KDM-commit message and 63 bytes for KDM-reveal message. They are broadcast over a 15-minute period and can be processed on sensor nodes only using 15 one-way function evaluations and two hash operations on average.

The signature overhead for broadcast authentication is almost minimal as a public key technique. The base station only needs to broadcast 64 bytes of signature for typical broadcast messages and a sensor node only requires a single squaring modulo RSA-512 and some hash operations for message decoding. Since modular squaring can be further speeded up with our special modulus, the overall signature verification could be done in a few tens of milliseconds with moderately optimized implementations of AES and modular arithmetic. [3] The message expansion due to signing in our scheme is almost comparable to

---

[3] For RSA-512, it is reported that on MicaZ mote, modular multiplication takes 8.15msec in [10] and squaring (without modular reduction) takes 3.87msec in [36]. Thus we can expect that modular squaring in our scheme would take far less than 10msec only with a moderately optimized implementation.

the most compact ECC signatures. The proposed scheme will be particularly efficient for applications requiring many full packet transmissions, such as code dissemination (e.g., see [13]), as each full packet incurs only 20-byte signature overhead and can be signed and transmitted independently, without chaining.

Comparison with Existing Schemes. The proposed scheme overcomes all major drawbacks in existing schemes. Compared to μTESLA, it does not require any network-wise clock synchronization and provides strong resistance against DoS attacks. Compared to one-time signatures, it incurs almost negligible signature overhead and is equipped with a robust mechanism for public key update. Our scheme also has other desirable features inherited from digital signatures (see Section 5), together with additional proactive security feature.

5   Extensions

Several extensions can be considered to the proposed scheme (details will be given in the full paper due to the space limitation in the present version):

- *Supporting mobile users*: There also exist mobile users in many sensor network applications (e.g., fire-fighters, rescue officials, tanks/soldiers, etc.), who query neighbor nodes to collect necessary data while moving along the network. Thus we also need to support broadcast authentication by such mobile users. It is quite easy to support mobile users in our scheme since the base station can play the role of a trusted certification authority and issue a signed modulus to a mobile user as a simple certificate.
- *Three-level extension*: The present scheme has a two-level hierarchy, TESLA level for key management and RSA level for message authentication. We may also envision a three-level extension with two RSA levels. For example, the public key update protocol can be used to distribute RSA-640 of lifetime 1 day, which can then be used to distribute RSA-512 of lifetime 20 minutes. This extension may have some advantages over the basic scheme (without much loss of efficiency) due to the flexible use of two level RSA signatures. The low level RSA needs not have a pre-determined fixed lifetime (even can be activated on demand when necessary). The high level RSA is better suited to play the role of CA to mobile users (as well as the base station). The pre-specified bits P_i actually contain all parameters required for public key certification and thus a signed modulus can be used as a simple certificate.
- *Secure fail-over of base stations*: It is usually assumed that base stations cannot be compromised during the whole network lifetime. However, this assumption may not be realistic in practice [7]. In fact, a base station may fail due to attacks, such as physical compromise or destruction and persistent jamming, etc. Thus, to avoid a single point of failure, we may pre-deploy or deploy on demand some backup base stations which can be activated in case of existing base station failure. The proposed scheme can also be extended to support secure fail-over in this scenario.

6   Conclusion

We have presented a hybrid approach to broadcast authentication in wireless sensor networks, where the base station periodically distributes a short RSA modulus of limited lifetime using a robust TESLA-like protocol and uses short-lived Rabin signatures for efficient broadcast authentication. The proposed scheme is very efficient and robust, overcoming all major drawbacks in existing schemes. We thus believe that our approach could be a very practical and viable solution to broadcast authentication in large-scale distributed sensor networks.

References

1. Aoki, K., Franke, J., Kleinjung, T., Lenstra, A.K., Osvik, D.A.: A kilobit special number field sieve factorization. In: Kurosawa, K. (ed.) ASIACRYPT 2007. LNCS, vol. 4833, pp. 1–12. Springer, Heidelberg (2007)
2. Bellare, M., Rogaway, P.: The exact security of digital signatures - how to sign with RSA and rabin. In: Maurer, U.M. (ed.) EUROCRYPT 1996. LNCS, vol. 1070, pp. 399–416. Springer, Heidelberg (1996)
3. Black, J., Rogaway, P., Shrimpton, T.: Black-box analysis of block-cipher-based hash function constructions from PGV. In: Yung, M. (ed.) CRYPTO 2002. LNCS, vol. 2442, pp. 320–335. Springer, Heidelberg (2002)
4. Chang, S.M., Shieh, S., Lin, W.W., Hieh, C.M.: An efficient broadcast authentication scheme in wireless sensor networks. In: ASIACCS 2006 (March 2006)
5. Chen, J.-M., et al.: Improved factoring of RSA modulus. In: the 25th Workshop on Combinatorial Mathematics and Computation Theory (2008)
6. Coron, J.S.: Optimal security proofs for PSS and other signature shcemes. In: Knudsen, L.R. (ed.) EUROCRYPT 2002. LNCS, vol. 2332, pp. 272–287. Springer, Heidelberg (2002)
7. Deng, J., Han, R., Mishra, S.: Enhancing base station security in wireless sensor networks, Technical Report CU-CS-951-03, University of Colorado (2003)
8. Geiselmann, W., Steinwandt, R.: Special-purpose hardware in cryptanalysis: The case of 1,024-Bit RSA. IEEE Security & Privacy Magazine 5(1), 63–66 (2007)
9. Geiselmann, W., Shamir, A., Steinwandt, R., Tromer, E.: Scalable hardware for sparse systems of linear equations with applications to integer factorization. In: Rao, J.R., Sunar, B. (eds.) CHES 2005. LNCS, vol. 3659, pp. 131–146. Springer, Heidelberg (2005)
10. Gura, N., Patel, A., Wander, A.: Comparing elliptic curve cryptography and RSA on 8-bit CPUs. In: Joye, M., Quisquater, J.-J. (eds.) CHES 2004. LNCS, vol. 3156, pp. 119–132. Springer, Heidelberg (2004)
11. Huang, Y., He, W., Nahrstedt, K., Lee, W.C.: DoS-resistant broadcast authentication protocol with low end-to-end delay, In. In: IEEE INFOCOM 2008, pp. 1–6 (2008)
12. Hui, J.W., Culler, D.E.: Extending IP to low-power, wireless personal area networks. IEEE Internet Computing, 37–45 (July/August 2008)
13. Hyun, S., Ning, P., Liu, A., Du, W.: Seluge: Secure and dos-resistant code dissemination in wireless sensor networks. In: IPSN 2008 (April 2008)
14. Joye, M.: RSA moduli with a predetermined portion: Techniques and applications. In: Chen, L., Mu, Y., Susilo, W. (eds.) ISPEC 2008. LNCS, vol. 4991, pp. 116–130. Springer, Heidelberg (2008)

15. Krontiris, I., Dimitriou, T.: A practical authentication scheme for in-network programming in wireless sensor networks. In: REALWSN 2006 (2006)
16. Lenstra, A.K.: Generating RSA moduli with a predetermined portion. In: Ohta, K., Pei, D. (eds.) ASIACRYPT 1998. LNCS, vol. 1514, pp. 1–10. Springer, Heidelberg (1998)
17. Lenstra, A.K., Lenstra Jr., H.W.: The development of the number field sieve. LNM, vol. 1554. Springer, Heidelberg (1993)
18. Liu, D., Ning, P.: Multi-level μTESLA: Broadcast authentication for distributed sensor networks. ACM Trans. Embedded Computing Systems 3(4), 800–836 (2004)
19. Liu, D., Ning, P.: TinyECC: A configurable library for elliptic curve cryptography in wireless sensor networks. In: IPSN 2008, pp. 245–256 (2008)
20. Liu, D., Ning, P., Zhu, S., Jajodia, S.: Practical broadcast authentication in sensor networks. In: MobiQuitous 2005 (July 2005)
21. Naor, D., Shenhav, A., Wool, A.: One-time signatures revisited: Practical fast signatures using fractal merkle tree traversal. In: IEEE 24th Convention of Electrical and Electronics Engineers in Israel (November 2006)
22. Ning, P., Liu, A., Du, W.: Mitigating DoS attacks against broadcast authentication in wireless sensor networks. ACM Transactions on Sensor Networks (2007)
23. Passing, M., Dressler, F.: Experimental performance evaluation of cryptographic algorithms on sensor nodes. In: IEEE MASS 2006, pp.882–887 (2006)
24. Perrig, A., Canetti, R., Song, D., Tygar, D.: Efficient authentication and signing of multicast streams over lossy channels. In: IEEE Symp. on Security & Privacy (2000)
25. Perrig, A., Szewczyk, R., Wen, V., Culler, D., Tygar, D.: SPINS: Security protocols for sensor networks. In: MobiCom 2001 (July 2001)
26. Piotrowsi, K., Langendoerfer, P., Peter, S.: How public key cryptography influences wireless sensor node lifetime. In: ACM SASN 2006 (October 2006)
27. Poettering, B.: AVRAES: The AES block cipher on AVR controllers, http://point-at-infinity.org/avraes/
28. Rinne, S., Eisenbarth, T., Paar, C.: Performance analysis of contemporary lightweight block ciphers on 8-bit microcontrollers. In: SPEED 2007, Amsterdam, The Netherlands (June 2007)
29. Sastry, N., Wagner, D.: Security considerations for IEEE 802.15.4 networks. In: WiSE 2004 (October 2004)
30. Seys, S., Preneel, B.: Power consumption evaluation of efficient digital signature schemes for low power devices. In: IEEE WiMob 2005, pp. 79–86. IEEE, Los Alamitos (2005)
31. Shamir, A., Tromer, E.: Factoring large numbers with the TWIRL device. In: Boneh, D. (ed.) CRYPTO 2003. LNCS, vol. 2729, pp. 1–26. Springer, Heidelberg (2003)
32. Scott, M., Szczechowiak, P.: Optimizing multiprecision multiplication for public key cryptography. In: IACR ePrint, Report 2007/192 (2007)
33. Szczechowiak, P., et al.: NanoECC: Testing the limits of elliptic curve cryptography in sensor networks. In: Verdone, R. (ed.) EWSN 2008. LNCS, vol. 4913, pp. 305–320. Springer, Heidelberg (2008)
34. Uhsadel, L., Poschmann, A., Paar, C.: Enabling full-size public-key algorithms on 8-bit sensor nodes. In: Stajano, F., Meadows, C., Capkun, S., Moore, T. (eds.) ESAS 2007. LNCS, vol. 4572, pp. 73–86. Springer, Heidelberg (2007)

35. Wang, R., Du, W., Ning, P.: Containing denial-of-service attacks in broadcast authentication in sesnor networks. In: MobiHoc 2007, pp.71–79 (2007)
36. Wang, H., Li, Q.: Efficient implementation of public key cryptosystems on mote sensors. In: Ning, P., Qing, S., Li, N. (eds.) ICICS 2006. LNCS, vol. 4307, pp. 519–528. Springer, Heidelberg (2006)
37. Recommendation for Block Cipher Modes of Operation: The CMAC Mode for Authentication, NIST SP 800-38B (May 2005)
38. IEEE Std. 802.15.4-2003, http://standards.ieee.org/getieee802/download/
39. IEEE Std. 802.15.4-2006, http://standards.ieee.org/getieee802/download/
40. IEEE P1363a-2004, Stanard specification ofr public key cryptography: Additional techniques

Author Index